Handbook of Vanilla Science and Technology

Handbook of Vanilla Science and Technology

Second Edition

Edited by

Daphna Havkin-Frenkel and Faith C. Belanger
Rutgers University
NJ, USA

This edition first published 2019
© 2019 John Wiley & Sons Ltd.

Edition History
First edition published @ 2011 Blackwell Publishing Ltd.
9781405193252 Handbook of Vanilla Science and Technology

Registered Office(s)
John Wiley & Sons, Inc., 111 River Street, Hoboken, NJ 07030, USA
John Wiley & Sons Ltd, The Atrium, Southern Gate, Chichester, West Sussex, PO19 8SQ, UK

Editorial Office
The Atrium, Southern Gate, Chichester, West Sussex, PO19 8SQ, UK

For details of our global editorial offices, customer services, and more information about Wiley products visit us at www.wiley.com.

Wiley also publishes its books in a variety of electronic formats and by print-on-demand. Some content that appears in standard print versions of this book may not be available in other formats.

Library of Congress Cataloging-in-Publication Data
Names: Havkin-Frenkel, D. (Daphna), 1951- editor. | Belanger, Faith C., editor.
Title: Handbook of vanilla science and technology / edited by Dr Daphna Havkin-Frenkel, Rutgers University, NJ, US, Dr Faith C. Belanger, Rutgers University, NJ, US.
Description: Second edition. | Hoboken, NJ, USA : Wiley, [2018] | Includes bibliographical references and index. |
Identifiers: LCCN 2018014797 (print) | LCCN 2018015790 (ebook) | ISBN 9781119377313 (pdf) | ISBN 9781119377290 (epub) | ISBN 9781119377276 (cloth)
Subjects: LCSH: Vanilla. | Vanillin. | Vanilla industry.
Classification: LCC SB307.V2 (ebook) | LCC SB307.V2 H36 2018 (print) | DDC 664/.52–dc23
LC record available at https://lccn.loc.gov/2018014797

Cover Design: Wiley
Cover Images: Courtesy of Daphna Havkin-Frenkel

Set in 10/12pt WarnockPro by SPi Global, Chennai, India

Printed in Singapore by C.O.S. Printers Pte Ltd

10 9 8 7 6 5 4 3 2 1

Contents

List of Contributors

Paul Bayman
Departamento de Biologia
Universidad de Puerto Rico – Rio Piedras
San Juan, Puerto Rico

Faith C. Belanger
Department of Plant Biology
School of Environmental and
Biological Sciences
Rutgers University
New Brunswick
New Jersey, USA

Deborah Y.J. Booth
50 Clover Hill Road
Millington
New Jersey, USA

Richard J. Brownell Jr.
Vice President Vanilla Products
Virginia Dare Extract Company, Inc
New York City
New York, USA

Felix Buccellalto
Custom Essence Inc.
Somerset
New Jersey, USA

Kenneth M. Cameron
Department of Botany
University of Wisconsin
Madison
Wisconsin, USA

Alan Chambers
Tropical Research & Education Center
University of Florida
Homestead
Florida, USA

Paul Chavarriaga
International Center for Tropical
Agriculture (CIAT)
Cali, Colombia

Fang Chen
BioDiscovery Institute and Department
of Biological Sciences
University of North Texas
Denton
Texas, USA

Dawn Dean
Organic Vanilla Association
Barranco Village
Toledo District, Belize

Richard A. Dixon
BioDiscovery Institute and Department
of Biological Sciences
University of North Texas
Denton
Texas, USA

Richard Exley
Australian Vanilla Bean
Karama
Northern Territory, Australia

Nicola S. Flanagan
Departamento de Ciencias Naturales y
Matemáticas
Pontificia Universidad Javeriana
Cali, Colombia

Chaim Frenkel
Department of Plant Biology
School of Environmental and
Biological Sciences
Rutgers University
New Brunswick
New Jersey, USA

Kathryn E. Galasso
Tufts University
Medford
Massachutsetts, USA

Rebeca Alicia Menchaca Garcia
Centro de Investigaciones Tropicales
Universidad Veracruzana
Xalapa de Enriquez
Mexico

Carlos Javier Hernandez Gayosso
Universidad Tecnólogica
Puebla, Mexico

Nelle Gretzinger
249 Smith Street
Brooklyn
New York, USA

Thomas G. Hartman
Center for Advanced Food Technology
School of Environmental and Biological
Sciences
Rutgers University
New Brunswick
New Jersey, USA

Daphna Havkin-Frenkel
Department of Plant Biology
School of Environmental and
Biological Sciences
Rutgers University
New Brunswick
New Jersey, USA

and

Bakto Flavors
North Brunswick
New Jersey, USA

Sylvia M. Heredia
Department of Botany and Plant
Sciences
University of California
California, USA

Juan Hernández-Hernández
Instituto Nacional de Investigaciones
Forestales, Agrıcolas y Pecuarias
(INIFAP)
Veracruz, Mexico

Patrick G. Hoffman
PGH Consulting LLC
Freeland
Maryland, USA

Ivica Labuda
Georgetown University
Washington, D.C.
USA

Keun Joong Lee
Schering-Plough
Hialeah
Florida, USA

Pesach Lubinsky
Foreign Agricultural Service, USDA
Washington, DC
USA

Ana T. Mosquera-Espinosa
Departamento de Ciencias Naturales y
Matemáticas
Pontificia Universidad Javeriana
Cali, Colombia

Andrzej Podstolski
Faculty of Biology
Institute of Experimental Plant Biology
University of Warsaw
Warsaw, Poland

Andrea Porras-Alfaro
Department of Biological Sciences
University of Western Illinois
Macomb
Illinois, USA

Elida Varela Quiros
Las Dos Manos Vainilla Ltda.
Detras Iglesia San Juan de Naranjo
Alajuela, Costa Rica

Arvind S. Ranadive
Premier Vanilla, Inc.
East Brunswick
New Jersey, USA

Gustavo A. Romero-Gonzalez
Harvard University Herbaria
Cambridge
Massachusetts, USA

Francis P. Tangel
Flavor & Fragrance Specialties
Mahwah
New Jersey, USA

Stephen Toth
International Flavors & Fragrances Inc.
Union Beach
New Jersey, USA

Javier Tochihuitl Vazquez
Principal of the Career of Agroindustrial
Processes
Universidad Tecnológica de
Xicotepec de Juarez
Puebla, Mexico

Filip van Noort
Horticulture & Product Physiology
Group
Wageningen University and Research
Wageningen, Netherlands

Stephanie Zabel
Harvard University Herbaria
Cambridge
Massachutsetts, USA

Charles M. Zapf
Technical Innovation Center
McCormick & Co. Inc.
Hunt Valley
Maryland, USA

Preface

Vanilla is the world's most popular flavor. It is a universally appreciated flavor on its own and it also provides smoothness and body when used in combination with other flavors. The aroma of vanilla extract is intoxicating and those who work with any aspect of vanilla become intoxicated with the subject. The vanilla orchid is indigenous to Mexico and was first used and cultivated by the Totonac Indians. Now vanilla cultivation, extraction, analysis, and marketing are major international industries. This book has a chapter covering each of these aspects of vanilla, as well as chapters on the biology of vanilla and the potential for biotechnological production of vanillin. There have been many new developments in vanilla research since publication of the first edition of the *Handbook of Vanilla Science and Technology*. In this second edition of the book there are seven new chapters and four updated chapters. Currently, there is considerable interest in expanding vanilla cultivation to additional geographical regions as well as breeding vanilla for specific traits. There are also expanded efforts to understand the basic biology of vanilla plants, including genetics and the vanillin biosynthetic pathway. These exciting topics are discussed in this new edition.

Part I

Production of Vanilla – Agricultural Systems and Curing

1

Mexican Vanilla Production

Juan Hernández-Hernández

1.1 Introduction

The vanilla species of commerce, *Vanilla planifolia* G. Jacks, known as "Mexican" or "Bourbon" vanilla, is native to tropical forests of southeastern Mesoamerica (Portères 1954; Soto-Arenas 2003; Hágsater *et al.* 2005). By at least the nineteenth century, *V. planifolia* was introduced into other tropical countries in Asia and Africa from the original Mexican cultivated stock (Bory *et al.* 2008; Lubinsky *et al.* 2008). Vanilla was used in pre-Hispanic Mesoamerica for a variety of purposes: tribute, fragrance, cacao flavoring, medicinal, etc., and by numerous indigenous groups such as the Maya, Aztec, and Totonac. In this sense, vanilla is a gastronomic legacy that Mexico has imparted to the world.

Beginning in the mid- to late eighteenth century, the Totonac of the Papantla region of the state of Veracruz were the first and only vanilla exporters in the world for nearly 100 years, in part because of the exceptional quality of the vanilla that was produced. Gold medal prizes for Mexican vanilla were awarded in Paris (1889) and Chicago (1892) (Chávez-Hita and González-Sierra 1990), as Papantla was famed as, "the city that perfumed the world." Initially, Mexican vanilla production depended on harvesting the fruits from the wild, which were the result of natural pollination by bees that are endemic to the New World tropics.

The Mexican monopoly on vanilla fell apart with the discovery of a method for hand pollination of vanilla in Belgium in 1836. This knowledge enabled other countries to become vanilla producers. By 1870, French colonies in the Indian Ocean, especially Reunion and Madagascar, surpassed Mexico as the leading producer. Madagascar has retained the leading role in production since that time (Bruman 1948; Bory *et al.* 2008).

Although Mexico has lost its standing as the major vanilla exporter, it continues to be the center of origin and genetic diversity for this important orchid. Cultivation in Mexico endures to the present, mostly by the Totonac, who have continued to use their vanilla crop as a means to obtain cash, and because it is part of their historical and cultural fabric.

The area of vanilla production in Mexico is found between the coast and Sierra Madre Oriental on the Gulf, from sea level to a height of 700 m, where the climate is hot,

Handbook of Vanilla Science and Technology, Second Edition.
Edited by Daphna Havkin-Frenkel and Faith C. Belanger.
© 2019 John Wiley & Sons Ltd. Published 2019 by John Wiley & Sons Ltd.

humid, and tropical. Average temperatures are around 24 °C, relatively humidity is 80%, and average annual precipitation is 1,200 to 1,300 mm. A marked dry season occurs from March to June. In winter, there are humid, cool winds of low intensity called "*nortes*" that bring cool temperatures to the area, which is believed to stimulate the flowering in vanilla.

The state of Veracruz accounts for 70% of national production. Oaxaca and Puebla together produce most of the remaining 30%, and small quantities of vanilla are also supplied by San Luis Potosí, Hidalgo, Chiapas, and Quintana Roo. The municipality (*municipio*) of Papantla, located in northern Veracruz and inhabited by Totonac communities, is the largest producer in the country, and is the center of vanilla curing and commercialization.

An estimated 4,000 families are engaged in vanilla cultivation, mostly indigenous people, who exclusively sell green vanilla. Six private companies and four farmer cooperatives also exist, and participate in curing and selling of vanilla to national and international markets.

Annual production in Mexico varies from 80 to 200 tons of green vanilla (10–30 tons cured vanilla beans), depending on climatic conditions and the intensity of flowering, among other factors. In 2008/2009, according to estimates by the Consejo Nacional de Productores de Vainilla, 150 tons of green vanilla beans were produced (ca. 20 tons cured vanilla beans).

The principal limiting factors to vanilla production in Mexico are:

- drought and high temperatures, which occur during flowering and fruit development;
- the fungus *Fusarium oxysporum*, which causes mortality and reduces the productive life of individual cultivated areas (*vainillales*); and
- high production costs and low prices for vanilla.

1.1.1 The Mexican Vanilla Legend

The Mexican vanilla legend, which is an oral Papantla tradition, is compiled and interpreted by Professor J. Núñez-Domínguez (Curti-Diaz 1995):

> At the summit of a mountain close to Papantla, was the temple of Tonacayohua, the goddess of food and planting crops. During the reign of King Teniztli III, one of his wives gave birth to a daughter whose beauty was so great that she was named Tzacopontziza ("Bright Star at Sunrise"), and was consecrated to the cult of Tonacayohua.
>
> As time passed, a young prince named Zkatán-Oxga ("Young Deer") and Tzacopontziza fell in love, knowing that this sacrilege was condemned by death.
>
> One day, Bright Star at Sunrise left the temple to look for tortillas to offer to Tonacayohua, and fled with the young prince to the jagged mountains in the distance. Not before long, a monster appeared and surrounded them by a wall of flames, and ordered them to return.
>
> When the couple returned to the temple, a group of irate priests had been waiting for them, and before Zkatán-Oxga could say anything, the young lovers were shot with darts, and their bodies were brought to a temple where their hearts were removed, and their carcasses were thrown down into a canyon.

In the place where the bodies landed there was a herb, and its leaves started to wilt as if the scattered blood of the victims had scorched the plant like a curse. Sometime later a new tree began to grow, and within days its vigorous growth covered all the ground around it with its brilliant foliage.

When finally it stopped growing, next to its trunk began to grow an orchid that climbed and also was amazingly vigorous. Within a short amount of time, it had branched and covered the trunk of the tree with its fragile and elegant leaves, and protected by the tree, the orchid grew more until finally it took the form a woman lying in the embrace of her lover.

One day the orchid became covered with small flowers and the whole area was filled with an exquisite aroma. Attracted to the pleasant smell, the priests and the pueblo came to observe, and no one doubted that the blood of the young lovers had transformed into the tree and the climbing orchid.

To their surprise, the beautiful little flowers also transformed into large, thin fruits. When the fruits matured, they released a sweet, subtle perfume whose essence invoked the innocent soul of Bright Star at Sunrise and the most exotic fragrances.

This is how the vanilla was born, the one that is called *"Caxixanath"* (Recondite Flower), which is a sacred plant and a divine offering in Totonac temples.

1.2 Cultivation Methods

Vanilla is a hemi-epiphytic orchid that in cultivation needs a tree to provide physical support, shade, and organic material.

In Mexico, vanilla is cultivated in different settings:

- in environments similar to the natural habitat, i.e. a forest composed of mostly secondary vegetation (*"acahual"*), which is the "traditional" style;
- intercropped with other crops such as coffee or orange;
- "intensively", with *Erythrina* sp. or *Gliricidia sepium* as support trees; and
- "intensively", in shade houses.

1.2.1 "Traditional"/*Acahual*

Acahual refers to a secondary forest or fallow that is regenerating, in many cases following maize cultivation. These sites are where vanilla is primarily cultivated, and are very similar to the natural habitat of the species. Over 90% of vanilla growers, mostly from indigenous groups, use this setting, which is almost always less than 1 ha.

Species commonly encountered in *acahual* are used as support trees for vanilla. They include: "laurel" (*Litcea glaucescens*), "pata de vaca" (*Bahuinia divaricata*), "cojón de gato" (*Tabernaemontana* sp.), "cacahuapaxtle" or "balletilla" (*Hamelia erecta*), and "capulín" (*Eugenia capuli*), among others (Curti-Diaz 1995). A relatively low density of vanilla plants is cultivated without irrigation and with minimal overseeing. Consequently, yields are low, varying between 50 and 500 kg of green vanilla/ha, with an average yield of 200 kg/ha.

This "traditional" style of cultivation is also used where vanilla is intercropped with coffee, where the vanilla benefits from the abundant organic matter and shade typical of such *cafetales*. Support trees in this setting are trees that are used to provide shade to the coffee, such as *Inga* sp., or are species introduced to the site, such as *Erythrina* sp.

The advantage of the coffee-vanilla production system is that the grower diversifies his/her economic activities, obtaining two products from one site.

Establishing a "traditional" *vainillal* requires an initial investment of around $2,000 USD/ha, with maintenance costs typically totaling $1,500 USD per year.

1.2.2 Intensive System (Monoculture)

This system is normally practiced in deforested areas that have been used to cultivate another crop. The name of this system is "pure cultivation" (Chauds 1970), and the first step consists of planting support trees. After a year, when there is sufficient shade (50%), the vanilla is planted (Pennigton *et al.* 1954). This system is utilized by growers with more economic means, in lots of 0.5 to 2 ha per grower.

Support trees that are regularly used are *"pichoco"* (*Erythrina* sp.) and/or *"cocuite"* (*Gliricidia sepium*), two leguminous trees with the capacity to fix atmospheric nitrogen and that can be propagated clonally through cuttings. Per ha, 1,000 to 5,000 support trees are planted, as are 2,000 to 10,000 cuttings of vanilla (2 vanilla plants/support tree). The planting distances between trees are 1 × 2 m, 2 × 2 m, 1.5 × 2.5 m, and 3 × 3 m.

This system of vanilla cultivation has the advantage of relatively high yields, but generally only in the fourth or fifth year after planting (second or third harvest). After this time, yields decline drastically, most likely due to the difficulties of managing mature plants in such a confined space (especially for adequate shade and ventilation).

Yields of green vanilla beans vary from 1 to 2 tons per ha in rain-fed systems, and 2 to 4 tons per ha with a higher density of plantings (10,000 plants per ha) and with irrigation.

Establishing a monoculture of vanilla from a cleared area requires around $10,000 USD to cover the costs of establishing support trees and the high density of plantings. Maintenance costs per year average $7,500 USD.

1.2.3 Vanilla Cultivation in Existing Orange Groves

Orange trees are excellent support trees for vanilla, because their branches are durable and grow laterally and are able to support a good quantity and distribution of shoots (Figure 1.1). These features help mitigate the problem of the shoots shading out other shoots. The canopy of orange trees is capable of providing vanilla plants with sufficient sunlight throughout the year. In most systems with orange trees as supports, vanilla flowers in the second year.

This system is one of the best ventilated, with a low incidence of pests and diseases. Yields are higher and costs of production are lower because orange trees in coastal Veracruz have been extensively cultivated for decades.

Many of the vanilla growers started off cultivating oranges and continue to do so when managing vanilla. The vanilla plants are established when the orange grove is producing. Orange trees that are selected as supports have an average height of 4 m and a well-formed canopy. Dry branches (*"chupones"*) are pruned, as are those in the interior of the canopy that impede the spread of vanilla plants as they are growing or block out too much sunlight.

Figure 1.1 Vanilla vines growing on orange trees as a support.

Densities of orange tree plantings vary between 204 to 625 individuals per ha. Trees are spaced on a grid of 4 × 4 m, 5 × 5 m, 6 × 6 m, and 7 × 7 m, and 3 to 6 cuttings of vanilla are planted per orange tree, yielding a total of between 1,224 and 1,875 vanilla plants per ha.

Growers manage 1 to 5 ha and harvest 500 to 2,500 kg of green vanilla/ha, although most obtain 1 ton.

Establishing vanilla cultivation in an existing orange grove requires a minimum initial investment of $7,000 USD/ha. The orange trees represent an economically sustainable resource in the sense that they do not have to be purchased or planted. Annual maintenance costs average $6,000 USD/ha per year.

1.2.4 Shade Houses

This is the most recent and intensive form of vanilla management in Mexico. Its principal feature consists of substituting or complementing natural shade with artificial shade by means of shade cloth (black or red) of 50% luminosity, which is stretched above all the support trees at ca. 3 to 5 m high, at the four sides of the planted area. These systems are referred to as "shade houses". In size, they are usually on the order of 25 × 40 m (1,000 m^2) and some are up to 1 ha.

Shade houses most commonly feature artificial or "inert" support tress, such as concrete posts, or posts made from wood or bamboo. On occasion, living support trees, such as "*pichoco*" (*Erythrina* sp.) or "*cocuite*" (*Gliricidia sepium*), are used in lieu of or in combination with artificial supports. High planting densities are typical of this system, with 254 to 2,500 supports and 1,524 to 2,500 vanilla plants per 1,000 m^2.

Shade houses are appropriate on flat ground that has been deforested or on patios, and for use by growers with relatively more economic means. The initial investment is high, usually $10,000 USD per 1,000 m^2, with annual maintenance costs of $2,000 USD. For this reason, most shade houses in Mexico are subsidized by the government.

The first yields from shade houses have been variable, with the maximum thus far being 514 kg green vanilla per 1,000 m^2, from 1,524 vanilla plants. This value theoretically scales up to 5,140 kg green vanilla per ha, similar to yields obtained from shade house production systems in other countries.

Growers agree that shade houses provide for a system of better care and overseeing of vanilla plants, which tend to grow vigorously as a consequence. However, it is yet to be determined what the real outcomes and economic viability of this system of production are.

In whatever system of vanilla cultivation, the maximum yields occur in the fourth or fifth year following after planting (second or third harvest). After this time, production volume can be lower or higher, but after 9 years, yields steadily decline until productivity ceases almost completely by the twelfth year.

1.3 Vanilla Propagation Techniques

Vanilla is propagated almost entirely by stem cutting. The cuttings are procured from another grower or from a government agricultural entity. Cuttings are made from highly productive and vigorous individuals that have never produced fruits. The cutting itself should not be a flowering shoot and should have at least 3 nodes with viable axillary buds for producing new shoots from which the plant will grow. Cuttings should be free of damage or symptoms of pests/diseases so as to avoid future proliferation of disease. A best practice is to ensure that the cuttings are certified as virus-free. Cuttings are normally 6 to 8 nodes (80–20 cm long, 1 cm in diameter) in length. Longer or thicker cuttings form new vegetative and reproductive shoots more rapidly (Ranadive 2005), but are more difficult to deal with during planting, and are more expensive.

1.3.1 Preparation and Disinfection of Cuttings

Cuttings are prepared prior to planting. The three most basal leaves are removed by hand by twisting at the petiole and taking care not to tear into the stem where open wounds can facilitate the spread of pathogens.

In order to prevent stem rot, caused primarily by *F. oxysporum*, stem cuttings are disinfected prior to planting. The basal portion of the cutting is submerged for 2 to 5 minutes in a fungicidal solution. The solution may consist either of carbendazim (2 g/L) or Bordeaux mixture (1 kg lime + 1 kg copper sulfate in 100 L of water), the latter being less effective but authorized for the production of organic crops. Fungicidal solutions are handled with rubber gloves to avoid harmful exposure to the body.

After disinfection, cuttings are hung separately on a structure 1 to 1.5 m tall, in a shaded and well-ventilated area for a period of 7 to 15 days. The cuttings slightly dehydrate allowing for more flexible material for planting. Calluses form over areas of the cuttings that were damaged during leaf removal.

1.3.2 Establishing Cuttings – Timing

Cuttings are planted when support trees have developed sufficient foliage to prevent the young vanilla plants from being burned. With shade cloth, cuttings are planted immediately after the establishment of support trees. The best conditions for planting cuttings are in humid substrates during warm, dry months preceding the onset of the rainy season (Ranadive 2005). This timing favors a high percentage (>90%) of successfully established cuttings, since high temperatures are conducive to the emergence of new shoots and roots.

1.3.3   Establishing Cuttings – Planting

Cuttings are planted in the following manner. Adjacent to the support, a shallow ditch is dug 5 to 10 cm deep, into which the cutting is placed horizontally (but only the part that has had the leaves removed). The cutting is then buried with 3 to 5 cm of organic material and/or fertile soil or leaves, which will serve as a mulch and as a source of nutrients. The extreme basal end of the cutting (2–3 cm) is left uncovered to prevent rot (Wong *et al.* 2003; Ranadive 2005), especially when the substrate is humid. Some cuttings are established without making ditches, and are simply placed on top of a humid substrate.

Once planted, the rest of the cutting (with leaves, ca. 4–5 nodes) is positioned vertically on the support and fastened with bio-degradable material such as banana leaves, tree bark, or henequen fiber.

Under optimal conditions of humidity and temperature, and with vigorous, healthy cuttings, the first roots begin to emerge the first week after planting and the first shoots in about 1 month.

1.3.4   New Bud Formation and Root Growth

Warm temperatures stimulate both bud break and the longitudinal growth of shoots. In Mexico, most vegetative growth occurs in spring and summer (58–67.8 cm/month). In fall and winter, this rate of growth declines to 22 to 52.2 cm/month.

In general, growth is affected by humidity, nutrition, health, environmental conditions, etc. Vegetative growth during the first 2 years (3.97–5.94 m/year) is markedly less than when the plant is in the third and fourth years (7.49–7.63 m/year). After the fourth year, vegetative growth declines (5.74–6.8 m/year).

The first 2 years following establishment of the vanilla consist almost entirely of vegetative growth. By the third year, plants begin to flower and produce, when shoots have reached a minimal length of 10 m. The plants continue to produce from there on.

1.4   Irrigation

The main vanilla production region of Mexico – the Papantla area in northern Veracruz – characteristically suffers drought on an annual basis. The drought is most pronounced during the most critical season for vanilla, during flowering and pollination. Most growers in Mexico nonetheless cultivate vanilla in rain-fed systems.

The most frequent form of irrigation in Mexico is the use of micro-emitters to moisten the mulch layer where the vanilla roots are growing. One criterion for irrigation is to maintain at all times a moist layer of mulch without reaching saturation levels. During the dry season, watering is performed once to twice per week.

1.5   Nutrition

The primary source of nutrition for vanilla in cultivation is organic material (humus) that results from the natural decomposition of vegetable/animal residues (mulch), composting (via micro-organisms), or vermi-culture (worm-mediated breakdown of organic material).

1.5.1   Mulch

In addition to providing nutrients, the benefits of mulch are:

- it helps maintain soil humidity;
- it serves as a porous substrate, aiding soil aeration and permitting the unrestrained development of roots;
- maintains an adequate temperature; and
- decreases the incidence of weeds.

The most common mulch for vanilla is from decaying leaf litter derived from leaf fall, pruning, and from herbaceous plants in the *vainillal*.

The mulch should be 10 to 20 cm deep and laid down on either side of the support where the vanilla roots will grow. To prevent the loss of mulch from runoff from heavy rains, most prevalent in *vainillales* managed on slopes, borders are constructed out of trunks of wood, bamboo canes, rocks, or other materials. New applications of mulch are made when roots are observed growing out of the surface of the mulch, generally 2 to 3 times/year, and mostly in the hot/dry months, when mulch is carefully managed to prevent dehydration.

1.5.2   Building Compost

In addition to available natural organic material, the nutritional requirements of vanilla can also be met by developing a composting system.

Compost can be made from a diversity of primary organic materials, but it is best to use locally abundant resources. Fresh sawdust may contain substances that are toxic to plants, such as phenols, resins, terpenes, and tannins. Fresh manure or manure that has not decomposed adequately, can cause burning or root-rot and eventual mortality. When using either of these materials as fertilizer, it should be ensured that they are first well decomposed to avoid causing damage to the plant.

Compost is developed in many ways, but a simple and practical method for composting for vanilla, which gives good results, has been developed by growers in San Rafael, Veracruz. Vanilla plants are grown on orange tree supports, and are fertilized with a mixture of sheep manure and pine sawdust.

This compost is made by:

- mixing 70% pine sawdust with 30% dry sheep manure. The mixing is done on the ground with a shovel until the mixture homogenizes.
- applying water until 45 to 65% moisture is achieved. In practice, a grower decides when this percentage is arrived at by inspecting a small amount (a pinch) of the mixture in his hands. The water should not drop down onto the hands, but adhere to the mixture, and the moisture should be felt between the fingers.

- covering the mixture with plastic to protect it from the rain. High temperatures are generally not a problem, but should not exceed 65 °C, which could cause the death of the microorganisms responsible for breaking down the organic material. If this temperature is exceeded, the plastic cover is removed, and the compost is re-mixed (aerated) and water is also applied.
- turning the compost over every 15 days to accelerate decomposition and to maintain good aeration, especially during the initial stages of degradation, since the microorganisms (bacteria and fungi) depend on oxygen to live.
- Compost is ready to use in generally 3 months, when the compost pile cools and has the color and smell of earth; the best indicator is when young herbs start to germinate from the compost. At this stage, the compost should have about 30% moisture.

Composts are applied 1 to 2 times each year. Immediately after they are applied, growers irrigate in order to facilitate the absorption of the nutrients.

1.6   Weed Control

Between rows, weeding is performed with a hoe or machete. At the base of the plants themselves, weeds are carefully pulled out by hand in order to not disturb the shallow rooting structure of the vanilla plants. After removal, weeds that are annual herbs can be added to the mulch or composted and added later. Perennial weeds, such as *Commelina diffusa* and *Syngonium podophyllum*, are removed from the *vainillal* because they do not readily decompose. Weeds should be dealt with whenever they impede access to the vanilla plants and/or when support trees defoliate a disproportionate amount. In general, weeding is performed 3 to 4 times per year.

In shade houses, the rows between plantings are covered with milled "tezontle" (reddish, porous volcanic rock) or ground limestone, in order to prevent the growth of weeds.

1.7   Shade Management (Pruning of Support Trees)

In *vainillales* with living support trees such as *Erythrina* sp. or *Gliricidia maculata*, shade is controlled by periodic pruning, usually 2 or 3 times per year. Pruning should be timed to take place in the rainy season (July–November) to avoid the development of diseases in vanilla due to inadequate sunlight. Shade levels are between 30 to 50% during the rainy season. In dry and hot times of the year (March–June), which coincide with flowering/pollination and fruit development, support trees should have a denser canopy to provide 70 to 80% shade, which conserves humidity, prevents burning from intense sunlight, and decreases the incidence of young fruit drop.

Pruning is accomplished by removing the thicker central branches and leaving the laterals in order to achieve a canopy in the shape of a parasol that also maximizes the equitable distribution of vanilla shoots. Branches are pruned with either saws or machetes, down to about 40 cm from where they diverge from the trunk. The thinnest of the cut branches are broken into longitudinal pieces and placed at the base of the support as an additional source of organic material. Thicker branches are removed from the *vainillal* entirely. Over-pruning results in sunburn to the vanilla plants, and should be avoided.

With orange tree supports, shade management also consists of eliminating young buds, which impede the growth of the vanilla plant. Shoots of the orange tree are pruned when they over-shade the vanilla, which are generally the unproductive or dry/dead shoots. This pruning is generally performed once to twice per year following flowering and the harvest.

1.8   Shoot Management – Looping

The most common practice involving shoot management is "looping", i.e. re-directing a growing shoot over a branch and towards the ground once it reaches the height of the first branches of the support tree. This practice maintains the height of the vanilla at roughly 2 m, facilitating hand pollination and harvesting. Another consequence of looping is hormonal induction promoting flowering and new shoot formation (usually just below the height of the fork in the tree where the shoot is bent). Shoots are managed so that they are equally distributed among the branches of the support tree such that no one shoot shades out another.

1.9   Shoot Management – Rooting

Once a shoot has been looped and has reached the level of the ground, a portion of it, usually 2 to 3 internodes long, is buried, leaving the growing apical meristem uncovered. This practice promotes root formation at the buried nodes. The shoot apex is fastened back to the support tree to continue growing. Rooting of shoots is performed every instance a new shoot has reached ground-level, helping to maintain the vigorous growth of the plant, which obtains more nutrients and is more resistant to *F. oxysporum*. In this way, rooting helps counteract the mortality of plants due to pathogens (Hernández-Hernández 2005).

1.10   Main Vanilla Insect Pest

The *"chinche roja"* (*Tenthecoris confusus* Hsiao & Sailer [Hemiptera: Miridae]) is a small insect that passes through many life stages, including 4 instars. At the nymphal stage, it measures less than 5 mm in length (Figure 1.2). It is at this stage that it causes the most damage to vanilla. As an adult, the *chinche* measures 5 to 6 mm and is black and red, from where it gets its name (Pérez 1990; Arcos *et al.* 1991; Sánchez 1993).

The *chinche* is the single most damaging vanilla pest, causing tissue damage in the leaves, stems, and fruits. The wounds left by the *chinche* allow for the colonization of fungus and bacteria that cause rot, wilting, and defoliation.

The *chinche* is controlled when it is present at low population density, simply by killing them by hand in the early morning hours (when they are most present and least active). An organic control is prepared from 3 onions, 3 heads of garlic juiced in a blender, and a bar of pH neutral soap (in pieces), all dissolved in 40 L of water. The solution is left to sit for 48 hours and applied to the vanilla plants with a sprayer. Applied

Figure 1.2 The main vanilla insect pest, "chinche roja" (nymphal stage).

correctly, it is more effective at eliminating *chinches* than other insecticides (Hernández-Hernández 2008).

An alternative organic control is oil from the neem tree (*Azadirachta indica*). The dosage is 4 mL of neem oil per 1 L of water. Neem oil is a natural insecticide that is biodegradable and non-toxic to beneficial insects and to humans.

Vanilla also suffers herbivory from caterpillars that occasionally damage floral buds.

1.11 Main Vanilla Diseases

Root/stem rot (*Fusarium oxysporum* f. sp. *vanillae*) is a fungus that causes rotting of the roots, stems, and fruits, and plant mortality. It is found to some degree wherever vanilla is cultivated, principally where management is deficient and/or in plants that are bearing fruit. In Mexico, it is estimated to kill 67.4% of vanilla plants within 4 years of planting (Hernández-Hernández 2004).

When *Fusarium* infects the plant, it is very difficult or impossible to eliminate. Prevention is the best practice, and can be achieved by different techniques: using well-drained ground, planting only healthy and vigorous plants, ensuring the roots are always protected with a layer of organic material/compost, meeting nutritional requirements, looping and rooting shoots, avoiding over-pollination, regulating shade, and eliminating diseased plants or buds.

Fungicides may be applied during the rainy season, once to twice per month to prevent infection. Either carbendazim or Bordeaux mixture can be used, in the dosages indicated.

1.11.1 Anthracnose

This disease, caused by the fungus *Colletotrichum* sp., attacks leaves, fruits, stems, and flowers. It is identified by small, sunken spots that are dark brown. Infected fruits fall from the plant before they mature, and so overall yield decreases, sometimes by as much as 50%.

Anthracnose is prevented by ensuring that roots are healthy and that the plant is well-nourished. Fungicides can also be applied, such as inorganic copper oxy-chloride or Mancozeb at concentrations of 2 g/L in water or Bordeaux mixture. The application is done immediately after the cool winter winds (nortes) begin.

1.11.2 Rust

Rust (*Uromyces joffrini*) is identified by the presence of round pustules that are yellow-orange on the underside (abaxial side) of the leaves. As the rust develops, the pustules grow and merge together, eventually drying out the entire leaf. Rust is most frequently encountered in more traditional cultivation systems where there is little ventilation, excessive shade, and where precipitation is too great.

Plants infected with rust cease to develop, and so their productive capacity is reduced. Untreated, rust can defoliate entire plants or plantings.

When the symptoms of rust are first observed, growers immediately eliminate leaves, increasing the amount of light filtering to the plants. Bordeaux mixture or other products that contain copper are then applied weekly, at concentrations of 2.5 g/L of water. Infected leaves are taken out of the *vainillal* and buried.

1.11.3 Yellowing and Pre-mature Fruit Drop

Yellowing and fruit drop of immature fruits manifest at high temperatures exceeding 32 °C, and low relative humidity (<80%), during months of intense sunlight.

The fruit drop occurs 2 months after pollination, mostly in June, after a strong rainfall. Fruit drop varies from 15 to 90%, depending on the cultivation system.

In diseased fruits, two fungal species have been identified: *Fusarium incarnatum-equiseti* species complex and *Colletotrichum* sp. The *Fusarium* is the most commonly encountered, and is thus considered more responsible for causing fruit drop, but only under the environmental conditions cited earlier. In Mexico, these species have only recently been identified (Hernández-Hernández 2007). In India, other species of *Fusarium* have also been identified and reported to produced the same problem (Vijayan and Kunhikannan 2007), although *Colletotrichum vanillae* has also been isolated there as well (Anandaraj *et al.* 2005).

During flowering and fruit development, growers should eliminate the stressful conditions that lead to fruit drop, by maintaining 50% shade and by misting plants. Vanilla should not be cultivated in areas with poor ventilation since this raises temperatures, and leads to problems of stress and pathogen development.

1.12 Flowering and Pollination

In general, the first flowering, or "rehearsal" ("*ensayo*"), happens 3 years following planting. When *Citrus* sp. are used as supports, or when vanilla is cultivated in shade houses, flowering initiates in the second year since the plants tend to grow more vigorously as a result of more consistent shade and management.

The physiological cue to flower is promoted by climatic or mechanical stress. The principal stress in Mexico that induces flowering are the low temperatures of Autumn and Winter, when cool air masses known as "*nortes*" blow down unimpeded from the Arctic Circle, dropping temperatures to below 10 °C; the lower the temperature, the greater the expectation of a good flowering year. The cool temperatures "burn" the apical tip, killing it, and break the apical dominance of the plant while stimulating lateral floral buds to develop. The flowering season is March to May, with peak flowering occurring in April.

1.12.1 Percent of Flowering Plants

The percentage of plants that flower varies each year. The first flowering usually involves a low percentage (27.19%) of plants, but by the third year of flowering (fourth or fifth year after planting) this amount reaches 97.07%. After the third flowering, the percentage of plants that flower may increase or decrease. Heavy flowering in one year is generally followed by reduced flowering the following year, due principally to the low number of developed flowering shoots. There are also numerous other mitigating factors, such as the amount of light filtering through to the vanilla plants, the health of the plants, etc.

1.12.2 Natural Pollination

Mexico is one of few countries where it is possible to obtain vanilla beans through natural pollination, although it happens rarely, accounting for only about 1% of all fruits. The identity of the natural pollinator(s) of vanilla is unclear, and for a long time it has been said that bees (*Melipona beechii*), hummingbirds (*Cynniris* sp.), and bats pollinate vanilla. The preponderance of evidence favors the hypothesis that the most common pollinator is the shiny green orchid bee *Euglossa viridissima* (Soto-Arenas 1999a, 2003; Hagsater *et al.* 2005; Lubinsky *et al.*, 2006). These bees have been documented visiting vanilla flowers but their visits are irregular and their potential for effecting pollination even smaller, perhaps only just 1 fruit per 100 or 1,000 flowers (Soto-Arenas 1999a,b).

Other orchid bees, namely, individuals of *Eulaema* sp. ("*jicotes*"), frequently visit the flowers of *V. pompona* in northern Veracruz, Mexico (Figure 1.3). On rare occasions, they also effect pollination of the flowers (5%) while looking for nectar inside and at the base of the labellum. The mechanism by which these bees actually pollinate vanilla flowers is yet to be documented.

1.12.3 Hand Pollination

Inside the labellum of the vanilla flower, the part which attaches to and wraps around the column, is a tissue that flaps down from the column, called the rostellum. The rostellum hangs exactly in between the stigma (female organ) and the anther sac (male organ), and is considered to be a product of evolution selected to prevent self-fertilization. In hand pollination, pollen is manually moved from the anther sac to the stigma, bypassing the rostellum.

Figure 1.3 An *Eulaema* sp. bee on a flower of *V. pompona*.

Hand pollination is performed with a small, thin stick roughly the size and shape of a toothpick, but can be made from bamboo, bone, spines, or other materials (Figure 1.4). The method of hand pollination consists of:

i) Use a toothpick or similar tool to make a longitudinal slit in the labellum on the side opposite of the column to reveal the reproductive structures.

Figure 1.4 Hand pollination of a vanilla flower.

ii) With the same end of the toothpick, lift underneath the rostellum and flip vertically so that the anther sac can hang down unimpeded over the stigma lobes.

iii) Gently press the anther to the stigma until the two stick together and then remove the toothpick.

Hand pollination is performed from 7 am to noon, or a little bit later when it is overcast, but never when the flowers have already closed or withered. Hand pollination should be conducted by able and experienced people. Women are more commonly involved in the task. An experienced person pollinates 1,000 to 1,500 flowers per 5 to 7 hour period (ca. 4 flowers/minute), assuming that the plants are in the same area. The first flowers in the raceme that are pollinated yield longer and straighter fruits, while the last flowers to open characteristically produce smaller and curved fruits that have less value.

Hand pollination is a daily task for a period of 3 months. Per hectare, 300 to 600 days of work are required to carry out pollination, depending on the abundance of flowers, their location, efficacy of the pollinator, and distance between plants.

1.12.4 Quantity of Flowers to be Pollinated

In general, 6 to 8 flowers per raceme are pollinated to ensure obtaining a minimum of 4 to 5 fruits of acceptable quality (pollination is not 100% successful). Obtaining 100 to 120 fruits per plant requires 8 to 5 flowers per raceme to be pollinated. These approximations are rough since much depends on environmental conditions, the position and vigor of the plants, as well as the biological characteristics of the clone or cultivar. Vanilla growers determine the amount of flowers to be pollinated by considering pricing as well. Over-pollination leads to an abundance of many smaller fruits of lesser value that increase the cost of pollination and exert a heavy cost on the plants. Over-pollination is also associated with major fluctuations in production volume from year to year (Hernandez 1997).

1.12.5 Fruit Development

Immediately following hand pollination, pollen tubes begin their germination and growth and eventual fertilization of the ovules. The ovary quickly begins to enlarge and assume a strong, dark green aspect as it orientates itself downward. The maximum length and diameter of the fruit is achieved 45 days after hand pollination (Figure 1.5). Afterwards, growth ceases, and the fruit enters into a period of maturation lasting roughly 7 to 8 months.

1.13 Harvesting

The harvest in Mexico begins on December 10 of each year, in respect of an agreement taken by growers, curers, and industrial manufacturers. Growers harvest their entire crop in a single day, with the fruits at different stages of development. These stages can be significantly different, since flowering occurs over at least a 3-month time period. The heterogeneity in harvested fruits effects attempts at dehydrating the beans during the curing process, since immature fruits lose water more quickly than mature fruits.

Figure 1.5 Developing vanilla fruits.

The ideal is for fruits to be harvested only when they have reached a ready stage for commercialization, that is, when the distal tip of the bean changes color from green to yellow. This transition normally occurs 8 to 9 months following pollination.

1.13.1 Harvesting Practices

In order to avoid rapid dehydration, the whole bundle or raceme of fruits is harvested with hand shears. The central stalk of the inflorescence, the rachis, remains attached. Harvested fruits are placed in baskets or plastic crates to prevent mechanical damage, which can lead to pathogen infection. The fruits are also kept in well ventilated and shady areas.

After harvesting, it is customary to prune shoots that have already flowered. These shoots will not produce again (or as much) unless they retain buds. The pruning is performed with a knife or blade that is disinfected prior to use in a solution of 1 part bleach to 6 parts water.

The removal of "spent" shoots serves to eliminate unproductive parts of the plant that occupy space and deplete the plant's energy resources. Their removal facilitates the maintenance of adequate ventilation and light conditions for the plant. Some of these spent shoots may serve as cuttings to start new plants if they retain meristematic tissue.

1.13.2 Preventing Theft

Mexico has taken some actions to prevent theft:

i)   Each grower should have a permit to transport and sell vanilla. The permits can be obtained direct from SAGARPA, from the Consejo Nacional de Productores de

Vainilla, from regional government offices, or from local officials. Officials may con-fiscate vanilla from a person who cannot present their permit. Middle-men are noti-fied that they should not buy vanilla from a grower who does not present his/her permit, since the vanilla could have been stolen. In practice, middle-men do make purchases without permits, since they can obtain more vanilla for a cheaper price.

ii) Growers have sought out and receive help from state security forces to protect and transport vanilla (via horse-back escorts or helicopters). This happens when the price of vanilla is high, so when the risk of theft is high.

1.14 Green Vanilla Commercialization

The majority of vanilla growers in Mexico sell non-value-added, green vanilla to mid-dle-men and processors who cure and export the cured vanilla beans. The two cities of Papantla and Gutiérrez Zamora, both in Veracruz, serve as the centers of vanilla curing and export. Green beans are sourced from growers in the state of Veracruz, as well as from Puebla and Oaxaca.

1.14.1 Prices

Prices for green vanilla are set by curers-exporters who consider world prices, supply and demand, costs for curing and exporting, etc., in order to ensure a profit. In recent years, vanilla growers have been forced to sell green vanilla at a loss, on average $4 USD/kg. One exception are growers who sell to the Consejo Nacional de Productores de Vainilla (Asociación de Vainilleros), at a fixed price of $8 USD/kg (2008–2009 har-vest), for beans that are larger and better quality than average. Growers are paid only after the vanilla is cured and sold. Some growers have also sold green vanilla to private companies, for as much as $12 USD/kg, but for individually harvested beans longer than 20 cm.

1.15 Curing

The curing process allows for the development of aromatic compounds and flavor in vanilla beans that can be used in different industries and applications.

In Mexico, curing is accomplished in a traditional, artisan style that includes ovens, and sun curing of vanilla beans laid out on mats of woven palm (*"petates"*) to facilitate cellular breakdown and dehydration (Figure 1.6). The entire process lasts 3 to 5 months (January–May), and consists of:

i)   *Selection and "despezonado"*: Beans are detached from the rachis, or *"pezón"*, and sorted by size and type. The type classes are "entire", "split" (i.e. when the vanilla beans have opened), "painted/spotted" (fruits infected by *Colletotrichum* sp.), and *"zacatillo"* (i.e. small and curved beans). Each class is cured separately, because of the differences in quality.

ii) *Cellular breakdown in ovens, or "killing"*: This step terminates the cellular processes of the beans, and among other consequences, prevents beans from opening

Figure 1.6 Sun curing of vanilla beans on mats of woven palm.

further. The fruits are placed in wooden boxes or inside folded *petate* mats, and placed in ovens from 24 to 48 hours at a temperature of 60°C. Afterwards, the fruits are removed and placed in larger "sweat boxes" for usually 18 to 24 hours (but sometimes as long as 48 hours) to receive their first sweat. The sweat-boxes are capped with matting and *petates* to prevent heat loss so that the beans continue to sweat. In recent years, some curers have replaced the oven method with the Bourbon process of killing beans in hot water, as is used in Madagascar.

iii) *Sun curing and successive sweating*: The fruits are removed from the sweat boxes and placed on *petates* on a patio with full sun for 3 to 4 hours, during which they are allowed to reach a maximum temperature of 50 to 55°C. Immediately afterwards the beans are returned to the sweat-boxes and once more are insulated with a covering of *petates* in order to conserve heat and allow the beans to gradually lose water. The following morning, usually between 9 to 10 a.m, the beans are taken out of the boxes and repositioned on the patio in full exposure to the sun. This cycle of sun curing followed by sweating is repeated until the beans reach a 30% humidity content and a dark brown color, usually after 11 cycles for younger, less mature fruits and 24 cycles for fully mature fruits.

iv) *Classification of cured beans*: Due to the fact that the curing process is not uniform, beans are re-classified according to how they feel and look. This is usually done after 8 to 11 cycles of curing. The beans are grouped according to their thickness (thick, intermediate, or thin), which is an indicator of moisture content. Once sorted and separated, these groups receive different amounts of curing/sweating. When curing is finished, the beans are re-classified again, this time according to

Figure 1.7 Classification of cured vanilla beans.

thickness, flexibility, and color. The classification scheme includes three categories, "supple/raw", "bland", and "dry", indications of the progress of the curing.

v) *Conditioning*: Beans classed as "dry" are no longer cured, but instead placed on wooden racks ("*camillas*") so that they continue to gradually develop flavor and aroma. The beans are also inspected at this point to verify that they were adequately cured. If the beans show indications of colonization by fungus, their moisture content is too high, and the beans are returned to the sun to be dry further. Conditioning lasts 30 to 45 days, with every 15-day period serving to mark another round of inspection.

vi) *Classification*: Beans that show no problem of developing fungus are classified by length and quality (color, sheen, flexibility, and aroma) (Figure 1.7).

1.15.1 Yield Ratio of Green/Cured Vanilla

The normal yield ratio of green to cured vanilla is 5:1. In other words, 5 kilos of green vanilla are needed to produce 1 kilo of cured vanilla. This ratio varies according to weight, size, and maturity of the green vanilla beans.

1.16 Grading

Cured vanilla is classified as either "whole", "split", or "*picadura*" ("chopped"). *Picadura* refers to beans that have been cured from immature, small, or damaged fruits or were improperly cured beans. For whole and split beans, five categories have been established in Mexico (Galicia *et al.* 1989; Curti-Diaz 1995):

i) *Extra*: Thick beans, flexible and lustrous, dark brown "chocolate" color, sweet and delicate aroma, with a vanillin content greater than 2.5% of dry weight. These beans are harvested at the optimal time and are well cured.
ii) *Superior*: Similar to "extra", but less thick and lustrous, with a vanillin content between 2.25 and 2.29%.
iii) *Good*: Flexible and lustrous, sweet aroma, dark brown color with red longitudinal streaks, and a vanillin content of between 2 and 2.24%.
iv) *Medium*: Little flexibility/sheen, light aroma, dark brown with light edges, with a vanillin content between 1.75 and 1.99%.
v) *Ordinary*: No flexibility/sheen, weak aroma, light brown with dark edges, with a vanillin content between 1.5 and 1.74%.
vi) *Picadura*: Lowest quality beans, both physically and in aroma. Sold in small pieces of about 1 cm for use in extracts.

In practice, this grading system may or may not be used in lieu of standards set by other countries and/or standards set by the buyer such as "gourmet", "splits", "small", "chopped", etc.

1.16.1 Packing

Mexican vanilla is traditionally shipped in bulk, wrapped in wax paper, and packaged in cardboard boxes (Figure 1.8). "Extra" or "gourmet" vanilla is also sold in rolls called "mazos".

Figure 1.8 Packaged cured vanilla beans.

1.17 Buyers

The principal buyers of Mexican vanilla are international companies such as Aust Hatchman, McCormick, Eurovanille, Vanipro, Coca-Cola, Vanilla Saffron Imports, International Flavors & Fragrances (IFF), Nielsen-Massey Vanilla, and Dammann & Co., among others. Most of these are based in the United States, France, Germany, and Canada. Within Mexico there are also business that buy vanilla for extract manufacture and for re-sale.

1.18 Export Volume

The majority of Mexican vanilla is destined for export. In the past 3 years, since the price has been less than $50 USD/kg, not all of the vanilla in Mexico has been sold, and has remained in warehouses until prices improve.

Usually, Mexico annually exports 20 to 30 tons of cured vanilla, about 1% of total annual supply worldwide. The United States is the number one buyer of Mexican vanilla, followed by Germany, France, Japan, and Canada. About 5% of the supply of Mexican vanilla is sold within Mexico for extracts and for making handicrafts.

1.19 Prices

The price for cured vanilla is set by international companies, and is normally similar to the price in Madagascar. In the last 3 years, these companies have offered less than $50 USD/kg, except in some instances where small quantities of gourmet beans have been sold for $80 USD/kg. Mexico does not enjoy a different price for its quality of vanilla since the international companies/brokers re-sell the Mexican vanilla to the same markets where vanilla from other countries is also sold.

1.20 Aromatic Profile

The aroma of Mexican vanilla is described as intense, sweet, lightly spicy, with tobacco notes. The vanillin content is generally 2%. The characteristic aroma of Mexican vanilla is due to the presence of vanillin as well as other volatile compounds that, while present at low concentrations, nevertheless strongly impact the overall flavor of the beans.

One study found that Mexican vanilla contains 65 volatile compounds, predominantly acids and phenolics (Pérez-Silva *et al.*, 2006). Another study (Hartman 2003) identified 61 volatile compounds, 11 of which were unique to Mexican vanilla: hexanoic acid, vanillyl methyl ketone, methyl eicosanoate, 4-butoxy-3-methyl-2-butanone, methoxy-methylacetate, 4-hexen-1-olacetate, 3-ethyl-3-methylpentane, 2,4-dimethyl-1-heptanol, 4-methylene-2-oxethanone, 2-methyl-3-ethylpentane, and 2-ethyl-1,3-dioxolane. In comparison to vanilla from other countries, Mexican vanilla tends to have greater concentrations of acetic acid and less anisyl compounds (Black 2005).

Mexican vanilla is preferred in the international markets for gourmet uses and for household consumption because of its exquisite taste and aroma. It differs in its aroma and taste from other countries because of its unique compounds and in the curing method it receives.

1.21 Summary

Growers of vanilla in Mexico have started to organize themselves in national and state associations in accordance with legal and judicial frameworks in order to obtain economic resources from the government. Growers have also sought out from the government technical assistance, help with establishing their own curing facilities and organizations (in which growers receive a better price by selling a value-added product), and in linking directly to external markets. In other words, growers have been trying to break the traditional commercialization scheme. As part of this initiative, some growers have also been promoting shade-house cultivation, subsidized by the government, that they hope will produce higher yields because of the high density of plantings and increased overseeing and technology.

A typical feature of vanilla growers in Mexico is that personal investment in time and resources directly correlates with good prices for vanilla. When prices fall, growers decrease their own investments, to the extent of abandoning vanilla cultivation altogether, as is happening currently. This is the main factor that explains why the volume of Mexican vanilla production has been so low for the last 50 years. The interest to cultivate vanilla in Mexico among growers is strong, but the price factor and fluctuations in international demand are the prime determinants for the increase or decrease in Mexican vanilla production.

Few scientific/technical studies in Mexico have addressed how vanilla cultivation can improve, mainly because of a lack of government funds since vanilla does not represent a crop of major socio-economic or political importance in Mexico. There remain few institutions that conduct vanilla research, most of which are thesis projects by university students.

Only the Instituto Nacional de Investigaciones Forestales, Agrícolas y Pecuarias (INIFAP) has two full-time vanilla researchers who have contributed fundamentally to the technological improvement of vanilla cultivation, and to capacity-building, via work-shops and courses for growers. The majority of the applied knowledge in vanilla cultivation is the product of cumulative experience of growers, from generations of transmitting knowledge from fathers to sons.

INIFAP and other institutions have made commitments to establish a germplasm repository and to identify cultivated material, but the lack of funding has made it difficult to realize such advances.

References

Anandaraj, M., Rema, J., Sasikumar, B. and Suseela Bhai, R. (2005) *Vanilla* (extensión pamphlet) Indian Institute of Spices Research.

Arcos, C.G., Jiménez, H.R. and Pérez, A.V. (1991) Dinámica poblacional y ciclo biológico de *Tenthecoris confusus* Hsiao and Sailer en vainilla. 4$_a$. *Reunión anual del INIFAP en el estado de Veracruz (resultados y avances de investigación)*. Resumen de trabajo.

Black, J.M. (2005) Vanilla bean volatile analysis, origin and species. Flavor and Fragrance Research, LLC. Presentation (conference) from Vanilla 2005. *The International Symposium on the Vanilla Business.* Veracruz, Mexico. 15–16 November. www. Spectraflavor.com/pdfs/vanilla-GCO-presentation.pdf. Site accessed August 3, 2009.

Bory, S., Lubinsky, P., Risterucci, A.M. et al. (2008) Patterns of introduction and diversification of *Vanilla planifolia* (Orchidaceae) in Reunion Island (Indian Ocean). *American Journal of Botany*, 95, 805–815.

Bruman, H. (1948) The culture history of Mexican vanilla. *The Hispanic American Historical Review*, 28, 360–376.

Chauds, P. (1970) Les Derniers résultats obtenus par la recherche agronomique dans la culture du vanillier (a la Réunion). *Plantes a éspices*, 25, 101–112.

Chávez-Hita, A. and González-Sierra, J. (1990) Papantla Veracruz: Imágenes de su historia. Archivo General del estado de Veracruz ISBN-968-6171-32-0. Litográfica Turmex. S.A. de C.V. México, D.F.

Curti-Diaz, E. (1995) Cultivo y beneficiado de la vainilla en México. *Folleto técnico para productores. Organización Nacional de Vainilleros Indígenas.* Papantla, Veracruz.

Galicia, R., Garcia, M. and Curti-Diaz. E. (1989) Anteproyecto de norma oficial mexicana - Especias y Codimentos - *Vanilla fragans* (Salisbury) Ames o *Vanilla planifolia* (Andrews) Entera en estado seco. Comisión Nacional de Fruticultura. Memorias, Primera Semana Nacional de la Vainilla. pp. 93–112.

Hagsater, E., Soto-Arenas, M.A., Salazar-Chávez, G.A., Jiménez-Machorro, R., López-Rosas, M.A. and Dressler, R.L. (2005) *Las orquídeas de México: Orquídeas y gente.* Instituto Chinoin, México D.F. pp. 38–71.

Hartman, T.G. (2003) *Composition of vanilla beans from different geographical regions.* Presentation in Vanilla: First International Congress, Princeton, NJ, November 11–12.

Hernández, A.M. (1997) *Crecimiento y reproducción de Vanilla planifolia (Orchidaceae) en Usila*, Oaxaca. Tesis de maestría en ciencias. UNAM.

Hernández-Hernández, J. (2004) Tecnología para producir vainilla. *Memoria técnica no. 12. Día del Productor Agropecuario y Forestal.* Campo Experimental Ixtacuaco. pp. 1–21.

Hernández-Hernández, J. (2005) Comparación de dos sistemas de producción intensiva de vainilla. resultados finales. In: *Avances en la Investigación Agrícola, Pecuaria, Forestal y Acuícola en el Trópico Mexicano 2005.* Libro Científico No. 2. Veracruz, México pp 81–94.

Hernández-Hernández, J. (2007) Bad weather and hurricane Dean: the beginning and end of México's vanilla production collapse of 2007. *Presentation in Vanilla 2007, International Congress.* Jamesburg, NJ, 6–8 November.

Hernández-Hernández, J. (2008) Manejo integral de plagas y enfermedades en vainilla. *Revista Agroentorno* 96, 21–25.

Lubinsky, P., Van Dam, M. and Van Dam, A. (2006) Pollination of vanilla and evolution in Orchidaceae. *Lindleyana*, 75, 926–929.

Lubinsky, P., Bory, S. Hernández-Hernández, J., Kim, S.C. and Gómez-Pompa, A. (2008) Origins and dispersal of cultivated vanilla (*Vanilla planifolia* Jacks. (Orchidaceae). *Economic Botany*, 62, 127–138.

Pennigton, C., Jiménez, F.A. and Theis, T. (1954) A comparison of three methods of vanilla culture in Puerto Rico. *Turrialba*, 4, 79–87.

Pérez, P.E. (1990) *Dinámica poblacional e identificación de la chinche roja en el cultivo de la vainilla (Vanilla planifolia A.) en el municipio de Papantla, Ver. Tesis de Licenciatura.* Universidad Autónoma del Estado de Morelos. Cuernavaca, Mor.

Pérez-Silva, A., Odoux, E., Brat, P., et al. (2006) GC-MS and GC-olfactometry analysis of aroma compounds in a representative organic aroma extract from cured vanilla (*Vanilla planifolia* G. Jackson) beans. *Food Chemistry*, 99, 728–735.

Portères, R. (1954) Le genre *Vanilla* et ses espèces. In: Lechevalier P. (Ed) *Le vaniller et la vanille dans le monde*, Lechevalier, P. (ed.), Encyclopédie Biologique XLVI. Paris. France. pp. 94–290.

Ranadive, A.S. (2005) Vanilla cultivation. In: *Vanilla: First International Congress*. Allured Publishing Corp., Carol Stream, IL, pp. 25–31.

Sánchez, M.S. (1993) Manual de producción de vainilla en el estado de Veracruz. *Folleto para productores No. 6. Campo Experimental Papantla*. CIRGOC-INIFAP.

Soto-Arenas, M.A. (1999a) Conservation of the genetic resources of vanilla. *Journal of the Canadian Orchid Congress* 11 (Abstract).

Soto-Arenas, M.A. (1999b) Fitogeografia y recursos geneticos de la vainillas de Mexico. *Project J101*. Comision Nacional para el conocimiento uso de la biodiversidad, Mexico City, Mexico.

Soto-Arenas, M.A. (2003) Vanilla. In: *Genera Orchidacearum Vol. 3 Orchidoideae (Part 2). Vanilloideae*, Pridgeon, A.M., Cribb, P.J., Chase, M.W. and Rasmussen, F.N. (eds), Oxford University Press, Oxford, pp. 321–334.

Vijayan, A.K. and Kunhikannan, C. (2007) Assessing the risk of aflatoxin in vanilla for export from India to USA. Workshop. Indian Cardamom Research Institute and Institute of Forest Genetics and Tree Breeding. http://www.angrau.net/participantsPresents.htm (accessed August 14, 2009).

Wong, C., Wong, M. and Grisoni, M. (2003) Culture de la vanilla. *Fiches techniques/technical leaflets*. 37 p. (audiovisual material).

2

Vanilla Diseases

Juan Hernández-Hernández

2.1   Introduction

Diseases are some of the main factors that damage and reduce vanilla production, as well as the productive period of a plantation. The conditions of temperature and humidity under which vanilla grows tend to favor the development of pathogens, mainly fungi. The incidence of diseases is higher in traditional culture systems, plantations in the stage of production, and in older plantations. This chapter describes the main diseases of vanilla in Mexico and presents guidelines for prevention and control. It is always better to prevent a disease than to try to control the damage. When necessary, it is important to use chemical control in a rational way to avoid environmental contamination and also to respect the norms for production of healthy foods. Some environmental conditions leading to damage of vanilla plants are also discussed.

2.2   Root and Stem Rot (*Fusarium oxysporum* f. Sp. *Vanillae*)

This disease is caused by the fungus *Fusarium oxysporum* f. sp. *vanillae*, also known as *Fusarium batatatis* var. *vanillae* Tucker (Childers and Cibes 1948; Bouriquet 1954; Childers *et al.* 1959; Ben Yephet *et al.* 2003; Ranadive 2005; He 2007).

2.2.1   Description

F. oxysporum f. sp. *vanillae* is the most harmful fungal pathogen of vanilla, causing root and stem rotting and consequently the death of the plants. The fungus lives in the ground and is difficult to eliminate. Lesions in the roots are initially brown, which is followed by a blackening, and finally the infected tissue dries out (Figure 2.1). In general, a plant with root rotting, will also display apical rotting, stop producing new buds or shoots and, therefore, its growth pauses. A plant with root rotting does not die quickly, since it develops new roots from the aerial part of plant, which grow to the ground and, if they find sufficient moisture and organic matter, can survive for a considerable time.

Handbook of Vanilla Science and Technology, Second Edition.
Edited by Daphna Havkin-Frenkel and Faith C. Belanger.
© 2019 John Wiley & Sons Ltd. Published 2019 by John Wiley & Sons Ltd.

Figure 2.1 Stem rot (left), and a diseased plant exhibiting symptoms caused by *Fusarium oxysporum* f. sp. *vanillae* (right).

However, if there is not enough moisture, the stem dehydrates showing longitudinal cracking, the leaves wilt and become yellow, and the plant finally dries out and dies (Curti-Diaz 1995; Hernández-Hernández 2005). Stem rot disease begins with a dark lesion that extends longitudinally, eventually covering the stem, and the plant dries out (Loredo 1990).

Heavy and prolonged precipitation, deficient soil drainage, excess shade, poor ventilation, drought stress, low nutrients, and high plant density are favorable conditions for root and stem rot disease. Plants that are not well rooted, with inadequate nutrition, over-pollinated, under drought stress, or planted at high densities, are the most susceptible to the disease. The rotting of roots is observed when there is high moisture in the soil; however, the number of dead plants exhibiting the disease symptoms is higher under drought conditions.

2.2.2 Damage

The fungus causes varying degrees of damage and production losses in vanilla plantations throughout the world. In Mexico, it killed 67% of plants in a 4-year-old plantation and also infected 15% of the total fruit production (Hernández-Hernández 2005). This fungus has been the main limiting factor for some vanilla producing countries, such as Puerto Rico, Costa Rica, China, and currently Madagascar (Anonymous 2008a,b).

2.2.3 Control

When the fungus infects the plant, it is difficult to eradicate the disease. Therefore prevention of the disease through cultural methods is recommended. Some of the cultural practices that are recommended for disease prevention are:

- Use land with good drainage.
- Use healthy and vigorous cuttings.
- Maintain a 10 cm cover of mulch over the roots.
- Keep plants well-nourished.
- Avoid overcrowding plants by maintaining appropriate distance between plants (1.5–2.0 m × 2.0–2.5 m).

- Avoid excess shade and excess sunlight.
- Prune plants to remove infected parts.
- Avoid over pollination.
- Sterilize any new planting areas.

Some chemical agents (carbendazim fungicide) have some effectiveness against *Fusarium*. However, the treatments are very expensive and not practical. In addition, they contaminate the ground.

For these reasons, other strategies of control are the use of essential oils (i.e. clove and cinnamon oil), tolerant or resistant plants, and biological control micro-organisms (*Trichoderma harzianum Bacillus* spp., and *Pseudomonas fluorescens*).

Although *V. planifolia* is susceptible to *Fusarium*, some other *Vanilla* spp. are resistant. Resistant hybrid plants, the product of crosses and backcrosses between (*V. planifolia* × *V. pompona*) × *V. planifolia* were developed in Madagascar. These plants were called "*Tsy taitry*", which means "nonsusceptible" (Anonymous 1995; Grisoni *et al.* 1997). The hybrid plants are very vigorous and produce very heavy fruits larger than 10 mm in thickness and 20 to 30 cm in length, but were not commercially cultivated in Madagascar.

2.3 Black Rot (*Phytophtora* Sp.)

2.3.1 Description

This fungus is very aggressive, it can attack any part of the plant and kill it in only a few days. The disease is distinguished by watery injuries of greenish to blackish color and causes general rotting of the infected tissue. One week after the infection, fine (thin) white filaments, the mycelium of the fungus, are observed (Wong *et al.* 2003; Anandaraj *et al.* 2005). Damage begins in the apical part of the plant and extends to the stem, leaves, aerial roots, and the rest of the plant. However, damage can be restricted to immature fruits or to specific plant parts. The disease can be confused with that caused by *Fusarium*, but *Phytophtora* is less aggressive and it differs in the formation of a mycelium on the injury and production of pin-head sized conidia (Anandaraj *et al.* 2005). The favorable conditions for the development of the disease are prolonged rains, poor soil drainage, excess shade, high plant density, and deficient control of weeds.

2.3.2 Damage

The disease causes high losses in production due to rotting, falling of fruits, and loss of plants.

2.3.3 Control

The incidence of the disease can be diminished by using the appropiate distance in between plants, from 1.5 to 2.0 meters between plants and from 2.0 to 2.5 meters between rows. The tutors should be pruned to allow from 30 to 50% or more of sunlight. Weeds should be controled. Infected plant parts should be removed and burned. Wong *et al.* (2003) recommend the monthly application of the following mixtures: Fosetyl-Al

(2.5 g/L of water) + Carbendazim (2.0 g/L of water) and the mixture of Metalaxyl (2.5 g/L of water) + Benomyl (2.0 g/L of water).

2.4 Anthracnose (*Colletotrichum* Sp.)

2.4.1 Description

The fungal pathogen *Colletotrichum* sp. attacks leaves, fruits, stems, and flowers. Characteristic of the disease are the small sunken dark coffee spots, irregular in color (Figure 2.2). It damages the leaves and the stem during the time called "nortes", the season characterized by cold air and moderate rain (Curti-Diaz 1995; Hernández-Hernández 2005). In general the symptoms develop on the first five young leaves of the apical part of the plant.

Fruit damage (Figure 2.3) is pronounced during the humid and warm months. Although the symptoms are similar to those on the leaves, the pathogens can be considered different species or forms of the fungus, because they appear in different climatic conditions (Hernández-Hernández 2005). An excess of shade and high density of plants favors anthracnose development, as well as root rot and stem rot.

Figure 2.2 Anthracnose on leaves: initial, intermediate, and end stage.

Figure 2.3 Damaged fruits, "pintos", caused by *Colletotrichum* sp.

2.4.2 Damage

Damage of leaves and stems results in a reduction of new growth. Infected fruits fall prematurely before reaching their commercial maturity and the yields fall significantly, up to 50%.

2.4.3 Control

Anthracnose attack can be prevented by maintaining healthy root systems and adequate plant nutrition. Also, it can be prevented by applying any fungicide that contains copper oxychloride or mancozeb, in concentrations of 2 g/L with water or Bordeaux mixture (1 kg of lime + 1 kg of copper sulphate in 100 liters of water) before or immediately after the arrival of "*norte*". To avoid burns it is important not to apply copper compounds on days with intense sunlight or during flowering and development of fruit. Young leaves and fruits affected with antracnose must be removed and buried outside the plantation to avoid further infection sources.

2.5 Rust (*Uromyces* Sp.)

2.5.1 Description

Rust is charactherized by the presence of yellow-orange spots on the leaves (Figure 2.4). As the disease advances, the pustules coalesce, eventually resulting in completely dried leaves. This fungal disease is more frequent in traditional production systems with little ventilation and excess shade, and in very rainy places.

Figure 2.4 Symptoms of *Uromyces* sp. on a vanilla leaf.

2.5.2 Damage

Plants affected by rust stop growth and development. Therefore the disease eliminates the productive capacity and if it is not controlled in time, the resulting defoliation of the plants can destroy the vanilla plantation.

2.5.3 Control

Infected leaves should be removed and burned as soon as symptoms are observed. Also, it is important to increase the amount of light within the plantation and to make weekly applications of Bordeaux mixture, or other products that contain copper, in concentrations of 2.5 g/L of water.

2.6 Rotting of Recently Planted Cuttings

2.6.1 Description

According to Dequaire (1976), this rotting can be caused by *Fusarium oxysporum* and *Rhizoctonia solani* (syn. *Corticium solani*), but the primary causal agent is not known. Days after planting, cuttings exhibit rotting of the underground section, which advances towards the upper part of the stem. In some cases, soft rotting is observed as well as the formation of white-cottony mycelium at the base of the stem (Figure 2.5).

2.6.2 Damage

The percentage of damaged cuttings varies from 5 to 50%, depending on the quality and health of the cuttings, the season of planting, and the type of land. Thus, more damage

Figure 2.5 Rotting caused by *Fusarium oxysporum* and *Rhizoctonia solani*.

is observed when cuttings are planted in rainy months. Poorly drained land presents the greater percentage of rotting. Also, the percentage of damage increases when very young cuttings are used, they are not disinfected, or the soil was already contaminated with fungi. Rotted cuttings will not produce roots or vegetative growth.

2.6.3   Control

Only healthy cuttings should be planted and they should be desinfected with carbendazim (2 g/L of water). Planting should be done during the less rainy months, but after an irrigation or rain. If the cuttings become infected by the fungus, they should be replaced by new healthy cuttings and the soil should be drenched with carbendazim (2 g/L in water).

2.7   Yellowing and Shedding of Young Fruits

2.7.1   Description

In Mexico, the yellowing and shedding of young fruits happens 2 months after pollination and with greater intensity in June, after heavy rain. Intense sunlight with high temperatures (>32 °C) and low relative humidity (<80%) are characteristic of May to June and favor infection. The fallen fruits (Figure 2.6) are of normal size, but of yellow color and smaller weight, without the floral remainder (corola). The color on the inside is coffee and with tender white seeds. After fruits fall, or even before, rotting appears in the apical part and continues throughout the fruit.

In Mexico, *Fusarium incarnatum-equiseti* species complex and *Colletotrichum* sp. have been isolated from yellowing fruits (Hernández-Hernández 2007). However,

Figure 2.6 Yellowing and falling of fruits: diseased fruits (left) and healthy fruits (right).

Fusarium is found most frequently, which is why it is considered the probable causal agent. It develops when environmental conditions are appropriate. In India, *Fusarium* sp. has been reported as causing the same problem (Vijayan and Kunhikannan 2007), although *Colletotrichum vanillae* has also been found (Anandaraj *et al.* 2005).

2.7.2 Damage

The damage is more severe in plantations exposed to high sunlight and with poor ventilation, for example, in plantations under plastic mesh (shade-house) with temperatures of 45 °C. In these conditions, up to 90% of the fruits can fall. In plantations where the vanilla is grown on tutors of orange trees at a spacing of 5 × 5 m between trees and 7 × 7 m between rows, which results in intermediate shade and greater ventilation, losses have been quantified around 50% of fall of yellow fruits. On the other hand, in intensive systems with high densities of *Erythrina* sp. or *Gliricidia sepium* tutors (1.5 × 2.5 m) and therefore better shade, the fall of fruits has been lower than 15%. Also, the damage is more severe in the border plants, which are not protected from the sun, and in fruits without the remaining floral parts (corolla), since this favors dehydration and attack by pathogens

2.7.3 Control

During the flowering stage and development of the fruit, the conditions conducive for the development of the disease should be controled. It is important to provide the crop with greater than 50% shade and sufficient irrigation. Also, vanilla should not be cultivated in spaces with poor ventilation, since the temperature is increased, which can cause major damage. In India, it is also recommended to apply any of the following fungicides: Methylic Tiofanato (0.2%) or the mixture of carbendazim + mancozeb (0.25%), during the time of flowering and pollination, with intervals of 15 to 20 days to prevent the development of the mentioned fungi (Anandaraj *et al.* 2005).

2.8 Viral Diseases

Vanilla is also affected by viral diseases, mainly in the plantations of French Polynesia and India, where they represent a serious problem. In Mexico, there are no scientific reports of damage by viruses (Hernández-Hernández 2008). However, Soto-Arenas (2006) reported the presence of some symptoms of virus in Veracruz, which may be limiting the vanilla production. The damage caused by viruses can be difficult to distinguish, since some plants do not exhibit clear symptoms or are asymptomatic. The viruses most common in vanilla, according to Pearson *et al.* (1991), are described in Sections 2.8.1 to 2.8.4.

2.8.1 Cymbidium Mosaic Virus (CYMV)

The plants infected with the virus are generally asymptomatic, but occasionally they exhibit mild chlorosis in the leaves of *V. planifolia* and *V. tahitensis*. The virus is transmitted through the sap and dispersed through propagation material. It is not known if it is transmitted by a vector. The virus was first reported in the vanilla producing region of the South Pacific (French Polynesia). It has since been found in vanilla plots in many countries, such as Madagascar, Reunion Island, and India (Grisoni et al. 2010).

2.8.2 Vanilla Mosaic Virus (VMV)

The virus causes distortion of the leaf and mosaic lesions in *V. planifolia*, *V. pompona*, and *V. tahitensis* (Figure 2.7). It is transmitted by the sap and is spread by using infected cuttings in the establishment of the crop. Tests of transmission have shown that this virus can be transmitted by aphids (*Myzus persicae*). This virus occurs mainly in the islands of French Polynesia, where *V. tahitensis* is the cultivated species.

2.8.3 Vanilla Necrosis Potyvirus (VNPV)

Plants infected with VNPV exhibit distorted young leaves with chlorotic spots and necrotic lesions in leaves and mature stems, eventually resulting in defoliation and death of the plant. It is transmitted by the sap and dispersed by propagation material. It has been reported in *V. planifolia* cultivated in Tonga, Fiji and Vanuatu.

2.8.4 Odontoglossum Ringspot Virus (ORSV)

Plants infected by ORSV are generally asymptomatic, although sometimes small spots on the leaves of *V. planifolia* and *V. tahitensis* are observed. The virus is reported in the

Figure 2.7 Typical symptoms of distortion and mosaic lesions on leaves of *V. tahitensis* infected by vanilla mosaic virus.

producing region of the South Pacific. The virus is transmitted through the sap and dispersed through propagation material. The transmitting vector is not known.

2.8.5 Prevention of Viral Diseases

It is important to use healthy certified cuttings, control the insect vectors (aphids), and eliminate weeds and other crops around the plantation that can be reservoirs for the virus, for example, *Commelina difusa*, *Cucurbita maxima*, *Physalis angulata*, *Momordica charantia*, watermelon, and pumpkin (Wong *et al.* 2003; Anandaraj *et al.* 2005). Plants with virus symptoms must be removed from the plantation and burned. The movement of infected cuttings from one region to another must be avoided.

2.9 Damage by Adverse Climatic Factors

In addition to the direct damage caused by diseases, vanilla culture is affected by environmental conditions, which can significantly affect the efficiency of vanilla production.

2.9.1 Natural Pruning of the Apical Buds

2.9.1.1 Description
During the winter, when temperatures of around 7 °C extend for more than 1 hour, the terminal shoots are burned. They initially exhibit a light brown color, and later with the humidity of rains or dew they began to rot and finally dry out, becoming a dark color (Figure 2.8).

Figure 2.8 Apical bud showing damage from cold (natural pruning).

2.9.1.2 Damage

A plant without apical buds no longer grows and therefore it must develop a new bud. For small young plants less than 2 years old, "natural pruning" is not recommended because it delays plant growth. However, in mature plants, the "pruning" serves as an indicator that the plant underwent stress by the cold and that it is going to bloom.

2.9.1.3 Control

In small plants, the damage can be minimized by maintaining high moisture levels in the ground and in the mulch, as well as with natural or artificial shade of around 50% during winter time.

2.10 Damage from Sunburn

2.10.1 Description

Initially, a yellowing in the leaves is observed and later some leaves dry completely (Figure 2.9).

2.10.2 Damage

Sunburn of plants occurs frequently in the intensive production systems where *Erythrina* sp. and *Gliricidia sepium* tutors are used. Serious sun damage can be observed when these species are not pruned at the suitable time (they will shed their leaves in winter), or if they are over-prunned, or if their foliage is damaged by disease. Sun damage is pronounced only in the leaves or stems that received direct sunlight, and the plant

Figure 2.9 Yellowing and sunburn of leaves in a vanilla plantation with deficient shade.

can be total or partially burned. Burned leaves will not recover because their photosynthetic capacity is diminished and therefore the growth of the plant is affected. This condition predisposes the plant to pathogen attacks.

2.10.3 Control

Prune plants at the recommended time to avoid total defoliation and water them during dry periods to accelerate development of new foliage. In addition, in some cases it is necessary to apply chemical control for certain pests. Plants can be covered with some material (banana leaves, grass, etc.) to provide shade and to avoid burns. Also, when little shade is available and there are intense sunny days, plants can be covered with plastic mesh. Although this measure adds an additional cost, it could be justifiable since it protects the plants from sun damage.

2.11 Hurricanes

Hurricanes can cause total losses to vanilla plantations, mainly in the producing regions of the Indian Ocean (Madagascar, Reunion Island, and the Comoros) and Indonesia. In Mexico, these natural phenomena appear in the period of August to October, but do not always affect vanilla plantations. However, a major disaster occurred in 2007 when Hurricane Dean severely affected the plantations located in the coastal zone where it made landfall (Figure 2.10). The damage was mainly to the mesh coverings used for shade (Hernández-Hernández 2007). In order to mitigate the damage, it is necessary to establish supports and tutors that are able to resist the effects of hurricanes. Also, curtains of trees can be established, using for example Australian pine, that serve as wind

Figure 2.10 Plastic mesh used to provide shade were destroyed by hurricane "Dean", in the region of Tecolutla, Veracruz, Mexico, August 22, 2007.

barriers. After a hurricane, the main activity is to repair the mesh used for shade to protect the plants from sunburn, to raise the plants and to apply fungicides, as described previously, as preventive measures against fungal diseases.

References

Anandaraj, M., Rema, J., Sasikumar, B. and. Suseela Bhai, R. (2005) *Vanilla* (extension pamphlet). Rajeev P. and Dinesh, R. (eds), Indian Institute of Spices Research. Kochi, India.

Anonymous (1995) *Vanille: Manitra Ampotony, Tsy Taitry, deux variétes prometteuses.* FOFIFA/Centre National de la Recherche Appliquée au Développemnt Rural. http://www.fofifa.mg/res_van.htm. Site accessed July 27, 2009.

Anonymous (2008a) *Madagascar hit by deadly vanilla-killing fungus.* http://news.mongabay.com/2008/1208-vanilla.html, site accessed July 27, 2009.

Anonymous (2008b) *Une maladie affecte la vanille. L'Express de Madagascar. Economie.* http://www.lexpressmada.com/index.php?p=display&id=21294&search=vanille. Site accessed July 27, 2009.

Ben Yephet, Y., Dudai, N., Chaimovitsh, C. and Havkin-Frenkel, D. (2003) *Control of vanilla root rot disease caused by* Fusarium. Vanilla 2003, November 11–12, Princeton, NJ.

Bouriquet, G. (1954) *Le vanillier et la vanille dans le monde.* Encyclopedie Biologique XLVI. Lechevalier, Paris VI, pp. 459–491.

Childers, N.F. and Cibes, H.R. (1948) *Vanilla culture in Puerto Rico. Cir. No. 28.* Federal Experiment Station in Puerto Rico (USDA). Mayaguez, Puerto Rico.

Childers, N.F., Cibes, H.R. and Hernández-Medina, E. (1959) Vanilla – The orchid of commerce. In: *The Orchids, A, Scientific Survey*. Withner, C.L. (ed.), The Ronald Press Co., New York, pp. 477–508.

Curti-Diaz, E. (1995) *Cultivo y beneficiado de la vainilla en México*. Organización Nacional de Vainilleros Indígenas. Papantla, Veracruz.

Dequaire, J. (1976) L'amelioration du vanillier a Madagascar. *Journal d'Agriculture Tropicale et de Botanique Appliquee*, 23, 140–158.

Grisoni, M., Come, B. and Nany, F. (1997) *Project de reliance de la vanilliculture dans la region du SUVA*. Compte Rendu de Mission a Madagascar. La Reunion.

Grisoni, M., Pearson, M. and Farreyrol, K. (2010) In: *Virus diseases of vanilla*, Vanilla. Odoux, E. and Grisoni, M. (eds.), CRC Press Taylor & Francis Group. USA. pp. 97–123.

He, X-H. (2007) Bio-control of root rot disease in vanilla. PhD thesis, University of Wolverhampton, Wolverhampton, UK.

Hernández-Hernández, J. (2005) Comparación de dos sistemas de producción intensiva de vainilla. resultados finales. In: *Avances en la Investigación Agrícola, Pecuaria, Forestal y Acuícola en el Trópico Mexicano*, Libro Científico No. 2, 81-94. Veracruz, México.

Hernández-Hernández, J. (2007) Bad weather and hurricane Dean: the beginning and end of México's vanilla production collapse of 2007. *Vanilla 2007*, November 6-8, International Congress, Jamesburg, NJ, USA.

Hernández-Hernández, J. (2008) Manejo integral de plagas y enfermedades en vainilla. *Revista Agroentorno*, 96, 21–25.

Loredo, S.X. (1990) Etiología de la necrosis del tallo de vainilla (*Vanilla planifolia* Andrews) en Papantla, Veracruz. *Tesis de Maestría en Ciencias*. Colegio de Postgraduados. Montecillos, México.

Pearson, M.N., Jackson, G.V.H., Zettler, F.W. and Frison, E.A. (eds) (1991) FAO/IBGR Technical guidelines for the safe movement of vanilla germplasm. *Food and Agriculture Organization of the United Nations*, Rome/International Board for Plant Genetic Resources, Rome.

Ranadive, A.S. (2005) Vanilla cultivation. In: *Vanilla, First International Congress*. Allured Publishing Corporation, Carol Stream, IL, pp. 25–31.

Soto-Arenas, M.A. (2006) La vainilla: Retos y perspectivas de su cultivo. *Biodiversitas*, 66, 2–9.

Vijayan, A.K. and Kunhikannan, C. (2007) Assessing the risk of aflatoxin in vanilla for export from India to USA. *Workshop. Indian Cardamom Research Institute and Institute of Forest Genetics and Tree Breeding*. http://www.angrau.net/participantsPresents.htm (accessed August 14, 2009).

Wong, C., Wong, M. and Grisoni, M. (2003) *Culture de la vanilla*. Fiches techniques/technical leaflets. (audiovisual material).

3

Vanilla Production in Costa Rica

Elida Varela Quirós

3.1 Introduction

Costa Rica, a country located in Central America, was discovered by Christopher Columbus in 1502 during his fourth trip. As a result of the conquest by a European country, the culture of Costa Rica is strongly influenced by Spain. Currently, less than 1% of the total population of over 4 million belongs to indigenous tribes. Costa Rica is considered one of the most stable nations in Latin America since its army was abolished in 1949. The current president Oscar Arias won the Nobel Peace Prize Award in 1987 for his support of the Central American peace process. The education is free and obligatory. The official religion is Catholicism; however, there is total freedom of religion. The official language is Spanish and the national currency is the colon. The national flower is the purple orchid (*Guarianthe skinneri*). Costa Rica is divided into 7 Provinces: San Jose (the capital of Costa Rica), Heredia (flowering province), Cartago (the vegetables province), Limón (the Caribbean province), Puntarenas (the Pacific province), Guanacaste (the dry province) and Alajuela (the Mangos province). Costa Rica is divided by a rugged highland landscape from south to north, creating an Atlantic and a Pacific zone with different weather conditions. The volcanoes and the beach area along the country's edge create different climatic zones from one town to another. Costa Rica has two seasons: the dry season and the rainy season. The dry season occurs from December to June in the Pacific zone and from February to May in the Atlantic zone. This is why it is possible to find cloudy forest, dry forest, and rain forests in Costa Rica. With only 51,100 square kilometers, Costa Rica is one of the countries with the most biodiversity in the world. Costa Rica is a biological bridge of botanical and zoological species, where the North and South American fauna and flora converge. There are more than 10,000 plant species, 800 butterfly species, 500 mammalian species, and 850 bird species in this small country. Costa Rica has protected around 25% of its territory under categories such as national parks, biological reservoirs, and national refuges of wild life.

Handbook of Vanilla Science and Technology, Second Edition.
Edited by Daphna Havkin-Frenkel and Faith C. Belanger.
© 2019 John Wiley & Sons Ltd. Published 2019 by John Wiley & Sons Ltd.

3.2   History of Vanilla Production in Costa Rica

The first reference to vanilla production in Costa Rica dates from 1987: "In General, there are about 20 hectares of vanilla plantations at different stages in several parts of Costa Rica" (Ocampo 1987). "There were also four vanilla plantations owned by foreign investors," Ocampo wrote. There is not much information about the first plants of vanilla and where they came from. There was one vanilla hectare planted in Upala, Alajuela, and one more hectare planted in Aguirre, Puntarenas, in 1986. The vanilla hectare planted in Upala was owned by a foreigner who had to travel to Europe before the vanilla harvest. He took some vanilla beans from his plantation from previous harvests to Europe. The beans were liked so much, that he returned, planning to cultivate more vanilla plants. When he came back, however, the vanilla plantation was damaged and he could not continue the vanilla cultivation.

3.2.1   The First Phase of Large-scale Cultivation in Costa Rica

Based on the high vanilla bean prices in the 1990s, and in an effort to develop a better standard of living for rural communities and to preserve buffering zones around the biological reserves and national parks, national and international institutions came together to support vanilla cultivation in Quepos and Puerto Jimenez. Both towns are located in the southern part of Costa Rica. The Biological Reserves of El Nara and Los Santos in Puntarenas were selected to develop this project. A group of farmers that have sustainable agriculture crops, such as beans and corn, came together and planted *Vanilla planifolia* in the boundary zones of the biological reserves. These farmers were the pioneers in the cultivation of vanilla. However, the fungal pathogen *Fusarium oxysporium* f sp. *vanillae* infected the vanilla plantations in 1993 (Ramírez *et al.* 1999). As a result, most of the vanilla plantations disappeared.

In 1995, the University of Costa Rica, together with the Agriculture Department in Quepos and the foundation Holland-Costa Rica (FundeCooperación), initiated a project on the organic development of vanilla cultivation. This was the first step in the development and research of vanilla cultivation in Costa Rica. Unfortunately, on July 30, 1996, Costa Rica sustained huge losses from hurricane Cesar, with 24 persons dead, 6 people missing, 2,875 evacuated, almost 22.83 million dollars of losses in road infrastructure, 7.3 million dollars in hospital installations, 16 bridges in bad condition, 7 drinking water installations affected, and 5 electricity systems suspended (Zúñiga 1996). The direct loss to the agriculture sector was about 1.16 million dollars, nearly 1.2% of the annual production, with about 354 hectares of crops affected. The most affected areas were Quepos, Parrita, and Puerto Jimenez, where 99% of the vanilla plantations were damaged (Marín-Gonzalez 2003). According to the Agriculture Department, almost 33 hectares had been planted with *V. planifolia* in Aguirre, Parrita, Puerto Jimenez, and Garabito (Guzmán-Diaz 1997).

3.2.2   The Second Phase of Vanilla Cultivation in Costa Rica

The Agricultural Microbiology Laboratory of the Agricultural Investigation Center (CIA) at the University of Costa Rica, along with the support of the Interamerican Bank for Development (BID), the National Institute of Biodiversity (INBio), and the private support of the La Gavilana Company, initiated a project to restore the vanilla plantations in Quepos. They found some beneficial micro-organisms, which improve the resistance of the plant to *Fusarium*. The University of Costa Rica produced bio-fertilizer products for use not only in vanilla plants but also in some other crops in the country (Marín-Gonzalez 2003). A few of the farmers in the region have small organic vanilla plantations, for which the most important product is tourism and the second product is the sale of the vanilla beans. As an example of good agricultural practice, the Villa Vanilla spice farm owned by Henry Karczynski, a pioneer of vanilla cultivation in Costa Rica, has the first Demeter* certified biodynamic vanilla farm. This is a sustainable system where the vanilla beans are grown using traditional organic farming. Villa Vanilla gives tours, and conferences and vanilla beans are sold to the public.

3.2.3 The Third Phase

By the year 2000, the vanilla bean price was increasing in the international market, which attracted foreign investors to create large-scale vanilla production companies. The private industry production began with traditional vanilla plantations supported by live tutors, such as a Melina tree (*Gmelina arborea*), *Gliricidia sepium*, and *Erytrina* spp. Soon, these companies would face the same problems as encountered before, disease infection, lack of water, and lack of shade in summer since the tutors were deciduous trees, which drop all the mature leaves in the dry season. But, this time, there was sufficient financial support to apply technological solutions to the process of production.

As one of the solutions to the problem of disease, a new vanilla cultivar, called "Vaitsy", from the Investigation and Research Institute of Agriculture in Madagascar was brought to Costa Rica. The proper species designation and the history of development of "Vaitsy" in Madagascar are not known. It is possible that it is the *V. planifolia* × *V. pompona* interspecific hybrid from Costa Rica, described in Chapter 21 by Belanger and Havkin-Frenkel. It was given to scientists at the Costa Rican Polytechnic University in Santa Clara, San Carlos, where the tissue culture propagation of vanilla was developed. Private laboratories obtained the technology and the mother plant to continue vegetative reproduction of the new vanilla cultivar. The first new vanilla plants of "Vaitsy" were planted with amazing results. The plants had good *Fusarium* resistance and produced large beans, up to 26 cm long, with a 30 g average weight per green bean.

The planting of the new cultivar was not the only technological advancement. The intensive vanilla cultivation under shade cloth was also improved. More than 46 hectares of vanilla were planted in a sophisticated system using organic mediums, less use of pesticides, improved cultural practices such as drainage systems and water irrigation systems and biological control of diseases. The vanilla production was on the way. But the future is unpredictable. After 4 years of practical experience in production, the vanilla bean price began to drop in 2004. By 2005, when Costa Rica had the first large vanilla bean harvest, the vanilla price dropped dramatically. Marketing vanilla beans was difficult and the profit was not enough. Some vanilla companies abandoned the plantations and went into more profitable businesses. Some vanilla plantations still struggle to survive.

Curiously, in 2006, despite the hard times for marketing, the National University of Costa Rica supported vanilla plantations using live tutors. Native trees such as pochote (*Bombacopsis quinatum*), neem (*Azadirachta indica*), teak (*Tectona grandis*), and others are being used (Paniagua 2006). The general objective of the program is to improve non-traditional crops as an economical alternative and support reforestation in the province of Guanacaste. Some other farmers in different parts of the country began planting vanilla as an economical alternative in hard times. In 2008, a national vanilla organization (Vanilla Foundation) was created to support small vanilla farmers. Currently, there are nearly 20 hectares of vanilla among the small farmers, about 20 hectares under shade cloth, and about 10 hectares for tourism purposes located in buffer zones of national parks, biological reserves, and private forest areas.

Currently, the author is in charge of 16 hectares of vanilla planted under shade cloth at Las Dos Mamos Vanilla Limitada in the northern part of Costa Rica, close to Nicaragua. Most of the population around the farm is unemployed. The only source of employment for many families is the vanilla company. This is why we struggle to be profitable, reduce costs, and use environmentally friendly organic cultural practices. Soon we will obtain organic certification for our plantation. History has taught us that vanilla plants require organic matter in the soil and good agricultural practices to thrive. When these are provided, the vanilla plant will produce long, heavy, and aromatic vanilla beans (Figure 3.1).

Figure 3.1 Vanilla beans.

3.3   Vanilla Production – The Traditional System

There are two types of production in Costa Rica; an intensive vanilla cultivation under net houses and the traditional method in open fields using live trees as tutors (Figure 3.2). The traditional system is used by farmers in small-scale cultivation, normally from 0.5 to 2 hectares. The supporting live tutors used are the *Erythrina* spp., guaba, or melina trees. Often, these trees are deciduous and loose their leaves in the dry season when the vanilla plant needs more shade. If the vanilla plant is exposed to direct sunlight, their leaves are burned. Research shows that a later pruning of the tree makes it produce new leaves near to the summer time. These young leaves will not drop as would happen with mature leaves and the vanilla plant can get 50% shade in the dry season. The pruning strategy is to open the live tutor and get as many branches as possible. In September, the first apical meristem is cut, and within a month two branches will have formed. Each branch is cut again and so on until a tree shaped like an umbrella is formed. In addition, the pruned material is used as organic fertilizer for the vanilla plants. In Cost Rica, the orange tree is not used as a vanilla tutor as in Mexico, although there are orange plantations around the vanilla production zones.

   Another disadvantage of the traditional vanilla cultivation system is the water supply. Building an irrigation system is expensive for small farmers. Also, the live tutors make long distance irrigation systems impractical. Individual irrigation is needed for each tree. The organic material is also expensive to obtain in the dry season. On the other hand, the live tutor can supply the organic matter for the vanilla plant in the rainy period. The hours of hand work needed are less than in an intensive system. A single worker can maintain 4 hectares of vanilla plantation in an open field. In summary, there are three main agricultural practices required for successful vanilla cultivation in the open field: shade in the dry period, organic material, and an adequate water supply. There are some other agricultural practices also needed in vanilla cultivation, such as

Figure 3.2 Traditional vanilla plantation.

good drainage and air circulation. The steps establishing a traditional vanilla plantation in Costa Rica are:

- preparing the land;
- planting the live tutor (tree)/building a net house;
- maintenance of the tree, cleaning around the base of the tree, fertilization, land cleaning;
- pruning the tree to an open branched tree;
- applying organic fertilizers at the base of the tutor;
- planting the vanilla plant at the base of the tutor; it should be planted opposite from the sun;
- maintainance of the vanilla plant with weeding, disease control, pest, fertilization, organic matter applications, and water supply;
- guiding the vanilla stem up and down from the tutor trunk and their branches;
- pollinating the vanilla flowers;
- harvesting the vanilla beans;
- processing the vanilla beans;
- selecting and packing of vanilla beans.

Preparation of the land is usually done by tractor. After the trees are planted, the cleaning is done by a weeding machine, as well as by hand around the base of the tree. The planting of the trees is done after an application of organic material. The density of trees is about 2,500 trees per hectare. The first pruning is done when the trees are about 1 m tall. When these trees are 3 years old, the lower branches will be about 2 m above the ground, which is a good height for hand pollination.

The application of 2 to 3 kg of organic matter around the tutor is needed before planting the vanilla plant. This organic matter must provide good drainage, aeration, and nutrients such as nitrogen, phosphorus, and calcium to the vanilla plant. It is very important that there is enough organic matter, both to supply nutrition for the plant needs and to retain enough water for the plant. At least two applications of organic matter per year are recommended. One organic application at the beginning of the dry period should retain enough water to support the plant. The other application should be 6 months later and provide enough drainage and good aeration for the plant roots. Some examples of organic medium used are sawdust, decomposing leaves, cane chaff bagasse, wood chips, rice hulls, and a mixture of organic cane products and rice hulls. The application of organic matter is recommended from December to February and from June to August in Costa Rica.

The maintenance of the vanilla plants includes some applications of organic fertilizer, biological control of fungi, and insect repellents. The guiding of the vanilla stem is up and down from the base of the tree to the top branches. For optimum blooming, the plants should have about 60% shade. Pollination is done in the dry season, from February to June, in the northern part of Costa Rica and from December to May in the Pacific coast region. The blooming is induced by water stress since Costa Rica does not have low temperatures because it is located near the equator. However, after induction of blooming, the plants need to be provided with adequate water. Because the blooming time is in the summer, irrigation is needed to protect the plant from stress and to prevent pollen drying. In addition, the young beans should be protected from lack of water, otherwise they turn yellow and eventually drop. The species *V. tahitensis* is very sensitive to lack of water. On the other hand, the vanilla cultivar "Vaitsy" is less sensitive to lack of water. The harvest is done 9 months after pollination, from December to February in Costa Rica. The vanilla beans are harvested when they are not only mature but ripe, which is indicated by a yellowish color at the tip of the bean. If the vanilla beans are left on the vine too long, they can split and are less valuable.

3.4 Vanilla Production – The Intensive System

The other cultivation system existing in Costa Rica is the intensive vanilla cultivation (Figure 3.3). The pioneers of this system were the foreign investors who implemented an intensive vanilla cultivation under shade netting. The number of vanilla plants per hectare is from 2,000 to 6,000. However, if the plant density is too high, the vanilla plants will not flower or the blooms will be scarce. The more plants per hectare, the lower the yield per hectare. The main density of the plant biomass should have good light exposure, with good air circulation and good drainage. There should be enough space between beds and plants to allow the vanilla plant to fully grow. A growth of 15 m

Figure 3.3 Intensive vanilla production system.

per year, for 5 years, depending of the cultural practices and organic matter used, is to be expected. Since it is an intensive system, it needs more hand labor for weeding, guiding of the stem, and pollination than in a traditional system.

The infrastructure is another difference between a traditional and an intensive system. The initial investment in the construction of net house beds and tutors is expensive, as is the maintenance of these structures for more than 7 years. Because of these high costs in the initial construction, the materials must be of high quality to last more than 7 years, or at least up to the first harvest. The net house post, wires, and tutors must be chosen well, otherwise structure maintenance will be expensive. Another difference of an intensive system is the irrigation system. It must have an irrigation system in the summer period to avoid the fall of the vanilla beans and stressing of the vanilla plant.

3.5 Propagation

There are several vanilla cultivars and species cultivated in Costa Rica. The common vanilla species used are *V. planifolia, V. tahitensis, V. pompona*, and native species from Costa Rica. As discussed in Sectiom 3.2.3, a common cultivar used is "Vaitsy", although the proper species designation is not known. These different vanillas have different physiological and morphological characteristics, such as vanillin content, bean size, stem thickness, and disease resistance (Table 3.1).

There are two propagation methods for vanilla plants in Costa Rica. The conventional way is to take a stem cutting of the vanilla plant 1 m long and put it in a planting medium. Placing a stem node under the medium in a dark and moist location promotes root growth. This vegetative vanilla propagation method has the advantage that in 3 years the vanilla plant could be blooming. This method reduces, from 1 to 2 years, the time before harvest compared to the *in vitro* method. If there is not enough vegetative material for planting, or if there is the need to decontaminate the plantation from a disease, then tissue culture is a good option. The tissue culture method, where small plantlets are produced from a mother plant, is done in private laboratories. In Costa Rica there are three private laboratories that generate tissue-culture derived vanilla plants. The plantlets are given to the vanilla farmer when they reach 3 cm in length (Figure 3.4). Before

Table 3.1 Vanilla cultivar differences.

Characteristics	"Vaitsy"	*V. planifolia*	*V. tahitensis*
Stem thickness, inches	4	2	1
Flower size	Large	Medium	Small
Bean size, cm	From 16 to 27	From 16 to 20	From 13 to 18
Growth rate, m per year	10	5	15
Water stress resistance	Medium	Low	High
Fusarium resistance	Medium	Low	High
Anthracnose resistance	High	Medium	Low

Figure 3.4 Tissue culture derived vanilla plantlets.

planting, these plantlets must first be acclimatized in a separate area provided with 80% shade until they are strong enough to endure rain, sun, pests, and diseases (Figure 3.4). This is a very sensitive stage where significant plant losses were reported in other countries (Chin 2004). The medium used for acclimatization is very important in the reduction of plant losses. The medium usually used is peat moss, rice hulls, or ornamental red stone. After 15 days, the small plants are fertilized and a bactericide can be applied. After 3 months, the plants are ready for planting inside the net house or in the open field. Around 80% survival of acclimatized vanilla plants was reported in some farms in Costa Rica. It can take about 4 to 5 years for the plants to bloom. Since the young plant is soft and succulent, it is more attractive to pests such as crickets and worms and some other insects. The most common disease is *Erwinia xansatinova* at this young stage.

3.6 Diseases and Pests

The most common vanilla pests are worms (*Plusia aurifera*, *Agrotis* sp.), crickets (*Stenacris* sp.), thrips (*Chaetanaphothrips* sp.), slugs (*Vaginulus* sp.), and mice (Figure 3.5 (left)). The common diseases of vanilla plants in Costa Rica are *Fusarium oxisporum* f. sp. *vanillae*, *Phytophthora* sp., anthracnose (*Colletotrichum vanillae*), *Xanthomonas* sp., and *Mycospharella* sp. (Figure 3.5 (middle and right)). The most important means of control are good cultural practices, such as good drainage, good air circulation, adequate organic matter, low vanilla plant density, some insect repellents, and biological control micro-organisms.

Figure 3.5 Damage to vanilla plants from mice (left), *Phytophthora* sp. (middle), and *Fusarium* sp. (right).

3.7   Vanilla Bean Processing

The vanilla pod curing is done in three steps: scalding, sunning/sweating, drying and conditioning. The most common method used is the Bourbon method, where the beans are submerged in hot water for several minutes in the killing process to stop the cellular metabolism of the pod. Sunning and sweating are done by putting the vanilla pods in the sunlight, until they are hot. They are then wrapped in sheets and put into an airtight container overnight. This process is repeated for several weeks until the bean has a moisture content of only 30%. Finally the vanilla pods are conditioned indoors for several months. During this period the pods are selected by size, cultivar, and quality. Then, the vanilla beans are ready to be packed for sale. Costa Rica does not yet have classification standards for vanilla beans size and quality. The classification used by Las Dos Manos Vanilla Ltda. is shown in Table 3.2.

The vanilla beans from Costa Rica are larger than those from other locations. We have had some 28 cm long, with a weight of 14 g per pod. The beans are fleshy, oily, and dark brown brilliant. The fragrance is sweet, woody, and vanillin-like.

3.8   Conclusions

In 16 years of vanilla cultivation in Cost Rica, a lot of new knowledge has been obtained. We know now that organic materials are very important for the vanilla plant. The intensive vanilla cultivation system works if the plant density is low and cultural practices such as drainage, plant guiding, and pest control are done adequately. The acclimation of tissue culture plants is a very sensitive stage of the plant. Since vanilla cultivation is environmentally friendly, it can be located near national parks and biological research preserves. Because of the intensive labor forces needed, the vanilla companies are a very important source of employment for many Costa Rican communities.

Table 3.2 Las Dos Manos Vanilla Ltda. Vanilla bean classification.

Grade	Color and Brilliance	Aroma	Size, cm	Vanillin, %[a]	Water content, %	No. beans per kilogram
Mini	Dark brown	Sweet	14–15	2	20	
Small	Dark brown, brilliant	Sweet	16–18	2	20	155
Medium	Dark brown, brilliant	Sweet	19–22	2	25	90
Large	Dark brown, brilliant	Sweet	23–26	2.5	25	78
Premium	Dark brown, brilliant	Sweet	27–up	2.5	25	

a)   Vanillin content determined by the spectrophotometric method.

References

Chin, C-K. (2004) Vanilla propagation and micro-propagation. *Vanilla Science and Technology Conference*. Rutgers, The State University of New Jersey, New Brunswick, NJ, August 2–6.

Guzmán-Diaz, G. (1997) *Cuadros Estadísticos sobre 23 Actividades Agrícolas y Pecuarias*. Ministerio de Agricultura y Ganadería. San Jose (Costa Rica). 27 p.

Marín-Gonzalez, R. (2003) Microorganismos Benéficos. *Appl. Revista Crisol.* 11–12.

Paniagua, V.A. (2006) *Respuesta en Crecimiento y producción de la Vainilla en Condiciones de Cultivo Orgánico, mediante el uso de cuatro especies forestales como tutores*: Instituto de Investigación y Servicios Forestales (INISEFOR) Universidad Nacional (UNA).

Ocampo, S.R. (1987) *Seminario sobre El Cultivo de Especias en Costa Rica*. Colegio de Ingenieros Agrónomos.

Ramírez, C., Rapidel, B. and Mattey, J. (1999) Principales Factores Agronómicos Restrictivos en el Cultivo de La Vainilla y su Alivio en la Zona de Quepos, Costa Rica. *XI Congreso Nacional Agronómico, pp.* 309–313.

Zúñiga, J.X. (1996) *Emergencia! Urge Revisión Total*: Cruz Roja Costarricense, Appl. La Nación.

4

Atypical Flowering of *Vanilla planifolia* in the Region of Junín, Peru[*]

Juan Hernández-Hernández

Mexico is one of the origin centers of the *Vanilla planifolia* G. Jacks Ex Andrews, whose processed fruits yield vanillin and hundreds of other compounds. It is used as an intensifier for flavor and aroma of foods, beverages, ice creams, fragrances, confectionery, pharmaceutical products, chocolate, baked goods, etc. *Vanilla planifolia* is the main commercial species of vanilla cultivated in the world, since it is preferred by both the agroindustry and the gourmet market, principally for its higher vanillin content. *V. tahitensis* is cultivated on a smaller scale in French Polynesia and Papua New Guinea.

In March 2011, the company Fundo San Rocco, EIRL (San Rocco Farm), of Lima, Peru, imported 3000 cuttings (propagation material) of *Vanilla planifolia* Cv. "Mansa" from Papantla, Veracruz, Mexico and planted them on a plot of land located in the following geographic coordinates: Southern Latitude 11°05′19.74″, Western Longitude 75°23′34.27″ and height 1124.2 meters above sea level, near the town of San Ramon, in the province of Chanchamayo, Junín, Peru, 371 kilometers from the capital city of Lima. This is the first vanilla plantation established in Peru, and it is at a higher altitude than vanilla is usually planted; the majority of the vanilla plantations in Mexico and in other countries are found under 700 meters above sea level.

The cuttings were planted under a greenhouse known as a "shade house" in an area of 3000 square meters completely covered in shading cloth (red) of 50% luminosity (Figure 4.1). For training or support for the vanilla, bamboo posts were used, and an aerial sprinkling system was installed. The planting beds were 1 meter wide and were filled with organic material (leaves and decomposing wood). To avoid dragging or loss due to rain, bamboo stalks were placed around the rows of plants, and vetiver grass (*Vetiveria* spp.) was planted as a living barrier on both sides of the rows of plants.

The establishment and management of the crop was carried out based on the training and technical recommendations provided by the National Institute of Forestry, Agriculture, and Livestock Investigations (Instituto Nacional de Investigaciones Forestales, Agrícolas y Pecuarias – INIFAP – SAGARPA – México (Hernández, Chapter 1 in this volume), which are described below.

---

[*]Note: The manuscript was translated by Athalia Sachowitz.

Handbook of Vanilla Science and Technology, Second Edition.
Edited by Daphna Havkin-Frenkel and Faith C. Belanger.
© 2019 John Wiley & Sons Ltd. Published 2019 by John Wiley & Sons Ltd.

Figure 4.1 Vanilla cultivation in a "shade house".

4.1 Preparation of the "Mother" Plant (Cuttings)

With the objective of making the planting of the cuttings easier, the three most basal leaves were removed by hand by twisting at the petiole and taking care not to tear into the stem where open wounds can facilitate the spread of pathogens.

To prevent rotting of the cuttings, principally caused by the fungus *Fusarium oxysporum* f. sp. *vanillae*, they were disinfected before being planted in the final site by submerging them in a container for 2 to 5 minutes with a fungicide solution of Carbendazim at a dose of 2 grams per liter of water.

4.2 Planting Method

The cutting was planted in the following way: adjacent to the tutor support, a surface trench 5 to 10 cm deep was opened with a spade, then the leafless part of the cutting (3 nodes) was placed horizontally (lying down) and covered with a layer of 3 to 5 cm of fertile soil and vegetable residues, which functioned as a mulch and source of nutrients. Once planted, the rest of the leafy cutting (4 to 5 nodes) was placed vertically on its corresponding tutor support and was tied with agave fiber thread. The size of cuttings was 80 cm in length, and they were physically healthy; that is to say without symptoms of phytosanitary damage. Initially, the cuttings were planted at a distance of 1 m between them and had a high percentage (almost 100%) of successful establishment.

4.2.1 Weed Control

The weeds in the rows of the vanilla plantations were manually controlled with hoes and "machetes", while at the base of the plants themselves, weeds are carefully pulled out by hand in order to not disturb the shallow rooting structure of the vanilla plants. The weed residues were placed at the base of the tutor, so they worked as cover or mulch.

4.2.2 Shoot Management – Looping

This practice consisted in detaching the vanilla guide (apical part of the vine), each time it reached the top of the tutor, to interrupt its growth upwards and redirect it towards the ground, in order to keep the vanilla plant at a height no greater than 2.0 m. That is to say, once a plant reached 2 meters in height, it was looped in order to ensure that it would not get too tall. The vines were distributed over the tutor without being crowded, to avoid the shading between them.

4.2.3 Shoot Management – Rooting

Once a shoot has been looped and has reached the level of the ground, part of it (two to three apical internodes long) were covered with vegetative cover or wet compost, to promote its rooting; the tip that remained free was tied to the same tutor with agave fiber thread, to promote its growth upwards. This practice was carried out every time a vine reached the ground, thus the plant remained vigorous, better nourished and acquired greater tolerance against the fungus *F. oxysporum*, resulting in several rooted vines, at least in the first years.

4.3 Nutrition

The residue of naturally decomposed vegetable matter covering the soil (mulch) usually constitutes the best source of humus or organic matter and essential nutrients required for the development of vanilla cultivation. As such, decaying wood residues were applied with a thickness of 10 to 20 cm, with a width of 50 to 100 cm, starting smaller at the beginning of the planting and growing larger as the root system of the plant grew.

Dry vetiver grass was also applied to the beds to conserve soil moisture.

Plant residues were applied to the base of the tutors, covering the area where the root of the vanilla grows. Each time the roots were observed growing on the soil surface, this material was incorporated to the base of the plant. No chemical fertilizers were applied.

In general the plants of the San Rocco Farm were observed to have very thick stems, around 12 mm in diameter, probably due to the quality of the luminosity provided by the red mesh and the good nutrition of the plants, as indicated by the results of the foliar analysis undertaken in different plants (Table 4.1).

4.4 Irrigation

Sprinklers were used to irrigate the vanilla vines, with the sprinklers moistening the organic matter (vegetation cover) or compost, where the roots of vanilla plants are

Table 4.1 Results of foliar analysis in vanilla plants grown in the San Rocco Farm, 2012.

No. of plants	N (%)	P (%)	K (%)	Ca (%)	Mg (%)	S (%)	Na (%)	Zn (ppm)	Cu (ppm)	Mn (ppm)	Fe (ppm)	B (ppm)	MS (%)
6	2.16	0.26	2.48	3.65	0.23	0.11	0.03	18	90	57	57	38	5.74
58	2.38	0.31	3.10	3.34	0.26	0.10	0.03	16	27	54	55	33	4.92
129	2.21	0.32	3.15	3.49	0.28	0.12	0.03	15	48	61	62	37	4.99
149	2.07	0.29	3.20	3.87	0.26	0.09	0.03	22	66	73	55	26	6.50
107	2.02	0.26	3.30	3.40	0.26	0.10	0.04	17	39	22	47	38	5.09

found. A useful criterion for watering the vanilla plants was to keep organic matter moist at all times, but without saturating it. It should be remembered that the roots of vanilla are of superficial growth; therefore, it is better to water several times with little water, than to water once with a lot of water.

4.5 Pests, Disorders, and Diseases

4.5.1 Vanilla Pest

At the San Rocco farm, no damage was observed from the "chinche roja" ("red orchid bug") *Tenthecoris confusus* (Hemiptera: Miridae), which is the main pest of vanilla in Mexico. The "chinche roja" is a small red insect approximately 6 mm in length, that feeds on the sap of the leaves, stem and fruits of vanilla. If it is not controlled in a timely manner, the orchid bug can end the plantation in a short time.

4.5.2 Diseases

A common disease of vanilla, *Fusarium oxysporum* f. sp. *vanillae,* which causes rotting of roots and stems and consequently the death of plants, was present in the vanilla plantation (Figure 4.2). Also, damage caused by *Colletotrichum gloesporioides* was observed, which is manifested by dark brown spots sunk in leaves, stems and fruits (Hernández, this volume, Chapter 2). The rate of infection in plants was quantified as 3% and 1% of plants affected by these fungi, respectively. To prevent infection by these fungi, it was recommended to use different fungicides such as copper oxychloride, carbendazim, among others, mainly in the months with abundant rains. Sanitation practices were also carried out, which consisted in eliminating all parts of the stem and damaged leaves, and some cases the whole plant, to avoid the advance of the disease in other plants.

4.5.3 Intense Solar Radiation

Damage from the intense radiation of the sun, which consists of a yellowing of the leaves and stems, was observed mainly in the rows of the plants that were more exposed to the sun, since the red mesh with 50% of shade did not seem to provide sufficient cover for plants at certain times of the year. Approximately 75% of plants were visually estimated to have yellowing due to sun damage. To reduce this damage, it was recommended to provide more shade during days with intense radiation, by placing additional mesh.

4.5.4 New Pest

In the plantation of vanilla in the San Rocco Farm, a new pest appeared, which caused superficial lesions on the epidermis of the developing fruits, affecting their external quality, in approximately 30% of fruits (Figure 4.3). The fruits were inspected in the field with a conventional magnifying glass and the presence of very small mites of transparent white color were found, as the probable cause. This damage has also been seen in Mexico, but only in some fruits.

Figure 4.2 Damage caused by *Fusarium oxysporum* f. sp. *vanillae.*

Figure 4.3 Superficial lesions on the epidermis of the beans.

4.5.5 New Disease

A new disease was also observed in 1% of plants. The symptoms of the new disease were small circular lesions on leaves and stems (Figure 4.4). Samples of diseased tissues were analyzed by a phytosanitary diagnostic laboratory, which isolated a bacterium of the

Figure 4.4 New disease in vanilla plants.

genus *Pectobacterium* sp., but the studies lacked the pathogenicity tests to confirm that this disease was the causal agent.

4.6 Flowering Period

In vanilla-producing countries, located in the northern latitude, including Mexico, India and China, vanilla flourishes mainly in the spring season, from March to May, but with greatest intensity during the month of April. In the producing countries of Madagascar, Comoros, Indonesia, Papua New Guinea, and French Polynesia, flowering occurs during the fall season, from September to December, because they are located geographically in the southern latitude. Normally, *V. planifolia* blooms only once a year.

4.6.1 Atypical Vanilla Bloom in Peru

Something surprising, that is not common to observe in *V. planifolia* cv. "Mansa", is the production of inflorescences in a constant way throughout the year, but with greater intensity in the months of September to December (as normally happens in the producer countries of vanilla located in southern latitudes). While in Mexico, this species and cultivar blooms once a year, in Peru the same species had multiple inflorescences characterized by the appearance of one to seven lateral or secondary inflorescences in the rachis of the main inflorescence. This characteristic was found in 10% of the plants (Figure 4.5).

This atypical flowering may be due to the fact that the plants are being cultivated in a geographical region different from most of the vanilla producing regions and therefore have special climatic conditions such as intense solar radiation, alternating presence of

Figure 4.5 Multiple inflorescences on *Vanilla planifolia* cultivated in Peru.

low and high temperatures or length of day and night, among other factors, which may be causing plant stress (Diez *et al.* 2017) and the stimulation of continuous flowering. It is necessary, however, to carry out corresponding studies to really determine what factors are inducing this constant flowering. In the plantations of Mexico, this type of inflorescence is occasionally present at a very low percentage of less than 1%.

It has also been reported in Uganda that two blooms may occur per year, due to the fact that there are two periods of drought that cause stress in the plants and therefore induce flowering.

For comparative purposes, the temperature and relative humidity data recorded at 12.00 p.m., at the San Rocco Farm planting and at C.E. Ixtacuaco-INIFAP Veracruz, Mexico, during the month of July 2014 are shown in Figure 4.6.

As can be seen, the relative humidity is lower and the temperature is more extreme during the day in the San Rocco Farm than in Veracruz, Mexico, at least for the month of July shown.

4.7 Hand Pollination

Each flower should be manually pollinated to allow adequate fruit production (Figure 4.7). It was recommended to pollinate 6 to 8 flowers to give at least five good quality fruits per main inflorescence and only four to five flowers in the secondary inflorescences. One should pollinate mainly the first flowers of each floral cluster because they are the ones that give the best and biggest fruits. The pollination work was carried out on a constant basis during most of the year, but mainly in the months of October to December. Ultimately, it was decided not to pollinate the flowers that were produced in the middle of the year (January to June), to avoid weakening of the plants.

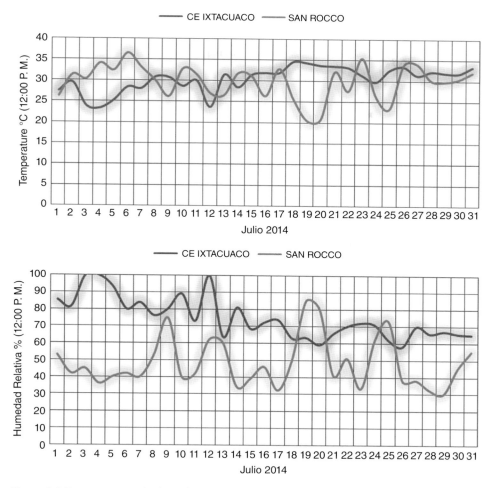

Figure 4.6 Temperature and relative humidity data, recorded in C.E. Ixtacuaco-INIFAP, Veracruz Mexico (20°02′19″ N, 97°05′49″ W, altitude 92 m) and in the San Rocco EIRL Farm (11°05′19″S, 75°23′34″W, altitude 11, 24 m), San Ramón Chanchamayo, Junín, Peru.

The flower buds left over from each cluster were manually removed before they opened. The instrument for pollination and hands were both disinfected with chlorine or alcohol daily to prevent the spread of disease. The pollinated flowers were tagged using different colors for periods of three months to identify pollination dates.

4.8 Harvesting

The fruits were harvested as they reached their full physiological maturity, which was noticed when the apex or tip of the fruit changed from a green to a yellow tone, which occurred eight or nine months after the pollination of the flowers. When all the flowers were pollinated, four harvests were made, but currently only two a year are carried out.

Figure 4.7 Hand pollination of vanilla flowers.

At the plantation of the San Rocco Farm, full fruit production occurred in 2014, with a yield of 1300 kilos of green vanilla, which is considered a good yield. In the following crops of 2015 and 2016, the yield decreased by 30 to 40%, as happens in the majority of the plantations of Mexico.

4.9 Vanilla Curing

Vanilla was processed according to the traditional method (Mexican), which consisted mainly of sunning and sweating the fruits. In order to do this, San Rocco Farm staff were trained by an expert "master beneficiator" or "master processor" from Mexico. In this Farm, there are favorable conditions for the processing of vanilla, since there are no prolonged days with drizzle or clouds that interrupt the process, as happens in Mexico. The favorable conditions are reflected in the excellent physical quality of the processed pods.

Quality: seventy percent of the fruits harvested in 2016 were over 16 cm in length. The processed vanilla was of excellent quality (glossy, flexible, dark brown) (Figure 4.8) and had a good aroma with a vanillin content of 1.5 to 1.7%.

4.10 Final Comments

The plantation of the San Rocco Farm has unique peculiarities, which are not common in other plantations, such as the production of multiple inflorescences and constant flowering, which can have an impact on the productivity of vanilla.

Figure 4.8 Vanilla beans processed by the traditional Mexican method.

This productive behavior can be considered as beneficial, if for example it exceeds fruit yields in comparison to when a single harvest is normally carried out per year, or it can be a disadvantage. In either case, the increase exists and production costs are increased in order to achieve several harvests and the processing of the pods during the year. This suggests the value of undertaking a study to determine the agronomic and economic advantages and disadvantages of vanilla production in agroecological conditions in Peru.

References

Diez, M.C., F. Moreno and E. Gantiva. (2017) Effects of light intensity on the morphology and CAM photosynthesis of *Vanilla planifolia* Andrews. *Revista Facultad Nacional de Agronomia* 70 (1), 8023–8033.

Hernández, H.J. (2011a) Mexican vanilla production. In: Havkin-Frenkel, D. and F.C. Belanger. *Handbook of Vanilla Science and Technology*, 1st edn. Wiley-Blackwell Publishing Ltd., Oxford, pp. 3–24.

Hernández, H.J. (2011b) Vanilla diseases. In: Havkin-Frenkel, D. and F.C. Belanger. *Handbook of Vanilla Science and Technology*, 1st edn. Wiley-Blackwell Publishing Ltd., pp. 26–39.

5

Vanilla Production in the Context of Culture, Economics, and Ecology of Belize

Nelle Gretzinger and Dawn Dean

5.1 Introduction

Vanilla flourished wild in the damp shade of Central America's lowland forests long before humans discovered its tantalizing aroma and undertook its cultivation. Today, Belize boasts an astounding density of natural vanilla populations in which several species of vanilla are represented. In some cases these may be wild, or may be relic cultigens of the now extinct Manché Chol Maya agriculture.

Certainly the present-day Maya word for vanilla, *ché si'bik* (Dawn Dean, personal communication), derives from a bygone era. Franciscan friar Bartolomé Fuensalida visited the Yucatec Maya town of Lucú in 1618 (Thompson 1988) and remarked upon the vanilla he found there, referring to it as *cizbiques* (McNeil 2006). This is almost certainly a Spanish translation of a Yucatec word. Fuensalida was fluent in Yucatec and it was also the language spoken by the Itzà, who controlled trade of vanilla across a large swath of Mexico and Central America in the sixteenth and seventeenth centuries. In Choltí, the language of the Manché Chol, who cultivated vanilla for compulsory trade with the Itzà and the Spanish, the word is *chisbic* (Caso Barrera and Fernández 2006).

In areas contiguous to the historical vanilla growing regions of Belize, the preparation of cacao-based beverages that include vanilla has been recently documented. These beverages include the *chilate* of eastern Guatemala and the *tiste* from Copàn, Honduras (McNeil 2006). In southern Belize, the Kek'chi Maya still flavor their *cacau* with wild vanilla (Wilk 1997) when it is available.

Cultivation of the vanilla orchid in Belize, however, is no longer a skill passed from parent to child; the beans are merely considered a serendipitous find. Perhaps a hunter stumbles on them in the bush; perhaps a woman washing in the river searches for them, enticed by a delightful scent wafting on the breeze. A group of farmers in the Toledo District of Belize has begun cultivating vanilla and hopes to revive interest in this precious commodity that once helped fuel the region's economy.

Handbook of Vanilla Science and Technology, Second Edition.
Edited by Daphna Havkin-Frenkel and Faith C. Belanger.
© 2019 John Wiley & Sons Ltd. Published 2019 by John Wiley & Sons Ltd.

5.1.1 Toledo Agriculture and Socio-demographics Today

The southernmost district in Belize, Toledo is often called the "forgotten district" because it is the least-developed district with the highest poverty rate. Of the 27,000 people who live in the Toledo District, 78% are considered poor.

While impoverished, the Toledo District possesses a wealth of vibrant cultures: the Kek'chi Maya and Mopan Maya are the Mayan people present in Toledo District today. They are not descendants of the Manché Chol Maya, but rather immigrants (of several generations) to the area from Guatemala. While the demographics of Belize have been changing in recent years, so that Mestizo people are the majority of the population nation-wide, in the Toledo District they are still a relatively small ethnic group. People of East Indian descent also reside in the Toledo District, and they tend to be middle-men, the merchants and truckers and value-adders rather than primary producers. A culture born in the New World, the Garifuna people are descendants of Africans and Arawak Indians. While they have traditionally been farmers and fishermen, their culture is undergoing transformation. Many of today's Garinagu (the plural of Garifuna) are poor, but as a cultural group they are perhaps more highly educated than any other group in Belize. Finally there is the Creole culture, generally deemed to be the dominant culture of the country:

> Many people, especially Kek'chi and Mopan people in the western portion of the district, are subsistence farmers who grow most of their own food in addition to salable crops. Kek'chi people come from the cool highlands of Guatemala, and as such their agricultural approach leans more heavily on field crops, than does that of the Mopan Maya who hail from a climate similar to Toledo's, and whose agriculture is more attuned to the lush diversity of the humid tropics. The majority of people living in the Toledo District grow at least a small portion of their own food, but the trend is increasingly towards purchase rather than production of foods and household goods. In the Toledo District, as around the world, agriculture is being transformed into an energy and input intensive commodity driven industry.

Numerous projects, both governmental and non-governmental, have been promoted and run with an eye toward improving the economic, environmental, and social situations present in the Toledo District.

5.1.2 Maya Mountain Research Farm

The Maya Mountain Research Farm (MMRF), a registered non-governmental organization (NGO), is a training center and demonstration farm located in rural Toledo District. MMRF promotes sustainable agriculture and food security, with an emphasis on diversity and integration of the food-producing process into a natural ecological system:

> MMRF's mission statement: To research and demonstrate – within an ecosystem context – locally appropriate alternative technologies and sustainable agricultural techniques that promote and ensure food security, economic security, and environmental conservation, and to transfer this information to people in Toledo District and the rest of Belize and to other interested persons.

MMRF's premise is that truly sustainable agriculture must not only ensure the rendering of ecological services and food security, but also must be economically attractive to the farmers, while allowing them to retain their cultural and family roles. To fulfill this objective, MMRF looked into high value crops that could be integrated into agro-ecological systems, and selected vanilla as the best candidate.

5.1.3 Agro-ecological Systems

An agro-ecological system is an agricultural system, the structure of which replicates the diversity, resilience, and interconnectivity of the ecosystem that would naturally be present in that place. Species composition is comprised of:

i)   *Primary species*: plants useful to the agriculturalist, and
ii) *Secondary species*: plants that support those plants, which are useful to the agriculturalist.

It is important to emphasize that within an agro-ecological system, nearly all species will fulfill multiple functions.

MMRF is located on benthic and limestone soils, at an elevation of 100 to 430 feet, in the foothills of the Maya Mountains, in the humid semi-tropics. Tall rainforest is the natural ecosystem in this locale. Using the descriptors I and II, plants appropriate for, and used in, MMRF's agro-ecological system are:

i)   **Primary species**
   - Timber species: cedar (*Cedrella odorata*), mahogany (*Swietenia macrophylla*), ramon nut (*Brosimum alicastrum*), samwood (*Cordia Alliadore*);
   - Fruit species: anona (*Anona muricata*), bananas (*Musa spp.*), breadfruit (*Artocarpus alitilis*), breadnut (*Artocarpus camansi*), cacao (*Theobroma cacao*), cashew (*Anacardium occidentale*), guava (*Psidium guajava*), pineapple (*Ananas comosus*), plantains (*Musa paradisiacal*), tamarind (*Tamarindus indica*);
   - Semi-cultivated foods: jippy jappa palm (*Carludovica palmate*), pacaya palm (*Chamaedorea tepejilote*), ramon nut (*Brosimum alicastrum*);
   - Spices: allspice (*Pimenta doica*), black pepper (*Piper nigrum*), ginger (*Zingiber officinale*), hot pepper (*Capsicum spp.*), nutmeg (*Myristica spp.*), turmeric (*Curcurma longa*);
   - Leafy greens: chaya (*Cnidoscolus chayamansa*), collaloo (*Amaranth spp.*)
   - Ground foods: cassava (*Manihot esculenta*), dasheen (*Xanthosoma spp.*), yam (*Dioscorea spp.*)
   - Legumes: bri-bri (*Inga spp.*), peanuts (*Arachis hypogaea*), pigeon pea (*Cajanus cajan*)
   - Medicinal plants: jack-ass bitters (*Neurolaena lobata*), polly redhead (*Hamelia coccinea*), sorosi (*Anurophorus sorosi*)

ii) **Secondary species**
   - Plants that attract pollinators: *Bauhinia spp.*, bukut (*Cassia grandis*), flamboyant tree (*Senna magnifolia*), hibiscus (*Hibiscus rosa-sinensis*), Pride of Barbados (*Caesalpinia pulcherrima*)

- Plants that give shade: the above listed timber species, chicle (*Manilkara zapota*), *Spondias* spp.
- Plants whose deep taproots suck nutrients out of the sub-soil and deposit them as leaf litter: *Erythrina* spp.
- Plants that support trellising vines: madre de cacao (*Glyricidia sepium*);
- Plants that supply nutrients such as nitrogen: *Arachis pintoi*, bukut (*Cassia grandis*), guanacaste (*Enterolobium cyclocarpum*);
- Plants that protect against erosion: lemongrass (*Cymbopogon citrates*) and vetiver (*Chrysopogon zizanioides*), which can be planted in broad terraces across hillsides; maidenhair ferns and begonias that stabilize steep riverbanks; and Ficus trees whose roots secure seasonally submerged river edges.

The modern agricultural paradigm, which is displacing traditional land use and food production systems around the world, places more value on production levels than farmers' standard of living, environmental sustainability, or food quality. Imported agro-chemicals and seeds jeopardize local food security when farmers neglect endemic landraces in favor of imported hybrid seeds. Unsustainable farming practices undermine the ability of natural ecosystems to supply the ecological services past generations took for granted.

In contrast to this, agro-ecology conserves natural resources, and supports the surrounding ecosystem in providing ecological services such as regulation of river fluctuation, biodiversity preservation, erosion control, air purification, soil and water retention, and the creation of wildlife habitats. Agro-ecological systems support food security by offering a broad base (seasonally, nutrient-wise, and as a contingency plan when other food supplies are disrupted) of foods that can be directly consumed at home.

The inherent diversity of food resources in an agro-ecological system ensures food security in the event of natural or man-made disasters. In 2001, when Hurricane Iris hit the Toledo District, most of the fruit bearing trees in its path were either broken off at ground level, lost their branches, or lost their fruit. The field crops blew down and molded in the ensuing rainstorms. Dried staples stored inside homes were lost when the roofs protecting them blew off. However, crops such as tubers, nopales, plantains, and bananas, and in a week or two the perennial leafy greens that returned, were available for food. This underscores the importance of relying on a diversified food source.

5.1.4 Maya Mountain Research Farm Vanilla Cultivation and Introduction Project

Before launching a vanilla project with local farmers, MMRF spent over 2 years doing research. Liaisons were initiated, by MMRF staff, with producer groups in other countries. Literature reviews were conducted and visits were paid to vanilla farms in both Mexico and Guatemala. Staff also conducted field research, both biological and ethnobotanical, and cultivated wild-collected specimens of vanilla on-site at MMRF. Through these avenues information was gathered on such things as cultivation techniques, typical vanilla farmer demographics, production modalities, market approaches, international production level fluctuation, hybridization, and micro-propagation.

When MMRF began their vanilla cultivation and introduction project in August of 2007, only one person in Belize was commercially cultivating vanilla. This was Cyrila Cho of San Felipe Village (more can be read about her in the sidebar on page 65).

In San Antonio Village, the Ah family was also cultivating vanilla; however, their production was and still is for home consumption (see sidebar on page 67). Both of these farms were entirely reliant on wild vanilla vines, both *Vanilla planifolia* and *V. odorata*. The farmers had not transplanted the vines, but were tending them *in situ*.

It was decided that the vanilla cultivation and introduction project would focus on quality over quantity, and intensively train and include only 11 persons. These 11 persons are to be the seed for the next generation of Toledo District vanilla farmers. Rather than work in one village, the project selected 11 persons from 10 different villages. Participants were hand-selected to represent the 6 different ethnic groups present in the Toledo District, to be geographically distributed widely throughout the district, and to be experienced, established farmers with good land tenure, on a wide variety of soils. Both men and women were selected, with an age range of 18 to 64. It was considered that by choosing a group this diverse, and working closely with them, that vanilla could be introduced throughout the district by them.

This project was initiated with a 3-day workshop held at MMRF, led by Dawn Dean. The vanilla vines growing at MMRF were the first such plants the farmers had seen. Pictures of the Vallejo vanilla farm in Veracruz, Mexico were shown, along with the insights Victor Vallejo suggested for the fledgling group regarding vanilla cultivation, such as his technique for "footing" the vines to constantly keep young growth on the vines.

In October 2007, farmers received their vanilla vines. ForesTrade in Coban, Guatemala, donated 282 cuttings of *V. planifolia* to this project and these were distributed to the farmers (Figure 5.1). By the time these vines arrived, 1 original member of the 11-member selected vanilla farmer group was employed off his farm, and no longer wished to be involved. Also, more people had heard about the project and were interested in vanilla cultivation by then. The 282 vines were distributed among 32 farmers; the most any farmer received was 26, and several farmers received only 1 vine.

5.1.5   The Belize Organic Vanilla Association

In December 2007, at a meeting held in Barranco Village for the 11-member selected farmer group, the Toledo District Agriculture Officer, Mr Barry Palacio, was invited to speak. He addressed the issue of markets and marketing, an issue with which the Toledo District has always had difficulty, due to its small size and distance from domestic markets and ports. It was his advice that the farmers form an association as a vehicle from which to work together to access internal and external markets, lobby government, apply for grants, and as a self-regulatory mechanism for product quality. After thorough discussion with Mr Palacio and among themselves, the farmers unanimously agreed that forming a production-based organization was in their best interest and initiated an election. The 7 persons elected to serve as the founding members and on the board are as follows: Eugenio Ah – chairperson; Egbert Valencio – vice chairperson; Dawn Dean – secretary/treasurer; and as councilors Tereso Sho, Ophelia Chee Sanchez, Irma Gonzalez, and Constance Ramclam. The association was registered the following week, in the capital of Belmopan, as the Organic Vanilla Association, OVA.

5.1.6   OVA Description and Goals

From the beginning, OVA has been a farmer-based organization, unified around the goal of producing vanilla for an economic return, but always aware of the potential

Figure 5.1 Michaelyn Bachuber of ForesTrade, Guatemala prepares vines for donation to MMRF vanilla project.

environmental ramifications of their venture, particularly as relates to organic production and sourcing of wild vines. Regular meetings are held, wherein OVA members tour
one another's farms. The organization's bylaws are not ratified to date, but include
membership criteria pertaining to environmental issues:

> OVA's mission statement: OVA is a farmer organization whose goal is to produce
> organic vanilla. We will use research and education to help us promote vanilla
> and the vanilla industry. We will be socially just. We will encourage community
> involvement in OVA and vanilla cultivation. We will cultivate vanilla and grow
> our organization in a way that safeguards ecological resources, while addressing
> the economic needs of Southern Belize.

Market analysis, conducted largely by Nelle Gretzinger, has from the beginning shown
that successful entry into the vanilla market could only come about by specialization
and innovation. Unique approaches to value-adding are being investigated, as well as
indirect methods of marketing Belizean vanilla, such as eco-tourism and gastro-tourism.
Organic cultivation has been tirelessly promoted, as has a healthy respect for the wild

vanillas found in Belize. The members of OVA have positioned themselves as the guardians of this resource, which is invaluable both to the scientific community and the nascent vanilla industry in Belize.

5.1.7 Innovative Vanilla Plantation Establishment Method Pioneered by OVA Members Nicasio and Ophelia Chee Sanchez

One of the obstacles to introducing any perennial crop is that the farmer must make a significant investment over several years to establish and care for plants that are not yet productive and hence offer no financial return. OVA members Nicasio and Ophelia Sanchez have pioneered a new technique for establishing vanilla that addresses this problem; they intercrop the vanilla in a cornfield for the first few years. This method is broadly applicable, as corn is a staple food throughout Central America.

In Belize, hand-cultivated corn is grown in the following way: An area is cleared by machete and everything is chopped to ground level and left to rot where it falls. With a sharp pointed planting stick, a hole of 6 to 8 inches is made in the ground. The planting stick is gently wriggled out of the hole, so the sides of the hole remain intact. Five or six corn seeds are dropped in the hole, and the hole is left uncovered. These seed holes are spaced evenly in a diamond pattern with 5 to 6 feet between holes (corn plants). While the corn is always the dominant crop, sometimes other plants will be intercropped with the corn. The most common interplant is local pumpkin, a cucurbit that dries and stores well. Collaloo, a leafy amaranth grown for its edible leaves, is another common interplant, as are tomatoes. Volunteer wild edible greens, tomatillos, and medicinal herbs also frequently pop up in the cornfield. These are noticed and tended. The cornfield is "cleaned" again 6 to 8 weeks after planting, by machete chopping all weeds to the ground. Further tending of the corn is unnecessary until harvest time (6 months after sowing), but if intercrops are used, these may receive additional attention. It is in this context that the Sanchezes established their trial vanilla plantation.

At the same time as planting the cornfield, they put sticks of madre de cacao (*Glyricidia sepium*), a leguminous tree, in the ground to serve as tutors for the vanilla. The vanilla cuttings were also put in the field at the same time. To create shade for the vanilla, little thatch teepees of palm leaf (the Sanchezes used coconut and cohune leaf) were tied to each madre de cacao stick over the vanilla (Figure 5.2). These teepees are durable enough to last through the first corn harvest and well into the second cornfield, which is planted on the same site immediately subsequent to the first harvest. By the time the second cornfield is ready for harvest, the thatch teepees are rotted down, and the madre de cacao sticks have thrown enough branches to shade the vanilla. A third cornfield can be planted in the same spot, but by the third harvest, the madre de cacao and vanilla will have grown sufficiently to dominate the space, and it is unlikely that a fourth cornfield will be feasible (Figure 5.3). The leaf detritus from the corn plants is heaped at the base of the vanilla vines.

By intercropping the vanilla with corn, the labor necessary to tend the fledgling vanilla plantation is greatly reduced, as it is a by-product of maintaining the cornfield. Also of note is that the land itself is productive for the interim years between planting and harvest. While the Toledo District is not a land-poor region, this is a serious consideration in other regions where corn is a staple crop.

Figure 5.2 Thatch teepee shading newly planted vanilla vine.

5.1.8 Wild/Relic Vanilla Stands in Toledo District

The area within the Toledo District currently identified as having the highest density of vanilla plants, and which also hosts the greatest species diversity within the genus, is in the lands surrounding Barranco Village, and in the nearby Sarstoon and Temash National Park (STNP). In some places, 3 or more species exist within 100 yards of one another. The proximity and diversity of the vanilla present in these locales is to such an extent that we must logically question if the plants are not relic cultigens from historical cultivation. The Sarstoon and Temash rivers are located well within what had been Manché Chol territory and cacao and vanilla were once intensively cultivated in the areas surrounding these rivers (Caso Barrera and Fernández, 2006).

Barranco Village, located on the coast just north of the mouth of the Temash River, is a small Garifuna community, established between 1820 and 1830 (Wilk 1997), and at its peak was home to approximately 600 people. Because no recollection today exists with the elders of the community (who are only fourth, or in some cases the fifth, generation of Garifuna people residing in Barranco) regarding vanilla cultivation, it may be said with certainty that the vanilla plants are not relic cultigens of Garifuna agriculture. The next logical question is whether these plants are evidence of past land use by either Kek'chi or Mopan Maya. Once again the spoken record denies this

Figure 5.3 Mr and Ms Sanchez in corn/vanilla field where vanilla and madre de cacao tutors are beginning to dominate the space.

possibility. Taking one more step backwards through history, it bears considering if these plants attest to colonial era vanilla cultivation by the Manché Chol. The evidence suggests that this is not the case. In other areas previously inhabited by the Maya, pottery shards and house mounds are to be found. Some coastal areas of southern Belize are rife with these sorts of evidence of historical Maya occupation. Present-day Barranco residents do not find these archaeological records of previous Maya occupation, which would indicate that the Garifuna people were the first settlers of this land. That suggests that these vanillas, found so abundantly in the bush just a mile from human settlement, are in fact wild.

Within the Golden Stream Corridor Preserve (GSCP) there are many populations of *V. planifolia*, *V. odorata*, and likely other *Vanilla* spp. as well, the dissimilarity between leaves being more than could be expected from mere phenotypic difference. The buffer communities for the GSCP are well-established Kek'chi and Mopan Maya communities. Corn is central to Maya culture, and corn-based agriculture, as practiced in southern Belize, requires a large amount of land. Not only does corn supply less calories per unit of land than most other staples, but corn cultivation quickly depletes the soil, necessitating significant fallow periods after only a few harvests. As such, large tracts of land are required to support a Mayan community. Therefore, it comes as no surprise that informal verbal investigation has not turned up any leads on wild vanilla plants existing in the areas surrounding the GSCP.

Both *V. planifolia* and *V. odorata* were found in the community lands immediately adjacent to Jordan Village. While corn cultivation is a mainstay of the agriculture practiced by community members of Jordan Village, some of the land near the village is unsuitable for corn cultivation due to stony outcroppings or its low-lying nature. Therefore, these lands are untouched despite their proximity to settlement. In 2005, dozens of individual vanilla plants still survived within 5 minutes walk of the village. It must be noted that while Jordan is a recently established village (Wilk 1997), its location on the Moho River puts it in the vicinity of cacao cultivation that was remarked upon by Dominican friar Joseph Delgado when he journeyed to the area in 1677 (Thompson, 1988; McNeil, 2006).

Private parcels of land abutting the road that joins Punta Gorda Town with Boom Creek Village (also on the Moho River) commonly host a variety of vanilla species. More *V. odorata* has been found in these vicinities than any other species.

5.1.9 Possibility of Wild Superior or Useful Genotypes/Species

Preliminary evidence suggesting spontaneous natural hybridization in vanilla has been documented from Belize (Lubinsky *et al.* 2008b). Natural vanilla hybrids could have superior pollination rates, aroma, or vigor. Intentional hybridization could result in further potential for the vanilla industry by using these genetic resources to breed in traits of particular agronomic interest such as higher vanillin content, indehiscent fruit, and resistance to pathogens such as *Fusarium*. It could also help by inserting some genetic diversity into the plant stock currently utilized to initiate cultivation, which is generally acknowledged to derive from a very limited gene pool (Soto Arenas 1999; Lubinsky, 2003; Lubinsky *et al.* 2008a). However, in the United States and Europe, governmental regulations restrict the use of vanilla products in food to the two species *V. planifolia* and *V. tahitensis*. There are no such restrictions regarding the use of vanilla products for fragrances.

5.1.10 Dr Pesach Lubinsky's Research in Belize and Regarding
Vanilla tahitensis

Dr Lubinsky, who has done extensive research on the genetic diversity of vanillas in Mesoamerica, collected vanilla specimens with Mr Sylvano Sho in Blue Creek Village, Toledo District, among other locations in Belize. He found numerous unique *V. planifolia* clones as well as specimens of *V. odorata*. When he left Belize, he left these accessions in the care of the Belize Botanic Gardens, which is located in the Cayo District and managed by Heather Duplooy. Several of these accessions were utilized in Dr Lubinsky's work on the origin of *V. tahitensis*.

The origin of *V. tahitensis* was, until recently, a riddle without a very satisfying answer. It is not indigenous to the islands of French Polynesia, nor to the Philippines, the locale from which it was purportedly introduced to Tahiti by Admiral Hamelin in 1848 (Correll 1953; Portères 1954). *V. tahitensis*, in fact, has never been found in the wild (Portères 1954; Lubinsky *et al.* 2008a).

Much of the vanilla cultivated around the world today, predominantly *V. planifolia*, followed the same conduit. Cacao based beverages, frequently made with vanilla, became all the rage amidst the Spanish elite in Mexico by the mid-sixteenth century: Chocolate could be imbibed on certain fasting days imposed by the Catholic Church (Kourí 2004) and was substantial enough to assuage hunger. When cacao and vanilla in tandem were exported from Mexico to Spain, their consumption quickly became entrenched in the culture there. It was not long before knowledge of these New World novelties spread to other European countries. Then everyone wanted to grow vanilla and so cuttings of vanilla vines were transported to botanical collections in Europe and throughout the tropical colonies of the various European countries.

All of this vanilla dispersal originated with an Atlantic Ocean crossing, from Mexico to Spain, but the most direct route to the Philippines would have been across the Pacific. Spanish galleons did indeed ply this route, departing from Acapulco and sailing to Manila, from the latter half of the sixteenth century into the early nineteenth century. To the Philippines they carried silver and returned to Mexico with porcelain, ivory, silk cloth, and spices from Asia. In addition, they carried plants to the Philippines.

To transport vanilla plant stock from Veracruz, the cradle of the vanilla trade (Kourí 2004), to port in Acapulco would have meant a tortuous crossing of the Sierra Madre mountains or the swamps and dense jungle of the Isthmus of Tehuantepec. It would have been more practical then to collect plant stock on the Pacific littoral, specifically from Guatemala where cacao and vanilla were cultivated. Large amounts of Guatemalan cacao were transported overland via mule routes in the late sixteenth century to markets in Oaxaca, Puebla, and Mexico City (Coe and Coe 1996). Thomas Gage described, in the 1630s, the cacao cultivation of the Suchitepéquez region of Guatemala and the spices grown there that were used to flavor cacao-based beverages, which included vanilla (Thompson 1969). Francisco Fuentes y Guzmán's *Recordación florida*, from 1690, describes the methods used to plant vanilla in Yzquintepeque (Escuintla), Guatemala. The gathering of vanilla plant stock from a location so distant from Veracruz could have resulted in the collection of material that was genetically divergent from that which could be found in Veracruz (Lubinsky *et al.* 2008a).

Theories have existed for some time as to the origin of *V. tahitensis*, some suggesting that it is a hybrid of *V. planifolia* and another species, such as *V. pompona* (Portères 1954) or a *V. pompona-V. odorata* complex (Portères 1951; Soto Arenas 1999). Studies have since confirmed that while *V. tahitensis* and *V. planifolia* are very closely related genetically, the genetic distance between *V. tahitensis* and *V. pompona* is nearly three times greater than that between *V. tahitensis* and *V. planifolia* (Bory 2004; Duval *et al.* 2006). Furthermore, it has been noted that the dimensions of the *V. tahitensis* leaf are exactly intermediate between those of *V. planifolia* and *V. odorata* leaves (Soto Arenas, 2006). With all this in mind, Dr Pesach Lubinsky set out to determine the parentage of *V. tahitensis* by subjecting several different species of vanilla to DNA analysis. Dr Lubinsky thought, based on the morphology of the *V. tahitensis* leaf and flower, that it looked like something in between *V. planifolia* and *V. odorata* (Lubinsky 2007). To test his theory, Dr Lubinsky analyzed the DNA of several *V. tahitensis* accessions from French Polynesia and 40 vanilla accessions, of varying species including *V. planifolia* and *V. odorata*, which had been collected from southern Mexico down to Ecuador (Lubinsky *et al.* 2008b).

His initial analysis of the resulting data showed that *V. planifolia* and *V. odorata*, while only distantly related to each other, were both closely related to *V. tahitensis*. When he analyzed the chloroplast DNA, inherited exclusively from the "mother" plant, he discovered that *V. planifolia* was the "mother" of *V. tahitensis* and that it did indeed result from the hybridization of *V. planifolia* and *V. odorata* (Lubinsky *et al.* 2008b). It is interesting to note that the *V. planifolia* and *V. odorata* accessions from Belize proved to be very closely related to *V. tahitensis* on the phylogenetic tree that was developed from Lubinsky's DNA work.

Furthermore, the relatively short branch lengths on the phylogenetic tree indicate that *V. tahitensis* is a relatively new hybrid, perhaps 500 or 600 years old. Lubinsky speculates that it came about in 1350 to 1500. Those dates precede the arrival of the Spanish in Mexico and Central America, who were responsible for documenting the cacao and vanilla cultivation they found there. It is quite likely, however, that the cultivation, including that of the Manché Chol Maya, had been in existence for some time before the Spanish arrived. Since the discovery of hand pollination was not documented as having occurred until several centuries later, we must assume that the hybrid we have come to know as *V. tahitensis* was not engineered by man. So, it is possible then that somewhere in the lowlands of Belize, shaded by the bush and watered by abundant rains, resides the progenitor of *V. tahitensis*, a naturally occurring hybrid of *V. planifolia* and *V. odorata*.

5.1.11 Manché Chol

It is possible that the plethora of seemingly wild vanilla found today in southern Belize is vestigial, left behind by the Manché Chol.

Hernán Cortés traversed Chol territory in 1525 (Jones 1998), cutting across what is now the southwest corner of Belize, at the tail end of a journey that originated on the gulf coast of Mexico in the southern part of what is now the state of Veracruz (Dobson 1973). His chronicle of the *entrada*, a lengthy letter to Emperor Charles the V, includes several references to the cacao he came across in the region. Cortés was well aware of the value placed on cacao by the indigenous peoples he encountered on his travels, having noted in an earlier letter to Charles the V that "they use it as money throughout the land and with it buy all they need" (Pagden 1971). However, he had no idea of the role that cacao would play, in tandem with vanilla and annatto, in sustaining the local economy as the Spanish vied for domination of the Southern Maya Lowlands.

In the sixteenth and seventeenth centuries, the Spanish attempted to subdue, by forced relocation and conversion to Catholicism and by use of the *encomienda* system, the Maya peoples who inhabited the Southern Maya Lowlands, an area made up of the southern parts of Campeche and Quintana Roo in Mexico, the Petén in Guatemala, and Belize. The Itzà, who inhabited the central Petén, determinedly fought this fate and managed to retain their independence until the end of the seventeenth century (Caso Barrera and Fernández 2006). Theirs was the last independent Maya polity (McNeil 2006).

The Itzà elite consumed, for ritual purposes, great quantities of cacao-based beverages. While they grew a small amount of the three important ingredients for chocolate, cacao (*Theobroma cacao*), annatto (*Bixa orellana*), and vanilla (*V. planifolia*), it was only enough for local consumption on a small scale. The central Petén was, because of its soil and climate, an inhospitable place for growing cacao (McNeil 2006). The Itzà found a way to surmount this problem and controlled the production and trade of cacao, annatto, and vanilla in a large area of Mexico and Central America right up until they succumbed to Spanish domination in 1697 (Caso Barrera and Fernández 2006; McNeil 2006).

The Chontal Maya of Acalán, which translates as "Place of Canoes" (Henderson 1997), were excellent seafarers, in control of extensive maritime trade routes that stretched east around the Yucatan peninsula and all the way down the coast to the important trading center of Nito on the Gulf of Honduras (Coe 2005). They traveled these enormous distances to engage in the trade of luxury goods, including cacao, which they produced, feathers, jaguar pelts, and slaves. In the wake of the Spanish conquest of the Yucatan, however, their trading activity ceased (McNeil 2006). The Itzà stepped into the breach and reassembled the Chontal exchange system and resumed use of their trade routes (Caso Barrera and Fernández 2006).

Control of this trading system meant that the Itzà were assured an uninterrupted supply of cacao for their personal consumption. It must also have been very lucrative: Numerous Maya fled south from the Spanish incursion on the Yucatan to resettle in locations close to Itzà territory (Jones 1998), thus creating new outlets for trade. The Itzà went to any means necessary to maintain their power and control of their extensive trading system and to protect their territory from the advance of the Spanish. They bullied their neighbors, enslaving them, raping their women, and sacrificing a hapless few who were fool enough to offer aid to the Spanish. In 1630, they viciously attacked the Manché Chol, ultimately inciting the Chol to revolt against Spanish domination (Jones 1998). They warred with the Lacandón for control of the Salinas de los Nueve Cerros, the only source of salt in the region. They then used their control of this precious resource to force the Lacandón, and the Manché Chol, to exchange their valuable commodities, including cacao, for salt (Caso Barrera and Fernández 2006; McNeil 2006). It is the Manché Chol, who lived south and east of the Petén, with whom we are most concerned, as much of their territory was within what is present-day southern Belize.

With its numerous fertile river valleys, Manché Chol territory was ideally suited to growing cacao. In their orchards, called *pakab* in the Choltí language, the Chol grew great quantities of cacao, annatto (Thompson 1988), and vanilla (Caso Barrera and Fernández 2006; McNeil 2006). In 1620, the Dominican friar Gabriel de Salazar made a circuit around Central America that took him, among other places, the length of Belize, along the shoreline and through Chol territory (Feldman 2000; Caso Barrera and Fernández 2006; McNeil 2006). It is evident from Salazar's observations that cacao was important in the region, as he remarked that the Chol would "cast a spell for a cup of chocolate" (Feldman 2000). Salazar noted the large cacao and annatto orchards in the Chol villages along the coast of Belize: Yaxhal, Paliac, Campin, and Tzoité. Chol territory continued, tracing a crescent shape, south and west away from these settlements to the towns of Manché, Chocahau, Yaxhà, and Yol (all in present-day Guatemala). From these villages, the Manché Chol would transport their precious cargo to the Itzà capital of Noh Petén (Caso Barrera and Fernández 2006).

It was not only the Itzà who forced the Manché Chol into trade. The Spanish got in on the action too, extorting cacao, annatto, and vanilla from the Chol in exchange for overpriced metal tools and other wares. In fact, The Chol, surrounded by the Itzà to the northwest, the Yucatec to the north, and the Kek'chi and the Spanish of Verapaz to the southwest, managed to engage in trade with all their neighbors, some forcibly and some voluntarily. This attests to the value of the resources in the possession of the Manché Chol and illustrates that they must have intensively produced cacao, vanilla, and annatto in order to be able to supply everyone around them (Caso Barrera and Fernández 2006; McNeil 2006).

In 1689, the Manché Chol were rounded up by the Spanish and forcibly relocated to the Valley of Urrán in the Guatemala highlands (Caso Barrera and Fernández 2006; McNeil 2006). The terrain was absolutely foreign to them: J.E.S. Thompson made the acute observation that it was like banishing "Sicilians to the remoter highlands of Scotland" (Thompson 1970). It was not long before they started to perish. In 1699 it was noted by Marcelo Flores, a Spanish Captain, that some Chol still occupied what had been their lands in eastern Guatemala and southern Belize. However, by 1710 there were only four Manché Chol left in the town of Belén in the Valley of Urran (Thompson 1988; Caso Barrera and Fernández 2006). Their ultimate disappearance meant the loss of their acumen with regard to the cultivation of vanilla.

When the Spanish forcibly relocated the Lacandón in 1695 and defeated the Itzà in 1697, the cacao based trading network collapsed. While Kek'chi and Mopan Maya continued to grow cacao in what had been Chol territory, the trade never returned to its prior level of importance.

5.2 Discussion

To date, very little vanilla related research has been done in Belize. *V. planifolia* is a rare endemic to Belize and is sometimes sympatric with other species of vanilla that produce a scented fruit, such as *V. pompona*, *V. odorata*, *V. insignis*, and *V. hartii* (Soto Arenas 1999). Owing to the infrequency with which vanilla flowers in the wild (Childers *et al.* 1959), positive species identification is extremely difficult, and in the case of Belize has been very limited. Individual populations can be easily destroyed by insensitive land use, and natural reintroduction rates are low. Due to the dearth of definite information regarding vanilla (species present, distribution, preferred growing conditions), sustainable conservation and use patterns cannot now be established.

Because the wild/relic vanillas of Belize may represent lost varieties (Lubinsky *et al.* 2008a), it is imperative that they be located, and that specimens be collected and then identified both morphologically and genetically. With this, in addition to species identification, patterns of hybrid origin and overall phylogenetic relationship can be established. Relative rarity of each *Vanilla* spp. can also be determined. Specimen deposition in both domestic and foreign herbaria would logically follow.

Vanilla habitat within Belize needs to be mapped and described. This description should consist mostly of biotic elements such as associated vegetation and proximity to waterways, but should also include measurements of humidity, characterization of soils, consideration of elevation and slope direction, proximity to human settlements both present and historical, historical logging records, etc.

Using GIS, a method for predicting expected locations of vanilla, should then be created and tested. Actual output of this component would be dependent on whether vanilla is ascertained to be either a phytosociologically selective, exclusive, or preferential species.

Finally, the establishment of an *ex situ* conservation gene bank would contribute to: conservation of the genetic and morphological diversity of Belizean vanilla, ease of further research due to the plants' proximity to one another, and enablement of site manipulation to induce flowering, and cultivar selection and breeding use to aid the burgeoning vanilla industry in Toledo.

Construction of the information base outlined in Section **5.1.4** would allow for forest policy recommendations to be produced regarding conservation, sustainable use of wild-harvested vanilla for cultivation purposes, and an information base and intellectual guidelines for protected area management organizations.

While the disappearance of intensive vanilla cultivation by the Chol Maya may be mourned, it is nonetheless proof that the crop is well suited to the climate and geography of southern Belize. As such, vanilla agriculture and its trade could easily be revived in the region. With the introduction of cultivation made by MMRF and the establishment of OVA, the seeds of this renaissance have been sown. However, the aforementioned research is necessary to ensure that the Belizean vanilla industry flourishes and that the country's extant vanilla resources are safeguarded.

Acknowledgments

We owe a debt of gratitude to Daphna Havkin-Frenkel and Faith Belanger for inviting us to contribute to this volume. We would like to thank the Geoffrey Roberts Trust for providing the financial support that facilitated the research for and writing of this chapter. We would also like to thank Christopher Nesbitt of Maya Mountain Research Farm and Pesach Lubinsky for their generosity in sharing their wealth of knowledge with us. We are grateful to Egbert Valencio and Mauricio Ah, who led us to the locations of numerous wild vanilla vines, and to Adelina Caliz and Cyrila Cho, who spoke to us of the role vanilla plays in their culture. Photos are used with the permission of the individuals in the photos.

Interview with Adelina Caliz

In the village of Mafredi (population 160), Adelina Caliz, a Kek'chi woman in her mid-50s, farms a piece of land near her house that she has worked her entire adult life. What is grown on this property is for the consumption of Ms Caliz's family, which includes her husband Burton and her grown daughter Carla, who works the family farm and lives at the site with her husband and young son. Adelina sells at the market in Punta Gorda 3 or 4 days a week and the market farm is a couple of miles away from the Caliz home. The family also raises turkeys, chickens, and pigs and keeps parrots and rabbits as pets.

Burton and Adelina are dogmatically, emphatically organic. They like to grow vegetables that nobody else has and specialize in hard-to-grow or rare vegetables, such as lettuces or unusual gourds, and uncommon tree fruits. The Caliz family makes most of its money from farming and they count, especially, on their income from allspice production. They are possibly the largest producers of allspice in the Toledo District.

When we arrive at her home, Ms Caliz is in the midst of making dukunu and there is a 5-gallon bucket of stripped corn cobs sitting next to her. She's cut her finger and is happy to take a break from her work to talk with us. We come to the subject of vanilla via coffee, which her parents used to grow. "We used to have a lot of coffee. People from villages all around would buy it. If they didn't have money, they'd bring two baby chickens or a five or six pound pig."

According to Adelina, vanilla was frequently used to flavor coffee, going right into the hot water used to make the beverage. Her mother would collect whole, young plants from the wild when she found them and replant them in a shady part of the yard, wrapping them around sticks or coffee trees.

"We call them the wild ones. They grew by the creek side, around the cohune [palm]. When the pods were ripe, you could smell it." The beans, collected only from the variety of vanilla with dark green, sword-shaped leaves, were gathered when they were scented and "mauve colored," but before they split. They were then sun dried, to eliminate all moisture, and stored in bottles for later use in everything from coffee to cacao to sweets, such as bread pudding and stewed pumpkin. For inclusion in the cacao beverage, the vanilla beans were toasted on a comal (a type of griddle used extensively by the Maya), along with cacao beans, to get them crisp enough that they shattered when they were subsequently ground.

Interview with Cyrila Cho

Cyrila Cho is a Kek'chi Maya woman who is originally from Laguna Village, but now resides in San Felipe (population 350). Most Kek'chi women are very reserved, but Cyrila is outgoing and talkative. Possessed of an entrepreneurial spirit, she has a business grinding corn into masa for the community with her family's electric corn grinder. For a few years now, she has also had a chocolate making business. She is aided in this endeavor by her husband and two grown children, Abelina and Juan. Abelina assists with production and Juan purchases cacao from a handful of other farmers in the village. The chocolate business is growing and more cacao is needed than Cyrila and her family currently grow. It should be noted that Juan pays a higher price for cacao than is normally paid in the area. Juan is progressively community minded and wants to expand the chocolate making business to help establish a more lucrative market for the cacao grown by local farmers.

Cho's Chocolate uses traditional Maya methods to ferment, roast, and grind their cacao. Their products include cocoa powder, made by simply grating the dried and fermented cacao beans, baking chocolate, used locally for making a cacao beverage, and a sweetened chocolate bar. The flavoring for the chocolate bar is dependent upon what is available at any given time, but can include allspice, cloves, and vanilla. When vanilla is used, the vanilla bean is ground right along with the cacao.

Cyrila discovered Ché Si'bik (the Maya word for vanilla) in the bush behind her house in San Felipe when she cleared an area for planting. She was able to recognize the vine because she recalls her parents using vanilla. "My father have a farm and he have a vanilla. Since my father and my mother raised me they use it with cacao because it smell sweet."

The wild vanilla vines were left to grow after the area surrounding them was cleared. These plants self-pollinate and produce beans that are sun-dried for use in the chocolate bar recipe. Two different species of vanilla are used in the chocolate, V. planifolia and the beans from an unidentified vanilla vine that has a thin, narrow dark leaf.

In October 2007, Cyrila received 10 V. planifolia cuttings from MMRF to plant in her yard. As of January 11, 2009, when we visited with Ms Cho, six of those vines were still alive and appeared to be thriving. They were planted, appropriately enough, in an area in which several cacao trees were also planted. In the yard, there are also about a dozen fruit trees and many small plants that are used as herbs, both for seasoning food and as traditional medicines. Cyrila tells us enthusiastically, "I like plants. By my house, I have cabbage, cucumber. In my farm I have watermelon. Right now we plant beans and corn."

According to Cyrila, "Some of them [the women in the village] have vanilla, not all. Some of them want plant. I teach them." As we depart, she smiles and tells us that should OVA get more plants, she would like some more to plant in her yard.

Interview with Egbert Valencio

Egbert Valencio, 34 years old and the youngest of 9 children, could easily have remained in Belize City where he lived for several years, but his devotion to his cultural heritage was such that he determined to return to his native village of Barranco. His decision was influenced by a Garifuna youth movement born in Belize City to repopulate the oldest Garifuna village in Belize. Many of the people who returned to this small fishing village on the Caribbean coast simultaneous with Egbert remained only a short while, but Egbert has stayed and is raising his family here. He is the only one of his siblings currently living in Barranco and he resides with his father Raymond, who is the principal fisherman for the village, his mother Lucille, his wife Emelda, and their infant daughter Lumar.

For the last 4 years, Mr Valencio has worked as a ranger, and sometimes as a boat captain, for SATIIM. (SATIIM's mission is "linking biodiversity management with the physical and cultural survival of the indigenous people who surround the Sarstoon and Temash Nature Preserve.") He has become a well-respected member of the community, sitting on the Village Council and teaching environmental classes at the Barranco primary school.

About a mile outside the village of Barranco, Egbert owns a 10-acre parcel of land that abuts land owned by his brothers and sisters. Eventually, he will build a home on this property to house his family. In the meantime, he actively farms the land and has planted nutmeg, coconuts, cassava, agaves, and a wide variety of fruit trees.

When Mr Valencio speaks about vanilla, his eyes sparkle. He is a champion of the nascent vanilla industry in Toledo and is the vice-chair for OVA, the Organic Vanilla Association. When he is out in the field for SATIIM, Egbert, an acute observer of the native flora, frequently spots wild vanilla vines, sometimes in flower and sometimes with beans on them. We visit his farm where he's established a small plantation with plants received from OVA and about 50 plants gathered in the wild from a nearby village. The latter plants are thriving and already developing racemes at 14 months of age. When we walk the perimeter of his farm, he shows us the vanilla that he discovered growing there. Based on his observations and our own, it would appear that there are as many as five different species growing at the periphery of the farm. When we leave his farm, we are elated and of a particularly healthy vanilla vine Mr Valencio says "That plant was green. It was smiling. No," he pauses briefly before speaking again, "it was laughing!"

Interview with Mauricio Ah

Mauricio Ah gives us language lessons in Mopan as we hike from his house, in San Antonio (population 954), on to the land he tends with his two sons, Eugenio and Emerygildo. While he is clearly in his mid-60s, he is spry and moves at a rapid clip through the bush, swinging his machete as he goes. We walk for 20 minutes before we come to the first, seemingly, wild vanilla vine that Mauricio discovered growing on his land a few years ago.

It had been attached to a tree that fell down, so a rudimentary trellis now supports it, constructed from tree limbs found in the vicinity. When the support tree fell, it triggered the vanilla to flower. Now there are two beans on this vine, showing the queue de serin typical of ripening *V. planifolia* beans. Mr Ah pockets them after I tell him that I can give him and his wife, Priscilla, a lesson in curing when we get back to his house. The vine has well-developed racemes as well, indicating that it will flower again shortly.

A few yards away is another vine, the leaves of which are considerably narrower than those on the first vanilla, prompting us to speculate that it is probably a different species. Mr Ah tells us that this vanilla produced 12 beans last year and that those beans had a scent to them when they were still green and on the vine. It also flowered last year and while we are able to locate a withered raceme on the vine, it is devoid of beans.

On our trek back to the Ah house, Ms Dean and I are suddenly hit in the nose by a heavenly scent. We exchange looks, but not a word passes between us. In the course of our interviews we have heard from numerous sources, mostly Maya, that if you encounter the scent of vanilla in the bush, you must never say that you smell something sweet, for if you do, you will not find the vanilla that's emitting the scent.

To our left, about 3 feet from the trail we're on, is a very small clearing in which we immediately find a vanilla vine that, judging by appearances, is a *V. planifolia*. We find no beans, or flowers, or even racemes on this vine. To the right of the path is a tangle of various vines that leads to a creek, about 12 feet away. The abundance of razor vines is daunting, but Mr Ah gamely attacks it with his machete. Each whack of the machete serves to exaggerate the scent now permeating the air. We can now make out a vanilla vine and, revealed by a few more swings of the machete, we discover that the vine has three green beans hanging from it. Several inches from the green beans are two dried, and split, black beans. With the discovery of the black beans, we have located the source of the intoxicating aroma.

References

Bory, S. (2004) *Diversité des vanilliers cultivés – Utilisation de marqueurs microsatellites et AFLP.* Mémoire de DEA, ENSAM, Université de Montpellier II, Montpellier, France.

Caso Barrera, L. and Fernández, M.A. (2006) Cacao, vanilla and annatto: Three production and exchange systems in the southern Maya lowlands, XVI-XVII centuries. *Journal of Latin American Geography*, 5, 29–52.

Childers, N. F, Cibes, H.R. and Hernández-Medina, E. (1959) Vanilla – The orchid of commerce, in *The Orchids: A Scientific Survey*, Withner, C.L. (ed.), The Ronald Press Co., New York, pp. 477–508.

Coe, S.D. and Coe, M.D. (1996) *The True History of Chocolate*. Thames & Hudson, New York.

Coe, M.D. (2005) *The Maya*, 7th edn. Thames & Hudson, New York.

Correll, D.S. (1953) Vanilla: Its botany, history, cultivation and economic importance. *Economic Botany*, 7, 291–358.

Dobson, N. (1973) *A History of Belize*. Longman Caribbean, Trinidad and Jamaica.

Duval, M-F., Bory, S. Andrzejewski, S., et al. (2006) Diversité génétique des vanilliers dans leurs zones de dispersion secondaire. *Les Actes du BRG*, 6, 181–196.

Feldman, L.H. (ed. and trans.) (2000) *Lost Shores, Forgotten Peoples: Spanish Explorations of South East Maya Lowlands*. Duke University Press, Durham, NC.

Henderson, J.S. (1997) *The World of the Ancient Maya*. 2nd edn. Cornell University Press, Ithaca, NY.

Jones, G.D. (1998) *The Conquest of the Last Maya Kingdom*. Stanford University Press, Stanford, CA.

Kourí, E. (2004) *A Pueblo Divided: Business, Property, and Community in Papantla, Mexico*. Stanford University Press, Stanford, CA.

Lubinsky, P. (2003) Conservation of wild vanilla. In: *First International Congress on the Future of the Vanilla Business*. Princeton, NJ, November 11–12, 2003.

Lubinsky, P. (2007) Elucidating the evolutionary origins of Tahitian vanilla. In: *Vanilla 2007*. Jamesburg, NJ, November 6–8, 2007.

Lubinsky, P., Bory, S., Hernández, J.H., Kim, S-C. and Gomez-Pompa, A. (2008a) Origins and dispersal of cultivated vanilla (*Vanilla planifolia* Jacks.[Orchidaceae]). *Economic Botany*, 62, 127–138.

Lubinsky, P., Cameron, K.M., Molina, M.C. et al. (2008b) Neotropical roots of a Polynesian spice: The hybrid origins of Tahitian Vanilla. *American Journal of Botany*, 95, 1040–1047.

McNeil, C.L. (ed.) (2006) *Chocolate in Mesoamerica: A cultural history of cacao*. The University Press of Florida, Gainesville, FL.

Pagden, A.R. (ed. and trans.) (1971) *Hernán Cortés: Letters from Mexico*. Grossman Publishers, New York.

Portères, R. (1951) Observations sur le vanillier de Tahiti. *Bulletin de la Société Botanique de France*, 98, 126–127.

Portères, R. (1954) Le genere Vanilla et ses espèces. In: *Le vanillier et la vanille dans le monde*, Bouriquet, G. (ed.), Éditions Paul Lechavalier, Paris, France, pp. 94–290.

Soto Arenas, M.A. (1999) *Filogeografia y recursos genéticos de las vainillas de México*. Project J101, CONABIO, Mexico City, Mexico.

Soto Arenas, M.A. (2006) Vainilla: Los retos de un cultivo basado en una especie amenazada con una historia de vida compleja. In: *Congreso Internacional de Productores de Vainilla. Papantla*, Veracruz, Mexico, May 26–28, 2006.

Thompson, J.E.S. (ed.) (1969) *Thomas Gage's travels in the New World*. University of Oklahoma Press, Norman, OK.

Thompson, J.E.S. (ed.) (1970) *Maya History and Religion*. University of Oklahoma Press, Norman, OK.

Thompson, J.E.S. (ed.) (1988) *The Maya of Belize: Historical chapters since Columbus*. Cubola Productions, Benque Viejo del Carmen, Cayo, Belize.

Wilk, R.R. (1997) *Household ecology: Economic change and domestic life among the Kekchi Maya in Belize*. Northern Illinois University Press, DeKalb, IL.

6

Conservation and Sustainable Use of Vanilla Crop Wild Relatives in Colombia

Nicola S. Flanagan, Paul Chavarriaga, and Ana Teresa Mosquera-Espinosa

6.1   Introduction

6.1.1   Low Genetic Diversity in the Vanilla Crop

Vanilla is one of the most economically important crops for low-altitude humid tropical regions. The crop species, *Vanilla planifolia* G. Jacks, was domesticated in Mexico, and from there distributed widely in tropical regions from the early 19th century. The principal cultivating countries are now Madagascar, Indonesia and Papua New Guinea (FAOStat; data for 2014). The vanilla crop is almost exclusively vegetatively propagated, leading to an extremely narrow genetic base, in both the primary center of origin in Mexico and Central America, as well as the secondary center of origin in the paleotropics (Schluter *et al.* 2007; Besse *et al.* 2004; Bory *et al.* 2008a; 2008b; Divakaran *et al.* 2008; Verma *et al.* 2009). The crop has a high susceptibility to disease (Hernandez-Hernandez 2011), including the principal fungal pathogen, *Fusarium oxysporum* f. sp. *radicis-vanillae* that causes root and stem rot (Koyyappurath *et al.* 2015a), and this is a major limiting factor in vanilla crop production systems (Xiong *et al.* 2015). However, despite its economic importance vanilla has received relatively little research in terms of genetic resources.

6.1.2   The Importance of Crop Wild Relatives for Agriculture

Crop wild relatives (CWR) constitute an essential resource for combating the genetic bottleneck associated with domestication. These taxa are separated into the primary, secondary and tertiary gene pools. The primary gene pool includes the genetic diversity held within the crop species, in both cultivated varieties and wild populations. The secondary and tertiary crop gene pools comprise congeneric species that are more or less closely related to the crop, and harbor a broader spectrum of potentially useful traits for improving agricultural production and maintaining sustainable agroecosystems (Hajjar and Hodgkin 2007; Heywood *et al.* 2007). Although access to these traits in many crops may be limited by physiological obstacles to gene introgression between species, the high inter-specific compatibility in orchids offers a favorable prospect for vanilla breeding.

Handbook of Vanilla Science and Technology, Second Edition.
Edited by Daphna Havkin-Frenkel and Faith C. Belanger.

Research is urgently needed into vanilla CWR in order to harness the novel genetic variation necessary to overcome present agronomic challenges, and to develop new market products. Given the high disease index in the vanilla crop, the characterization and introgression of pest and disease resistance genes is a priority for breeding programs. Exploration of the primary gene pool has identified accessions with increased resistance to *Fusarium* (Koyyappurath *et al.* 2015b), and the "Handa" variety, which shows increased resistance to *Fusarium oxysporum* f. sp. *radicis vanillae* has been patented (https://www.google.com/patents/US20170013762).

Crop wild relatives from the secondary gene pool of cultivated vanilla have also proved of value, with hybrids between *Vanilla planifolia* and the related species *V. pompona* having an increased resistance to *Fusarium* infection (Belanger and Havkin-Frenkel 2011). In terms of diversifying the aroma product market, *Vanilla* × *tahitensis*, a hybrid between *V. planifolia* and *V. odorata* (Lubinsky *et al.* 2008), possesses distinct organoleptic properties that are valued in the gourmet market. Further traits that should be explored include those that may mitigate the abiotic stresses predicted under global scenarios of climate change (Dempewolf *et al.* 2016). At the same time, a concerted effort is also needed to develop a comprehensive strategy for the conservation of vanilla genetic resources (Roux-Cuvelier and Grisoni 2010), at the national (Azofeifa-Bolaños *et al.* 2014; Flanagan and Mosquera-Espinosa 2016) and the international level (Castañeda-Álvarez *et al.* 2016), so as to meet crop breeding requirements in the future. Here we review research into vanilla CWR, and identify priorities for research, and actions necessary for conservation and sustainable use for these valuable genetic resources in Colombia.

6.2   Vanilla Crop Wild Relatives

6.2.1   Phylogenetic Diversity Within the Genus *Vanilla*

The genus *Vanilla* comprises approximately 120 species distributed in tropical and subtropical regions across the Americas, Africa and Asia. Soto Arenas and Cribb (2010) divided the genus into two subgenera. The subgenus *Vanilla* is distributed in the Neotropics and contains 17 species, including the genus type species *Vanilla mexicana*. This basal clade in the genus phylogeny, also called the Membranaceous clade, has a center of diversity in Southern Brazil, where two new species have been recently described (Pansarin *et al.* 2012; Pansarin and Miranda 2016). The second, larger subgenus, *Xanata*, is further split into two sections, *Xanata* and *Tethya*. The crop species, *Vanilla planifolia* falls within the section *Xanata*, which contains 43 species distributed within the neo-tropics. The commercial, hybrid taxon *Vanilla* × *tahitensis*, derived from *Vanilla planifolia* and *Vanilla odorata* parental species (Lubinsky *et al.* 2008) is included in this group. In the section *Tethya* a further 59 species are reported (Soto Arenas and Cribb 2010), distributed principally in the paleotropics, with a group of five species occurring across the Caribbean islands.

6.2.2   The Secondary Gene Pool for Vanilla

Not all *Vanilla* species produce aromatic fruits. Those that do fall within the subgenus *Xanata*, section *Xanata*, which is commonly known as the "aromatic clade". This group of species, being those most closely related to the crop species *Vanilla planifolia,* represents the secondary gene pool of the vanilla crop. Research efforts for conservation and utilization of vanilla genetic resources should be primarily concentrated on this clade of species. The remaining species within the genus, being more distantly related to the crop species, represent members of the tertiary gene pool, which, nonetheless, may possess traits of interest in crop breeding programs, and from whom gene transfer to the crop species is still likely.

The neotropical region represents the natural center of distribution of the secondary gene pool of the vanilla crop, with members of the section *Xanata* distributed from Northern Argentina to Florida. Soto-Arenas and Cribb (2010) divide the section *Xanata* into six morpho-groups. The largest of these groups is that containing *Vanilla planifolia,* in which the authors recognized 16 species, including *Vanilla* × *tahitensis.* Recent taxonomical work has added a further six species to this group, predominantly from South America. Koch *et al.* (2013) described *V. labellopapillata* from the Brazilian Amazon region, and Sambin and Chiron (2015) described *V. inornata* and *V. aspericaulis* from French Guiana. From the Atlantic forest of Brazil de Fraga *et al.* (2017) have described *V. capixaba* and *V. paulista.* Additionally, Azofeifa-Bolaños *et al.* (2017) have described *Vanilla sotoarenasii,* a species closely related to *Vanilla planifolia,* found in Costa Rica.

The section *Xanata* contains five more groups: the *Vanilla pompona* group; the *Vanilla hostmanii* group; the *Vanilla trigonocarpa* group; the more basal *Vanilla palmarum* group; and *Vanilla penicillata.* Together, these groups contain a further 21 recognized species (Soto Arenas and Cribb 2010; Molineros-Hurtado *et al.* 2014), bringing the total number of taxa in the secondary gene pool of vanilla to 43 species.

6.2.3   Vanilla Diversity in Colombia

Based on a revision of material in major international herbaria, Soto Arenas and Cribb (2010) reported eleven species for Colombia, ten of which fall in the aromatic clade. Herbaria in Colombia were, however, not consulted. Ortiz-Valdivieso (2015) lists two further species, *Vanilla palmarum* and *V. penicillata* from the Amazon region. Recent field work has identified natural populations of twelve species, including four new reports for the country: *V. bicolor, V. cribbiana, V. helleri* and *V. phaeantha,* and a confirmation of wild populations of *V. planifolia* (Molineros-Hurtado 2012). Additionally a new species, *Vanilla rivasii* has been described in the *V. hostmanii* group (Molineros-Hurtado *et al.* 2014). All these additional reports constitute species in the aromatic clade. Also reported for the country are four species belonging to the Membranaceous clade: *V. guianensis, V. inodora, V. methonica* and *V. mexicana,* although these lack associated reference specimens, and need botanical verification (Ministerio de Ambiente y Desarrollo Sostenible (MADS) y Universidad Nacional de Colombia (UNAL), 2015). These reports bring the total number of species for Colombia to 22, of which 18 fall within the aromatic clade, distributed across the six recognized clades (Table 6.1; Figure 6.1). Three of these aromatic species, *Vanilla columbiana, V. espondae* and *V. rivasii* are endemic to the country.

Taxonomic work for the genus is hampered due to the rarity of plants in natural habitat, their low flowering frequency and ephemeral flowers, lasting generally no more than 24 hours. Thus, when encountered, the plants tend to be in the sterile state, so

Table 6.1 Register of *Vanilla* species in Colombia.

Vanilla species	Subgeneric classification	Habitat type: Distribution in Colombian regions	Citation for presence in Colombia
Vanilla bicolor Lindl.	Subgenus *Xanata*, Section *Xanata*, *V. palmarum* group	Humid Tropical Forest: Chocó, Amazonia	Molineros-Hurtado, 2012
Vanilla calyculata Schltr.	Subgenus *Xanata*, Section *Xanata*, *V. pompona* group	Tropical Dry Forest: Valle del Cauca, Valle del Magdalena, Caribbean	Soto Arenas and Cribb, 2010
Vanilla columbiana Rolfe	Subgenus *Xanata*, Section *Xanata*, *V. pompona* group	Tropical Dry Forest: Valle del Magdalena	Soto Arenas and Cribb, 2010
Vanilla cribbiana Soto Arenas	Subgenus *Xanata*, Section *Xanata*, *V. hostmanii* group	Humid Tropical Forest: Chocó	Molineros-Hurtado, 2012
Vanilla dressleri Soto Arenas	Subgenus *Xanata*, Section *Xanata*, *V. hostmanii* group	Humid Tropical Forest: Chocó	Soto Arenas and Cribb, 2010
Vanilla espondae Soto Arenas	Subgenus *Xanata*, Section *Xanata*, *V. trigonocarpa* group	Tropical Dry Forest: Valle del Magdalena. Humid Tropical Forest: Amazonia	Soto Arenas and Cribb, 2010
Vanilla helleri Hawkes	Subgenus *Xanata*, Section *Xanata*, *V. planifolia* group	Humid Tropical Forest: Chocó	Molineros-Hurtado *et al.*, unpublished data
Vanilla hostmanii Rolfe	Subgenus *Xanata*, Section *Xanata*, *V. hostmanii* group	Humid Tropical Forest: Amazonia	Soto Arenas and Cribb, 2010
Vanilla guianensis Splitg.	Subgenus *Vanilla*	Not registered	MADS and UNAL, 2015
Vanilla inodora Schiede	Subgenus *Vanilla*	Not registered	MADS and UNAL, 2015
Vanilla methonica Rchb.f and Warsz.	Subgenus *Vanilla*	Humid Tropical Forest: Amazonia	Soto Arenas and Cribb, 2010
Vanilla mexicana Mill.	Subgenus *Vanilla*	Not registered	MADS and UNAL, 2015
Vanilla odorata C. Presl	Subgenus *Xanata*, Section *Xanata*, *V. planifolia* group	Humid Tropical Forest: Chocó. Andean	Soto Arenas and Cribb, 2010
Vanilla palmarum Lindl	Subgenus *Xanata*, Section *Xanata*, *V. palmarum* group	Humid Tropical Forest: Amazonia	Ortiz-Valdivieso, 2015
Vanilla penicillata Garay and Dunst	Subgenus *Xanata*, Section *Xanata*, *V. penicillata* group	Riparian Forests: Llanos	Ortiz-Valdivieso, 2015
Vanilla phaeantha Rchb.	Subgenus *Xanata*, Section *Xanata*, *V. planifolia* group	Riparian Forests Llanos, Caribbean	Molineros *et al.*, unpublished data

Table 6.1 (Continued)

Vanilla species	Subgeneric classification	Habitat type: Distribution in Colombian regions	Citation for presence in Colombia
Vanilla planifolia Andrews	Subgenus *Xanata*, Section *Xanata*, *V. planifolia* group	Humid Tropical Forest: Chocó	Molineros-Hurtado 2012
Vanilla pompona Schiede	Subgenus *Xanata*, Section *Xanata*, *V. pompona* group	Tropical Dry Forest: Valle del Cauca	Soto Arenas and Cribb, 2010
Vanilla riberoi Hoehne	Subgenus *Xanata*, Section *Xanata*, *V. planifolia* group	Humid Tropical Forest: Amazonia	Soto Arenas and Cribb, 2010
Vanilla rivasii Molineros-Hurtado et al.	Subgenus *Xanata*, Section *Xanata*, *V. hostmanii* group	Humid Tropical Forest: Chocó	Molineros-Hurtado *et al*, 2014
Vanilla sprucei Rolfe	Subgenus *Xanata*, Section *Xanata*, *V. trigonocarpa* group	Humid Tropical Forest: Amazonia	Soto Arenas and Cribb, 2010
Vanilla trigonocarpa Hoehne	Subgenus *Xanata*, Section *Xanata*, *V. trigonocarpa* group	Humid Tropical Forest: Chocó	Soto Arenas and Cribb, 2010

hindering species identification. Continued exploration of *Vanilla* diversity in the country is necessary, and will likely result in further species being discovered. This work shows that Colombia is an important center of diversity for the *Vanilla* genus, and in particular for the secondary gene pool of this crop (Flanagan and Mosquera-Espinosa 2016).

6.3 Vanilla Species in the Wild

6.3.1 Vanilla Species are Rare in the Wild

Vanilla species are uncommon in the wild, even in undisturbed habitats (Soto Arenas and Dressler 2010), and populations known in Colombia often comprise only a few or even a single individual (Molineros-Hurtado 2012; Flanagan and Mosquera-Espinosa 2016). *Vanilla* plants flower infrequently, and when flowering does occur, natural pollination events in non-autogamous species are rare. Establishment of new plants through seed germination occurs rarely, and wild plants commonly undergo vegetative propagation, with vine division likely occurring through natural forest dynamics, such as tree fall. The small populations, together with low levels of genetic diversity resulting from vegetative propagation (Schluter *et al.* 2007; Gigant *et al.* 2014), make these species extremely vulnerable to population, and subsequent species extinction (Soto-Arenas 2003). The ecology of *Vanilla* species in the wild is still underexplored, yet knowledge of ecological and physiological processes and traits in wild populations is of great value for the development of effective strategies for conservation and sustainable use.

Figure 6.1 Colombian wild *Vanilla* species. **A** – *V. planifolia*; **B** – *V. odorata*; **C** – *V. phaeantha*; **D** – *V. rivasii*; **E** – *V. pompona*; **F** – *V. trigonocarpa*; **G** – *V. calyculata*; **H** – *V. espondae*; **I** – *V. bicolor*. Photos: A–C, E–G, Nicola Flanagan. D, I – Francisco Molineros Hurtado. H – Nicolas Gutierrez Morales.

6.3.2 Reproductive Biology of Vanilla Wild Species

6.3.2.1 Pollinators

The essential need for manual pollination is arguably the principal cost in vanilla cultivation. This is a consequence of a highly specialized pollination syndrome, dependent on a small number of specific pollinator species. In the neo-tropics, Pansarin *et al.* (2014) observed that the membranaceous species, *Vanilla edwallii*, produce floral fragrances to attract male bees in the genus *Epicharis* (Apidae: Centridini), however wider observations suggest that most species in the *Xanata* section are pollinated by euglossine bees (Apidae: Euglossini), the so-called orchid bees (Pansarin 2016). Euglossine pollination is generally considered to be reward-based pollination, with male bees visiting flowers to collect aromatic compounds for use in courtship displays (Roubik and Hanson 2004). Nonetheless, observations of fragrance-collecting behavior by euglossine bees in *Vanilla* species are rare, and food-deceptive pollination is an alternative mechanism, that may also explain the low natural fruit set seen in most *Vanilla* species (Lubinsky *et al.* 2006; Pansarin 2016).

Roubik and Ackerman (1987) observed bees in the genera *Euglossa*, *Eulaema* and *Eufriesea* with pollen from *Vanilla planifolia*, *V. pompona*, and other unidentified species. The smaller *Euglossa* bee species are reported as pollinators of *Vanilla planifolia* (*Eug. viridissima*) and *V. trigonocarpa* (*Eug. asarophora*) (Soto-Arenas and Dressler 2010). However, given the size of *Vanilla* flowers, the larger bees in the genus *Eulaema* are probably the most common pollinator species, and have been observed pollinating *V. pompona* ssp. *grandiflora* in Peru (*Eulaema meriana*: Lubinsky *et al.* 2006), *V. cribbiana* (*Eulaema* sp.) and *V. insignis* (*Eul. polychrome*) in Mexico (Soto-Arenas and Dressler 2010), and *V. bahiana* in N.E. Brazil (Mendoça dos Anjos *et al.* 2017). A diversity of bees in the genera *Euglossa* and *Exeretes*, as well as *Melipona*, and hummingbirds, were observed visiting *V. planifolia* (Lubinksy *et al.* 2006) and *V. hartii* (Soto-Arenas and Dressler 2010) in southern Mexico, but did not affect pollination, and males of both *Eulaema nigrita* and *Eufriesia violacea* were observed visiting the fragrant flowers of *V. dubia* in southeastern Brazil (Pansarin and Pansarin 2014). No definite registration has been made for *Vanilla* pollinators in the country; however Colombia is the most species-rich country for euglossine bees, with more than 113 registered (Ramirez *et al.* 2002).

Euglossine bees are native to the neo-tropics, and the absence of the specific pollinator species of *V. planifolia* hindered the establishment of the vanilla crop outside Mexico, until a manual pollination method was devised by Edmund Albius on the island of Reunion in 1841. Due to the low natural fruit set in *Vanilla planifolia*, manual pollination was quickly adopted in Mexico, where it remains essential in vanilla plantations today. In Colombia, a fairly high natural fruit set has been observed in wild species (Molineros-Hurtado 2012), however, with the increasing impact of ecosystem transformation on euglossine bee populations (Aguiar *et al.* 2015), natural pollination of wild *Vanilla* plants is likely to decline.

6.3.2.2 Autogamy

A key trait of interest for vanilla crop breeding is autogamy, offering the possibility to reduce the costs of manual pollination. In *Vanilla* flowers, the rostellar flap towards the apex of the column presents a physical barrier to self-pollination, keeping the pollen separate from the stigma. However, in some species the stigmatic fluid is sufficiently abundant for it to reach the pollinia, thereby inducing pollination (van der Pijl and Dodson 1966). Stigmatic leak has been observed in a small number of species in the tertiary gene pool (Soto-Arenas and Dressler 2010; Gigant *et al.* 2011, 2014, 2016), as well as *V. bicolor* and *V. palmarum*, in the most basal clade within the *Xanata* section (van Dam *et al.* 2010; Householder *et al.* 2010). Occasional spontaneous self-pollination has also been reported in *V. planifolia* (Soto-Arenas 2003), but the physiological basis is unstudied.

Further analysis of pollination syndromes and floral anatomy is required across different *Vanilla* species in order to understand possible mechanisms for autogamy in these species, and thereby develop strategies for its incorporation into commercial cultivars, so as to reduce the cultivation costs. Effective cross-pollination, although infrequent, is essential to maintain genetic variation, and so population viability in the wild. Thus, there is a vital need to understand pollination biology in wild *Vanilla* species, and so also take measures that ensure the conservation of pollinator species in native habitats.

6.3.3 Mycorrhizal Interactions

Vanilla species, like all orchids, require an association with mycorrhizal fungi in order to provide both the organic carbon and inorganic phosphorus and nitrogen nutrients necessary for seed germination, and subsequent plant growth, nutrition and defense in adult plants (Dearnaley *et al.* 2012). Although this interaction was considered as parasitism of the orchid upon the fungus, it has been shown in one temperate terrestrial orchid species that carbon may also be passed from orchid to fungus (Cameron 2006, 2008).

Across the orchid family different levels of specificity have been seen in their mycorrhizal fungal associations. The few studies of *Vanilla* orchid mycorrhizae to date reveal no clear picture regarding specificity of interactions. Fungal strains from three sexual genera, *Thanatophorus, Ceratobasidium* and *Tulasnella*, all in the Order Cantharellales, were isolated from both wild and cultivated adult plants of *V. planifolia*, and two other species in Puerto Rico and Costa Rica, suggesting a more general interaction between adult *Vanilla* plants and fungi (Porras-Alfaro and Bayman 2007; Bayman *et al.* 2011). In terms of function, however, a more specific interaction was observed, with only isolates of *Ceratobasidium* proving effective for promoting seed germination in *V. planifolia*. These data support the hypothesis that orchid mycorrhizal associations may change during plant growth and development.

In Colombia, both *Ceratobasidium* and *Tulasnella* species have been isolated from adult plants of several wild *Vanilla* species distributed across both dry and humid tropical forest habitats (Mosquera-Espinosa *et al.* 2010; 2012). Further research is required to fully characterize the inter and intra-population diversity of mycorrhizal fungal interactions across wild species, so as to understand the ecological networks in which these interactions are situated, and also to determine mycorrhizal function during the initial life cycle stages of seed germination and plantlet growth. This knowledge has potential applications for *ex situ* conservation and propagation efforts, as well as in sustainable cultivation practices. The use of *in situ* seed baits offers a promising technique to isolate those fungi with which the seeds associate for germination under natural conditions (Brundrett *et al.* 2003; Zettler *et al.* 2011).

6.3.4   Further Interactions with the Microbiome

In addition to interactions with mycorrhizal-forming fungi, *Vanilla* species also associate with a diversity of other endophytic and rhizospheric microorganisms, both fungi and bacteria. Endophytes are highly diverse and ubiquitous in plants, residing within the tissue of the plant without causing any visible manifestation of disease. A complex network of signaling interactions exists between the plant host and the endophyte community, finely balanced between pathogenic and beneficial outcomes, and dependent on the plant genotype colonized, amongst other factors (Rodriguez and Redman 2008). Plant signaling pathways may promote fungal gene expression, with the production of bioactive secondary metabolites by fungi *in planta*, but not in fungal cultures *in vitro* (Kusari *et al.* 2012), thereby co-opting endophytes to increase fitness in the plant hosts, such as the enhanced herbivore deterrence in grasses conferred by the fungi in the family Clavicipitaceae (Bacon and White 2000). In *Vanilla planifolia*, the bacterium, *Bacillus amyloliquefaciens*, with known *in vivo* antifungal properties in other plant species, has been reported (White *et al.* 2014). In addition to the endophytic community, the soil microbial community likely also plays a role in disease suppression in plants. A recent study by Xiong and co-workers (2017) showed that those soils associated with a lower incidence of *Fusarium* wilt disease had a higher fungal diversity, and lower bacterial diversity, with the fungal genus *Mortierella* having a particularly strong association with disease suppression.

The microbiome has also been seen to play a role in aromatic traits. Khoyratty and co-workers (2015) found that the presence of *Pestalotiopsis microspore* in maturing pods was correlated with a significant increase in the absolute amounts of vanillin and other flavor compounds, thereby raising the intriguing question regarding the importance of endophyte biotransformation in flavor complexity, and how therefore, agronomic practices which may alter the endophytic community impact on flavor development. A study of *V. planifolia* accessions in Mexico, however, found no association of aroma with environmental variation (Salazar-Rojas *et al.* 2011). More in depth studies of the three-way interactions plant-microbiome-environment are clearly needed.

In our studies of wild *Vanilla* populations, in both humid and dry tropical forest ecosystems, we have encountered a low to absent disease incidence, with no symptoms of *Fusarium* infection being observed (Mosquera-Espinosa *et al.* 2012). As wild populations will be adapted to the local conditions, including the local microbial communities, the study of the microbiomes of wild *Vanilla* plants is a clear strategy to understand the role of the *Vanilla* microbiome in stress tolerance, as well as to identify potential microbial bio-control agents to counter disease in the vanilla crop. In a study of endophytic fungi of five *Vanilla* species in Colombia, we have found a high diversity of fungi, with close to 150 Operational Taxonomic Species (OTUs) from eleven Ascomycete and four Basidiomycete orders (Mosquera-Espinosa and Flanagan, unpublished data). The evaluation of the symbiotic function of this microbial diversity is now required.

6.3.5 Bioclimatic and Biophysical Adaptations

Vanilla planifolia is cultivated in Mexico in the humid tropical forests along the Caribbean seaboard from sea level to 700 m.a.s.l., with average temperatures around 24°C, relative humidity at 80%, and average annual precipitation between 1200 and 1300 mm. Flowering is associated with the onset of the marked dry season in March (Hernandez-Hernandez 2011). These local conditions in Mexico provide general guidelines for new ventures in vanilla cultivation around the tropics.

However, across their neotropical distribution, the species within the aromatic clade of *Vanilla* are native to a variety of biomes with differing bioclimatic and biophysical conditions. An understanding of these adaptive patterns and the incorporation of abiotic adaptations into crop breeding programs may facilitate the expansion of vanilla cultivation beyond those regions currently considered optimal, as well as offering an important strategy to mitigate the impact of predicted global climate change on vanilla cultivation.

In Colombia, *Vanilla* species occur naturally up to 1200 m.a.s.l. in biomes ranging from the pluvial forests in the Pacific coastal and Amazon regions to sub-xerophytic conditions. Species adaptation to different climatic regimes appears not to be phylogenetically correlated. The conditions found in the tropical dry forests in the inter-Andean valleys most closely resemble those recommended for the vanilla crop, with annual precipitation between 1,000 and 1,500 mm, a relative humidity of 70–80%, and mean temperatures of 24 to 28°C, although these regions present two notable dry seasons (IDEAM 2017). In such regions the species *V. calyculata, V. odorata* and *V. pompona* are found. In contrast, the Chocó biogeographic region along the Pacific coast, possibly the wettest place of Earth, has up to 8,000 mm annual precipitation, relative humidity up to 90%, a mean temperature of 26 to 28°C, and little or no seasonality (IDEAM 2017), conditions quite distinct to those of the vanilla-growing region in Mexico. In this region wild populations of *Vanilla planifolia* have been identified, together with seven other species from four different groups in the aromatic clade (see Table 6.1).

Under predicted climate change scenarios in Colombia (IDEAM *et al.* 2015), in both the Chocó pacific region, and inter-Andean valleys, a light to moderate increase in rainfall is expected, while in the Amazon and Caribbean regions a reduction of up to 20% is predicted by 2070. There is very little understanding of the tolerance of the vanilla crop to changing patterns of precipitation. In terms of temperature shifts, a recent study has predicted that due to expected temperature increases of up to 2.6°C by 2050, the optimal distributions of orchid populations in tropical dry forest in Colombia will increase in altitude by between 250 m and 650 m (Reina-Rodriguez *et al.* 2017). The biology of *Vanilla* species, as long-lived plants with infrequent plantlet establishment, indicates that they likely have a very low ability to naturally shift distributions within the expected timeframe of climate change.

The natural distributions of *Vanilla* species also vary with respect to edaphic conditions. The soils of the inter-Andean valleys and Andean foothills range from loamy haplustoll soils on the alluvial plains to volcanic hapludands in the foothills, while soils on the Pacific coast are acidic, nutrient-deficient ultisols. This association of different Colombian species to different substrates reflects that seen in Central America where *V. planifolia* is limited to calcerous or sedimentary substrates, while *V. insignis* may be found on substrates of volcanic origin (Soto-Arenas 2003). There is very little understanding of the role of biophysical adaptations in the vanilla crop, and clearly much more work is needed here.

6.4   Conservation of Vanilla Crop Wild Relatives

6.4.1   Threats to Conservation

The genetic resources held within the vanilla crop wild relatives in Colombia are subject to severe conservation threats, due both to intrinsic biological traits of these species, including the low population densities, specialized pollination and the obligate interaction with mycorrhizal fungi outlined above, and to extrinsic anthropogenic activities.

The tropical forests below 1200 m.a.s.l. that provide the natural habitat for *Vanilla* species are under threat from deforestation for landscape conversion to agriculture, mining and urbanization (González *et al.* 2011). While the humid lowland forests in both the Amazon and the Pacific coastal region still retain considerable areas, deforestation rates in these areas are increasing (IDEAM 2015). For those species native to tropical dry forest the outlook is most serious. Only 4% of the original cover of tropical dry forest remains in Colombia (Pizano and García 2014), and several *Vanilla* species are known only from a single locality.

In addition to habitat degradation, *Vanilla* species may also be subject to the pressure of collection from wild populations. Soto Arenas and colleagues documented the disappearance of known populations of *Vanilla planifolia* in Mexico, Guatemala and Belize, ostensibly for establishment of plantations, leaving fewer than 30 known wild *V. planifolia* plants (Schluter *et al.* 2007). In Colombia, *Vanilla* species are currently not much sought-after compared with other, ornamental orchid species. However, as interest grows in its cultivation, it is likely that vines will be increasingly extracted from the wild. With species determination being problematic due to infrequent flowering, it is probable that both commercial and non-commercial species will be targeted, as has also been seen in Central America and Mexico. Thus, there is an urgent need for effective strategies for *Vanilla* species conservation that integrate *in situ*, *ex situ* and *circa situm* actions (Figure 6.2), before the scarce populations disappear from the wild.

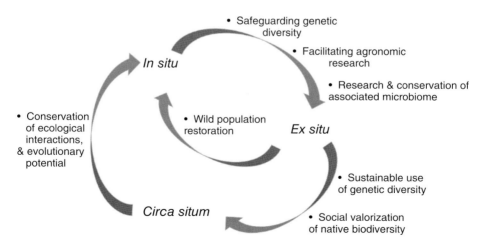

Figure 6.2 An integrated strategy for the conservation and sustainable use of vanilla crop wild relatives. Simultaneous *in situ*, *ex situ* and *circa situm* actions complement each other, and maximize our ability to conserve genetic resources while also facilitating research and sustainable use.

6.4.2 Conservation *In situ*

In situ conservation facilitates the maintenance of a sufficient population size to retain genetic variation and viability, as well as enabling conservation of the ecological interactions in which the species are involved, thereby also contributing to the conservation of other species in inter-dependent ecological networks. Moreover, *in situ* conservation permits ongoing adaptive change in response to environmental variation, so maximizing persistence of the species in the long-term (Maxted *et al.* 1997). Thus, protection of the natural habitat is the single most important conservation action for most plant species.

As of 2016, close to 28 million hectares (25%) in Colombia are included under different modalities of protection, with half of this comprising the National Parks system. However representation across the diverse biomes in the country varies. In the Amazon region, more than 25% of the territory is protected, while for the tropical dry forest habitats in the inter-Andean valleys and the Caribbean region this drops to less than 1 percent (Forero-Medina and Joppa 2010). Given the small population sizes for wild *Vanilla* species, even in those regions that have larger protected areas, the existence of a protected area network is not sufficient in itself, and additional actions for *in situ* population management are necessary. As well as augmentation of existing populations, population translocations are recommended to new localities within the species historic range, particularly to more secure protected areas (Flanagan *et al.* 2012). Efforts to include *Vanilla* CWR populations within protected areas must also take into consideration the predicted future distributions of species under climate change projections (Reina-Rodriguez *et al.* 2017). Thus, assisted migration of species and establishment of populations beyond the current native range must also be considered (Liu *et al.* 2015), particularly for those species native to the inter-Andean valleys. Long term monitoring of the size and health of both *Vanilla* populations and also their key ecological interactions is recommended.

6.4.3 Conservation *Ex situ*

A significant *ex situ* collection of crop wild relatives is an essential complement to *in situ* conservation, not only to safeguard these valuable genetic resources, but also for plant breeding activities (Smith 2016), through molecular and agronomic characterization for traits such as autogamy, enhanced flowering frequency, unique organoleptic quality, indehiscence, shortening or eliminating maturation by curing, as well as disease resistance. In Colombia an *ex situ* collection has been established for *in vitro* and in-greenhouse conservation, as well as germplasm characterization. To date the collection holds 14 species with 27 accessions from diverse biomes of the country. In-greenhouse, the vines are maintained under 80% shading mesh on organic leaf litter, collected principally from leguminous trees such as *Inga spuria*. This is topped-up 2–3 times per year, and plants are also fertilized each month with a generic fertilizer containing N-P-K at 20-30-10, as well as trace elements. In the greenhouse the plants are subject to the climate profile of the region (28–>32°C; 70–80% R.H.). Although these conditions differ from native conditions, some species have flowered within a period of 18 months, opening the possibility of establishing an *ex situ* breeding program with Colombian vanilla CWR under controlled greenhouse conditions.

The in-greenhouse collection has been replicated *in vitro*, with meristem/explants from a diversity of species established under standard growing conditions at 28 ± 2°C, 16/8 h photoperiod, with a photosynthetic flux of 120–150 μmol m^{-2} s^{-1}, and without moisture regulation (Figure 6.3). For maintenance, explants are seeded on 50 ml of Orchimax$^{©}$ basal medium with activated charcoal, (Duchefa cat # O0262), with propagation carried out every four months, depending on the species (Sanchez-Chamorro *et al.*, unpublished data). These conditions are optimal for multiplication and research

Figure 6.3 *In vitro* germplasm collection of vanilla crop wild relatives. **A** – Propagation of different *Vanilla* species from axillary buds under *in vitro* conditions. From left to right, *V. calyculata, V. rivasii, V. odorata, V. phaeantha, V. pompona,* and *V. planifolia.* **B** – General view of the clonally propagated collection. **C** – Immature seeds of F$_1$ interspecific hybrids. The approximate size of a seed is 350 × 260 μm. **D** – F$_1$ interspecific hybrids sown from seed *in vitro.* **E** – Eight month-old-plants of the same F$_1$ interspecific hybrid.

purposes, however for long-term conservation it is necessary to establish slow-growth conditions in order to minimize costs (Bello-Bello *et al.* 2015).

The *in vitro* system developed to preserve the diversity of Colombian vanillas also facilitates propagation of the wild species from seed, thereby facilitating the establishment of a bank of hybrids and breeding lines, and so accelerating breeding efforts. Hybridization efforts are limited by the necessity of two species flowering simultaneously, an uncommon event given the infrequent and ephemeral flowering in *Vanilla*. However, this objective will be facilitated through the establishment of pollen conservation actions (Hawkes *et al.* 2012).

6.4.4 Conservation *Ex situ* of the Vanilla Microbiome

Given the importance of the microbiome for plant health, any long-term *ex situ* CWR conservation strategy for *Vanilla* genetic resources must also contemplate the simultaneous conservation of microbial symbionts. While protocols do exist for asymbiotic *in vitro* germination of *Vanilla* seed (Menchaca *et al.* 2011) the application of mycorrhizal fungi will likely enhance the success of transfer of plantlets to *ex vitro* conditions. Cryopreservation techniques have been developed to store orchid mycorrhizal fungi either together with the seed (e.g. Wood *et al.*, 2000), or separately (Batty *et al.* 2001). Isolates of *Tulasnella* and *Ceratobasidium* genera stored at −80°C for up to 24 months retained or even improved their ability to promote *in vitro* seed germination (Ercole *et al.* 2013). Once an understanding is gained of the important component taxa of the vanilla microbiome, research will be needed to optimize protocols for their long-term conservation and application *ex situ*.

6.4.5 Conservation of *Circa situm* and Sustainable Use

Circa situm conservation refers to those actions carried out within altered agricultural landscapes, such as agroforestry systems or home gardens that are outside natural habitats but within the native geographical range of a species (Dawson *et al.* 2013) (Figure 6.4). *Circa situm* actions are often successfully promoted in combination with the sustainable use of the genetic resources. To date, there has been little attention given to the potential for cultivation and commercialization of wild aromatic *Vanilla* species, however it is likely that some wild species, including *V. odorata*, one of the parental species of *V.* × *tahitensis*, will present favorable organoleptic and agronomic traits. Thus, vanilla CWR not only represent valuable genetic resources for crop improvement, but also, in and of themselves, are a source of new products in a diversified fragrance and aroma market.

In Colombia, cultivation of promising *Vanilla* species offers the forest-dependent communities whose territories contain these plants a potential income for sustainable livelihood improvement. *Vanilla* plants can be easily incorporated into agroforestry systems, in either humid or dry tropical forest regions (Flanagan *et al.* 2012; Flanagan and Mosquera-Espinosa 2016), thereby also minimizing the loss of beneficial ecological interactions, including those with pollinators and microorganisms. A diversified market offers an opportunity to consolidate a more constant value chain, which could also be based on certification for organic or sustainable agricultural production, fair-trade or "biodiversity-friendly" cultivation.

Figure 6.4 Actions for *circa situm* conservation and sustainable use in rural communities.
A – A community germplasm bank for Colombian Pacific *Vanilla* species. **B** – *Vanilla phaeantha* in a home garden where it is easily accessible for pollination and harvesting for household use. **C** – Community fencing, and **D** – population augmentation of *Vanilla calyculata* in native forest relicts. **E** – A community educational activity. **F** and **G**. Cultivation of *Vanilla rivasii* within secondary forest stands by Don Alvaro Rivas, a small-scale farmer on the Colombian Pacific coast, and namesake of the species *V. rivasii*.

Raising awareness in local communities of the potential value of these wild genetic resources will encourage community conservation actions for *Vanilla*, including community germplasm collections (Flanagan *et al.* 2012; Mosquera-Espinosa *et al.* 2012), as well as promoting the more biodiverse agro-forestry ecosystems in which the vanilla is cultivated. Nonetheless, there is no tradition of vanilla cultivation in Colombia, and the implementing of programs for sustainable production and on-farm conservation requires technical training and accompaniment in effective eco-agricultural practice, for at least the duration of the initial crop establishment period of two to three years, or longer.

6.5 Biotechnological Approaches for Vanilla Genetic Resource Conservation and Utilization

6.5.1 Characterization and Utilization of Genetic Diversity

6.5.1.1 DNA Barcoding

Actions for both the conservation and sustainable utilization of *Vanilla* genetic resources are hindered by difficulties in identifying plants found in the wild, due to their infrequent flowering. DNA barcodes offer a straightforward alternative for species identification through the use of diagnostic genetic variation in the DNA sequence of specified loci. A number of loci have been suggested as candidate barcode loci in plants, however the *matK* locus in the chloroplast genome has been recommended for orchid diagnostics (Lahaye *et al.* 2008). Species differentiation is also possible with the *rRNA-ITS* locus in the nuclear genome (Soto-Arenas and Dressler, 2010). Azofeifa-Bolaños *et al.* (2017) found two nucleotide substitutions in the *ITS* locus distinguishing the newly-described species *V. sotoarenasii* from *V. planifolia*. These two loci comprise the majority of sequences for the *Vanilla* genus in the public databases, a necessary resource for DNA barcoding, and therefore probably comprise the most useful barcoding loci. Nonetheless, further work is needed to complete the genetic data available for all known *Vanilla* species. This work is on-going for those Colombian species for which we have unambiguous identifications based on floral morphology (Molineros-Hurtado *et al.*, unpublished data).

6.5.1.2 Genomic Characterization of Vanilla

In order to fully exploit the genetic diversity present in vanilla crop wild relatives, an understanding of their genomic diversity is essential. Subsequent to the sequencing of the chloroplast genome of *Vanilla planifolia* (Lin *et al.* 2015), sequencing of the nuclear genome of *Vanilla shenzhenica* (syn. = *V. somai*), a member of the tertiary gene pool, is proceeding (http://english.agri.gov.cn/news/dqnf/201502/t20150212_25013.htm).

An *ex situ* collection of the genetic diversity in the secondary gene pool will clearly facilitate the task of identifying useful genetic variation (Henry 2013), however the genome size and occurrence of somatic endo-replication of *Vanilla* genomes means this work is not straightforward (Trávníček *et al.* 2015). Variation in genome size encountered within cultivars of *Vanilla planifolia* and *V. × tahitensis* (Bory *et al.* 2008c; Lepers-Andrzejewski *et al.* 2012) is also being seen in our study of accessions of different species in the secondary gene pool (Camila Gonzalez *et al.*, personal communication). Nonetheless, comparative whole genome re-sequencing across the secondary gene pool of *Vanilla* is an increasingly necessary and viable venture, and these efforts must continue.

6.5.2 Application of Microorganisms in Vanilla Cultivation

Studies of mycorrhizal function in *Vanilla* show that isolates of *Ceratobasidium* are effective for promoting *V. planifolia* seed germination and growth and survival of *in vitro* plantlets of a commercial cultivar (Porras-Alfaro and Bayman 2007). The application of mycorrhizal fungi not only enhances plant nutrition but may also promote plant defenses under cultivation conditions. *Ceratobasidium* spp. obtained from terrestrial roots of different orchid species in Colombia, including *Vanilla* spp., have shown potential as biocontrol agents against the pathogen *Rhizoctonia solani* in rice (Mosquera-Espinosa *et al.* 2013). In other crops, studies show both a direct action of the biocontrol agent on the *Fusarium* pathogenic fungus through competition for space and nutrients, and an indirect action, through induction of systemic resistance in the host (e.g. Burns and Benson 2000; Sneh *et al.* 2004).

Other microorganisms have also shown positive results as bio-control agents in vanilla diseases. Both soil-derived and endophytic fungi, including the genera *Trichoderma*, *Mortierella* and *Emericella* not only suppress *Fusarium* and other vanilla diseases, but, in some cases, also promote plant growth. Other rhizosphere bacteria, including *Pseudomonas fluorescens*, *Staphylococcus xylosus*, *Serratia* sp. and *Stenotrophomonas* sp. are associated with disease suppression, in particular when inoculated in conjunction with a bio-control fungus (Duijff *et al.* 1999; Radjacommare *et al.* 2007; Talubnak and Soytong 2010; Sandheep *et al.* 2012, 2013; Ravindran and Shaike 2013; Adame-García *et al.* 2016; Xiong *et al.* 2017). The natural populations of *Vanilla* wild species in Colombia, which show a very low incidence of disease, offer a natural reservoir for further bioprospecting for potential beneficial microbial diversity (Mosquera-Espinosa *et al.* 2012).

6.6 An Integrated Strategy for Conservation and Sustainable Use of Vanilla Crop Wild Relatives

6.6.1 A Colombian National Strategy for Vanilla CWR

Vanilla offers an exceptional opportunity for the development of a high-value chain for rural livelihood improvement in a sustainable and participatory manner, inclusive of gender. In particular, in Colombia, the crop has great potential to support rural development subsequent to the peace agreement with armed rebels, as potential areas for cultivation, coincide with the transition zones for the resolution of the armed conflict of Colombia.

However, in this country little government attention has been focused on the exploration, conservation and sustainable use of novel plant genetic resources. Colombia is one of two countries on the continent that has not submitted a national report on the state of the nation´s plant genetic resources to the FAO global report in 2010.

A review of endangered species published in 2014 (Resolution 192, Ministerio del Ambiente) listed 70 orchid species as critically endangered or endangered. The genus *Vanilla* was not assessed, and the lack of population data for these species hinders a comprehensive threat assessment. To date, only a small number of small populations are known, and it is probable that most *Vanilla* species in Colombia fall within a higher category of risk. To improve protection for populations of *Vanilla* species native to Colombia, it is also important to assess these species for listing in the IUCN Red List of Threatened Species. Currently, not one species from the aromatic clade of *Vanilla* is listed in the IUCN.

Nonetheless, it is important that conservation policy actions do not restrict the ability to develop a comprehensive research program for sustainable utilization of these species. Such research must include continued genomic, metabolomic, ecological and agronomical study of native *Vanilla* species, thereby informing effective conservation strategies, in addition to facilitating efficient use of genetic diversity in crop improvement programs.

In Colombia, indigenous and afro-descendent communities living in rural areas have autonomous jurisdiction over their traditional territories. The majority of the Pacific Chocó region falls within these collective jurisdictions, which also include autonomous determination for the use of the biological resources within the territories. In areas with collective territories, any actions for the conservation and sustainable use of the native vanilla resources must be carried out in a participative manner with local communities.

6.6.2   International Strategy for Conservation of Vanilla CWR

At the international level, Appendix II of the CITES Convention listing species that may become threatened with extinction if trade is not closely controlled includes almost the entire Orchidaceae family, including the "fruits, and parts and derivatives thereof, of naturalized or artificially propagated plants of the genus *Vanilla*". Equally, seeds, seedpods, and pollen (including pollinia), and tissue cultures obtained *in vitro* are also exempt. While these exemptions are important for plant breeding they may hinder conservation efforts, particularly given the small, sparse populations of *Vanilla* we have found in the wild. Artificially propagated plants are often hard to verify, and, with increased attention on *Vanilla* wild populations for commercial ventures, we suggest it may be necessary to limit this exemption to only the major cultivated species of *V. planifolia, V. pompona,* and *V. × tahitensis,* or those plants of certifiable artificial origin.

Clearly, vanilla is a crop of international relevance, and it is now important that a joint effort be established amongst those countries that commercialize or hold important genetic resources for vanilla in order to leverage funding from the international agencies that promote agricultural research for sustainable livelihood improvement.

References

Adame-García, J., M. Luna-Rodríguez and L.G. Iglesias-Andreu (2016) Vanilla rhizobacteria as antagonists against *Fusarium oxysporum* f. sp. *vanillae*. *International Journal of Agricultural Biology* 18, 23–30.

Aguiar, W.M.D., Sofia, S.H., Melo, G.A. and Gaglianone, M.C. (2015) Changes in orchid bee communities across forest-agroecosystem boundaries in Brazilian Atlantic Forest landscapes. *Environmental Entomology*, 44(6), 1465–1471.

Azofeifa-Bolaños, J.B., Paniagua-Vásquez, A. and García-García, J.A. (2014) Importance and conservation challenges of *Vanilla* spp.(Orchidaceae) in Costa Rica. *Agronomía Mesoamericana*, 25(1), 189–202.

Azofeifa-Bolaños, J.B., Gigant, L.R., Nicolás-García, M., Pignal, M., Tavares-González, F.B., Hágsater, E., Salazar-Chávez, G.A., Reyes-López, D. Archila-Morales, F.L., García-García, J.A., da Silva, D., Allibert, A., Solano-Campos, F., del Carmen Rodríguez-Jimenes, G., Paniagua-Vásquez, A., Besse, P., Pérez-Silva, A. and Grisoni, M. (2017) A new vanilla species from Costa Rica closely related to *V. planifolia (Orchidaceae) European Journal of Taxonomy* 284, 1–26.

Bacon, C.W. and White, J.F. (2000) Physiological adaptations in the evolution of endophytism in the Clavicipitaceae. *Microbial Endophytes*, 237–261.

Batty, A.L., Dixon, K.W., Brundrett, M. and Sivasithamparam, K. (2001) Long-term storage of mycorrhizal fungi and seed as a tool for the conservation of endangered Western Australian terrestrial orchids. *Australian Journal of Botany* 49, 619–628.

Bayman, P., Mosquera-Espinosa A.T. and Porras-Alfaro, A. (2011) Mycorrhizal relationships of *vanilla* and prospects for biocontrol of root rots. In: *Handbook of Vanilla Science and Technology*, 1st edn. Daphna Havkin-Frenkel and Faith C. Belanger (Eds). Wiley-Blackwell, Oxford, pp. 266–279,

Belanger, F.C. and Havkin-Frenkel, D. (2011) Molecular analysis of a *Vanilla* hybrid cultivated in Costa Rica. In: *Handbook of Vanilla Science and Technology*, 1st edn. Daphna Havkin-Frenkel and Faith C. Belanger (Eds). Wiley-Blackwell, Oxford, pp. 256–265.

Bello-Bello, J.J., García-García, G.G. and Iglesias-Andreu, L. (2015) Conservación de vainilla (*Vanilla planifolia* Jacks.) bajo condiciones de lento crecimiento *in vitro*. *Revista Fitotecnia Mexicana* 38(2), 165–171.

Besse, P., Da Silva, D., Bory, S., Grisoni, M., Le Bellec, F. and Duval, M.F. (2004) RAPD genetic diversity in cultivated vanilla: *Vanilla planifolia*, and relationships with *V. tahitensis* and *V. pompona*. *Plant Science* 167(2), 379–385.

Bory, S., Lubinsky, P., Risterucci, A.M., Noyer, J.L., Grisoni, M., Duval, M.F. and Besse, P. (2008a) Patterns of introduction and diversification of *Vanilla planifolia* (Orchidaceae) in Reunion Island (Indian Ocean) *American Journal of Botany* 95(7), 805–815.

Bory, S., Da Silva, D., Risterucci, A.M., Grisoni, M., Besse, P. and Duval, M.F. (2008b) Development of microsatellite markers in cultivated vanilla: Polymorphism and transferability to other vanilla species. *Scientia Horticulturae* 115(4), 420–425.

Bory, S., Catrice, O., Brown, S., Leitch, I.J., Gigant, R., Chiroleu, F., Grisoni, M., Duval, M.F. and Besse, P. (2008c) Natural polyploidy in *Vanilla planifolia (Orchidaceae) Genome* 51(10), 816–826.

Brundrett, M., Scade, A., Batty, A.L., Dixon K.W. and Sivasithamparam, K. (2003) Development of *in situ* and *ex situ* seed baiting techniques to detect mycorrhizal fungi from terrestrial orchid habitats. *Mycological Research* 107, 1210–1220.

Burns J and Benson M. (2000) Biocontrol of damping-off of *Catharanthus roseus* caused by *Pythium ultimum* with *Trichoderma virens* and binucleate *Rhizoctonia* fungi. *Plant Diseases* 84, 644–648.

Cameron, D.D., Leake, J.R. and Read, D.J. (2006) Mutualistic mycorrhiza in orchids: evidence from plant–fungus carbon and nitrogen transfers in the green-leaved terrestrial orchid *Goodyera repens*. *New Phytologist* 171(2), 405–416.

Cameron, D.D., Johnson, I., Read, D.J. and Leake, J.R. (2008) Giving and receiving: measuring the carbon cost of mycorrhizas in the green orchid, *Goodyera repens*. *New Phytologist* 180(1), 176–184.

Castañeda-Álvarez, N.P., Khoury, C.K., Achicanoy, H.A., Bernau, V., Dempewolf, H., Eastwood, R. J. and Müller, J.V. (2016) Global conservation priorities for crop wild relatives. *Nature Plants* 2, 16022.

van Dam, A.R., Householder, J.E. and Lubinsky, P. (2010) *Vanilla bicolor* Lindl. (Orchidaceae) from the Peruvian Amazon: auto-fertilization in *Vanilla* and notes on floral phenology. *Genetic Resources and Crop Evolution* 57(4), 473–480.

Dawson, I.K., Guariguata, M.R., Loo, J., Weber, J.C., Lengkeek, A., Bush, D., Cornelius, J., Guarino, L., Kindt, R., Orwa, C., Russell, J. and Jamnadass, R. (2013) What is the relevance of smallholders' agroforestry systems for conserving tropical tree species and genetic diversity in *circa situm, in situ* and *ex situ* settings? A review. *Biodiversity and Conservation* 22(2), 301–324.

Dearnaley, J.D.W., Martos, F. and Selosse, M.A. (2012) 12 Orchid mycorrhizas: molecular ecology, physiology, evolution and conservation aspects. In: *Fungal Associations*. Springer Berlin Heidelberg, pp. 207–230.

Dempewolf, H., Eastwood, R. J., Guarino, L., Khoury, C. K., Müller, J. V., and Toll, J. (2014) Adapting agriculture to climate change: a global initiative to collect, conserve, and use crop wild relatives. *Agroecology and Sustainable Food Systems*, 38(4), 369–377.

Divakaran, M., Jayakumar, V.N., Veena, S.S., Vimala, J., Basha, A., Saji, K.V., Nirmal Bau, K. and Peter, K.V. (2008) Genetic variations and interrelationships in *Vanilla planifolia* and few related species as expressed by RAPD polymorphism. *Genetic Resources and Crop Evolution* 55(3), 459–470.

Duijff, B.J., Recorbet, G., Bakker, P.A.H.M., Loper, J.E. and Lemanceau, P. (1999) Microbial antagonism at the root level is involved in the suppression of *Fusarium* Wilt by the combination of nonpathogenic *Fusarium oxysporum* Fo47 and *Pseudomonas putida* WCS358. *Phytopathology* 89, 1073–1079.

Ercole, E., Rodda, M., Molinatti, M., Voyron, S., Perotto, S. and Girlanda, M. (2013) Cryopreservation of orchid mycorrhizal fungi: A tool for the conservation of endangered species. *Journal of Microbiological Methods* 93, 134–137.

Flanagan, N.S., Otero J.T., Molineros F.H., Mosquera-Espinosa A.T, et al. (2012) *Aprovechamiento sostenible de Recursos Biológicos Nativos del Distrito de Manejo Integrado de Atuncela*. Guía Práctica. Cartilla de los Recursos Biológicos del Distrito de Manejo Integrado de Atuncela. Proyecto de CVC No 0051 y la Pontificia Universidad Javeriana-Cali, 31 pp.

Flanagan, N.S. and Mosquera-Espinosa, A.T. (2016) An integrated strategy for the conservation and sustainable use of native vanilla species in Colombia. *Lankesteriana* 16(2), 201–218.

Forero-Medina, G. and Joppa, L. (2010) Representation of global and national conservation priorities by Colombia's protected area network. *Plos One* 5(10), e13210.

Fraga de, C.N., Couto, D.R. and Pansarin, E.R. (2017) Two new species of *Vanilla* (Orchidaceae) in the Brazilian Atlantic Forest. *Phytotaxa* 296(1), 63–72.

Gigant, R. L., Bory, S., Grisoni, M. and Besse, P. (2011) Biodiversity and evolution in the Vanilla genus. In: *The Dynamical Processes of Biodiversity-Case Studies of Evolution and Spatial Distribution*. Grillo, O. and Venora, G. (Eds) pp. 1–27. [online]. Website http://www.intechopen.com/books.

Gigant, R.L., De Bruyn, A., Church, B., Humeau, L., Gauvin-Bialecki, A., Pailler, T., Grisoni, M. and Besse, P. (2014) Active sexual reproduction but no sign of genetic diversity in range-edge populations of *Vanilla roscheri* Rchb. f.(Orchidaceae) in South Africa. *Conservation Genetics* 15(6), 1403–1415.

Gigant, R.L., Rakotomanga, N., Goulié, C., Da Silva, D., Barre, N., Citadelle, G., Silvestre D., Grisoni, M. and Besse, P. (2016) Microsatellite markers confirm self-pollination and autogamy in wild populations of *Vanilla mexicana* Mill.(syn. *V. inodora*) (Orchidaceae) in the Island of Guadeloupe. In: *Microsatellite Markers*. I. Abdurakhmonov (Ed.) InTech. DOI: 10.5772/64674.

González, J.J., Etter, A.A., Sarmiento, A.H., Orrego, S.A., Ramírez, C., Cabrera, E., Vargas, D., Galindo, G., García, M.C. and Ordoñez, M.F. (2011) *Análisis de tendencias y patrones espaciales de deforestación en Colombia*. Instituto de Hidrología, Meteorología y Estudios Ambientales-IDEAM. Bogotá D.C., Colombia, 64 pp.

Hajjar, R., and Hodgkin, T. (2007) The use of wild relatives in crop improvement: a survey of developments over the last 20 years. *Euphytica* 156(1–2), 1–13.

Hawkes, J.G., Maxted, N. and Ford-Lloyd, B.V. (2012) *The ex situ conservation of plant genetic resources*. Springer Science and Business Media.

Henry, R.J. (2013) Sequencing of wild crop relatives to support the conservation and utilization of plant genetic resources. *Plant Genetic Resources* 12(S1), S9–S11.

Hernández-Hernández, J. (2011) *Vanilla* diseases. In: *Handbook of Vanilla Science and Technology*, 1st edn. Daphna Havkin-Frenkel and Faith C. Belanger (Eds). Wiley, Oxford, pp. 26–39,

Heywood, V., Casas, A., Ford-Lloyd, B., Kell, S. and Maxted, N. (2007) Conservation and sustainable use of crop wild relatives. *Agriculture, Ecosystems and Environment* 121(3), 245–255.

Householder, E., Janovec, J., Mozambite, A.B., Maceda, J.H., Wells, J., Valega, R. and Christenson, E. (2010) Diversity, natural history, and conservation of *Vanilla* (Orchidaceae) in Amazonian wetlands of Madre De Dios, Peru. *Journal of the Botanical Research Institute of Texas*, 227–243.

IDEAM (Instituto de Hidrología, Meteorología y Estudios Ambientales de Colombia) (2015) Consulted 2-01-2017. http://www.ideam.gov.co/web/sala-de-prensa/noticias/-/asset_publisher/96oXgZAhHrhJ/content/aumenta-deforestacion-en-colombia-para-2014.

IDEAM, PNUD, MADS, DNP, CANCILLERÍA (2015) *Nuevos escenarios de Cambio Climático para Colombia 2011-2100. Herramientas Científicas para la toma de decisiones – Enfoque Nacional – Departamental: Tercera Comunicación Nacional de Cambio Climático*. ISBN 978-958-8902-55-5.

IDEAM (Instituto de Hidrología, Meteorología y Estudios Ambientales de Colombia) (2017) *Atlas Interactivo*. Consulted 30-04-2017 http://atlas.ideam.gov.co/visorAtlasClimatologico.html

Koch, A.K., Fraga, C.N.D., Santos, J.U.M.D. and Ilkiu-Borges, A.L. (2013) Taxonomic notes on *Vanilla* (Orchidaceae) in the Brazilian Amazon, and the description of a new species. *Systematic Botany* 38(4), 975–981.

Khoyratty, S., Dupont, J., Lacoste, S., Palama, T.L., Choi, Y.H., Kim, H.K., Payet, B., Grisoni, M., Fouillaud, M., Verpoorte, R. and Kodja, H. (2015) Fungal endophytes of *Vanilla planifolia* across Réunion Island: isolation, distribution and biotransformation. *BMC Plant Biology* 15(1), 142.

Koyyappurath, S., Atuahiva, T., Le Guen, R., Batina, H., Le Squin, S., Gautheron, N., Edel Hermann, V., Peribe, J., Jahiel, M., Steinberg, C., Liew, E.C.Y., Alabouvette, C., Besse, P., Dron, M., Sache, I., Laval, V. and Grisoni, M. (2015a) *Fusarium oxysporum* f. sp. *radicis-vanillae* is the causal agent of root and stem rot of vanilla. *Plant Pathology* 65, 12445.

Koyyappurath, S., Conéjéro, G., Dijoux, J.B., Lapeyre-Montès, F., Jade, K., Chiroleu, F., Gatineau, F., Verdeil, J.L., Besse, P. and Grisoni, M. (2015b) Differential responses of *Vanilla* accessions to root rot and colonization by *Fusarium oxysporum* f. sp. *radicis-vanillae*. *Frontiers in Plant Science* 6, 1125.

Kusari, S., Hertweck, C., and Spiteller, M. (2012) Chemical ecology of endophytic fungi: origins of secondary metabolites. *Chemistry and Biology* 19(7), 792–798.

Lahaye, R., Van der Bank, M., Bogarin, D., Warner, J., Pupulin, F., Gigot, G., Maurin, O., Duthoit, S., Barraclough, T.G. and Savolainen, V. (2008) DNA barcoding the floras of biodiversity hotspots. *Proceedings of the National Academy of Sciences* 105(8), 2923–2928.

Lepers-Andrzejewski, S., Causse, S., Caromel, B., Wong, M. and Dron, M. (2012) Genetic linkage map and diversity analysis of Tahitian Vanilla (×, Orchidaceae) *Crop Science*, 52(2), 795–806.

Lin, C.S., Chen, J.J., Huang, Y.T., Chan, M.T., Daniell, H., Chang, W.J. and Liao, C.F. (2015) The location and translocation of ndh genes of chloroplast origin in the Orchidaceae family. *Scientific Reports* 5, 9040.

Liu, H., Ren, H., Liu, Q., Wen, X., Maunder, M. and Gao, J. (2015) Translocation of threatened plants as a conservation measure in China. *Conservation Biology* 29, 1537–1551.

Lubinsky, P., Van Dam, M. and Van Dam, A. (2006) Pollination of *Vanilla* and evolution in Orchidaceae. *Lindleyana* 75(12), 926–929.

Lubinsky, P., Cameron, K.M., Molina, M.C., Wong, M., Lepers Andrzejewski, S., Gómez-Pompa, A. and Kim S.C. (2008) Neotropical roots of a Polynesian spice: the hybrid origin of Tahitian vanilla, *Vanilla tahitensis (Orchidaceae) American Journal of Botany* 95, 1040–1047.

Maxted, N., Ford-Lloyd, B.V. and Hawkes, J.G. (Eds) (1997) *Plant conservation: The in situ approach*. Chapman and Hall, London, 446 pp.

Menchaca, R., Ramos, J.M., Moreno, D., Luna, M., Mata, M., Vázquez, L.M. and Lozano, MA. (2011) Germinación *in vitro* de híbridos de *Vanilla planifolia* y *V. pompona*. *Revista Colombiana de Biotecnología* 13(1), 80–84.

Mendoça dos Anjos, A., Barberena, F.F.V.A. and Pigozzo, C.M. (2017) Biologia reprodutiva de *Vanilla bahiana* Hoehne (Orchidaceae). *Orquidário* 30(3–4), 67–79.

Ministerio de Ambiente y Desarrollo Sostenible y Universidad Nacional de Colombia (MADS and UNAL) (2015) *Plan para el estudio y la conservación de las orquídeas en Colombia*. Textos: Betancur, J., Sarmiento, H., Toro-González, L. and Valencia, J. Universidad Nacional de Colombia. Facultad de Ciencias. Instituto de Ciencias Naturales; Coord. Técnica: Higuera Díaz, Diego – Minambiente. Dirección de Bosques, Biodiversidad y Servicios Ecosistémicos Bogotá D.C.: Colombia. Ministerio de Ambiente y Desarrollo Sostenible; Universidad Nacional de Colombia, 336 pp.

Molineros-Hurtado, F.H. (2012) *Caracterización morfológica y filogenia del género Vanilla en el distrito de Buenaventura – Valle del Cauca (Colombia)*. Tesis de Maestría en Ciencias Biológicas, Línea de Investigación Recursos Fitogenéticos Neotropicales. Universidad Nacional de Colombia, sede Palmira, 157 pp.

Molineros-Hurtado, F.H, González-Mina, R.T., Flanagan, N.S. and Otero, J.T. (2014) *Vanilla rivasii* (Orchidaceae), a new species from the Colombian pacific region. *Lankesteriana* 13(3), 353–357.

Mosquera-Espinosa, A.T., Bayman, P. and Otero, J.T. (2010) *Ceratobasidium* como hongo micorrízico de orquídeas en Colombia. *Acta Agronómica* 59, 316–326.

Mosquera-Espinosa, A.T., Otero, J.T., Molineros, F., Vásquez, E. and Flanagan, N.S. (2012) Bioprospección de los recursos nativos de la orquídea *Vanilla* spp. presente en el Valle del Cauca. *Acta Agronómica* 52(Especial), 37–38.

Mosquera-Espinosa A.T., Bayman, P., Prado, G., Gómez-Carabalí, A. and Otero, J.T. (2013) The double life of *Ceratobasidium*: orchid mycorrhizal fungi and their potential for biocontrol of *Rhizoctonia solani* sheath blight of rice. *Mycologia* 105, 141–150.

Ortiz-Valdivieso, P. (2015) Vanilla. In: *Catálogo de plantas y líquenes de Colombia.* Bernal R., Gradstein S.R. and Celis M. (Eds) Instituto de Ciencias Naturales, Universidad Nacional de Colombia, Bogotá. http://catalogoplantascolombia.unal.edu.co

Pansarin, E.R., Aguiar, J.M.R.B.V. and Ferreira, A.W.C. (2012) *A new species of Vanilla (Orchidaceae: Vanilloideae) from São Paulo, Brazil.* Brittonia (Bronx, N.Y.), 64, 157–161.

Pansarin E.R. and Miranda M.R. (2016) A new species of *Vanilla* (Orchidaceae: Vanilloideae) from Brazil. *Phytotaxa* 267, 084-088.

Pansarin, E.R., Aguiar, J.M.R.B.V. and Pansarin, L.M. (2014) Floral biology and histochemical analysis of *Vanilla edwallii* Hoehne (Orchidaceae: Vanilloideae): an orchid pollinated by *Epicharis* (Apidae: Centridini) *Plant Species Biology* 29, 242–252.

Pansarin, E.R. and Pansarin, L.M. (2014) Floral biology of two Vanilloideae (Orchidaceae) primarily adapted to pollination by euglossine bees. *Plant Biology* 16(6), 1104–1113.

Pansarin, E.R. (2016) Recent advances on evolution of pollination systems and reproductive biology of Vanilloideae (Orchidaceae) *Lankesteriana* 16(2), 255–267.

van der Pijl, l. and Dodson, C.H. 1966. *Orchid Flowers: Their Pollination and Evolution.* Fairchild Tropical Garden and Miami University Press, Coral Gables, Florida.

Pizano, C. and García H. (2014) *El Bosque Seco Tropical en Colombia.* Instituto de Investigación de Recursos Biológicos Alexander von Humboldt (IAvH) Bogotá, D.C. Colombia, 213 pp.

Porras-Alfaro, A. and Bayman, P. (2007) Mycorrhizal fungi of *Vanilla*: diversity, specificity and effects on seed germination and plant growth. *Mycology* 99(4), 510–525.

Radjacommare, R., Usharani, R. and Samiyappan, R. (2007) Genotyping antibiotic producing fluorescent pseudomonads to select effective rhizobacteria for the management of major vanilla diseases. *Annals of Microbiology* 57, 163–170.

Ramírez, S., Dressler, R.L. and Ospina, M. (2002) *Abejas euglosinas* (Hymenoptera: Apidae) de la Región Neotropical: Listado de especies con notas sobre su biología. *Biota Colombiana* 3(1)7–118.

Ravindran, A.S. and Shaike, J.M. (2013) Biocontrol of Rhizoctonia rot of Vanilla (*Vanilla planifolia*) using combined inoculation of *Trichoderma* sp. and *Pseudomonas* sp. *Acta Biologica Indica* 2(1), 292–299.

Reina-Rodríguez, G.A., Mejía, J.E.R., Llanos, F.A.C. and Soriano, I. (2017) Orchid distributions and bioclimatic niches as a strategy to climate change in areas of tropical dry forest in Colombia. *Lankesteriana* 17(1) 17–47.

Rodriguez, R. and Redman, R. (2008) More than 400 million years of evolution and some plants still can't make it on their own: plant stress tolerance via fungal symbiosis. *Journal of Experimental Botany* 59(5), 1109–1114.

Roubik, D.W. and Hanson, P.E. (2004) *Orchid bees of Tropical America: Biology and field guide*. San Jose, Costa Rica: INBIO.

Roubik, D.W. and Ackerman, J.D. (1987) Long-term ecology of euglossine orchid-bees (Apidae: Euglossini) in Panama. *Oecologia* 73(3), 321–333.

Roux-Cuvelier, M. and Grisoni, M.I. (2011) Conservation and movement of *Vanilla* germplasm. In: *Vanilla*. Odoux, E. and Grisoni, M. (Eds) CRC Press, Boca Ratón, FL, pp. 31–41.

Salazar-Rojas, V.M., Herrera-Cabrera, B.E., Delgado-Alvarado, A., Soto-Hernández, M., Castillo-González, F. and Cobos-Peralta, M. (2011) Chemotypical variation in *Vanilla planifolia* Jack. (Orchidaceae) from the Puebla-Veracruz Totonacapan region. *Genetic Resources and Crop Evolution*, 59, 875–887.

Sambin, A. and Chiron, G.R. (2015) Deux nouvelles espèces de *Vanilla* (Orchidaceae) de Guyane française. *Richardiana* 15, 306–316.

Sandheep, A.R., Asok, A.K. and Jisha, M.S. (2012) Biocontrol of Fusarium wilt of Vanilla (*Vanilla planifolia*) using combined inoculation of *Trichoderma* sp. and *Pseudomonas* sp. *International Journal of Pharmarcy and Biological Science* 3(3), 706–716.

Sandheep, A.R, Asok, A.K. and Jisha, M.S. (2013) Combined inoculation of *Pseudomonas fluorescens* and *Trichoderma harzianum* for enhancing plant growth of vanilla (*Vanilla planifolia*) Pak. *Journal of Biological Sciences* 16(2): 580–584.

Schluter, P.M., Soto-Arenas, M.A. and Harris, S.A. (2007) Genetic variation in *Vanilla planifolia* (Orchidaceae) *Economic Botany* 61, 328–336.

Soto-Arenas, M.A. (2003) *Vanilla*. In: *Genera orchidacearum: Orchidoideae*. Pridgeon, A.M., Cribb, P.J., Chase, M.W. and Rasmussen F.N. (Eds) Oxford University Press, New York, pp. 321–334.

Soto-Arenas, M.A. and Cribb, P. (2010) A new infrageneric classification and synopsis of the genus *Vanilla* Plum. Ex Mill. (Orchidaceae: Vanillinae) *Lankesteriana* 9, 355–398.

Soto-Arenas, M.S., Dressler, R.L. (2010) A revision of the Mexican and Central American species of *Vanilla* Plumier ex Miller with a characterization of their ITS region of the nuclear ribosomal DNA. *Lankesteriana* 9, 285–354.

Smith, C. (2016) Keeping a finger on the pulse: Monitoring the use of CWR in crop improvement. In: *Enhancing Crop Genepool Use: Capturing Wild Relative and Landrace Diversity for Crop Improvement*. N. Maxted, M. Ehsan-Dulloo and B.V. Ford-Lloyd (Eds). CABI, Boston, MA, pp. 78–86.

Sneh, B., Yamoah, E. and Stewart, A. (2004) Hypovirulent *Rhizoctonia* spp. isolates from New Zealand soils protect radish seedlings against Damping-off caused by *R. solani*. *New Zealand Plant Protection* 57, 54–58.

Talubnak, C. and Soytong, K. (2010) Biological control of vanilla anthracnose using *Emericella nidulans Journal of Agricultural Technology* 6, 47–55.

Trávníček, P., Ponert, J., Urfus, T., Jersáková, J., Vrána, J., Hřibová, E., Dolezel, J. and Suda, J. (2015) Challenges of flow-cytometric estimation of nuclear genome size in orchids, a plant group with both whole-genome and progressively partial endoreplication. *Cytometry Part A* 87(10), 958–966.

Verma, P.C., Chakrabarty, D., Jena, S.N., Mishra, D.K., Singh, P.K., Sawant, S.V. and Tuli, R. (2009) The extent of genetic diversity among *Vanilla* species: comparative results for RAPD and ISSR. *Industrial Crops and Products* 29(2), 581–589.

White, J.F., Torres, M.S., Sullivan, R.F., Jabbour, R.E., Chen, Q., Tadych, M., Irizarry, I., Bergen, M.S., Havkin-Frenkel, D. and Belanger, F.C. (2014) Occurrence of *Bacillus amyloliquefaciens* as a systemic endophyte of vanilla orchids. *Microscopy Research Technique* 77, 874–885.

Wood, C.B., Pritchard, H.W. and Miller, A.P. (2000) Simultaneous preservation of orchid seed and its fungal symbiont using encapsulation-dehydration is dependent on moisture content and storage temperature. *CryoLett* 21, 125–136.

Xiong, W., Zhao, Q., Zhao, J., Xun, W., Li, R., Zhang, R., Wu, H. and Shen, Q. (2015) Different continuous cropping spans significantly affect microbial community membership and structure in a vanilla-grown soil as revealed by deep pyrosequencing. *Microbial Ecology* 70(1), 209–218.

Xiong, W., Li, R., Ren, Y., Liu, C., Zhao, Q., Wu, H. Jousseta, A. and Shen, Q. (2017) Distinct roles for soil fungal and bacterial communities associated with the suppression of vanilla Fusarium wilt disease. *Soil Biology and Biochemistry* 107, 198–207.

Zettler, L.W., Corey, L.L., Richardson, L.W., Ross, A.Y. and Moller-Jacobs, L. (2011) Protocorms of an epiphytic orchid (*Epidendrum amphistomum* A. Richard) recovered *in situ*, and subsequent identification of associated mycorrhizal fungi using molecular markers. *European Journal of Environmental Sciences* 1, 108–114.

7

The History of Vanilla in Puerto Rico: Diversity, Rise, Fall, and Future Prospects

Paul Bayman

7.1 Introduction

Puerto Rico was once an important center for *Vanilla* production and research. Production died off in the 1950s because of root rot disease, high labor costs and a general decline in agriculture. Could and should *Vanilla* production return to Puerto Rico? I review wild *Vanilla* in Puerto Rico, the history of its cultivation and abandonment, and prospects for renewing the industry. Since the barriers to renewal are not unique to Puerto Rico, this discussion should also be relevant to other countries that are interested in establishing or strengthening their *Vanilla* programs.

7.2 Diversity of Wild Vanilla in Puerto Rico

7.2.1 Species and Distributions

Seven *Vanilla* species grow wild in Puerto Rico (Ackerman 1995). Another four species have been found elsewhere in the Greater Antilles, but not in Puerto Rico (Ackerman 2014).

Of these seven species, two (*V. planifolia* and *V. pompona*) belong to the Neotropical, leafy, fragrant clade as defined by Cameron (2011, and Chapter 20 in this volume). Four belong to the Caribbean leafless subclade, and are found mostly in the relatively dry forests of western Puerto Rico: *V. dilloniana*, *V. barbellata*, *V. claviculata* and *V. poitaei* (though *V. poitaei* is widely distributed) (Ackerman 1995). This Caribbean leafless subclade is apparently derived from an African subclade, and may have been originally transported to the Caribbean by hurricanes (Cameron 2011). The remaining species, *V. mexicana*, is assigned by Cameron to the Membranaceous clade, basal to the rest of the genus. It is impressive that the three major clades of *Vanilla* are represented on one small island. However, of these species only *V. planifolia* and *V. pompona* are of commercial interest (Childers *et al.* 1959; Cameron 2011).

Some of these species may have escaped from cultivation. However, only *V. planifolia* and *V. pompona* are widely cultivated and the others have received limited attention as

Handbook of Vanilla Science and Technology, Second Edition.
Edited by Daphna Havkin-Frenkel and Faith C. Belanger.
© 2019 John Wiley & Sons Ltd. Published 2019 by John Wiley & Sons Ltd.

ornamentals. The simplest assumption is that their presence is not due to intentional human introduction.

7.2.2 Flowering, Pollination, and Fruit Set

Flowering is often inconsistent on wild *Vanilla* plants in Puerto Rico. Some large patches of *V. planifolia* in El Yunque National Forest show no signs of ever having flowered; reproduction is mostly vegetative.

Gardeners report mixed results with *V. planifolia* and *V. pompona*: some say their plants flower every year, usually in the dry season from February to April, whereas others have had large, healthy plants for years and a single flower has never been seen. The reason for this variation is not clear: it may be due to genetic differences in the plants, environmental differences, or the way in which the plants are grown. Flowering is encouraged when the shoot tip bends over towards the ground, presumably because nutrients accumulate in the apex (Childers *et al.* 1959).

Fruit set is even more irregular than flowering. The presumed pollinators of *Vanilla* are male euglossine bees, which are not found in Puerto Rico (Ackerman 1995, 2014). Honeybees and *Centris* bees may pollinate these and other species occasionally (J.D. Ackerman, personal communication). *Melipona* bees, ants and hummingbirds are reported to visit *Vanilla* flowers in other areas, but reports of successful pollination are scarce (Cameron 2011). Less than 1% of flowers are thought to form fruits without manual pollination (Childers *et al.* 1959). Even in *V. planifola*'s native range in Mexico, where wild pollinators are found, natural pollination is about 1% (Hernández-Hernández, Chapter 1 in this volume).

However, hybrids were reported between two species in the Caribbean leafless clade (*V. claviculata* and *V. barbellata*) in several sites in Puerto Rico. Hybridization was inferred from floral morphology and isozyme data (Nielsen and Siegismund 1999; Nielsen 2000). Hybridization implies both simultaneous flowering and cross-pollination of two sympatric species. It has also been reported elsewhere between species of *Vanilla* (Cameron 2011). A hybrid, which may be natural, is described by Belanger and Havkin-Frenkel (Chapter 21 in this volume).

Low natural fruit set and scarcity of pollinators means that commercial *Vanilla* in Puerto Rico (and elsewhere) needs to be hand-pollinated (Childers *et al.* 1959; Cameron 2011). Manual pollination is labor-intensive, and since flowers only last a day, requires daily attention during flowering season. This was a major limitation in Puerto Rico, where labor is more expensive than in most other *Vanilla*-producing countries (Childers *et al.* 1959; Carro-Figueroa 2002).

7.3 Rise and Fall: The History of Vanilla Cultivation in Puerto Rico

Vanilla planifolia is indigenous to Puerto Rico according to McClelland (1919), but Ackerman (2014) suggested it may have been introduced from Mexico. The following history is based on Childers and Cibes (1946, 1948) and Correll (1953). *Vanilla* was first brought to Puerto Rico as an ornamental by "a relative of Sr. Miguel Morell from Utuado" (translated from Childers and Cibes 1946). Material for commercial production was introduced in 1909 from two sources: Mexico and the US Plant Introduction Garden, Florida (Correll 1953). These plants were propagated at the Federal Experiment Station (later renamed the Tropical Agricultural Experiment Station, or TARS) in Mayagüez by McClelland and distributed to growers.

Coffee and other crops were hit hard by several hurricanes in the early 1900s, including the Category 5 San Felipe II in 1928. *Vanilla* was viewed as a supplementary crop that could help coffee growers recover from hurricane damage (Childers and Cibes 1946). *Erythrina berteroana* (Fabaceae, coral tree) was distributed along with *Vanilla* for support, shade and nitrogen fixation. Technical support was also provided.

Commercial production increased steadily until 1944, when production reached 6000 lbs (Pennigton *et al.* 1954). Acreage likewise increased, from 106 acres in 1937 to 434 acres in 1940, distributed among 168 farms (Childers and Cibes 1946). (For comparsion, coffee at that time occupied about 100,000 acres (Childers and Cibes 1948). At this point, Puerto Rico was the ninth largest *Vanilla* producers worldwide (Figure 7.1), as well as an important center for research. The farms were mostly in the Cordillera Central mountain range of west-central Puerto Rico, where coffee was the main industry (Figure 7.2). Growers organized a cooperative, called the Cooperativa de Cosecheros de Vainilla, and established a curing facility in Castañer. This allowed uniformity of processing and grading, which facilitated export. Trade disruption during WWII led to higher prices for domestic vanilla, and stimulated growers to increase planting and pollination. In 1944–5 over 6,000 lbs of cured vanilla and 2,000 lbs of unprocessed fruits were exported.

However, production then fell to only 350 lbs in 1953. Why this sudden decline? The main cause was *Fusarium* root rot (discussed below). By 1945 40–50% of plants were infected and growers were beginning to abandon the crop (Pennigton *et al.* 1954). Different cultivation systems were tested to reduce susceptibility to the disease; results were mixed.

Experiments on *Vanilla* cultivation were mentioned every year in the reports of the Federal Experiment Station until the early 1950s. After that time it is hard to find any

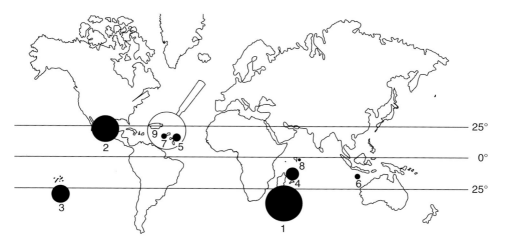

Figure 7.1 Map of *Vanilla* production from 1941–44, showing Puerto Rico as the ninth largest producer worldwide. (From Childers and Cibes1948.)

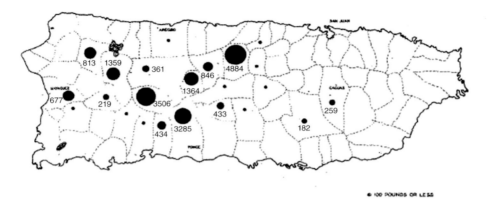

Figure 7.2 Map showing distribution of *Vanilla* production in Puerto Rico in 1941–1944. Production was concentrated in the central mountain range of western Puerto Rico, where coffee is grown. (From Childers and Cibes 1948.)

mention of it, and agronomists working in Puerto Rico in the 1960s have no recollection of research on *Vanilla*.

A combination of root rot diseases, high labor costs and a general abandonment of agriculture in favor of industrialization led to the decline and fall of the crop. Each of these factors is discussed below, with emphasis on diseases.

7.4 Socioeconomic Factors Contributing to the Decline of Vanilla

This decline of *Vanilla* in Puerto Rico was part of a general decline in agriculture. In the post-WWII era the new Puerto Rican Commonwealth government promoted industrialization as the key to economic growth (Carro-Figueroa 2002). Many people abandoned agricultural areas to move to San Juan and New York. Agriculture came to be viewed as backward and unsophisticated.

As a consequence, farm labor became scarcer. Fewer people were willing to work in agriculture (Carro-Figueroa 2002). This remains a major challenge for coffee growers, who need labor for harvesting (Fain *et al.* 2017). Since *Vanilla* is a labor-intensive crop, especially for pollinating, harvesting and curing, availability of labor is an obstacle to re-establishment of the crop. (See also section 7.6, below.)

7.5 Diseases and Decline

Since *Vanilla* has not been grown commercially in Puerto Rico for the last fifty years, there are no recent studies on its diseases under local conditions. The information presented here is based on a review by Childers and Cibes from 1948, and more recent reports from other places.

7.5.1 *Fusarium* Root and Stem Rot (RSR)

The most serious disease in *Vanilla* production is root and stem rot (RSR). The disease was originally noticed in Puerto Rico in 1918 and the pathogen described by Tucker in 1924. As RSR spread, some growers started to abandon *Vanilla* (Childers and Cibes 1946). By 1945 40–50% of plants on many *Vanilla* farms in Puerto Rico were infected (Childers and Cibes 1948).

RSR is now widespread, occurring in Madagascar, Indonesia, India, China, Mexico, Colombia, Reunion Island and French Polynesia – wherever *Vanilla* is grown (Ramírez-Mosqueda *et al.* 2015; Santa *et al.* 2012; Bhai and Dhanesh 2008). In Mexico it can kill two-thirds of *Vanilla* plants within four years of planting (Hernández-Hernández, Chapter 1 in this volume). In Indonesia RSR is considered the principal impediment to increasing production, causing losses of up to 80% (Pinaria *et al.* 2010, 2015).

7.5.1.1 The Pathogen

Root and stem rot is caused by *Fusarium oxysporum* f. sp. *vanillae* (formerly called *F. batatatis* var. *vanillae*). For readers who are not mycologists, this taxonomy merits explanation: **f. sp.** or **forma specialis** is an intraspecific taxon defined by the host plant on which a pathogen is found (Webster and Weber 2007), similar to a pathovar in microbiology. Since *F. oxysporum* attacks many hosts, it includes many formae speciales, which are not necessarily mutually exclusive or phylogenetically distinct.

Two recent studies estimated relationships among *F. oxysporum* f. sp. *vanillae* isolates from various countries (in each case based on sequences of two genes) and tested isolates for pathogenicity. In one study, *F. oxysporum* f. sp. *vanillae* isolates from Indonesia, Mexico and Reunion formed three separate clades (Pinaria *et al.* 2015). Each clade included at least three other formae speciales as well; one clade included isolates from *Vanilla* in various countries while the other two only included isolates from Indonesia. This result suggests that *F. oxysporum* f. sp. *vanillae* is polyphyletic, that *F. oxysporum* pathogens from other hosts can attack *Vanilla*, and that virulence factors may be transferred between isolates (Pinaria *et al.* 2015).

Similarly, in the second study isolates of *F. oxysporum* f. sp. *vanillae* formed several clades which also included isolates of other formae speciales (Koyyappurath *et al.* 2016). Also, as in the previous study, consistency of biogeographic patterns was seen, but each country had isolates in several different clades. Similar patterns of polyphyly were found for *F. oxysporum* f. sp. *vanillae* in Colombia, based on rDNA sequences (Santa *et al.* 2012) and for *F. oxysporum* f. sp. *cubense*, a serious pathogen of banana (O'Donnell *et al.* 1998).

These results have several implications. First, no cultivars of *V. planifolia* or *V. tahitensis* are known to be resistant to *Fusarium oxysporum* f. sp. *vanillae*, but efforts are underway to develop some. However, if the pathogen varies among countries and includes pathogens from other host plants that are able to cross over to attack *Vanilla*, a cultivar resistant in one country may not be resistant when grown in another, because different pathogens may be present. Second, even though RSR was first identified in Puerto Rico, distribution of *Vanilla* germplasm from Puerto Rico was probably not responsible for the spread of the disease worldwide. *F. oxysporum* is ubiquitous and attacks many plants; wherever *Vanilla* is planted on a large scale some genotypes of this species will probably be able to attack it. Eliminating and escaping the pathogen are not practical options.

However, there are few if any reports of RSR on wild *Vanilla* plants. This suggests that the disease is density-dependent, and that high fruit set in *Vanilla* farms makes plants more susceptible by depleting resources. This was reported for Puerto Rican Vanilla during WWII, when scarcity pushed up prices and growers responded by overpollinating, making plants more susceptible to RSR (Childers and Cibes 1948).

7.5.1.2 Symptoms of RSR

The primary symptom of RSR is that roots turn brown and dry out (Childers and Cibes 1948; Hernández-Hernández, Chapter 2 in this volume). An initial response to loss of roots may be production of new roots, which often become infected as well. Subsequently stems and leaves become flaccid, turn yellow, and then shrivel.

Plants tend to become diseased after fruiting, suggesting that depletion of resources may make them more susceptible. Overpollination driven by high *Vanilla* prices during World War II was at least partly responsible for the spread of RSR, which led to the elimination of the crop in Puerto Rico (Childers and Cibes 1948).

The pathogen appears to be opportunistic in the sense that healthy plants are relatively resistant. Sufficient water, drainage, shade, spacing and mulch help keep plants healthy. Preventing infection is much more effective than treating it (Childers and Cibes 1948).

7.5.1.3 Other *Fusarium* Species

F. solani was also reported as a causal agent of RSR in Puerto Rico (Alconero and Santiago 1969), but was considered a weak or secondary pathogen in Indonesia (Koyyappurath *et al.* 2016). *Fusarium* species in general are common in *Vanilla*, as they are in many plants: twelve species were isolated from rotting *Vanilla* stems throughout Indonesia (Pinaria *et al.* 2010). Of these species *F. oxysporum* was the most common and the only one that caused disease when inoculated on *Vanilla* shoot cuttings. These results suggest that the other *Fusarium* species are weak pathogens, endophytes or secondary colonizers of diseased tissue (Pinaria *et al.* 2010). *Fusarium* species have a similarly wide range of interactions in maize and other crops (Bacon *et al.* 2008).

7.5.2 Other Diseases and Pests

Insect pests of *Vanilla* were not a serious problem in Puerto Rico (Childers and Cibes 1948). Chickens can scratch and damage roots and stems. Some gardeners report rats gnawing through *Vanilla* stems.

Other diseases of *Vanilla* have been reported from other places, but not from Puerto Rico. The most important are discussed by Hernández (Chapter 2 in this volume).

7.5.3 Possible Solutions to RSR

7.5.3.1 Biological Control

There is interest in both bacteria and fungi as agents of biological control of *F. oxysporum* f. sp. *vanillae.*

Several bacteria isolated from *V. planifolia* roots in Mexico were tested for antagonism against *F. oxysporum* f. sp. *vanillae in vitro* (Adame-García *et al.* 2014). Isolates of

Staphylococcus xylosus, Serratia sp. and *Stenotrophomonas* sp. significantly reduced colony size. However, *in vivo* tests were not reported. A range of bacteria have been isolated from *Vanilla* roots, some of which may serve as biofertilizers or antagonists of pathogens (Álvarez López *et al.* 2013).

As in maize, it is possible that some *Fusarium* species might protect *Vanilla* plants from *F. oxysporum* f. sp. *vanillae*. Biocontrol may be most effective when the protective agent and the pathogen are closely related, because the same environmental conditions may favor them and they may occupy similar niches in the host. For example, non-toxigenic isolates of *Aspergillus flavus* are effective at protecting crops from isolate of the same species that produce aflatoxins (Cotty *et al.* 1994).

7.5.3.2 Mycorrhizæ
Mycorrhizal fungi of *Vanilla* have been isolated and identified (Porras-Alfaro and Bayman 2007; Mosquera-Espinosa *et al.* 2013, Johnson *et al.* 2016; González-Chávez *et al.* 2017, Bayman *et al.*, Chapter 22 in this volume). Most of the fungi were *Ceratobasidium* and *Thanatephorus*, whose asexual stages are often assigned to the form-genus *Rhizoctonia*.

These genera include serious plant pathogens as well as mycorrhizal fungi of orchids. Some isolates of *Ceratobasidium* were successful at protecting rice seedlings from a pathogenic isolate of *R. solani* (causal agent of sheath blight) both *in vitro* and in greenhouse experiments (Mosquera-Espinosa *et al.* 2013).

A similar strategy might be useful in protecting *Vanilla* from *Fusarium* root rot. It could offer a double benefit: inoculation with a mycorrhizal fungus can promote plant growth as well as protect against pathogens (Ordoñez *et al.* 2012; Bayman *et al.*, Chapter 22 in this volume). The improved nutrition provided by mycorrhizæ could strengthen the plants' defenses against pathogens.

7.5.3.3 Chemical Control
Several fungicides have been used to control RST (Fouche and Jouve 1999). Plant extracts have been used to decrease prevalence of RST in Indonesia (Suprapta and Khalimi 2009).

7.5.3.4 Breeding
No cultivars of *V. planifolia* or *V. tahitensis* have been shown to have absolute resistance to *Fusarium oxysporum* f. sp. *vanilla*, but efforts are underway to develop some (Ramírez-Mosqueda *et al.* 2015; Koyyappurath *et al.* 2015). Chambers (Chapter 11 in this volume) discusses development of a *Vanilla* breeding program in Florida. *Fusarium*-resistant hybrids were developed in Puerto Rico in the 1930s, but apparently never put in commercial production (Childers and Cibes 1948). These hybrids have apparently been lost. More recently, germplasm of *Vanilla planifolia* and hybrids was tested for resistance to RSR and mechanisms of resistance were studied (Koyyappurath *et al.* 2015); the amount of lignin deposition on the hypodermis, which forms a barrier to entrance of the pathogen, was related to resistance. A recent plant patent application describes a *V. planifolia* cultivar, "Handa", which is stated to be resistant to RSR (Grisoni and Dijoux 2017).

7.5.3.5 Cultural Control
F. oxysporum can survive in the soil for years and re-infect new plants, due at least in part to production of chlamydospores resistant to adverse environmental conditions

(Ploetz 2006). Preventing the establishment of the RSR is critical, because the pathogen is hard to eradicate.

Well-spaced plants are less susceptible to RSR than more crowded plantings. Agroforestry systems, described by Flanagan *et al.* (Chapter 6 in this volume) are ideal in this respect. As noted in section 7.5.1, heavy fruit set makes plants more susceptible to RSR.

7.6 Future Prospects

Currently *Vanilla* commands record prices on the world market, encouraging expansion of production. However, the high price is a double-edged sword. Production will increase, leading eventually to an oversupply and price crash. This cycle of fluctuation in prices has happened before (see Brownell, Chapter 14 in this volume).

As for dealing with root and stem rot, *Vanilla* growers in Puerto Rico have resources today that were unavailable to their predecessors sixty years ago: agricultural extension agents in every municipality, diagnostic tools and a range of plant protection options. Integrated pest management has never been used on *Vanilla* in a systematic manner, but could be developed (Ordoñez *et al.* 2012).

At present *Vanilla* is not grown commercially in Puerto Rico. However, agriculture is becoming popular again. A new crop of farmers is motivated by concerns about food security, lack of employment opportunities and patriotism (Associated Press 2016). Many are focusing on artesanal and sustainable farming. *Vanilla* would fit well with their vision.

High labor costs (and U.S. labor laws) put Puerto Rico at a disadvantage relative to other *Vanilla*-producing areas. However, labor intensity and RSR may be related in a sense that has not yet been explored. Hybrids between *V. planifolia* and *V. pompona* (or other species) could be developed that combine improved resistance to RSR with increased attractiveness to pollinators. Such hybrids would not require intensive labor for pollination, and lower fruit load would reduce susceptibility to RSR. This would be a low-input, low-yield strategy, compatible with agroforestry, agroecotourism and coffee and cacao farming.

Seventy years have past since the last review article about *Vanilla* in Puerto Rico (Childers and Cibes 1946, 1948). Since that time (or shortly thereafter) there has been no commercial crop in Puerto Rico. Can it make a comeback?

Acknowledgments

I thank orchidologist James D. Ackerman for advice and comments.

References

Ackerman, J.D. (1995) *Orchid Flora of Puerto Rico and the Virgin Islands*. New York Botanical Garden Press, NY.

Ackerman, J.D. (2014) *Orchid Flora of the Greater Antilles*. New York Botanical Garden Press, NY.

Adame-García, J., Rodríguez-Guerra, R., Iglesias-Andreu, L.G., Ramos-Prado and Luna-Rodríguez, M. (2014) Variación patogénica de especies de *Fusarium* aisladas de *Vanilla planifolia* Andrews en Papantla, México. *I Seminario Internacional de Vainilla*, p. 186. Instituto de Investigación y Servicios Forestales, Universidad Nacional, Heredia, Costa Rica.

Álvarez López, C.L., Osorio Vega, N.W. and Marín Montoya, M. (2013) Identificación molecular de microorganismos asociados a la rizosfera de plantas de vainilla en Colombia. *Acta Biológica Colombiana* 18, 293–306.

Associated Press. (2016) Puerto Rico finds unexpected source of growth in agriculture. *Caribbean Business*, Sept. 28. http://caribbeanbusiness.com/puerto-rico-finds-unexpected-source-of-growth-in-agriculture/

Bacon, C.W., Glenn, A.E. and Yates, I.E. (2008) *Fusarium verticillioides*: managing the endophytic association with maize for reduced fumonisins accumulation. *Toxin Reviews* 27, 411–446.

Bhai, S. and Dhanesh, J. (2008) Occurrence of fungal diseases in vanilla (*Vanilla planifolia* Andrews) in Kerala. *Journal of Spices and Aromatic Crops* 17, 140–148.

Cameron, K. (2011) *Vanilla Orchids: Natural History and Classification.* Timber Press, Inc., Portland, OR.

Carro-Figueroa, V. (2002) Agricultural decline and food import dependency in Puerto Rico: a historical perspective on the outcomes of postwar farm and food policies. *Caribbean Studies* 30, 77–107.

Childers, N.F. and Cibes, H.R. (1946) El cultivo de la Vainilla en Puerto Rico. *Revista de Agricultura de Puerto Rico* 37, 1–14.

Childers, N.F. and Cibes, H.R. (1948) *Vanilla culture in Puerto Rico.* Puerto Rico Agricultural Experiment Station Circular No. 28.

Childers, N.F., Cibes, H.R. and Hernández-Medina, E. (1959) *Vanilla*-the orchid of commerce. In: *The Orchids: A Scientific Survey.* Withner, C.L. (Ed.) The Ronald Press Co., NY, pp. 477–508.

Correll, D.S. (1953) *Vanilla*: its botany, history, cultivation and economic import. *Economic Botany* 7, 291–358.

Cotty, P.J., Bayman, P., Egel, D. and Elias, K. (1994) Agriculture, *Aspergillus*, and aflatoxins. In: *The Genus Aspergillus.* Powell, K.A., Renwick, A. and Peberdy, J.F. (Eds) Plenum Press, NY, pp. 1–27.

Fain, S.J., Quiñones, M., Álvarez-Berríos, N.L., Parés-Ramos, I.K. and Gould, W.A. (2017) Climate change and coffee: assessing vulnerability by modeling future climate suitability in the Caribbean island of Puerto Rico. *Climatic Change* 143, 1–12.

Fouche, J.G. and Jouve, L. (1999) *Vanilla planifolia*: history, botany and culture in Reunion island. *Agronomie* 19, 689–703.

González-Chávez, M.C.A., Torres-Cruz, T.J., Albarrán-Sánchez, S., Carrillo-González, R., Carrillo-López, L.M. and Porras-Alfaro, A. (2018) Microscopic characterization of orchid mycorrhizal fungi: *Scleroderma* as a putative novel orchid mycorrhizal fungus in *Vanilla*. *Mycorrhiza* 28, 147–157.

Grisoni, M. and Dijoux, J.B. (2017) *Vanilla* variety named 'Handa'. *USPTO Publication US* 20170013762 P1.

Johnson, L., Gónzalez-Chávez, M.C.A., Carrillo-González, R., Porras-Alfaro, A. and Mueller, G. (2016) Amplicon sequencing reveals differences between root microbiomes of a hemiepiphytic orchid, *Vanilla planifolia* at four Mexican farms. *Inoculum* 67(4), 26.

84th Meeting of the Mycological Society of America, University of California, Berkeley. [Poster]

Koyyappurath, S., Atuahiva, T., Le Guen, R., Batina, H., Le Squin, S., Gautheron, N., Edel Hermann, V., Peribe, J., Jahiel, M., Steinberg, C., Liew, E.C.Y., Alabouvette, C., Besse, P., Dron, M., Sache, I., Laval, V. and Grisoni, M. (2016) *Fusarium oxysporum* f. sp. *radicis-vanillae* is the causal agent of root and stem rot of vanilla. *Plant Pathology* 65, 612–625.

Nielsen, L.R. (2000) Natural hybridization between *Vanilla claviculata* (W. Wright) Sw. and *V. barbellata* Rchb. f. (Orchidaceae): genetic, morphological, and pollination experimental data. *Botanical Journal of the Linnean Society* 133, 285–302.

Nielsen, L.R. and Siegismund, H.R. (1999) Interspecific differentiation and hybridization in *Vanilla* species (Orchidaceae). *Heredity* 83, 560–567.

Mosquera-Espinosa, A.T., Bayman, P., Prado, G., Gómez-Carabalí, A. and Otero, J.T. (2013) The double life of *Ceratobasidium*: orchid mycorrhizal fungi and their potential for biocontrol of *Rhizoctonia solani* sheath blight of rice. *Mycologia* 105, 141–150.

O'Donnell, K., Kistler, H.C., Cigelnik, E. and Ploetz, R.C. (1998) Multiple evolutionary origins of the fungus causing Panama disease of banana: concordant evidence from nuclear and mitochondrial gene genealogies. *Proceedings of the National Academy of Sciences USA* 95, 2044–2049.

Ordoñez, N.F., Díez, M.C. and Otero, J.T. (2012) La *Vanilla* y los hongos formadores de micorrizas. *Orquideología* 29, 56–69.

Pennigton, C., Jimenez, F.A. and Theis, T. (1954) A comparison of three methods of *Vanilla* culture in Puerto Rico. *Turrialba* 4, 79–87.

Pinaria, A.G., Liew, E.C.Y. and Burgess, L.W. (2010) *Fusarium* species associated with vanilla stem rot in Indonesia. *Australasian Plant Pathology* 39, 176–183.

Pinaria, A.G., Laurence, M.H., Burgess, L.W. and Liew, E.C.Y. (2015) Phylogeny and origin of *Fusarium oxysporum* f. sp. *vanillae* in Indonesia. *Plant Pathology* 64, 1358–1365.

Ploetz, R.C. (2006) *Fusarium*-induced diseases of tropical, perennial crops. *Phytopathology* 96, 48–652.

Porras-Alfaro, A. and Bayman, P. (2007) Mycorrhizal fungi of *Vanilla*: diversity, specificity and effects on seed germination and plant growth. *Mycologia* 99, 510–525.

Ramírez-Mosqueda, M.A., Iglesias-Andreu, L.G., Luna-Rodríguez, M. and Castro-Luna, A.A. (2015) In vitro phytotoxicity of culture filtrates of *Fusarium oxysporum* f. sp. *vanillae* in *Vanilla planifolia* Jacks. *Scientia Horticulturae* 197, 573–578.

Santa Cardona, C., Marín Montoya, M. and Díez, M.C. (2012) Identificación del agente causal de la pudrición basal del tallo de vainilla en cultivos bajo cobertizos en Colombia. *Revista Mexicana de Micología* 35, 23–34.

Suprapta, D.N. and Khalimi, K. (2009) Efficacy of plant extract formulations to suppress stem rot disease on *Vanilla* seedlings. *Journal of the International Society for Southeast Asian Agricultural Sciences* 15, 34–41.

Webster, J. and Weber, R.W.S. (2007) *Introduction to Fungi*, 3rd edn. Cambridge University Press, Cambridge.

8

Origins and Patterns of Vanilla Cultivation in Tropical America (1500–1900): No Support for an Independent Domestication of Vanilla in South America

Pesach Lubinsky, Gustavo A. Romero-González, Sylvia M. Heredia, and Stephanie Zabel

8.1 Introduction

The pan-tropical genus *Vanilla* Plumier ex Miller [Orchidaceae] comprises an estimated 100 to 107 species of monopodial, terrestrial, and hemi-epiphytic herbs with branching stems, half being endemics to tropical America (there are none that occur in Australia) (Portères 1954; Ackerman 2002; Cameron and Soto Arenas 2003; Mabberley 2008). Leaves of *Vanilla* can be fleshy, leathery, or absent; the flowers are showy and generally ephemeral (lasting fewer than 24 hours), and the fruits are elongate, deciduous berries with many exceedingly small seeds (Cameron and Soto Arenas 2003; Lubinsky 2007). In contrast to the vast majority of orchids, the seed coat in *Vanilla* is hard and generally dark brown or black (Childers *et al.* 1959), one of many special traits (like a pan-tropical distribution and aromatic fruits) characterizing this basal orchid lineage, estimated to have diversified over 65 million years ago, roughly contemporaneous with large-scale continental break-up (Ramírez *et al.* 2007). Low rates of natural pollination, ephemeral flowers, natural hybridization (Nielsen 2000), and the rarity of the plants themselves have contributed to a largely incomplete and still confusing taxonomy of the genus, which is poorly represented in the world's herbaria. There is a standing need for a comprehensive revision of this economically important genus, for students of natural history and breeders alike.

The two species of *Vanilla*, whose cured fruits are presently commercialized for flavor and fragrance, are *Vanilla planifolia* Jacks. and *V. tahitensis* J.W. Moore. Genetic studies support the hypothesis that both cultivars originated in Mesoamerica (Bory *et al.* 2008; Lubinsky *et al.* 2008a,b), even though *V. tahitensis* has actually never been found in the wild (Lubinsky *et al.* 2008b). The natural distribution of *V. planifolia*, an extremely rare species, is restricted to the lowland tropical evergreen forests of eastern Mexico and the Caribbean watersheds of Guatemala, Belize, and Honduras; it remains controversial whether *V. planifolia* is also native to South America (Cameron and Soto Arenas 2003; Hágsater *et al.* 2005). Besides *V. planifolia* and *V. tahitensis*, it is estimated

Handbook of Vanilla Science and Technology, Second Edition.
Edited by Daphna Havkin-Frenkel and Faith C. Belanger.
© 2019 John Wiley & Sons Ltd. Published 2019 by John Wiley & Sons Ltd.

that there are 25 to 30 other neotropical species, which possess aromatic fruits (Cameron and Soto Arenas 2003; Lubinsky 2007).

In Mesoamerica, the earliest historical evidence for the practice of vanilla cultivation ("*vainillales*"), as opposed to the gathering of wild vanilla fruits, is around the 1760s in the Colipa/Misantla and Papantla regions of north-central Veracruz (Fontecilla 1861; Bruman 1948; Kourí 2004). This cultivation, carried out predominantly by Totonac communities, served to provide nascent European demand for exotic luxury items from tropical colonies, such as cacao (*Theobroma cacao* L.) beverages spiced with vanilla and cinnamon, and later sugar (Kourí 2004; Coe and Coe 2007). Whether pre-Columbian cultivation (i.e. "domestication") of vanilla existed in Mesoamerica is unclear, as is the nature and extent of vanilla's pre-Columbian cultural importance (Hágsater *et al.* 2005; Lubinsky 2007). The antiquity of vanilla use as a cacao-beverage flavoring is probably nearly equivalent to that of the consumption of cacao beverages themselves (Hurst *et al.* 2002; McNeil 2006; Crown and Hurst 2009). Unfortunately, archaeo-botanical analysis of ancient Maya "chocolate pot" residues, which have positively confirmed the presence of theobromine (Henderson *et al.* 2007), are unlikely to show traces of vanillin, since the compound is relatively simple and breaks down readily (W.J. Hurst; personal communication). Lacking such evidence, there are only Contact-era references to vanilla that provide support for pre-Columbian vanilla use in Mesoamerica (see Section I). A fair judgment, given the available information, is that Totonac cultivation of vanilla arose for the purpose of commercial export, while the pre-Columbian Maya probably were first to experiment with sporadic vanilla cultivation, since the natural distribution of *V. planifolia* overlaps with what was once the principal region for cacao and achiote/annatto (*Bixa orellana* L.) cultivation and trade/tribute during the Late Postclassic (AD 1350–1500), namely, in the vicinities of the Soconusco, Lacandon, and Peten (Bergmann 1969; Sauer 1993; Caso Barrera and Fernández 2006; Lubinsky 2007). Maya vanilla cultivation was a possibility at least by 1699, when Marcelo Flores, a Spanish captain, remarked that in the vicinity of eastern Guatemala/southern Belize:

> ...there is a town... that belongs to the doctrine of the priests of Santo Domingo, which is the town of Belén, close to Rabinal. And in all of these localities there is evidence that there are Indians using these paths and trails at their own manner and habit, as is evidenced in the care and tidiness of their cacao and vanilla orchards and other fruits (Caso Barrera and Fernandez 2006).

Although there are many aromatic species of *Vanilla* in northern South America, and despite both historical and present-day confirmation of vanilla use in the region (see Figure 8.1, Table 8.1), there has never been an attempt to synthesize or characterize the nature of South American vanilla ethnobotany. Here, we provide a review of the relevant literature and specifically explore the possibility that vanilla may have been independently domesticated in South America. To have achieved a vanilla culture that was as elaborate as that which existed in New Spain in the late eighteenth century, at least two requirements had to have been met in South America: cultivation (planting), and post-harvest processing consisting of curing-fermentation. For convenience, in comparison, we have separated our review of the literature into three stages of vanilla history, more or less defined by technological advancements or changes that impacted

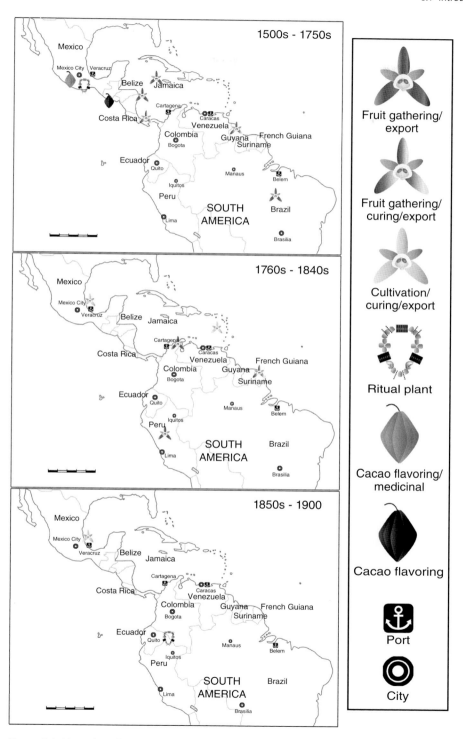

Figure 8.1 Map of vanilla uses in Meso, Central and Tropical South America, ca 1500-present.

Table 8.1 Vanilla uses in Meso, Central, and Tropical America, three periods during ca. 1500–present.

ca 1500 – 1750s (Precultivation)

Year	Place	Comments	References
1552	Aztec–Veracruz	Earliest document, records vanilla fruits ground up with other aromatic constituents and worn in an amulet necklace.	Bruman 1948
1580s	Colonial Mexico and Guatemala	Consumption of hot chocolate as a beverage, flavored with vanilla and sugar.	Sauer 1993
1619	Amazon Basin	*"Of aromatic things, what we have seen is those little pods, growing in trees, and when flavorful are black and very aromatic, and when mixed among clothing, leaves a long lasting aroma like a musk …"*	Patiño 2002
1630s	Pacific coast of Guatemala	*"[…] from the provinces of Soconusco and Suchitepéques, which are extreme hot, and subject to thunder and lightning, where growth scarce any remarkable commodity, save only cacao, achiote, "mecasuchil," vanilla and other drugs for chocolate"*	Thompson 1958
1640	Northeastern Nicaragua and Honduras	*"Englishmen spent a good deal of time at Cayos Miskitus and Caratasca Lagoon region procuring vanilla, silk grass, annatto"*	Offen 2000
1651	Mexico	Vanilla was described as curative and cacao-beverage flavoring.	Varey 2000
1655	Jamaica, Pomeroon	Jews settlers had secured a monopoly of the vanilla and pimento trades. The "Chocolate trade": cacao and vanilla by Jews, they figure out the curing procedure for vanilla	Fortune 1984; Arbell 1995
1660	Campeche and Tabasco–Mexico (New Spain) and Bocas del Toro (Bocca-toro), West Indies	Spaniards lay them up like tobacco stems, Indians cure them in the sun and sell (3 pence/ pod) them to the Spaniard who sleek them with oil. The vines grow plentifully. Sold by druggist to used among chocolate to perfume it.	Dampier 1776
1676	Bay of Campeche and coast of Veracruz	Chocolate became popular throughout Europe during 17th and demand for vanilla greatly exceeded the supply.	Sauer 1993
1699	Coast of South America	He said that he had traveled on the coast of South America, and he knew how to prepare vanilla extract.	Arbell 1995
1699	Eastern Guatemala/ Southern Belize	*"[…] there is evidence in the care and tidiness of their cacao and vanilla orchards […]"* Marcelo Flores, Spanish Captain.	Caso Barrera and Fernandez 2006
18 century	Venezuela	Vanilla was the first orchid mentioned from Venezuela. Potential for cultivation, growth habit	Romero-Gonzalez 1998
1700 soon after	Mosquito Coast, Colombia and Venezuela	Wild vanilla exported	Sauer 1993

Table 8.1 (Continued)

ca 1500 – 1750s (Precultivation)

Year	Place	Comments	References
1707 and 1725	Jamaica	*"although the long-term therapeutic use and value of plants from the Americas may have faded during the nineteenth century, their impact on European pharmacopoeia was immense"*	Sloane 1707–1725
1735	Pará-Brazil	*"...we imagine that [...], cacaos, and vanilla, are the only useful plants which the fruitful blossom of American presents?", "various useful articles as well from the rivers which fall into the Amazons, as from the river itself, such as [...], vanilla, sugar, coffee, and in abundance cocoa, which is the currency of the country"*	Condamine 1735–1745
1741	Santa Marta, Colombia	*"...in some of the hills there is wild vanilla, no cultivation, but very little is used and is only identified by scent..."*	Patiño 2002,
1743	Santo Domingo	François Geoffroy goes on to state, *"It is certain that the vanilla of Santo Domingo is not different from that of Mexico, which was described by Hernández, save for the color of the flowers and the aroma of the pods since the Mexican flower is black and the pod has a pleasant aroma"*	Étienne 1743
1750	Veracruz, Mexico	First record of planting vanilla, vainillales, among the Totonac communities.	Fontecilla 1861; Bruman 1948; Kourí 2004
		1800s were a golden era for Papantla	
Mid-1700s	Caribbean coast, Colombia	Mid-18 century, vanilla was one the main products of extraction near Providence Island	Patiño 2002

1760s – 1840s

Year	Place	Comments	References
1762–1764	French Guiana; Pará, Brazil	"That is the usual manner [used to process vanilla] by Galibi & Caribes naturals of Guiana, and by the Garipons escaped from Para, Portuguese colony in the banks of the Amazon river..."	Aublet 1775
1771	Jamaica	Vanilla appears in a table entitled "Duties payable upon importation into Great Britain on the following commodities, being of the produce of Jamaica"	Long 1774

(Continued)

Table 8.1 (Continued)

1760s – 1840s

Year	Place	Comments	References
1777–1788	Perú, Huánuco, Village of Pozuzo	"Vanilla gathering in different locations and selling by Indians. "The harvest in those forest is small because of the little value there"	Ruiz, 1998
1839	West Indies	Vanillons produced from V. pompona in Guadaloupe. "Vanilla was cultivated as early as 1839 on [...] Martinique and [...] Guadeloupe". Introduction of vanilla to Reunion from French Guiana	Purseglove *et al.* 1981; Weiss 2002; Arbell 1995

1850s-Present

Year	Place	Comments	References
1840s and 1860s	Réunion, West Indies	Edmond Albius discovered methods to effect hand pollination of vanilla flowers.	Ecott 2004; Kourí 2004
		Mexico was replaced by Réunion as the world's principal vanilla producer	
1878	Venezuela	In Venezuela vanilla was later used as a flavoring agent	Spence 1878
1889	Amazon Basin	"*Indigenous people wear one or more vanillas in necklaces for its fragrance*"	Patiño 2002
1894	Amazon Basin	"*[they] like very much a species, known in cities like Cartagena and Panamá as Bainilla, mixed with chocolate and Indigenous people wore around their neck...*"	Patiño 2002
1927	Maynas region, Perú, at the time Ecuador	"*As good as the best from Spain*"	Patiño, 2002,
1942	Putumayo river, Colombia	Richard E. Schultes acquires a necklace from Siona Indians	Botanical Museum 6836; Romero and Sabel in preparation
1988	Surinam	Fermented fruit is made into vanilla crystals, which are put into carapa oil (*Carapa guianensis*).	Heyde 1987; DeFilipps *et al.* 2004
		For blood circulation (circulatory problems), skin conditions (skin diseases).	
Present	Papantla, Veracruz, Mexico	New York had its suppliers from Mexico and, increasingly, from Baltimore, Michigan and New Jersey	See text

vanilla production specifically in the Mesoamerican region, as well as on a world-wide scale:

i) "Pre-cultivation" (ca. 1500–1750s), defined by an absence of cultivation; confusion in Europe over the correct botanical identity of vanilla; cacao exports to Europe being predominantly Mesoamerican in origin; and European frustration over how to cure vanilla;

ii) the "Papantla monopoly" (1760s–1840s), characterized by the initiation of cultivation in Veracruz, Mexico (but without artificial pollination); the decline of cacao cultivation in Mesoamerica and its expansion into South America; the establishment of Linnaean taxonomy, and the correct determination of *V. planifolia* as the vanilla orchid of commerce; the introduction of *V. planifolia* cuttings to Europe; and lamentations that cultivation in South America had unrealized potential;

iii) the "Vanilla revolution… and we've never looked back" (1850s–1900), set forth by the discovery of a practical method for artificial pollination of vanilla flowers, which allowed for the initiation of vanilla cultivation in Old World tropical colonies, especially in French-controlled Réunion; coinciding with the turbulent demise of Papantla as a significant vanilla producer; a sea-change in vanilla being used mainly in chocolate to being principally employed in ice-creams; and, in the United States, the near total replacement of vanilla beans by newly synthesized, "plain vanilla" imitation flavors.

Finally, in the last section of the paper, "The Vanilla Necklace", we discuss a unique use of vanilla beans as magic-ritual items for the Siona-Secoya of northern South America.

8.1.1   I. Pre-Cultivation, ca. 1500–1750s

During the colonial period in the New World (ca. 1500–1800), Europe's sea-borne empires (Portugal, Spain, England, the Netherlands, and France) sought out, and competed for, New World wealth by means of trade, settlement, and conquest. In addition to the extraction of precious metals, the commercialization of tropical plants provided one of the most profitable and enduring enterprises for Europe's colonial interests. The motivation to locate, identify, and capitalize on new and rare plants in all of Europe's tropical colonies ("colonial botany") was thus "big science" and integral to colonial endeavors as a whole, spawning the world's first multi-national trading companies such as the English East India Company, the French East and West India Companies (Compagnie des Indes Orientales/Occidentales), the Dutch West India Company, and the Dutch East India Company, or Verenigde Oostindische Compagnie (VOC) (Rich and Wilson 1967; Boxer 1969; MacLeod 2000; Brockway 2002; Schiebinger 2004; Schiebinger and Swan 2005). The scale by which Europe was able to marshal resources to these ends during the period was unprecedented, perhaps no better epitomized than by the production of sugar cane (an Old World plant) by African slave labor in tropical America for sale to Europe (Mintz 1984; Dunn 2000). Among other far-reaching consequences, the consumption in Europe of sugar and other stimulating plants of colonial derivation such as cacao, tobacco, coffee, and tea, would over time play an essential role in the "energizing" and enabling of an industrial workforce in the nineteenth century, along with staples such as maize and potato (Goody 1982).

Vanilla gained some notice in Europe, beginning in the sixteenth century. Columbus was credited for being the first to introduce vanilla beans into Europe while returning from his fourth voyage (Morren 1839; Smith *et al.* 1992; Weiss 2002); the apothecary Hugh Morgan recommended vanilla beans as a flavoring to his Queen, Elizabeth I, (Kourí 2004); Morgan sent beans to Flemish botanist C. Clusius in 1602; Clusius in *Exoticorum Libri Decem* (1605) would call them *"lobus oblongus, aromaticus"* (apparently on Morgan's suggestion) (Purseglove *et al.* 1981); W. Piso used "vanilla" for the first time in print while working in Brazil in 1658; shipments of vanilla were received in Cadiz, etc. (for full details, see Correll, 1953; Lubinsky 2007; but esp. Kourí 2004).

The development of European interest in vanilla was to a large extent the by-product of a greater interest in cacao (Kourí 2004; Coe and Coe 2007). Spanish priests-ethnographers, such as F. Bernardino de Sahagún, and royal physicians, such as Diego Durán, described the indigenous use of vanilla as a condiment/medicine in cacao beverages (de la Cruz 1940; Sahagún 1963; Durán 1994). An especially insightful example into Contact-era vanilla use is the work of Dr Francisco Hernández, royal physician to the King of Spain, whose monumental treatise on the medicinal flora of New Spain was disseminated in print in 1651. His entry on "Tlilxóchitl" (Nahuatl, "Black Flower") is the most detailed description of the native use of vanilla in Mesoamerica:

> ... the vanilla beans, smell like musk or balsam of the Indies, and they are black – hence the name. It grows in hot, moist places. They are hot in the third degree and are usually mixed with cacao as well as with mecaxóchitl ... Two vanilla beans dissolved in water and taken will provoke urine and menstruation, if mixed with mecaxóchitl. It hastens birth, expels afterbirth and a dead fetus. It strengthens the stomach, and expels flatulence. It heats and thins the humors. It invigorates the brain and heals fits of the mother. It is said that these vanilla beans are a similar remedy against cold poisons and against cold poisonous animal stings. It is also said to be one of the most aromatic plants in this region (Varey 2000; p. 167).

By the time Hernández's work was published, England was actively accumulating direct knowledge about the whereabouts and uses of vanilla in tropical America. In the 1630s, Thomas Gage reported on the presence of vanilla along the Pacific coast of Guatemala:

> The chief commodities which from along that coast are brought to Guatemala, are from the provinces of Soconusco and Suchitepéquez, which are extreme hot, and subject to thunder and lightning, where groweth scarce any remarkable commodity, save only cacao, achiote, "mecasuchil", vanilla and other drugs for chocolate (Thompson 1958).

Around the same time, in the 1640s, the English began to exploit wild vanilla populations in northeastern Nicaragua and Honduras (i.e. the Cayos Miskitus), specifically in the region around the Caratasca Lagoon, along with populations of silk grass and annatto (Offen 2000).

In the 1660s, the English pirate William Dampier recorded observing wild vanilla populations three times, twice in New Spain (near the present-day states of Campeche and Tabasco, Mexico), and lastly in the vicinity of Bocas del Toro, on the Caribbean side of the Costa Rica-Panama border. At "Bocca-toro", Dampier made his most extensive comments about vanilla, discussing aspects of curing and trade:

> The vanilla is a little pod full of black seeds... the Indians (whose manufacture it is, and who sell it cheap to the Spaniards) gather it, and lay it in the sun, which makes it soft; then it changes to a chestnut color. They press it frequently between their fingers, which makes it flat. If the Indians do anything to them besides, I know not; but I have seen the Spaniards sleek them with oil... These vines grow plentifully at Bocca-toro, where I have gathered and tried to cure them, but could not, which makes me think that the Indians have some secret that I know not of to cure them. I have often asked the Spaniards how they were cured, but I never could meet with any could tell me. One Mr Cree who spoke Spanish well ... and had been a privateer all his life, and seven years a prisoner among the Spaniards at Portobello and Cartagena ... could not find any of them that understood it... At, or near a town also, called Caibooca, in the bay of Campeachy, these pods are found. They are commonly sold for three pence a pod among the Spaniards in the West Indies, and are sold by the druggist, for they are much used among chocolate to perfume it. Some will use them among tobacco ... I never heard of any vanillas but here in this country, about Caibooca, and at Bocca-toro (Dampier 1776; pp. 369–370).

Dampier and the English were not alone in their belief that vanilla curing was an "Indian secret". Like the English, the Dutch had been on the lookout for commercial prospects in the New World. They became heavily involved in partnership with the English in the sugar cane trade in the Caribbean, as well as Guyana in northern South America (Davis 2006). Refugee Sephardim established communities in both places, often serving as intermediaries in the trans-Atlantic trade of specialized products such as cacao, pimento, and vanilla, after having been banned from the principal trade in sugar cane (Arbell 1995). For example, by 1655, Jamaican Jews had secured a monopoly on the island's pimento and vanilla trade (Fortune 1984), while in Guyana, Jews had apparently learned the curing of vanilla from the native population. A letter from Commander Beekman of Essequibo and Pomeroon [Guyana] to the headquarters of the Dutch West India Company, March 31, 1684, states:

> The Jew Salomon de la Roche having died some 8 to 9 months ago, the trade in vanilla has come to an end, since no one here knows how to prepare it, so as to develop proper aroma and keep it from spoiling. I have not heard of any this whole year. Little is found here. Most of it is found in Pomeroon, whither this Jew frequently traveled, and he sometimes used to make me a present of a little. In navigating along the river, I have sometimes seen some on the trees and picked with my own hands, and it was prepared by the Jew... I shall do my best to obtain for the company as much as shall be feasible, but I am afraid it will spoil, since I do not know how to prepare it... (Arbell 1995; p. 359).

In response, Com. Beekman was sent the following in August 21, 1684:

> As to the vanilla trade, which we recommend you carry on for the company, where you answer us saying this trade has come to an end through the death of a Jew, Salomon de la Roche ... a meager and poor excuse (Arbell 1995; p. 360).

The same pattern was seen in Curaçao. In 1699, Jean Baptiste Labat, a French missionary stationed in Martinique, wrote the following:

> ...a Jew who inherited Benjamin d'Acosta, who came from Curaçao to ask for the money due to his relative. He said that he had traveled on the coast of South America, and he knew how to prepare vanilla extract. I begged him to teach me how the Indians prepared the vanilla, how to dry it, and how to have the extract. I observed exactly the way he showed me and tried several times to prepare it with no results. I concluded that maybe the vanilla in Martinique was different from the one in Cayenne. But I think he had deluded me. It is not extraordinary to this sort of people... (Labat 1722; p. 3).

If vanilla processing was not straightforward or obvious in some instances, it did not prove to be an impediment to the export of vanilla from many regions during the seventeenth and early eighteenth centuries. In 1735, the French mathematician and surveyor Charles Marie de La Condamine, trekked across the Amazon from Peru to Brazil, and noted the export of vanilla along with other goods traded near Para for export to Lisbon:

> The commerce of Para direct with Lisbon ... enable those of the place whose circumstances are easy, to provide themselves with all the comforts of life. They receive European commodities in exchange for the produce of the country... all the various useful articles... from the rivers which fall into the Amazons, as from the river itself, such as clove-wood and the black nutmeg, sarsaparilla, vanilla, sugar, coffee, and in abundance cocoa, which is the currency of the country, ... (de la Condamine 1745; p. 249).

Vanilla was exported from many parts of tropical America to various countries in Europe. Comparisons were inevitably made between the quality of the varied products (Kourí 2004), and botanists were keen to identify if the multiple sources of vanilla represented multiple species. The namesake for the genus *Vanilla*, Charles Plumier (1646–1704), discussed three vanilla types he had observed in the Antilles in his *Plantarum Americanarum* (1758, originally published in 1693). Since Plumier's description antedated the establishment of the binomial system of Linnaeus (1735), his "generic concept" had to be shared with Philip Miller (1754), the first to describe the genus using the Linnaean system. It is not clear which, if any, of the three vanilla types Plumier mentioned were *V. planifolia*. At the time, the vanilla market offered more diversity of commercialized species than at any time prior or since.

By the beginning of the eighteenth century, vanilla, the "chocolate drug", had established its popularity in Europe as a desirable plant, and its reputation spread through publications, especially as a medicinal. The tradition of listing vanilla as part of New World *materia medica* was passed on from F. Hernández to the English, largely through the effort of Sir Hans Sloane's *Natural History of Jamaica* (ca. 1725), which included significant portions of Hernández's work. Such products as vanilla were recorded in standard trade data in England. In *The present state of His Majesties isles and territories in America* (London 1687), vanilla is enumerated among the island's products: "cacao, indigo, cotton, sugar, and *DRUGS*, which this Island produces in great abundance, as,

Guiacum, China-roots, Sarsaparilla, Cassiafistula, Tamerinds, Vinello's (i.e. vanilla) and *Achiots* or *Anetto* (Chabrán and Varey 2000)." The French also had a concept of vanilla influenced by Hernández. In 1743, François Geoffroy compared a *Vanilla* species he found in Santo Domingo to Hernández's description:

> It is certain that... [it] is not different from that of Mexico, which was described by Hernández, save for the color of the flowers and the aroma of the pods since the Mexican flower is black and the pod has a pleasant aroma (Étienne 1743; cited by López Piñero and Pardo Tómas 2000; p. 132).

In South America, Jesuit priests who served in the Orinoco River basin commented briefly on the natural abundance of vanilla in the region and the aromatic properties of its fruits (Rivero 1888; p. 4; Gumilla 1745, I: 366; 1993: 250; Caulín 1779: 18; Gilij 1780–1782, I: 176, 1965:168; Romero-González 1998). Caulín even cited a common local name for vanilla: "*Ekére-nuri*", meaning "lengua de Tigre" or "Jaguar's tongue", in reference to the leaf shape. While these commentators were aware of the economic value of vanilla, none made allusion to vanilla cultivation in the Orinoco.

8.1.2 II. Papantla Monopoly, 1760s–1840s

Growing demand for chocolate in Europe kept the cacao and vanilla markets surging in the early 1700s. Vanilla beans were arriving in Europe from throughout tropical America (Sauer 1993), all without the aid of cultivation. Instead, the commercialization of vanilla beans was based on fruit-gathering from wild populations of multiple species of *Vanilla*, which were subsequently cured in one fashion or another and then exported.

The onset of severe indigenous population declines in the seventeenth century along the Pacific littoral of Mesoamerica (Chiapas, Guatemala, El Salvador) contributed to a restructuring of the cacao trade, with increasing exports of cultivated cacao being produced with imported African slave labor in northern South America, in areas such as Ecuador (Guayaquil), Colombia, Venezuela, and the Guianas (Hussey 1934; Price 1976; Ferry 1981; Piñero 1988; Presilla 2001; Salazar 2004; Coe and Coe 2007; MacLeod 2008). While this emergent industry had the potential to also stimulate cultivation of local *Vanilla* species alongside cacao in the region, such would not be the case. One of the strongest mitigating factors obviating the need for South American vanilla cultivation was the establishment of *vainillales* in Veracruz in the 1760s. Veracruz was the only legal port of entry into Spain for over 300 years (Knight and Liss 1991). Cacao exports from Spanish ports in South America, such as Cartagena and La Guaira, were obligatorily shipped to Spain via maritime trade with Veracruz. The geographic origin for vanilla cultivation, near the all-important port of Veracruz, could therefore have not been better positioned to take advantage of the European market. For the next 100 years, it would be primarily botanists and natural historians who would comment on vanilla. In Europe, biological taxonomy based on a system on binomial nomenclature was becoming standardized by Linnaeus. He named the first *Vanilla* specimen he described as *Epidendrum vanilla* (= *V. mexicana*) in *Species Plantarum* (1753, II: 952). In his earlier *Materia Medica* (1749), Linnaeus listed vanilla's supposed curative attributes as, "calefaciens, corroborans, cephalica, diuretica, aphrodisiaca", used to treat: "melancholia, apoximeron". He gave no indication of how vanilla should be prepared in such instances but, in passing, noted that vanilla was used to make chocolate (Shrebero 1787, p. 234, entry 551).

During his Peruvian expedition of 1777–1788, Spanish botanist Hipólito Ruiz twice mentioned gathering of vanilla (of apparently different species) as secondary trade items. In the region of Cuchero, an area where quinoa (*Cinchona nitida*) bark collection predominated, Ruiz made mention of vanilla:

> [*Vanilla officinalis*, vainilla (vanilla)]... The Indians gather some of the fruits, or pods, which they take to Huánuco to sell, but the harvest in those forests is small because they have little value there (1998).

Ruiz observed a similar pattern in the village of Pozuzo:

> *Vanilla volubilis*, vaynilla (vainilla, vanilla)]... Indians gather the fruits for sale to traveling merchants... who arrive at Pozuzo [to] buy coca leaves, paying for this product with cloth of various kinds, ribbons, glass beads, and other trinkets that the local people use as adornments, for holidays and their drunken parties when they drink maize chicha (1998, p. 259).

Between 1762 and 1764, French botanist/agronomist Jean Baptiste Christophe Fusée Aublet provided relatively detailed coverage of vanilla botany and use in Cayenne (French Guiana). Like other commentators, Aublet noted that vanilla cultivation was, "... not estimated or searched by the inhabitants (p. 78)," but native *Vanilla* populations did occur in abundance, and local markets seemed to have existed. This is evidenced by one of the main topics Aublet discusses: curing. In a section titled, "To prepare the Vanilla, to turn its scent smooth, aromatic and marketable," Aublet gives a description of nearly the entire curing process:

> Once a dozen of Vanilla is assembled, more or less, one ties them together or threads them like beads on a string, by their peduncular end: in a large pot or any other pot suitable for cooking, full of clear and clean boiling water; when the water is boiling hot, dipping the Vanillas to whiten, which works in an instant; once done, one extends and ties from the opposite ends, the string where the Vanillas are attached or threaded, in such a way that they are suspended in open air, where sun-exposed, for a few hours during the day. The following day, with a paint brush or with the fingers, one coats the Vanilla with oil, to shrivel it up slowly, to protect it from insects, from flies that dislike oil, to avoid drying up the skin, and to become tough and hard, finally to keep away from external air penetration and to preserve it soft. One has to observe to wrap the pods with cotton yarn soaked in oil, so as to keep them from opening and to contain the three valves. While they are hung up, to shrivel up, from the superior end, there is a thin stream of overabundant viscous liqueur; one squeezes the pod slightly, to facilitate the passage of the liqueur: before squeezing, one soaks his hands in oil and repeat the pressure two or three times per day (Aublet 1776: pp. 83–84).

It is unclear whether Aublet was superimposing prior knowledge of vanilla curing onto his observations in Cayenne, or whether in fact what he was describing a local system of curing. At one point he explicitly says:

> Here is the way commonly used by of Galabis of Suriname and the Caribs naturalized from Guiana, and by the Garipons maroons from Para [a Portuguese Colony], at the river bank of the Amazon. I used varnished containers, although they only fire containers without varnish... (p. 84)

but fails to elaborate, and it is not obvious which "way" for curing he is making reference to, or for what purpose the "Galabis" and "Caribs" were curing vanilla. Likewise, Aublet states that a certain, unspecified practice for curing vanilla was similar to preparations used to, "preserve plums at Tours, Brignoles, Digne, etc. (p. 83)" and, "... the same for the raisins sent from Naples and Ciouta (p. 83)." Interestingly, Aublet's description contains elements of both the Bourbon process for curing practiced in Madagascar ("killing" the beans by means of scalding), and what Gage and other commentators had talked about a century before: the apparently widespread practice of using oil to prevent the fruits from dehydrating completely.

Aublet considered vanilla export a viable economic activity for French Guiana. His desire to develop vanilla as a crop most likely served as the motivation for his extended discourse, and he apparently took personally the fact that his advice was not being acted upon. In frustration, Aublet criticized the hypocritical colonial botanists who prepared shoddy viability studies of economically important plants:

> I do not understand why there are such a bold people, to propose the government department a crop that they fully ignore in their research; their Memoirs promise more than what the authors can demonstrate. Why do these men, with their well-digested written proposals, fail to comprehend how to put it in practice by themselves, since they presented it as so profitable? They support it as a common good, for the good of the government, that for years they have worked to further; but that patriotism is hiding a personal interest (p. 85).

Aublet may have been alluding to Pierre Poivre, a more charismatic and politically savvy botanist who championed the economic benefits of nutmeg cultivation. The two botanists were frequently at loggerheads (Spary 2005). In the 1770s, Poivre had successfully smuggled cloves and nutmeg from under the nose of the Dutch in the Moluccas, and in return, was rewarded by the French East India Company with promotion to the title *Intendant Général* for Mauritius (Île de France) and Réunion (Île de Bourbon).

Under Poivre's supervision, over 1,600 rare and notable plants were introduced to nurseries in Réunion, including cuttings of South American *Vanilla* (Ecott 2004). Around 1820, Pierre-Henri Philibert, captain of the *Le Rhône*, introduced to Poivre's collection *Vanilla* stems that were procured from French Guiana (Arditti *et al.* 2009). It is not clear what species were involved in this dispersal. Currently, there are an estimated 23 species of *Vanilla* in the Guianas: *V. acuta* Rolfe, *V. appendiculata* Rolfe, *V. barrereana* Veyret & Szlach., *V. bicolor* Lindl., *V. chamissonis* Klotzsch, *V. cristato-callosa* Hoehne, *V. fimbriata* Rolfe, *V. gardneri* Rolfe, *V. grandflora* Lindl., *V. guianensis*

Splitg., *V. hostmanni* Rolfe, *V. latisegmenta* Ames & C. Schweinf., *V. leperieurii* Portères, *V. marowyensis* Pulle, *V. mexicana* Mill., *V. odorata* C. Presl, *V. ovata* Rolfe, *V. palmarum* (Salzm. ex Lindl.) Lindl., *V. penicillata* Garay & Dunst., *V. planifolia* Andrews, *V. porteresiana* Veyret & Szlach., *V. surinamensis* Rchb. f., and *V. wrightii* Rchb. f. (Funk *et al.* 2007). Most of this taxonomy is suspect, and it remains possible that one of the species that was taken by Philibert was *V. tahitensis*, which could have evolved spontaneously from the hybridization of *V. planifolia* and *V. odorata* (Lubinsky *et al.* 2008b).

In any event, by 1839, the French had also introduced *V. pompona* from South America to Guadeloupe and Martinique (it would come to be known as "West Indies vanilla", or "vanillion") (Purseglove *et al.* 1981; Weiss 2002). Like *V. planifolia*, *V. pompona* is not native to the Antilles, but readily established itself after escaping from cultivation efforts in the late 1800s (Garay and Sweet 1974).

The English too were active in *Vanilla* dispersals at the time. Around 1800, glazed roof technology led to the proliferation of hot-houses for keeping exotic plants, and in a short time, horticultural societies were established, such as the Society for the Improvement of Horticulture (the forerunner of the Royal Horticultural Society) (Ecott 2004; for a discussion of the origins of American horticulture, see Pauly 2007). Cuttings of *V. planifolia* were introduced as horticultural rarities into the extensive private collection of George Spencer Churchill, the Marquis de Blandford (later to become the fifth Duke of Marlborough) at his estate, Whiteknights, at Reading. These cuttings would serve as both the lectotype of the species, described in 1808 in Paddington, as well as the genetic material for commercial clonal propagation in the Indian Ocean and Indonesia (Lubinsky 2007; Bory *et al.* 2008). The popular assumption is that Blandford's cuttings were shipped from Jamaica, where they could have been relicts of Spanish dominion over the island in the early seventeenth century (*V. planifolia* is not native to Jamaica). However, this assertion is probably impossible to validate, since Blandford's personal papers were destroyed (Cooke 1992).

In the early 1800s, German naturalist Alexander von Humboldt visited South America after also having toured through New Spain. In both places, he talked of vanilla. His comments from Venezuela echo Aublet's assertion that an abundance of native *Vanilla* could serve to develop an industry. Humboldt further described an aversion to vanilla:

> The Spanish, in general, dislike a mixture of vanilla with the cacao, as irritating the nervous system; the fruit, therefore, of that orchideous plant is entirely neglected in the province of Caracas, though abundant crops of it might be gathered on the moist and feverish coast between Porto Cabello and Ocumare; especially at Turiamo... The English and the Anglo-Americans often seek to make purchases of vanilla at the port of La Guayra, but the merchants procure with difficulty a very small quantity. In the valleys that descend from the chain of the coast towards the Caribbean Sea, in the province of Truxillo, as well as in the Missions of Guiana, near the cataracts of the Orinoco, a great quantity of vanilla might be collected; the produce of which would be still more abundant, if, according to the practice of the Mexicans, the plants were disengaged, from time to time, from the creeping plants by which it is entwined and stifled (Humboldt 1819, II:124; 1851, II:63; 1956, III:141).

Humboldt's claim of Spanish dislike of vanilla could refer either to Spaniards in Venezuela, another colony, or in Spain. At least one subsequent author has interpreted

Humboldt's words as particular to Venezuela (Patiño 2002, p. 539). If so, Humboldt is probably correct in attributing part of the failure to develop vanilla cultivation in South America to a cultural/social bias.

8.1.3 III. The Vanilla Revolution, 1850s–1900, "… and we've never looked back"

By the mid-nineteenth century, *V. planifolia* had been identified as "the vanilla of commerce" and was disseminated by cuttings throughout Europe and her Old World colonies, with the result that practically any tropical region, not just Papantla, was able to grow vanilla.

Just as soon as *V. planifolia* became available for others to exploit, a deluge of social and technological changes contributed to a massive re-organization of the vanilla business and global trade in general. The nineteenth century ushered in the end of the colonial period (*sensu stricto*) in the Americas, as well as social revolution in Europe. Former colonies won independence, including Mexico in 1821, and in relation to this, many regions outlawed bondage (for example, in the sugar-cane growing regions of the Caribbean and the Guianas). Emancipation, however, triggered a new era of importation of cheap labor, this time from India (Tinker 1974). Europe's nineteenth-century imperial ambitions also faced new, aggressive competition from the United States, which enacted the Monroe Doctrine in 1823, declaring all of the Americas as its personal sphere of influence for economic opportunity.

As the winds of change swept over many parts of the Americas, the first in a series of momentous events in the history of vanilla transpired: the implementation of artificial pollination. Both Charles Morren, a botanist from Belgium, in the 1830s (Arditti *et al.* 2009), and Edmond Albius, a boy and slave from Réunion, in the 1840s (Ecott 2004), discovered methods to effect hand pollination of vanilla flowers. The radical consequence of hand pollination of vanilla was that no longer would Papantla be able to maintain its monopoly. Very quickly, by the 1860s, Mexico was replaced by Réunion as the world's principal vanilla producer (Kourí 2004).

Almost simultaneously, vanilla had a new companion in addition to cacao: ice-cream. The manufacture of ice-cream on an industrial scale, beginning in the mid-nineteenth century in the United States, kept demand alive for Mexican vanilla, since both the Spanish and French had essentially stopped ordering (Kourí 2004). In terms of volume of production, the late 1800s were a golden era for Papantla (which was by then producing vanilla with the aid of hand-pollination), but just as soon as this prosperity materialized it was to vanish, following the isolation of vanillin in 1858 and its laboratory synthesis from pine bark in 1874. Lab/factory made vanillin, much cheaper and more stable price-wise than vanilla beans, has dominated formulations and the vanilla market since, with over 90% of vanilla products today, including "natural" ones, being derived from sources other than vanilla beans.

The vanilla business in South America would continue to be marginalized in the face of all these changes. New York had its suppliers from Mexico and, increasingly, from Baltimore, Michigan, and New Jersey, and France had its own vanilla supply from the Indian Ocean; the Seychelles, Réunion, and Madagascar. The market for vanilla flavor was essentially saturated. The only mentions of vanilla during this period in South America are manuals for cacao cultivation that also advocate for vanilla cultivation (Rossignon 1850, 1881, 1929; Díaz 1861, pp. 265–268; Martínez Ribón 1895; Ríos 1999; Salazar 2004). The manuals are written with superficial detail with regard to vanilla, do not mention hand-pollination of vanilla flowers and, like Aublet, fail to consider the market opportunities that may or may not have existed for South American vanilla. There are also some references to localized consumption of vanilla (Spence 1878).

8.2   The Vanilla Necklace

The Siona and Secoya, two closely linked Amazonian indigenous groups, occupy a small territory that extends across the converging borders of Ecuador, Peru, and Colombia. Traditionally, the Siona inhabited the territory around the Putumayo and Aguarico rivers, and the Secoya lived in the Santa Maria River region (Vickers and Plowman 1984). The two groups share strong cultural and linguistic parallels, both speaking similar dialects in the Western branch of the Tukanoan language family (Vickers 1989).

Their settlements are generally small in population and characterized by a low level of socio-political organization. After 5 to 20 years, villages are abandoned and groups migrate to new areas of the forest (Vickers 1989). Their subsistence consists of shifting cultivation of horticultural species and hunting, fishing, and gathering of wild food plants (Vickers and Plowman 1984). They also tend garden plots containing a diverse variety of species used for food, medicine, and handicrafts (Vickers 1989). Among Amerindians, the Siona-Secoya are regarded especially for their use of hallucinogenic plants in ceremonial contexts in order to divine the future, cure disease, commune with the spirit world, and perform sorcery (Matteson Langdon 1992).

Within the past century, the Siona and Secoya have joined together to live in common settlements. This shift was catalyzed by the outbreak of diseases that greatly reduced population sizes, and the increased need for defensive action against attacks by exploitative white *patrones* (Vickers 1989). Population estimates of the Siona-Secoya in Colombia and Ecuador vary between 400 (Paz y Miño *et al.* 1995) and 1,000 (Vickers and Plowman 1984).

In 1942, ethnobotanist R.E. Schultes acquired a necklace that came from a Siona village in Puerto Ospina, along the Putumayo River, in the department by the same name in Colombia (Figure 8.2; Botanical Museum # 6836). The necklace is about 140 cm in total length (ca. 65 cm long while hanging) and weighs approximately 238 g. It includes three pairs of snail shells, several beetle parts, three small bundles of tobacco, one short segment of bird bone, one small brass bell, and ten bundles of vanilla, the latter outnumbered only by the currently unidentified, cut-out seeds (the necklace will be described in detail separately; Romero *et al.*, in preparation). All ten bundles of vanilla are quite fragrant, indicating that the vanilla fruits from which they were made had been properly cured. The bundles vary in the way they were attached to each other: some join 3 to 7 fruits, folded lengthwise in 3.5 to 4.5 cm long segments; there are also single fruits, much thicker and longer, also folded lengthwise and tied in 7 cm long bundles. Some of the smaller vanilla bundles have been carefully "re-packaged" with strings of different texture and color, suggesting that the owners spent considerable time maintaining the integrity of the necklace.

Figure 8.2 Siona necklace (Harvard University, Botanical Museum 6836) from Puerto Ospina, Putumayo Department, Colombia. The arrow indicates one of the vanilla bundles. Scale bar = 10 cm. Photograph by G.A. Romero-Gonzalez.

The Siona have a common name for vanilla, *Semenquete* (Espada 1904; p. 40, cited in Patiño 2002; p. 538). If the Siona cured vanilla to be used in their artifacts, the practice may have eroded since there is hardly a trace of vanilla in the current, extensive ethnobotanical literature dedicated to the Amerindians of northern South America. Patiño (2002; p. 538) also cites the use of vanilla in Colombia as a perfume together with a necklace, which he says was for the purpose of magical protection.

As far as we have been able to ascertain, the Siona necklace described above is the only surviving artifact that tangibly shows the use of vanilla in a ceremonial context. Perhaps a more expansive search of ethnobotanical collections will reveal the existence of other such objects. Interestingly, the use of vanilla beans in necklaces is also documented from Mesoamerica. In the *Badianus Manuscript* (1552), it states that vanilla was worn in an amulet around the neck as a medicinal charm (Bruman 1948).

The magic-religious use of vanilla beans as pendants may have once been more widespread, but has doubtless decreased significantly due to processes of cultural erosion, and the concentration of rare knowledge among only a handful of remaining individuals that may now be reluctant to transmit it to anyone outside their group ("the custodians of botanical knowledge"; Hemming 2008; p. 18).

8.3 Summary

The origin of vanilla cultivation in Veracruz in the 1760s grew out of a context of European colonialism, when various tropical colonies worldwide served as staging areas for the establishment of plantation economies. First introduced to Europe as a rare luxury in the sixteenth century, vanilla reached its zenith as a plantation crop 300 years later in French colonies in the Indian Ocean (Smith *et al.* 1992). For most of its history, however, vanilla has been utilized in a strictly cultural context: as a flavoring agent, a ritual plant, and as a medicine. The advent of market economies and the need to supply an external entity was what engendered vanilla cultivation, as probably happened during the rule of the Maya and Aztec elite of Late Post-classic Mesoamerica (AD 1350–1500), and certainly with the explosive rise in demand for chocolate in Europe in the colonial period.

Other major commercial crops have had similar historical trajectories to vanilla. For example, coffee reached the status of an important cultivar only during the colonial period, despite a long history of relative low-intensity use (Smith *et al.* 1992). The pre-Columbian use of vanilla was based on discriminate and occasional harvesting from the tropical "healing forest", a use pattern that generally characterizes patterns of exploitation of non-staples and ritual plants (Schultes and Raffauf 1990; DeFilipps *et al.* 2004;). In contrast to *V. planifolia*, which is naturally rare and thrives in semi-isolation from its con-specifics, many of the areas where vanilla was gathered in the colonial period were lowland, swampy, seasonally-inundated habitats that tend to favor *Vanilla* species that live and reproduce in populations. For example, the species that Dampier mentions at "Caibooca" (Uxpanapa Delta, Tabasco, Mexico) were probably either *V. insignis* or *V. odorata* (at Bocas del Toro, Dampier may have observed *V. planifolia* or multiple species; R. Dressler, personal communication). In Peru, Ruiz observed trade in vanilla fruits that could have derived from species less rare than *V. planifolia*, such as *V. bicolor, V. grandiflora*, or *V. palmarum*. A localized, natural abundance of vanilla also provided fruits in Belize (see Gretzinger, Chapter 5, this volume), French Guiana (Aublet 1776/Humboldt 1819), and Honduras/Nicaragua (Offen 2000), and in none of these situations did knowledge of artificial pollination exist.

We can speculate on a number of factors that contributed to the absence of vanilla cultivation historically in South America. In addition to Veracruz being the best-situated locale for export to Europe, South America furthermore lacked a pre-Columbian tradition of drinking cacao-beverages (McNeil 2006). To this day, South American Amerindians prefer to consume the sweet aril surrounding cacao seeds, usually discarding the latter. The lack of reliable cultivars (Díaz 1861; p. 268), and pests and diseases, may also have stymied commercial production of vanilla in South America. There are many fungi that attack vanilla, including *Fusarium batatas* Wollenw. var. *vanillae* Tucker (Childers *et al.* 1959 [referred to as "batatatis"]; Duke 1993).

If vanilla cultivation was absent in pre-Columbian tropical America, vanilla curing was not. In addition to various Mesoamerican techniques for curing (i.e. in Papantla, Oaxaca (Teuitla), and the Maya area [Kourí 2004]), the Amerindians of the Guianas have cured vanilla since at least the mid-seventeenth century, while the Siona-Secoya of Colombia/Ecuador have also developed their own process. A "Guiana method" (Purseglove *et al.* 1981) and a "Peruvian process" (Ridley 1912) have been mentioned in the literature, with both employing the use of oil to help preserve the vanilla fruits as well as tying them up into bundles.

Arguably, the art and science of vanilla curing has yet to reach perfection. One of the major obstacles to the production of quality vanilla fruits today is the widespread harvesting of green fruits that are hardly redolent and not ripe. Early harvesting, which is now almost habituated in some countries, precludes the formation of a full complement of unique aromatic compounds in the curing and aging process (McGee 2004). It is not for nothing that vanilla fruits take nearly 11 months to reach the ripening stage.

Scholarly research will doubtless uncover more documents bearing testament to the history of vanilla use in colonial tropical America, but they are unlikely to diverge thematically from the core historical synthesis we have presented here. The records of vanilla use on hand are remarkably consistent with regard to time and space and, in most cases, for their brevity. Whatever emphasis may be given to the commercial history of vanilla, it is important also to not neglect the original uses for vanilla that were fashioned by various cultures. In this regard, we feel that the medicinal properties of *Vanilla*, which include treating severe headaches (Hernández-López 1988; Cano Asseleih 1997), use as a vermifuge (Lawler 1984; Atran *et al.* 2004), and use to relieve circulatory problems and skin diseases (Heyde 1987; DeFilipps *et al.* 2004), comprise an especially fruitful avenue for future research inquiry.

Acknowledgments

P.L. thanks R. Russell, C. Kelloff, J. Gasco, J. Carney, and R. Bussman for their support and input. G.A.R.-G. acknowledges S.M. Rossi-Wilcox for pointing out the necklace described in the text, and the Orchid Society of Arizona and the Massachusetts Orchid Society for their generous support, and S.Z. gives thanks to her co-workers at the Harvard University Herbaria.

References

Ackerman, J.D. (2002) *Flora of North America (Volume 26).* Oxford University Press, Oxford.

Arbell, M. (1995) The Jewish settlement in Pomeroon/Pauroma (Guyana), 1657–1666. *Revue des etude Juives,* CLIV, 343–361.

Arditti, J., Nagaraja Rao, A. and Nair, H. (2009) Hand-pollination of *Vanilla:* how many discoverers? In: *Orchid Biology: Reviews and Perspectives, X,* Kull, T. Arditti, J. and Wong, S.M. (Eds), Springer, Berlin, pp. 233–249.

Atran, S., Lois, X. and Ucan Ek' E. (2004) *Plants of the Petén Itza' Maya = Plantas de los Maya Itza' del Petén*. Museum of Anthropology, University of Michigan, Ann Arbor.

Aublet, M.F. (1775) Quatrieme Mémoire. Observations sur la nature de la Vanille, la maniere de la cultiver, & les moyens de la préparer pour la rendre commerçable, Mémoires sur divers objets intéressans. In: *Histoire des Plantes de la Guiane Française II*. Suppl. Pierre-François Didot jeune, Paris, pp. 77–94.

Bergmann, J.F. (1969) The distribution of cacao cultivation in pre-Columbian America. *Annals of the Association of American Geographers*, 591, 85–96.

Bory, S., Lubinsky, P., Risterucci, A.-M. et al. (2008) Patterns of introduction and diversification of *Vanilla planifolia* (Orchidaceae) in Reunion Island (Indian Ocean). *American Journal of Botany*, 95, 805–815.

Boxer, C.R. (1969) *The Portuguese Seaborne Empire, 1415–1825*. Alfred Knopf, New York.

Brockway, L.H. (2002) *Science and Colonial Expansion: the Role of the British Royal Botanic Gardens*. Yale University Press, New Haven.

Bruman, H. (1948) The culture history of Mexican vanilla. *Hispanic American historical review*, 28, 360–376.

Cameron, K.M. and Soto Arenas, M. -A. (2003) Vanilloideae. In: *Genera Orchidacearum, volume 3: Orchidoideae (Part 2), Vanilloideae*, Pridgeon, A.M., Cribb, P.J., Chase, M.W. and Rasmussen, F (Eds), Oxford University Press, Oxford, pp. 281–334.

Cano Asseleih, L.M. (1997) *Flora medicinal de Veracruz: inventario etnobotánico*. Universidad Veracruzana, Xalapa, Veracruz, Mexico.

Caso Barrera, L. and Fernández, M.A. (2006) Cacao, vanilla and annatto: three production and exchange systems in the southern Maya lowlands, XVI–XVII centuries. *Journal of Latin American Geography*, 52, 29–52.

Caulín, A. (1779) *Historia corografica, natural y Evangelica de la Nueva Andalucia [Provincias de Cumana, Nueva Barcelona, Guayana y Vertientes del Rio Orinoco]*. Jose Ramos, Madrid.

Chabrán, R. and Varey, S. (2000) Hernández in the Netherlands and England. In: *Searching for the Secrets of Nature: the Life and Works of Dr Francisco Hernández*, Varey, S., Chabrán, R. and Weiner, D. B. (eds), Stanford University Press, Stanford, CA, pp. 138–150.

Childers, N.F., Cibes, H.R. and Hernández-Medina. E. (1959) Vanilla- the orchid of commerce. In: *The orchids: a Scientific Survey*, Withner, C. L. (ed.), Ronald Press Company, New York, pp. 477–510.

Coe, S.D. and Coe, M.D. (2007) *The True history of Chocolate*. Thames and Hudson, London.

Cooke, I.K.S. (1992) Whiteknights and the Marquis of Blandford. *Garden History*, 20(1), 28–44.

Correll, D.S. (1953) Vanilla- its botany, history, cultivation and economic import. *Economic Botany*, 7, 291–358.

Crown, P.L. and Hurst, W.J. (2009) Evidence of cacao use in the Prehispanic American Southwest. *Proceedings of the National Academy of Sciences*, 106, 2110–2113.

Dampier, W. (1776) *The Voyages and Adventures of Capt. William Dampier. Wherein are Described the Inhabitants, Manners, Customs, … &c. of Asia, Africa, and America*. London.

Davis, D.B. (2006) *Inhuman Bondage: the Rise and Fall of Slavery in the New World*. Oxford University Press, New York.

DeFilipps, R.A., Maina, S.L. and Crepin, J. (2004) *Medicinal Plants of the Guianas (Guyana, Surinam, French Guiana).* Department of Botany, National Museum of Natural History, Smithsonian Institution, Washington, DC.

de la Condamine, C.-A. (1745) *Abridged Narrative of Travels Through the Interior of South America: from the Shores of the Pacific Ocean to the Coasts of Brazil and Guyana, Descending the River of Amazons.* Academy of Sciences, Paris.

de la Cruz, M. (1940) *The Badianus Manuscript, Codex Barberini, Latin 241, Vatican Library; an Aztec herbal of 1552.* Johns Hopkins Press, Baltimore.

Díaz, J.A. (1861) *El agricultor Venezolano, o lecciones de agricultura práctica nacional, I–II.* Imprenta Nacional, Caracas, Venezuela.

Duke, J.A. (1993) *CRC Handbook of Alternative Cash Crops.* CRC Press, Boca Raton, FL.

Dunn, R.S. (2000) *Sugar and Slaves: the Rise of the Planter Class in the English West Indies, 1624–1713.* University of North Carolina Press, Chapel Hill.

Durán, D. (1994) *The History of the Indies of New Spain: Translated, Annotated, and with an Introduction by Doris Hayden.* University of Oklahoma Press, Norman, OK.

Ecott, T. (2004) *Vanilla: Travels in Search of the Luscious Substance.* Penguin, London.

Étienne, F.-G. (1743) *Traité de la matière medicale... traité des végétaux. Section I. Des médicamens exotiques.* Paris.

Etkin, N. (2006) *Edible Medicines: an Ethno-pharmacology of Food.* University of Arizona Press, Tucson, AZ.

Ferry, R.J. (1981) Encomienda, African slavery, and agriculture in seventeenth-century Caracas. *Hispanic American historical Review,* 41, 609–635.

Fontecilla, A. (1861) *Breve tratado sobre el cultivo y beneficio de la vainilla.* Mexico City, DF.

Fortune, S.A. (1984) *Merchants and Jews: the Struggle for British West Indian Commerce, 1650–1750.* Latin American Monographs – Second Series, Center for Latin American Studies Book. University of Florida Press, Gainesville, FL.

Funk, V., Hollowell, T., Berry, P., Kelloff, C. and Alexander, S.N. (2007) Checklist of the plants of the Guiana Shield (Venezuela: Amazonas, Bolivar, Delta Amacuro; Guyana, Surinam, French Guyana). *Contributions from the United States National Herbarium,* 55, 1–584. Department of Botany, National Museum of Natural History, Washington, DC.

Garay, L.A. and Sweet, H.R. (1974) *Flora of the Lesser Antilles (Orchidaceae).* Arnold Arboretum, Harvard University, Jamaica Plain.

Gilij, F.S. (1780–1782, 1784) *Saggio di storia Americana I–IV.* L. Perego erede Salvioni, Rome.

Gilij, F.S. (1965) *Ensayo de historia Americana, I–III.* Fuentes para la Historia Colonial de Venezuela. Biblioteca de la Academia Nacional de la Historia, Caracas, pp. 71–73.

Goody, J. (1982) *Cooking, Cuisine, and Class: a Study in Comparative Sociology.* Cambridge University Press, New York.

Gumilla, J. (1745) *El Orinoco ilustrado, y defendido, historia natural, civil, y geografica de este gran Rio, y de sus Caudalosas Vertientes, I–II* (Segunda Impression, Revista, y Aumentada por su mismo Autor,y dividida en dos Partes). Manuel Fernandez, Madrid.

Gumilla, J. (1993) *El Orinoco ilustrado y defendido.* Fuentes para la Historia Colonial de Venezuela, Biblioteca de la Academia Nacional de la Historia, Caracas, Venezuela, p. 68.

Hágsater, E., Soto-Arenas, M.A., Salazar Chávez, G.A., Jiménez Machorro, R.L. López Rosas, M.A. and Dressler, R.L. (2005) *Orchids of Mexico*. Productos Farmacéuticos, Mexico City, Mexico.

Hemming, J. (2008) *Tree of Rivers: the Story of the Amazon*. Thames & Hudson, London.

Henderson, J.S., Joyce, R.A., Hall, G.R., Hurst, W.J. and McGovern, P.E. (2007) Chemical and archaeological evidence for the earliest cacao beverages. *Proceedings of the National Academy of Sciences*, 104(48), **18**, 937–18, 940.

Hernández-López, J.A. (1988) *Estudio sobre la herbolaria y medicina tradicional del municipio de Misantla, Vercruz*. Bachelor's thesis, Faculty of Sciences, UNAM, Mexico.

Heyde, H. (1987) *Surinaamse medicijnplanten* (2nd Edn). Westfort, Paramaribo, Surinam.

Humboldt, A. von. (1814, 1819, 1825) *Voyage de Humboldt et Bonpland, Première Partie, Relation Historique, I–III*. Dufour et Compie [I], N. Maze [II] and J. Smith [III], Paris.

Humboldt, A. von. (1851) *Personal narrative of travels to the equinoctial regions of America during the years 1799–1804, I–III*. Translated and edited by T. Ross. George Routledge & Sons Limited, London.

Humboldt, A. von. (1956) *Viaje a las Regiones equinocciales del Nuevo Continente, I–V*. Ediciones del Ministerio de Educación, Dirección de Cultura y Bellas Arrtes, Caracas, Venezuela.

Hurst, W.J., Tarka, S.S., Powis, T.G., Valdez, F. and Hester, T.R. (2002) Cacao usage by the earliest Maya civilization. *Nature*, 418, 289.

Hussey, R.D. (1934) *The Caracas Company, 1728–1784: a Study in the History of Spanish Monopolistic Trade*. Harvard University Press, Cambridge, MA.

Knight, F.W. and Liss, P.K. (Eds) (1991) *Atlantic port cities: economy, culture, and society in the Atlantic world, 1650–1850*. University of Tennessee Press, Knoxville, TN.

Kourí, E. (2004) *A Pueblo Divided: Business, Property and Community in Papantla, Mexico*. Stanford University Press, Stanford, CA.

Labat, J.-B. (1772) *Nouveau voyage au Isles de l'Amerique*. Paris.

Lawler, L.J. (1984) Ethnobotany of the Orchidaceae. In: *Orchid Biology: Reviews and Perspectives, III*, Arditti, J. (Ed.), Cornell University Press, Ithaca, NY, pp. 27–150.

López Piñero, J.M. and Pardo Tómas, J. (2000) The contribution of Hernández to European botany and materia medica. In: *Searching for the Secrets of Nature: the Life and Works of Dr Francisco Hernández*, Varey, S., Chabrán. R. and Weiner. D.B. (Eds), Stanford University Press, Stanford, CA, pp. 122–137.

Lubinsky, P. (2007) *Historical and evolutionary origins of cultivated vanilla*. PhD dissertation, University of California, Riverside.

Lubinsky, P., Bory, S., Hernández Hernández, J., Kim, S.-C. and Gómez-Pompa, A. (2008a) Origins and dispersal of cultivated vanilla (*Vanilla planifolia* Jacks. [Orchidaceae]). *Economic Botany*, 62, 127–138.

Lubinsky, P., Cameron, K.M., Carmen Molina, M., et al. (2008b) Neotropical roots of a Polynesian spice: the hybrid origin of Tahitian vanilla, *Vanilla tahitensis* (Orchidaceae). *American Journal of Botany*, 95(8), 1040–1047.

Mabberley, D.J. (2008) *Mabberley's Plant Book: a Portable Dictionary of Plants, their Classification and Ises* (3ed Edn), Cambridge University Press, Cambridge, UK.

MacLeod, M.J. (2008) *Spanish Central America: a Socioeconomic History, 1520–1720*. University of Texas Press, Austin, TX.

Macleod, R (Ed.) (2000) Nature and empire: science and the colonial enterprise. In: *Osiris* 15 (second series), University of Chicago Press, Chicago, IL.

Martínez Ribón, C. (1895) *Nuevo método para el cultivo del cacao, adicionado con un memorandum sobre los cultivos de la vainilla y el cacao, I–III.* Anales de la Junta Central de Aclimatación y Perfeccionamiento Industrial, Caracas, Venezuela.

Matteson Langdon, E.J. (1992) Dau: shamanic power in Siona religion and medicine. In: *Portals of Power: Shamanism in South America*, Matteson Langdon, E.J. and Baer, G. (Eds), University of New Mexico Press, Albuquerque, pp. 41–62.

McGee, H. (2004) *On food and Cooking: the Science and Lore of the Kitchen.* Scriber, New York.

McNeil, C.L. (Ed.) (2006) *Chocolate in Mesoamerica: a Cultural History of Cacao.* University Press of Florida, Gainesville, FL.

Miller, P. (1954) *The Gardener's Dictionary*, 4th Edition. John and James Rivington, London.

Mintz, S.W. (1985) *Sweetness and Power: the Place of Sugar in Modern History*, Viking Penguin, New York.

Morren, C. (1839) On the production of vanilla in Europe. *Annals of Natural History or Magazine of Zoology, Botany and Geology*, 3, pp. 1–9.

Nielsen, L.R. (2000) Natural hybridization between *Vanilla claviculata* (W. Wright) Sw. and *V. barbellata* Rchb.f. (Orchidaceae): genetic, morphological, and pollination experimental data. *Botanical Journal of the Linnaean Society*, 133, 285–302.

Offen, K. (2000) British logwood extraction from the Mosquitia: the origin of a myth. *Hispanic American Historical Review*, 80(1), 113–135.

Patiño, V.M. (2002) *Historia de la dispersión de los frutales nativos del Neotrópico.* Centro Internacional de Agricultura Tropical (CIAT), Cali–Colombia, Chapter 22, pp. 538–54.

Pauly, P.J. (2007) *Fruits and Plains: the Horticultural Transformation of America.* Harvard University Press, Cambridge, MA.

Paz y Miño C., Balslev, G.H. and Valencia, R. (1995) Useful lianas of the Siona-Secoya Indians from Amazonian Ecuador. *Economic Botany*, 49, 269–275.

Piñero, E. (1988) The cacao economy of the eighteenth-century province of Caracas and the Spanish cacao market. *Hispanic American Historical Review*, 68, 75–100.

Plumier, C. (1755–1760) *Plantarum Americanarum.* Viduam & Filium S. Schouten, Batavia.

Portères, R. (1954) Le genere *Vanilla* et ses espèces. In: *Le vanillier et la vanilla dans le monde*, Bouriquet, G. (ed.), Éditions Paul Lechevalier, Paris, pp. 94–290.

Presilla, M.E. (2001) *The New Taste of Chocolate: a Cultural and Natural History of Cacao with Recipes.* Ten Speed Press, Berkeley, CA.

Price, R. (1976) *The Guiana Maroons: a Historical and Bibliographical iIntroduction.* Johns Hopkins University Press, Baltimore.

Purseglove, J.W., Brown, E.G., Green, C.L. and Robbins, S.R.J. (1981) *Spices: Volume 2.* Longman, New York.

Rain, P. (2004) *Vanilla: the Cultural History of the World's Favorite Flavor and Fragrance.* Penguin, New York.

Ramírez, S.R., Gravendeel, B., Singer, R.B., Marshall, C.R. and Pierce, N.E. (2007) Dating the origin of the Orchidaceae from a fossil orchid with its pollinator. *Nature*, 448, 1042–1045.

Rich, E.E. and Wilson, C.H. (Eds) (1967) *The Economy of Expanding Europe in the Sixteenth and Seventeenth Centuries: Volume 4, the Cambridge Economic History of Europe.* Cambridge University Press, Cambridge.

Ridley, H.N. (1912) *Spices*. Macmillan, London.

Ríos, J. (1999) *Los libros del Hacendado Venezolano*. Colección V. Centenario del Encuentro entre dos Mundos, 1492–1992, No 10. Banco Central de Venezuela, Caracas.

Rivero, J. (1888) *Historia de las Misiones de los Llanos de Casanare y los Ríos Orinoco y Meta Escrita el año de 1736*. Silvestre y Compañía, Bogotá, Colombia.

Romero-González, G.A. (1998) De viaje por las orquídeas. *Imagen (Caracas)*, 31(2): 143–149.

Rossignon, J. (1850) *Manual del cultivo del café, cacao, vainilla y tabaco en la América Española y todas sus aplicaciones*. Enciclopedia Popular Mejicana, Libreria de Rosa y Bouret, Paris.

Rossignon, J. (1881) *Manual del cultivo del café, cacao, vainilla, y tabaco en la América Española* (3rd Edn). Librería de la Vda de Ch. Bouret, Paris.

Rossignon, J. (1929) *Nuevo manual del cultivo del café, cacao, vainilla, caucho, el arbol de la quina, el tabaco y el té*. Librería de la Vda de Ch. Bouret, Paris.

Ruiz, H. (1998) *The Journals of Hipólito Ruiz: Spanish Botanist in Peru and Chile, 1777–1788*. Translated by Schultes R.E. and von Thenen de Jaramillo-Arango, M.J.B. Timber Press, Portland, OR.

Sahagún, B. de. (1963) *General istory of the things of New Spain, Book 11 – Earthly things*. School of American Research and the University of Utah, Santa Fe, New Mexico.

Salazar, S. (2004) Cacao y riqueza en la provincia de Caracas en los siglos XVII y XVIII. *Tierra Firme*, 22(87), 293–312.

Sauer, J.D. (1993) *Historical Geography of Crop Plants: a Select Roster*. CRC Press, Boca Raton.

Schiebinger, L. (2004) *Plants and Empire: Colonial Bio-prospecting in the Atlantic World*. Harvard University Press, Cambridge, MA.

Schiebinger, L. and Swan, C. (Eds) (2005) *Colonial Botany: Science, Commerce, and Politics in the Early Modern World*. University of Pennsylvania Press, Philadelphia.

Shrebero, J.C.D. (1787) *Materia Medica*. Wolfgangum Waltherum, Lipsiae et Erlangae.

Schultes, R.E. and Raffauf, R.F. (1990) *The Healing Forest: Medicinal and Toxic Plants of the Northwest Amazonia*. Dioscorides Press, Portland, OR.

Smith, N.J.H., Williams, J.T., Plucknett, D.L. and Talbot, J.P. (1992) *Tropical Forests and their Crops*. Comstock, Cornell University Press, Ithaca.

Spary, E.C. (2005) Of nutmegs and botanists: the colonial cultivation of botanical identity. In: *Colonial Botany: Science, Commerce, and Politics in the Early Modern World*, Schiebinger, L. and Swan, C. (Eds), University of Pennsylvania Press, Philadelphia, pp. 187–203.

Spence, J.M. (1878) *The Land of Bolívar I-II*. Sampson Low, Marston, Searle & Rivington, London.

Thompson, J.E.S. (1958) *Thomas Gage's Travels in the New World*. University of Oklahoma Press, Norman, OK.

Tinker, H. (1974) *A New System of Slavery: the Export of Indian Labour Overseas, 1830–1920*. Oxford University Press, London.

Varey, S. (Ed.) (2000) *The Mexican Treasury: the Writings of Dr Francisco Hernández*. Translated by Chabrán, R., Chamberlain. C.L. and Varey, S., Stanford University Press, Stanford, CA.

Vickers, W.T. (1989) *Los Sionas y Secoyas, su adaptación al ambiente amazónico*. Ediciones ABYA-YALA, Quito, Ecuador.

Vickers, W.T. and Plowman, T. (1984) Useful plants of the Siona and Secoya Indians of eastern Ecuador. *Fieldiana*, 15, 1–63 (new series). Field Museum of Natural History, Chicago, IL.

Weiss, E.A. (2002) *Spice Crops*. CABI Publishing, New York.

9

Vanilla Production in Australia

Richard Exley

9.1 Introduction

Vanilla has a somewhat checkered association with Australia. Attempts to establish viable vanilla cultivation have occurred periodically over the past 130 years, yet commercial success has remained elusive. Today, a focus on premium quality and vertical integration from farm to market is providing hope that a niche industry may be successfully established.

9.2 History

It is hard to say when vanilla was first introduced to Australia. It is recorded as "growing in the Brisbane Botanic Gardens and at Bowen Park in 1866" and was distributed to northern localities in 1866, 1872, 1874, and 1885 (Bailey 1910). The origin of these first introductions was not recorded.

These first introductions seemed to have little results. It was not until 1901, when Howard Newport imported a crate of cuttings from Fiji, that any serious attempt to establish the commercial cultivation of vanilla was made (Newport 1916). It seemed that Northern Queensland was ideally suited to vanilla. At that time the demonstration and research plots at Kamerunga State Park, Cairns, clearly verified "that North Queensland possesses conditions that not only are eminently suitable, but in which this orchid will, and does, when properly treated, thrive and grow in a manner found in but few other parts of the world" (Newport 1916). In this connection, Ridley (1912) also pointed out that Queensland evidently has special advantages in the nature of its scrub lands for vanilla:

> "In the ordinary tropical forest the trees are of all sizes and so irregular in growth that it would be very troublesome to clear the undergrowth so that the trees could be connected by trellises or poles in a convenient way. So readily adapted woodland as the Australian bush appears is rarely found" (Ridley 1912).

Handbook of Vanilla Science and Technology, Second Edition.
Edited by Daphna Havkin-Frenkel and Faith C. Belanger.
© 2019 John Wiley & Sons Ltd. Published 2019 by John Wiley & Sons Ltd.

Sadly even Howard Newport's attempts to encourage commercial cultivation seemingly amounted to little and vanilla as a commercial crop in Australia fell into obscurity once again.

For many years it remained forgotten until Northern Territory Government/ Department of Primary Industry officers in the 1980s decided to informally conduct some field trials with vanilla. Unfortunately no records are available of the work on these trials, which were carried out at the Government's Coastal Plains Research Station. Again, although the anecdotal evidence suggested vanilla could be successfully grown and had considerable commercial merit, no real progress towards establishing a commercial industry was made. The failure of vanilla to capture the imagination of local farmers and horticulturists may be explained by the prevailing attitude of most Australian farmers and growers to vanilla, that it is a crop with a high labor requirement, requiring low cost labor, making it only suitable for the third world. This attitude is still widely prevalent today.

Little happened for another 20 years until around 2003, when vanilla again was explored as a niche crop with research taking place in Darwin (Northern Territory) and Cairns (far north Queensland). All of these new developments were focused on the production of premium products in a vertically integrated operation – growing, curing, packaging, and branding for retail/end user as an essential pre-requisite to capture the full product value in northern Australia. Capturing the full value of premium products is necessary to make vanilla cultivation and processing a viable, attractive, and sustainable industry in Australia. At least two small family operations and a larger commercial plantation currently exist around Cairns in far north Queensland. These plantations presently produce very limited quantities, marketed directly to local end users. The larger commercial operation is also an importer and distributor of quality vanilla beans from around the world and a wholesale nursery supplying *Vanilla planifolia* plant stock to the horticultural industry, nursery trade, and the public. No doubt there are others exploring the potential of vanilla in both Queensland and the Top End of the Northern Territory. In Darwin, work continues on the research, development, and refinement of a vanilla plantation system for the production and processing of premium vanilla beans, suited to Darwin's tropical climate.

9.3 Species

The species grown exclusively in the emerging commercial vanilla industry in Australia is *V. planifolia*, although some *V. tahitensis* was also grown in trials at the Northern Territory Coastal Plains Research Station. Both species are grown by hobbyists and orchid and tropical plant enthusiasts throughout northern Australia.

9.4 Climatic Regions of Australia Suitable for Vanilla

Vanilla is only commercially grown 20 degrees north or south of the equator within the hot humid tropics. Much of Australia south of 20 degrees is arid or semi-arid savannah, making it unsuitable for the cultivation of vanilla.

The specific conditions that *V. planifolia* prefers are:

- shade/filtered light around 50%;
- daytime maximum temperatures between 21 and 29°C;
- night-time minimum temperature of about 16°C;
- 65 to 85% relative humidity; and
- soil medium that is moist (but not soggy), well drained, and rich in mulch.

There are two distinct regions, which are climatically suitable for the cultivation of vanilla (Figure 9.1). The first is the Darwin region and the Top End of the Northern Territory, extending as far south as possibly Katherine and to the east Arnhem Land, up to 200 km from the coast, providing suitable ground water is available for irrigation. The other is a coastal strip of far north Queensland beginning from Mackay, stretching north to the tip of Cape York. The strip spreading from the coast inland to the ranges includes regional centers such as Cairns and Townsville.

9.5 Climatic Conditions in the Vanilla Growing Regions

Both the far north Queensland and Top End regions of Australia have a tropical climate, with hot humid summers and mild dry winters (http://www.bom.gov.au/weather/qld/cairns/climate.shtml; http://www.darwin.nt.gov.au/aboutdarwin/darwins_profile/climate.htm). Rains occur mainly during the summer months between December and March. The average annual rainfall is approximately 1,700–2,000 mm. The temperatures in these areas are typically uniform throughout the year with daytime temperature ranges of 23 to 33°C in summer and 18 to 30°C in winter.

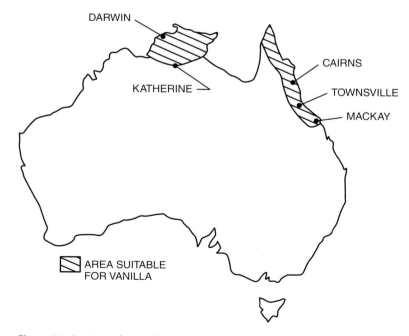

Figure 9.1 Regions of Australia suitable for cultivation of vanilla.

9.6 Soil and Nutrients

Growth of vanilla in a shade house system, with the use of raised mulch beds and timber posts and trellis supports, allows suitable conditions to be established and managed more readily. In Darwin, a typically two layer system of preparing the raised mulch bed for planting is used. The 2 to 3 inches thick (50–75 mm) bottom layer consists of 50% rich organic compost and 50% coarse hardwood wood chips (e.g. red gum or similar). The upper layer is either red gum wood chips or a blend consisting of a minimum of 80% red gum wood chips with up to 20% charcoal. This mixture provides the balance between moisture retention, aeration, and drainage in which the vines thrive.

9.7 Watering

As in the wild, *V. planifolia* is commercially grown in areas with a high annual rainfall (1,800–2,500 mm) spread throughout most of the year. It is important that the plants be kept moist. Vanilla prefers a well-drained growing medium, and does not like to be constantly wet. Drying out completely should be avoided. Vanilla requires a dry season of a minimum of 2 months to encourage flowering in the coming summer. In Darwin this period occurs June to August. The extended dry season in Darwin necessitates the need for irrigation, which also aids in maintaining appropriate humidity levels.

9.8 Fertilizing

V. planifolia benefits from foliar feeding. A weak fish emulsion based liquid fertilizer added to the water spray (irrigation system) is suitable. The weak (a quarter of the recommended strength) fish emulsion based liquid fertilizer provides a complete food for the vines and seems to work best if applied late in the day, ideally just after dusk.

Vanilla appears to have few insect pests in northern Australia, making it naturally suited to organic/bio-dynamic cultivation. The little damage very occasionally done by insects can be easily tolerated for the operation and business benefits that organic certified production may offer. To this end vanilla may be seen as friendly to native species of fauna due to its production potentially being chemical free. Its relatively small foot print means growers may be able to retain or replant native forests or woodlands, benefiting local wildlife and the climate.

9.9 Propagation

The propagation of all vanilla in Australia is predominantly done by cuttings from healthy vines. With limited stock available, consideration has been given to importing plant material (cuttings) from the Indian Ocean. This is a time consuming and costly exercise due to Australia's strict quarantine requirements. Import conditions require all cuttings to be fumigated and planted out in an approved quarantine facility at the importers expense for a period of 3 months.

9.10 Support

The most preferred method is the use of wooden posts 3 m in height to which the vine is tied up for the ease of access for pollination and harvesting (Figure 9.2). The vine is then allowed to grow straight upwards, emulating what it would do growing on a tree in the wild. When the vine has reached the top of the post, the upper half is carefully peeled off and draped over the round cross support set at a height of 1.2 to 1.5 m maximum. The vine is then looped to the mulch bed and allowed to climb to the top again so that the process is repeated. Some success was achieved with the use of an artificial post made of a drainage pipe covered with a non-woven geo-textile (Figure 9.3). These 3-m long pipes are attached to a tensioned steel cable in a way that their base can be buried in the raised mulch bed.

In Darwin, coco peat filled wire mesh tubes have also been tested as an alternative support to timber posts (Figure 9.4). While these proved very suitable for vine growth and support, they made training more difficult. Removing the portion of the vine to be looped without damaging roots proved difficult and increased the likelihood of damage to the vine if extra care was not taken in peeling it off the upper portion of the tube. Also, the mesh tubes (posts) did not have cross supports over which the vine could be draped, requiring tie off or spike to support the loop.

The benefits of using the wooden or artificial post systems in a shade house are:

- an increased yield per hectare through increased plant density;
- improved control over growing conditions, including the efficient application of irrigation and fertilization regimes; and
- greater efficiency in the operation of the plantation as the frequency of which the plants must be trained is reduced by the taller posts.

Another approach that may hold promise for implementation on small (hobby) mango farms of 5 to 5 acres is using established mango trees as supports. This has the

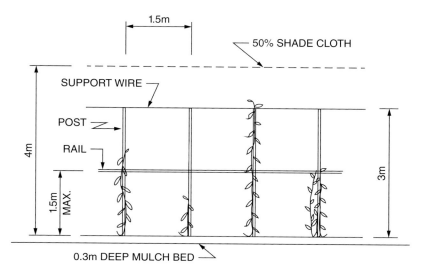

Figure 9.2 Diagram of wooden post method for growing vanilla.

Figure 9.3 Vanilla plant growing on a post made of drainage pipe.

advantage of diversifying and improving the sustainability of these small farms, which are at risk of being pushed out by larger corporate operations. However, the use of mango trees may present a two-edged sword, as mango trees in the Darwin region are at risk of the soil-borne fungus *Fusarium*, which has caused Mango Malformation Disease. This may be a serious threat to vanilla in the Top End, but good management and plantation design should eliminate this risk.

9.11   Light/Shade

Vanilla, like many orchids, is a shade loving plant. It prefers 50 to 60% shade and good to strong indirect light. While shade is necessary it should not be too overbearing, and an eastern exposure is preferable.

Figure 9.4 Vanilla plant growing on a coco peat filled wire mesh tube.

9.12 Spacing

Grown in its traditional woodland setting, the spacing of the vines is dictated by the spacing suitable for the support and shade trees. Typical spacing for the supporting trees is usually 2.5 to 3 m between trees and between rows. If vanilla cultivation is to be developed in an established mango plantation, spacing will be dictated by the existing trees. In a mature plantation this will be around 8 × 8 to 10 m. However, in the shade house environment, using timber posts for support, the plant spacing can be reduced. The inter plant spacing can be reduced to as little as 1 m, but 1.5 m is preferred, to allow the easy movement of workers among the crop. It is suggested that the row spacing be adequate to allow the mechanization of some plantation functions where possible; a minimum of 1.8 m, and preferably 2 m to accommodate a small vineyard style tractor or quad bike to assist in manual activities. This offers the best balance between increased yields and reducing the spacing to the point where it is a potential problem should a disease, mould, or pest occur in the plantation.

9.13 Training

Left to grow on their own, the vines will climb straight up the support to which they are attached, such as a tree. However to make the plant more accessible for workers to pollinate and harvest the beans, it should be trained to a height of not more than 1,500 mm. As the vine can grow up to 15 m in length, keeping it compact and accessible is achieved by looping the vine, which also aids in encouraging flowering. This is done by allowing the vine to grow up the support to twice the height of the desired level. Then carefully detaching the upper half of the vine and draping it over the cross support to the ground, with two nodes in contact with the mulch/humus and allowing it again to grow up the support. In doing this, the vine should be spaced around the support evenly.

9.14 Flowering, Fruit Set, Growth, and Maturation

9.14.1 Flowering

Flowering occurs once a year over a 2 to 3 month period. In Northern Australia (Darwin/ Cairns), flowering occurs in the period October to December at the beginning of the wet season, during the time locally known as the "build up".

9.14.2 Fruit Set (Pollination)

The hand pollination method perfected on the Ile de la Reunion by Edmond Albius in 1841 is today still the only reliable method used to pollinate the flowers. This is almost certainly the reason for vanilla's reputation as a high labor crop in Australia. This relatively simple but critical task must be carried out each morning during the flowering season. The flowers only remain viable for one day and the most reliable time to pollinate them is in the morning.

The description of the process for pollinating the flowers is:

- The lip is lowered and torn with a toothpick (or bamboo stick) to expose the column and anther.
- The rostellum is lifted up with a toothpick (or bamboo stick) and placed under the stamin.
- Gentle finger pressure is applied to bring the pollen and stigma into contact.

Care must be taken not to over pollinate the flowers on any given vine, as this can place excessive stress on the vine, ultimately resulting in a poor crop and failure of the vine. A rule of thumb in the industry is to pollinate around half the flowers and the practice is followed in most regions by pollinating only the flowers on the lower side of the raceme. This also allows the beans to hang straight down, aiding in the production of straight beans, an essential trait in the premium market.

9.14.3 Growth and Maturation

The growth of the beans occurs over the next 8 to 12 weeks after pollination. At the end of the flowering season, in late December to mid-January, small, crooked, and damaged beans are removed. The long process of maturation or ripening then begins, with the beans reaching maturity from late June through August.

9.15 Harvesting

Harvesting the beans is carried out 7 to 9 months after flowering and pollination when the head (or tip) of the bean (the free end) starts to turn yellow and the rest of the bean takes on a less green (paler) color, changing towards yellow. A delay in harvesting at this time will lead to a high number of split beans, which lowers their value considerably. Bringing it forward and picking too early leads to a poor quality bean in terms of the development of its aroma and flavor characteristics, particularly the vanillin content. This is due to insufficient glucovanillin development, the precursor to vanillin, which is released during curing. As a result, it is necessary to harvest the beans progressively by hand, as they ripen to achieve the best quality product.

9.16 Curing

9.16.1 Overview

The key to quality premium vanilla beans is the curing. It is during curing that the enzymatic process that converts the glucovanillin into vanillin occurs (this process is described in detail in the chapter 6 by Frenkel *et al.*). The appearance of a crust of fine white crystals on the cured vanilla beans is considered by many to be a sign of high vanillin content. The crust is in fact crystallized vanillin. The primary determinant of quality for cured vanilla beans is the aroma/flavor character. Other quality factors include general appearance, flexibility, length, and vanillin content. Traditionally, the visual indicators of a blemish free appearance, flexibility, and size characteristics have been important because there is a close correlation between these factors and the aroma-flavor quality.

There are two principle methods for curing vanilla, the Bourbon method and the Mexican method (see chapter by Frenkel *et al.*). Others exist, but they are hybrids or variations of the two systems. Australian growers presently predominantly use a variation of the Bourbon method adapted to the local conditions.

The final step is to package the beans for sale. Traditionally, 10 to 12 kg of vanilla bundles are packed into tin boxes lined with paper. However, Australian growers are focused on packaging and marketing direct to retail. Direct sales to the consumer market for home/gourmet cooking and baking requires packaging of 1 to 3 beans for this premium segment. The preferred packaging is vacuum-sealed plastic satchels, and glass or aluminium tubes.

References

Bailey, J.F. (1910) Introduction of economic plants into Queensland. Presidential Address – Royal Society of Queensland, February 26.

Newport, H. (1916) *Vanilla Culture for Tropical Queensland, 2nd Edn*, Anthony James Cumming, Government Printer – Brisbane.

Ridley, H.N. (1912) *Spices*. Macmillan and Co. Ltd, London.

10

Vanilla in Dutch Greenhouses:
A Discovery – From Research to Production

Filip van Noort

10.1   Introduction

Vanilla is one of the most well-known spices in the world. It is an orchid that grows as a vine, climbing up an existing tree, pole, or other form of support. It is mostly grown in the open, in (semi-)tropical countries, but is also grown in shaded houses with some technical support to increase vanilla cultivation.

   In the Netherlands many (semi-)tropical crops are grown under glass and, after many years of research and growing of these crops, it is now possible to achieve higher production than ever before. These higher levels have been achieved by being in control of as much (growing) parameters as possible, such as light, temperature, watering, fertilization, CO_2, etc.

10.1.1   Start of Research

In 2012 some in The Netherlands were thinking about "new" crops for greenhouses, and among others Vanilla seemed to be a promising, and also a challenging crop to consider. The research started by reading some "basic" vanilla-information in order to try to understand this plant, and getting some cuttings to begin growing vanilla on a small scale.

   From the literature we concluded that vanilla would be a challenging crop because it could take a lot of time for the plants to flower. There would be only a few hours for pollination and all the pollination would need to be done manually (labor is expensive in Holland). After pollination, there would be a 6–9 month wait for pod development and at the end one had to find a way to cure those pods. Additionally, there were questions about the growing system, fertilization, substrate, climate etc.

10.2   Review of Literature

Vanilla is an orchid, but unlike many other orchids, it is easy to grow from cuttings. Throughout the world three cultivars of Vanilla are grown for their pods: *V. planifolia*, *V. tahitensis* and *V. pompona* (Cameron 2011).

Handbook of Vanilla Science and Technology, Second Edition.
Edited by Daphna Havkin-Frenkel and Faith C. Belanger.
© 2019 John Wiley & Sons Ltd. Published 2019 by John Wiley & Sons Ltd.

Figure 10.1 Flowering Vanilla 2014.

Vanilla grows with temperatures at daytime between 20°–30°C and between 15°–25°C at night (Odoux and Grisoni 2010; Havkin-Frenkel and Belanger 2011). Vanilla uses the crassulacean acid metabolism (CAM) carbon fixation pathway. In CAM plants, the stomata are closed during the day, which helps the plants survive water stress. This influences the plant's requirements for light, temperature and water. Vanilla often grows under sun protection provided either by trees or other shading material. However, there is almost no information about the actual light levels or daily light requirement in relation to shading.

Vanilla produces adventitious aerial roots that are used for climbing, but that also play a role in survival by taking up water and nutrients. These roots need high humidity and water from above for good growth of the roots and therefore for the plants. In nature the plants grow as vines on trees and like a lot of other orchids, prefer tempered, diffuse light.

Flowering (Fig. 10.1) can be stimulated by climate and/or mechanical stress (looping and pinching). Different methods are used in different countries:

- In Uganda, Vanilla flowers twice, and light, temperature, and watering are considered to influence flowering.
- India uses a combination of methods to induce flowering – plants are pinched, watering is withheld, and more light is allowed by pruning the guide trees.
- In Mexico the plant tops will die in a cold wind and that stimulates flowering (Odoux and Grisoni 2010; Havkin Frenkel and Belanger 2011; Hernandez, Chapter 1 in this volume).

In The Netherlands, the vanilla seems able to start flowering at a length of 3 meters and one needs to "cull" or "loop" those plants to accelerate flowering (Fig. 10.2). Looping or culling is bringing the top of the plant back to the ground, sometimes re-rooting in the substrate, and then allowing the plant to grow back up. Looping should have an effect on the hormone balance and, in combination with other factors, this should have an effect on flowering. The re-rooting should be better for providing the plant enough

Figure 10.2 Different kind of looping systems in 2014 on a greenhouse table.

Figure 10.3 Vanilla in a high wire system in a greenhouse. Student is working on a rail cart.

water and fertilization as the plant matures. When flowering begins, they must be pollinated manually (Bhat and Sudharshan 2002).

10.3 Flowering

Two years after starting growing Vanilla in the greenhouse, two of the plants began to flower and that was promising. Around that point, Wageningen University at Bleiswijk received a subsidy to work on Vanilla and formed a group of people to work on growing Vanilla in greenhouses. This group was a mixture of researchers, growers of young plants, growers of tomatoes, and an ingredient consultant. This group decided to work

Figure 10.4 Vanilla with flowers and growing pods.

not only on growing Vanilla in greenhouses, but at the same time to work on the curing of the green pods, in order to make a final product.

10.3.1 Greenhouse

Growing Vanilla in a greenhouse allows the control of many parameters, such as light, temperature, watering, fertilization, CO_2 etc. However, one must determine the optimal combinations of growing parameters to obtain good Vanilla production and quality. The climate, screening (which influences light and temperature), fertilization and watering are controlled by computers, but the set points have to be found empirically.

10.3.2 Sustainability

Another important advantage of growing in greenhouses is that the production is sustainable, because the water and fertilization applied can be collected, cleaned and re-used, so there will be no loss of water and fertilization to the environment. Also it is easier to use biological control, because the predators are also in a protected environment and will stay in the greenhouse. The greenhouses themselves are also becoming more and more sustainable. There are screens in the greenhouse, not only for sun protection, but also to maintain higher temperatures. Some greenhouses are heated with warm wells, instead of being heated with gas or oil.

10.4 Varieties

Currently, different Vanilla varieties are grown, mostly *V. planifolia* from different origins, but mostly from Ugandan cuttings, and *V. tahitensis*, but also some *V. pompona*.

10.5 Propagation

The longer the cuttings, the sooner the plants will flower. Using two-node cuttings worked out very well and helped to grow enough plants for research. Single-node cuttings are also easily rooted, one only needs more time before the plants are big enough for flowering. So the recommendation of Umesha *et al.* (2011) to make 5-leaf cuttings to speed up growth was not used. Cuttings were grown, both with and without plastic, with the temperature just above 20°C, light levels around 200 µmol cm^{-2} s^{-1} and a humidity of 90%. The substrate should not be wet for too long, because roots won't appear or will die.

10.5.1 Cultivation

The growth of *V. planifolia* in a Dutch greenhouse can be categorized into four different stages. Towards the end of the year the induction of flowers starts. From the start of January, one can see the flower clusters developing. That will lead to actual flowering at the end of March and that lasts till the end of May. After this, the pod will mature and form seed for at least six months. Pods could be harvested from the end of November until February, depending on the time the pod was pollinated. At the end of the maturing phase the tip of the pod will turn yellow and then it is time for harvest and curing. As mentioned before, the advantage of cultivation in a greenhouse is the ability to control growth parameters and optimize the combination of these parameters for each stage of the plant's growth cycle.

Puthur (2005) reported the light preference of Vanilla and concluded that Vanilla grows between light levels of 300–800 µmol cm^{-2} s^{-1} and that 300–600 µmol cm^{-2} s^{-1} was most effective for vegetative growth. To control the light level, sun screens are used in greenhouses. The screens also make the light more diffuse, so the light will be "softer" on the top of the plants and will penetrate deeper into the crops. The day and night temperature for heating is 20°C and the windows open at 28°C. The temperature should not be above 30°C for extended periods, because as a CAM-plant, stomata are closed and the leaves are not able to cool themselves during the day.

An organic substrate is used for rooting. Good substrates should last several years, should not decompose very fast, and should be able to be dry and wet and everything in between. The substrate should also drain easily, because the roots of Vanilla don't do well when water-logged. Inorganic substrates such as rockwool and perlite may also be suitable, but these coarse substrates require more frequent watering than peat-based substrates.

The Vanilla is grown in pots on a gutter in order to collect the drain water. The watering is done with drip-irrigation and sprinklers from above. In a good Vanilla system one should be able to give water and fertilization at any time based on the plant's needs. Research on fertilization of Vanilla has not been done yet, so at the moment a general orchid fertilizer solution is used. Research on fertilization should be done in the near future.

10.5.2 Growing Systems

Currently, three growing systems are being compared. These growing systems are a high wire system such as used with tomatoes (Fig. 10.3), a short looping system (loops from about 1.5 m, no re-rooting) and a long looping system (loops from 4 m with re-rooting). A disadvantage of the high wire system is that with the long vines fruits will be hanging on the ground or in the dark. The disadvantage of the short looping system is that one will get a very bushy plant after only a few years of growing. At the moment the long looping system seems to have the most advantages for greenhouse growing in The Netherlands. For this system you need to have a radiator tube rail cart to be able to work at a height of 4 meters in the greenhouse for guiding the vine, pollination and checking the plants. The rail cart rides on heating pipes.

Production capacity is not clear, because our plants are quite young. There is some flowering on cuttings from 2015 but not more than four clusters per vine and many vines didn't flower (Fig. 10.4). The oldest plants are from 2014 and they gave serious production for the first time. The variation between plants is very big; there are plants without flowers, there are plants with >10 flowering clusters, there are flowering clusters with 1 pod, but also with 20 pods. So there is a lot of work to do on optimization of production and finding a good and productive variety.

10.6 Feasibility and Conclusions

A greenhouse will always be always more expensive than outside, because of all the equipment and the greenhouse itself. One needs at least a two–three year investment before earning any money. Labor for pollination manually is also a big issue when growing in a country with high wages. However, there are some expected benefits from greenhouses such as having almost full control of the light and climate. This should lead to optimal production and quality when there is enough understanding of the growing processes of the crop. This understanding includes watering (and re-use of water) and optimal fertilization. Tracking and tracing from products coming from greenhouses is easier. In time, after more research, is it my expectation that production in the greenhouse will be higher than outside, the pods will be longer, heavier and with good vanillin content. Another important issue is that new generation greenhouses are more sustainable, because they use almost no fossil energy. Last but not least, crop protection will be easier in greenhouses.

References

Bhat, S.S. and Sudharshan, M.R. (2002) Phenology of flowering in Vanilla (*Vanilla planifolia* Andr.) *Proceedings of Placrosym* XV, 128–133.

Cameron, K. (2011) *Vanilla Orchids: Natural History and Cultivation.* Timber Press. Portland, Oregon.

Castro-Bobadilla, G. and Garcia-Franco, J.G. (2007) Vanilla (*Vanilla planifolia* Andrews) crop systems used in the Totonacapan area of Veracruz, Mexico: Biological and productivity evaluation. *Journal of Food Agriculture and Environment* 5, 136–142.

Fouche, J.G. and Coumans, M. (1992) Four techniques for pollinating *Vanilla planifolia*. *American Orchid Society Bulletin* 61, 1118–1122.

Havkin-Frenkel, D. and Belanger, F.C. (2011) *Handbook of Vanilla Science and Technology.* Wiley-Blackwell, Oxford.

Korthout, H., Wang, M., Stehouwer, J.P., Graven, P., Dijkstra, T. and Rob Verpoorte, R. (2004) *Vanille als alternatief gewas voor de glastuinbouw: een haalbaarheidsstudie.* TNO Toegepaste Plantwetenschappen.

Lubinsky, P., Bory, S., Hernández Hernández, J., Seung-Chulkim, K. and Arture Gómez-Pompa, A. (2008) Origins and dispersal of cultivated Vanilla (*Vanilla planifolia* Jacks. [Orchidaceae]). *Economic Botany* 62, 127–138.

Odoux, E., and Grisoni, M., 2010. *Vanilla.* CRC Press, Boca Raton, Florida.

Osorio, A.I., Osorio, N.W., Diez, M.C. and F.H. Moreno, F.H. (2010) Effects of organic substrate composition, fertilizer dose, and microbial inoculation on vanilla plant nutrient uptake and growth. *Acta Horticulturae* 964, 135–142.

Puthur, J. (2005) Influence of light intensity on growth and crop productivity of *Vanilla planifolia* Andr. *General and Applied Plant Physiology* 31, 215–224.

Umesha, K., Murthy, G. and G.R. Smitha, G.R. (2011) Environmental conditions and type of cuttings on rooting and growth of vanilla (*Vanilla planifolia* Andrews). *Journal of Tropical Agriculture* 49, 121–123.

11

Establishing Vanilla Production and a Vanilla Breeding Program in the Southern United States

Alan H. Chambers

11.1   Introduction

Vanilla extract is the second most valuable spice globally, but the major growing regions are spatially separated from countries with the greatest consumption. Vanilla extract is commonly used around the world as a flavoring in desserts, a bitter-masking ingredient in chocolate and for other non-food uses. *Vanilla* cultivation is primarily concentrated outside its native range where individual flowers must be manually pollinated thus increasing production costs. *Vanilla* bean supply is also challenged by geopolitical instability, drought and disease that can impact global *Vanilla* bean supply and prices. Recently, large food companies have pledged to remove artificial ingredients from their products and will thus create an additional strain on the global supply of natural vanillin. Innovative solutions including step changes will be needed to meet the current and future demand for orchid-based vanillin.

Increasing *Vanilla* production especially within the native range is one potential solution to meet current and future challenges. Tropical fruit growers in the United States are showing increased interest in commercial *Vanilla* production, but lack foundational information for establishing profitable vanilleries. The University of Florida's Institute for Food and Agricultural Sciences supports the domestic production of agricultural commodities throughout Florida including in southern Florida at the Tropical Research and Education Center (TREC). Scientists at TREC are specifically dedicated to supporting local tropical fruit and vegetable growers by establishing optimized cultural practices, trialing varieties and releasing new cultivars with novel qualities. This chapter outlines a framework designed to support the establishment of commercial *Vanilla* production in southern Florida.

11.2   Southern Florida Climate

The southern Florida climate is subtropical, but is suitable for commercial production of tropical fruits. These include coconut palm, mango, citrus, avocado, guava, papaya, passionfruit, and many tropical ornamental species. Freezing temperatures can rarely

Handbook of Vanilla Science and Technology, Second Edition.
Edited by Daphna Havkin-Frenkel and Faith C. Belanger.

occur in the winter months for brief periods in specific microclimates. Some commercial plantings have irrigation for frost protection, though temperatures may not reach critical levels for many years at a time. Importantly, many of the key climactic factors are similar between southern Florida and that of the native *Vanilla* range in Mexico.

11.2.1 Average Temperatures

Monthly maximum, average and minimum temperatures as recorded at TREC by a Florida Automated Weather Network station (FAWN, 2017; Lusher *et al.* 2008) from 2006 to 2016 are shown in Figure 11.1. The weather station is located on a fully exposed and cleared ~10 acre plot on the TREC campus. The mean temperatures from monthly averages spanning 2006 to 2016 were 32.4 °C (maximum), 23.3 °C (average) and 12.2 °C (minimum). *Vanilla* species are regularly cited as growing only between 27° North and South latitude, and freeze events could be a concern for *Vanilla* production in southern Florida. December and January have the potential for localized freeze events that can last a few hours. *Vanilla* plantings will need to be empirically tested to determine vine responses to frost protection. Native species might also have adaptability traits for cold tolerance that could be incorporated into commercial quality cultivars through breeding. Otherwise, microclimates in production areas including tropical fruit plantings and shade house structures might not reach critical temperatures as might be expected in natural areas where native *Vanilla* species are endemic. For many tropical species, flowering is the critical time when freeze events can be the most damaging. *Vanilla* might naturally avoid this critical time as native species tend to start flowering around March (A. Chambers, *personal observation*).

11.2.2 Average Rainfall

Southern Florida experiences alternating wet and dry seasons like many tropical climates. Figure 11.2 shows the average rainfall by month over 10 years as recorded by the TREC FAWN weather station. The average total monthly rainfall for November through March is around 5 cm. Total rain increases from April to September with a high of around 20.6 cm in August and September gradually declining in October leading into the dry season. The dry season, reduced solar radiation or other environmental cues might be responsible for the induction of flowering in *Vanilla*, though native species have been observed to flower in the late summer months as well. Rainfall increases concurrently as *Vanilla* vines set seed capsules, and rainfall decreases as seed capsules reach maturity.

11.2.3 Average Solar Radiation

Vanilla species require reduced or filtered sunlight for optimal growth. This is provided either by tutor trees or by artificial shade houses usually targeting a 50% reduction in sunlight. The average monthly solar radiation as measured by the TREC FAWN weather station is shown in Figure 11.3. Solar radiation reaches its peak in April and May at around 241 w/m^2, and decreases to around 134 w/m^2 in December and January. The response of Vanilla vines to this large fluctuation in solar radiation will need to be empirically determined.

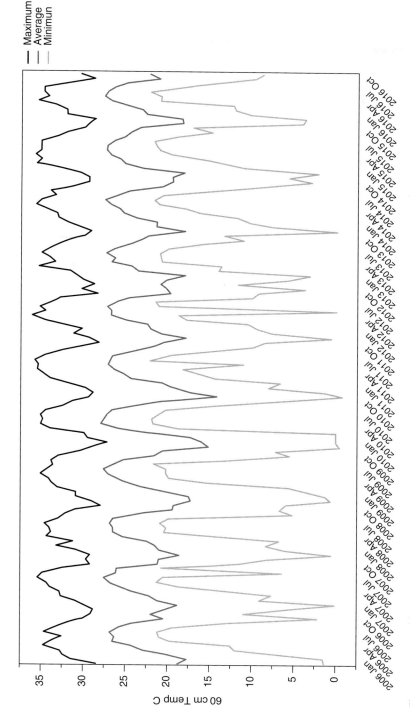

Figure 11.1 Monthly maximum (dark gray), average (medium gray) and minimum (light gray) temperatures as measured by a Florida Automated Weather Network station located at the Tropical Research and Education Center in Homestead, FL from 2006 to 2016. Temperatures were measured in °C at 60 cm above ground level in an unprotected location.

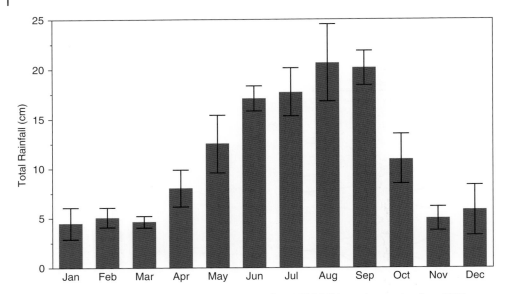

Figure 11.2 Average monthly rainfall as measured at the TREC FAWN weather station from 2006 to 2016.

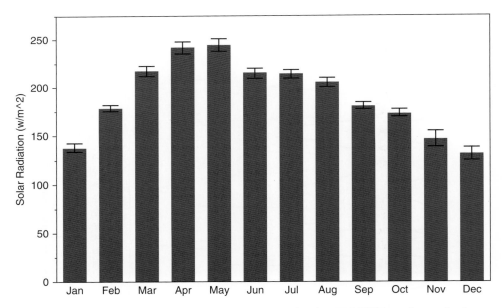

Figure 11.3 Average monthly total solar radiation measured at the TREC FAWN weather station from 2006 to 2016.

11.2.4 Major Weather Events

Many tropical areas are impacted by major tropical storms. Florida experiences an annual hurricane season from June to November when hurricanes are more frequent than during the rest of the year. Figure 11.4 shows maximum and average wind speed as recorded by the FAWN weather station at TREC. Major tropical storms are shown as increased maximum wind speed for the months shown. While major weather events are rare, extreme weather events could impact *Vanilla* tutors, growth structures and total production.

11.3   Native and Naturalized Vanilla Species of South Florida

There are four native *Vanilla* species in southern Florida, and also naturalized populations of *V. planifolia*. The native species are endangered and include *V. barbellata* (Wormvine Orchid), *V. dilloniana* (Leafless *Vanilla*), *V. mexicana* (Fuchs' *Vanilla*) and *V. phaeantha* (Leafy *Vanilla*). Each has been vouchered in at least one county in southern Florida and are mostly located in preserve lands. Naturalized populations of *V. planifolia* are also present in southern Florida on preserve lands and in the gardens of local orchid enthusiasts.

11.3.1   V. dilloniana

V. dilloniana or leafless *Vanilla* (Figure 11.5) is commonly thought to be extirpated in southern Florida. Online vendors and orchid enthusiasts can supply this species, though the exact origins of stock plants are largely unknown. Its morphological similarity with *V. barbellata* could lead to misidentification in natural settings. The application of genetic markers would be useful for identifying and conserving any remaining populations should they exist.

11.3.2   V. mexicana

V. mexicana or Fuchs' *Vanilla* (Figure 11.6) grows the farthest north of all the native species in Florida. It grows in fresh water tidal swamps among swamp bay (*Persea palustri*), cabbage palms (*Sabal palmetto*) and other tree species. This species has been observed to grow up to approximately eight feet, but is not a heavy climbing species. Natural seed capsule formation occurs on larger individuals, though the mechanism stimulating this has not been identified. *V. mexicana* tends to grow on the north side of supporting trees.

11.3.3   V. barbellata

V. barbellata or wormvine orchid (Figure 11.7) is a leafless species. Some natural populations spread both vertically up to 20 feet and horizontally over large areas. *V. barbellata* grows in hardwood hammocks among poison wood (*Metopium toxiferum*), swamp bay (*Persea palustri*) and other tree species. Evidence of flowering and the setting of seed capsules has not yet been observed.

11.3.4   V. phaeantha

V. phaeantha or leafy *Vanilla* (Figure 11.8) is another species that naturally flowers and develops seed capsules. Up to eleven seed capsules have been observed on a single

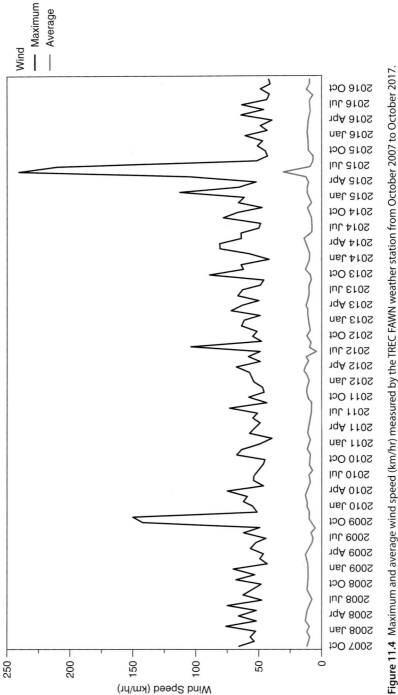

Figure 11.4 Maximum and average wind speed (km/hr) measured by the TREC FAWN weather station from October 2007 to October 2017.

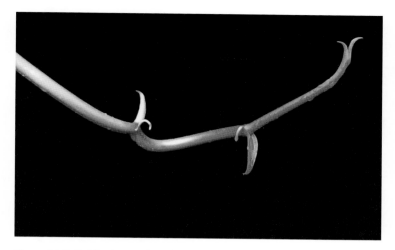

Figure 11.5 *V. dilloniana* collected in the United States and artificially propagated. This accession was provided by the Marie Selby Botanical Gardens, Tampa, FL, USA.

Figure 11.6 *V. mexicana* growing in a natural setting. Naturally set seed capsules are shown along with a close up of a single flower.

inflorescence. This species has been observed growing epiphytically as rising water levels cover and eventually lead to the complete rotting of the rooted base of individual plants. The plants can grow at least thirty feet up supporting trees. *V. phaeantha* density is estimated to be approximately one plant per 1–1.5 acres in some areas (Mike Owen, personal communication).

11.3.5   V. planifolia

V. planifolia (Figure 11.9) can be found in naturalized populations on preserve lands most likely from escaped introductions. The climbing vines can reach high into support

Figure 11.7 *V. barbellata* growing in a natural setting. This species is sprawling and a strong climber.

Figure 11.8 *V. phaeantha* growing in a natural setting. Natural seed capsule set is shown along with a mid-summer flower.

Figure 11.9 *V. planifolia* growing in a natural area. This is a strong climbing and spreading species in southern Florida, but probably does not pose a threat as an invasive species due to low natural seed set.

trees and spread across the understory. The origin and genetic diversity of naturalized *V. planifolia* in southern Florida is not currently known.

11.4 Establishing Vanilla Production in Southern Florida

All agricultural systems require optimization of cultural methods, and these methods require refinement over time. Differences in rainfall, temperature, solar radiation and soil conditions all require empirical testing at the variety level to maximize production. In southern Florida, native species could be a source of adaptability traits for local growing conditions and an important component of a viable breeding program, which has not yet been adequately studied.

11.4.1 Shade House Cultivation

At least two major growing systems could be viable for commercial *Vanilla* production in southern Florida. The first is the conventional crop monoculture. For *Vanilla* this would include shade house cultivation with optimized growing conditions. Shade house production of fruit and ornamental species is common in southern Florida especially for the large nursery industry. Routine monitoring of the varied pests and diseases are common for the diverse crops currently being grown. Optimizing these systems for *Vanilla* cultivation would require multi-year monitoring of plant growth and production. The advantage of the monoculture systems is the relative simplicity of growing a single species. Initially, growers would most likely only dedicate a portion of their growing area to *Vanilla* cultivation. This would reduce risk and provide invaluable information on *Vanilla* productivity in a commercial setting.

11.4.2 Tutor Tree Cultivation

An alternative production method could leverage existing orchards of palm or fruit trees as tutors in intercropping systems (Figure 11.10). Tutor trees could include palms, mango, lychee, longan, avocado and other evergreen tree species currently being grown commercially in southern Florida. The benefits of establishing a viable intercropping system with *Vanilla* could include supplemental or short term income to growers. This

Figure 11.10 Commercial, long-term palm and fruit tree plantings might provide suitable intercropping opportunities for local growers.

might especially be relevant in situations where orchard establishment requires many years to obtain a return on investment. Intercropping could also result in cultivation challenges. Growing *Vanilla* as an intercrop would require production practices suited to both primary and secondary crop species. Additionally, choice and timing of agricultural inputs such as fertilizer, pesticides and weed management would have to be optimized and could favor low input or organic production. Tree canopy management such as pruning, or harvesting operations could also impact *Vanilla* cultivation. Empirical research is needed in order to explore whether intercropping is feasible and what inputs and practices can be adjusted to favor the mixed system.

11.4.3 Substrate Considerations

Either of the options discussed above will have to rely of substrates to grow *Vanilla*. The agricultural lands in southern Florida are mostly comprised of calcareous soil with basic pH and negligible soil organic matter (Li 2001). *Vanilla* is grown globally on a number of substrates, and sourcing local materials is the most economical choice for establishing commercial plantings. Suitable substrates provide adequate moisture and the slow release of nutrients without becoming waterlogged, and that suppress pests and pathogens. Mulches from different tree species, landscaping debris, commercial green waste and other sources of organic material need empirical testing to identify the most suitable substrates for *Vanilla* cultivation, and must be investigated in conjunction with other aspects of optimized horticultural practices.

11.4.4 Local Economics and Niche Opportunities

Producing tropical fruits in the United States can be more costly than commodity imports depending on inputs, handling and processing. Historically, the United States has been the largest importer of cured *Vanilla* beans at ~1500-2000 Mt annually worth ~$52M USD with averages around $64 per kilogram (Champon Vanilla, 2009; Medina *et al.* 2009). It is unlikely that domestic growers will be able to compete with commodity imports from countries with lower production and regulatory costs. Still, there are many potential opportunities for domestic *Vanilla* growers including leveraging premium markets, niche quality opportunities and tourism.

Leveraging premium market segments could provide a viable option for domestic *Vanilla* growers. Vanilla extract is a premium product that is inherently attractive to domestic growers. Growing demand for local and organic markets could be leveraged to increase profitability. Specialty products from local growers and producers might resonate with domestic consumers and support a local markets. Another unknown opportunity is for ornamental *Vanilla*. *Vanilla* is a popular landscaping plant in southern Florida, and new varieties with novel qualities could open other market opportunities.

Vanilla extract quality is an active area of discussion primarily around counterfeit products that defraud consumers at premium prices. Beyond this active debate is the potential to deliver new and exciting sensory experiences through novel *Vanilla* genetics. The potential is exemplified by the historical spread and cultivation of the species. Few founding clones established what is now the major *Vanilla* cultivar(s) grown today (Bory *et al.* 2008b; Lubinsky *et al.* 2008). Additionally, some have argued that *Vanilla* should not be considered domesticated given the absence of plant improvement through breeding, and the similarities between wild and cultivated plants (Soto Arenas, 2009). Therefore, the true genetic potential of *Vanilla* is still enigmatic, and *Vanilla* science could be at the very foundation of new and exciting future possibilities to delight consumers with novel sensory attributes from *Vanilla*-based products.

Finally, agriculture in southern Florida is located between major urban areas and protected natural lands. This creates a market opportunity for agricultural tourism for crops like *Vanilla* that resonate with consumers. This business model has already been successfully employed domestically at the Hawaiian *Vanilla* Company (Paauilo, Hawaii) and The Vanillerie (Kailua-Kona, Hawaii), and also in other countries (Quirós, 2011). The revenue from agricultural tourism could offset production costs and provide year-round income for growers. Additionally, the popularity of *Vanilla* in general could increase support for conservation efforts protecting native and endangered *Vanilla* species.

11.5 Vanilla Breeding

Plant breeding is an iterative process that regularly improves upon the best varieties to enhance yield, quality and other traits of interest. The objective of plant breeding is to combine favorable genetics from multiple sources into a single improved cultivar. This would especially be useful for *Vanilla*, but *Vanilla* breeding is still underutilized globally. *Vanilla* orchids are challenged by biotic and abiotic factors that each increase the risks and costs associated with production. The few global production areas for *Vanilla* have historically been prone to severe weather events and geopolitical instability that can greatly impact supply. Adapting commercial-quality cultivars to local growing conditions could expand commercial *Vanilla* cultivation for domestic markets. *Vanilla* is also threatened by biotic stresses like pathogenic *Fusarium* and viruses. Ironically, the ease of asexual *Vanilla* propagation has also increased the risks associated with cultivation. Historically, clonal reproduction enabled traders to disseminate *Vanilla* globally, but also resulted in an extremely narrow genetic base in modern commercial plantings (Bory *et al.* 2008b; Lubinsky *et al.* 2008). The risks associated with clonally propagated species growing in monoculture is well documented including major examples like *Fusarium* wilt of banana (Ploetz, 2015) and citrus greening of sweet orange (McCollum *et al.* 2016). Innovative genetic solutions through plant breeding could play a significant role in the security of *Vanilla* production as demand for orchid-based *Vanilla* extract increases.

11.5.1 Establishing a Vanilla Breeding Program in the United States

Establishing a new *Vanilla* breeding program in support of domestic *Vanilla* production requires a few essential components. First, the acquisition of diverse *Vanilla* accessions including native species. The success of any breeding program is limited by the diversity within that program. Second, a deep understanding of the primary genetic pool is essential for *Vanilla*. Currently, *Vanilla* species are characterized by morphological and/or genetic markers (Bory *et al.* 2010), but this information is not suitable for predicting success for artificial hybridizations.

11.5.2   Acquiring Diverse Vanilla Accessions

Living collections of *Vanilla* accessions have been developed by multiple organizations around the globe. Some of these resources are identified in published sources (Bory *et al.* 2008b; Soto Arenas and Dressler 2009), and others can only be identified by direct inquires to botanical gardens or hobbyist growers. These materials can often carry material restrictions that can be burdensome to a breeding program. Overall, obtaining *Vanilla* germplasm can be challenging given international protections for threatened species, and import/export permit restrictions. The end result is that local collections often only benefit local directives.

A living collection of *Vanilla* germplasm for breeding purposes was initiated in 2016 at the Tropical Research and Education Center, University of Florida, Homestead, Florida, USA. As of August 2017, about 80 accessions from 22 species were obtained by petitioning domestic botanical gardens, online vendors, and hobbyist resources. Each of the four native Florida species including *V. barbellata*, *V. mexicana*, *V. phaeantha* and *V. dilloniana* were acquired through licensed vendors or from preserve lands after obtaining collection permits as part of conservation and genetics research. These foundational materials will be used to explore genomic diversity in *Vanilla* and also to create populations for genetic studies.

11.5.3   Creating Diversity in Vanilla

Creating diversity *de novo* is an alternative approach to obtaining diversity through artificial hybridization. Many tropical fruit species are similar to *Vanilla* in that they have limited diversity available in the primary, commercial species. In these cases, diversity can be induced through multiple methods including somatic mutations, chemical mutagenesis, ionizing radiation mutagenesis and inducing polyploidy (Ahloowalia and Maluszynski 2001).

Somatic mutations arising from tissue culture can be used as a means to capture novel diversity. *Vanilla* is a species that responds well to tissue culture and regeneration. *Vanilla* tissue culture can be routinely used for mass propagation of virus-free material for commercial plantings. An efficient method for identifying somatic mutants would include screening regenerated material from mass propagation for unique morphological traits. Normally these plantlets would be discarded, but could be tested for utility as part of a breeding program through resources that provide the seed material for commercial production. A more rapid approach would rely on historical mutations that have occurred naturally and were preserved in commercial plantings. This approach would require genomic testing of global, commercial *Vanilla* material to identify variants.

The use of chemical mutagens to create genetic diversity is common in plant genetic screens. EMS (ethyl methanesulfonate) is one common mutagen that results in DNA point mutations (G:C to A:T base pair conversions). For *Vanilla*, this would require a large number of viable seeds and space to grow out the resulting treated plants. Each accession requires optimization for chemical concentration and treatment time. A second, selfed generation is usually required for phenotypic evaluation of inbred crops, but treating a heterozygous species like *Vanilla* could result in morphological differences in the first generation.

Ionizing radiation similarly requires large numbers of viable seeds for treatment, but results in DNA lesions from double stranded breaks instead of point mutations. Sources of ionizing radiation can also be more challenging to obtain than chemical mutagens. The most efficient application of mutation breeding is for target traits that can be scored easily especially at the seedling stage. Seedling resistance to soil-borne pathogens is one example of an efficient target. Regardless of the method, mutation breeding can create useful diversity for a breeding program especially when there is a general constraint to available diversity.

Inducing polyploids is another way to increase diversity in a species. Naturally occurring polyploid accessions of *Vanilla* have been documented (Bory *et al.* 2008a; Duval *et al.* 2006; Lepers-Andrzejewski *et al.* 2011). Further work is needed to demonstrate the artificial induction of polyploids and the resultant changes in morphological and quality traits.

11.5.4 Identifying the Primary Gene Pool

The primary gene pool for *Vanilla* includes species and accessions that readily cross and produce viable and fertile offspring. This information is critical for breeding *Vanilla*, but is mostly unknown. A few hybrids have been mentioned that include *V. planifolia*, *V. pompona* and *V.* × *tahitensis* species with varying levels of supporting documentation (Bory *et al.* 2010; Rodolphe *et al.* 2008). Hybrids between the distantly related *V. planifolia* × *V. aphylla* ♂ (Divakaran *et al.* 2006; Minoo *et al.* 2008), reciprocal crosses between *V. planifolia* and *V. pompona* (Menchaca *et al.* 2011), and natural hybrids between *V. claviculata* and *V. barbellata* have been reported (Nielsen 2000; Nielsen and Siegismund 1999). These reports and studies are useful, and in general tend to support a claim that many interspecific crosses in *Vanilla* are possible. Still, a systematic, artificial hybridization program at a sufficient scale is lacking in *Vanilla* research. Numerous challenges need to be overcome in order to develop this foundational information. These include establishing mature living collections of diverse *Vanilla* species, conducting controlled artificial hybridizations, overcoming seed germination inhibition and multiyear monitoring of sufficient numbers of progeny to identify promising hybrids.

Breeding new cultivars of *Vanilla* would also require input from regulatory agencies and interested parties concerning species restrictions for *Vanilla* beans. Currently, only seed capsules from *V. planifolia* and *V.* × *tahitensis* are acceptable under the standard of identity for *Vanilla* beans (Food and Drug Administration, 2016). This definition might restrict the adoption of promising new *Vanilla* cultivars if plant breeding delivers on the promise of increased yield, improved sensory characteristics and other target traits by leveraging the diversity within the *Vanilla* genus. Otherwise, domestic growers could potentially seek lucrative, non-commodity markets outside the *Vanilla* standard of identity.

11.5.5 Target Traits

Target traits for a *Vanilla* breeding program would need to be identified by the major limitations impacting commercial production in southern Florida. All plant breeding programs, though, focus on key traits like yield, disease resistance and quality. Each of these depends on grower needs, available solutions and market opportunities. Both cultural and genetic solutions will play a role in improving target traits. A *Vanilla* breeding program would include traits like large, non-dehiscent beans and disease resistance for common pathogens. Additionally, the setting of seed capsules without manual pollination and beans that yield potent, higher quality flavor extracts are potentially innovative traits that should be investigated especially for production in southern Florida.

11.5.6 A Case for a Publically Available Vanilla Genome

A publically available *Vanilla* genome would open many research possibilities around the globe. The initial investment in a high quality genome would enable researchers to obtain relatively inexpensive genomics information for local *Vanilla* collections. Chloroplast or ITS sequences have been used to assess *Vanilla* diversity in the past, and this has been especially useful for overcoming variability in morphological markers and identifying unknown accessions. Genome-wide sequence data offers a new level of opportunity for diversity assessment. Perhaps the greatest impact of a *Vanilla* genome would be the ability to use next generation sequencing data and segregating populations to establish marker-trait associations and the identification of candidate genes for target traits like disease resistance and aroma quality of seed capsules. A publically available *Vanilla* genome would be the starting point of many potentially innovative research studies yet to be explored.

11.6 Conclusions

Vanilla is a crop with great potential, but also has many challenges. Price and supply fluctuations as well as biotic challenges and a lack of diversity all threaten commercial *Vanilla* production. Plant breeding is one potential solution to some of these challenges, though foundational work to support a breeding program needs additional research. Future research leveraging local *Vanilla* species and controlled hybridizations could lead to improved disease resistance, increased yield and higher quality *Vanilla*. Each of these traits could improve profitability for *Vanilla* growers. This is especially important for domestic growers that face high labor and regulatory expenses. *Vanilla* production in southern Florida could be commercially viable especially when combined with agrotourism, but more research into optimal production practices and deep economic analyses are needed.

References

Ahloowalia, B. and Maluszynski, M. (2001) Induced mutations – A new paradigm in plant breeding. *Euphytica* 118, 167–173.

Bory, S., Brown, S., Duval, M.-F. and Besse, P. (2010) *Evolutionary processes and diversification in the genus Vanilla*. Vanilla. CRC Press, Florida, US, 15–29.

Bory, S., Catrice, O., Brown, S., Leitch, I.J., Gigant, R., Chiroleu, F., Grisoni, M., Duval, M.-F. and Besse, P. (2008a) Natural polyploidy in *Vanilla planifolia* (Orchidaceae). *Genome* 51, 816–826.

Bory, S., Lubinsky, P., Risterucci, A.M., Noyer, J.L., Grisoni, M., Duval, M.F. and Besse, P. (2008b). Patterns of introduction and diversification of *Vanilla planifolia* (Orchidaceae) in Reunion Island (Indian Ocean). *American Journal of Botany* 95, 805–815.

Champon Vanilla (2009) *Vanilla Imports/Exports and Market Update January 2009.* Available from: http://vanillabean.com/vanilla_report.html.

Divakaran, M., Babu, K.N., Ravindran, P. and Peter, K. (2006) Interspecific hybridization in vanilla and molecular characterization of hybrids and selfed progenies using RAPD and AFLP markers. *Scientia Horticulturae* 108, 414–422.

Duval, M.-F., Bory, S. Andrzejewski, S., Grisoni, M., Besse, P., Causse, S., Charon, C., Dron, M., Odoux, E. and Wong, M. (2006) Diversité génétique des vanilliers dans leurs zones de dispersion secondaire. *Les Actes du BRG* 6, 181–196.

FAWN, U.o.F.I.f.F.a.A.S. (2017) FAWN: Florida Automated Weather Network.

Food and Drug Administration. (2016) CFR–Code of Federal Regulations, Title 21: US Department of Health & Human Services. Available from: https://www.accessdata.fda.gov/scripts/cdrh/cfdocs/cfcfr/CFRSearch.cfm?fr=169.3.

Lepers-Andrzejewski, S., Siljak-Yakovlev, S., Brown, S.C., Wong, M. and Dron, M. (2011) Diversity and dynamics of plant genome size: an example of polysomaty from a cytogenetic study of Tahitian vanilla (Vanilla× tahitensis, Orchidaceae). *American Journal of Botany* 98, 986–997.

Li, Y. (2001) *Calcareous Soils in Miami-Dade County.* University of Florida Cooperative Extension Service, Institute of Food and Agriculture Sciences, EDIS.

Lubinsky, P., Bory, S., Hernandez, J., Kim, S. and Gomez-Pompa, A. (2008) Origins and dispersal of cultivated vanilla (*Vanilla planifolia* Jacks. [Orchidaceae]). *Economic Botany* 62, 127–138.

Lusher, W.R., Jackson, J.L. and Morgan, K.T. (2008) The Florida automated weather network: Ten years of providing weather information to Florida growers. In: *Proc. Florida State Hort. Soc.* pp. 69–74.

McCollum, G., Baldwin, E., Gradziel, T.M., Mitchell, C.A. and Whipkey, A.L. (2016) Huanglongbing: Devastating Disease of Citrus. *Horticultural Reviews* 315–361.

Medina, J.D.L.C., Jiménes, G.C.R. and García, H.S. (2009) *Vanilla: Post-harvest operations.* Food and Agriculture Organization of the United Nations.

Menchaca, G., Rebeca, A., Ramos, P., José, M., Moreno, M., Luna, R., Mata, R., Vázquez, G., Miguel, L. and Lozano, R. (2011) In vitro germination of *Vanilla planifolia* and *V. pompona* hybrids. *Revista Colombiana de Biotecnología* 13, 80–84.

Minoo, D., Jayakumar, V., Veena, S., Vimala, J., Basha, A., Saji, K., Babu, K.N. and Peter, K. (2008) Genetic variations and interrelationships in Vanilla planifolia and few related species as expressed by RAPD polymorphism. *Genetic Resources and Crop Evolution* 55, 459–470.

Nielsen, L.R. (2000) Natural hybridization between *Vanilla claviculata* (W. Wright) Sw. and *V. barbellata* Rchb. f.(Orchidaceae): genetic, morphological and pollination experimental data. *Botanical Journal of the Linnean Society* 133, 285–302.

Nielsen, L.R. and Siegismund, H.R. (1999) Interspecific differentiation and hybridization in *Vanilla* species (Orchidaceae). *Heredity* 83, 560–567.

Ploetz, R.C. (2015) Fusarium wilt of banana. *Phytopathology* 105, 1512–1521.

Quirós, E.V. (2011) *Vanilla* production in Costa Rica. In: *Handbook of Vanilla Science and Technology* (pp. 40–49), First Edition. Edited by Daphna Havkin-Frenkel and Faith C. Belanger.

Rodolphe, G., Bory, S., Grisoni, M., Duval, M.-F. and Besse, P. (2008) Biodiversity and preservation of vanilla: present state of knowledge. *Genetic Resources and Crop Evolution* 55, 551–571.

Soto Arenas, M.A. (2009) *Recopilación y análisis de la información existente sobre las especies mexicanas del género Vanilla*. Reporte. México, DF: Herbario de la Asociación Mexicana de Orquideología, AC, Instituto Chinoín, AC.

Soto Arenas, M.A. and Dressler, R.L. (2009) A revision of the Mexican and Central American species of Vanilla Plumier ex Miller with a characterization of their ITS region of the nuclear ribosomal DNA. *Lankesteriana* 9, 285–354.

12

In vitro Propagation of Vanilla

Rebeca Alicia Menchaca García

The vanilla plant *(Vanilla planifolia* Andrews) is a tropical orchid the fruits of which provide a very popular flavoring. However, despite the fact that the vanilla crop comprises many plants, all of these individuals originate from the division of the stems of very few original individuals. For this reason, these plants are clones with a reduced genetic base. These individuals do not present the variation that would allow them to respond to attack by pests or diseases and present a low capability for adaptation to climatic change. In Mexico, the species *Vanilla planifolia* has been considered in danger of disappearance because of the reduced number of wild populations and the species also presents serious problems associated with the conservation of its natural habitat (FAO 1995; Soto-Arenas 2006; Duval *et al.* 2006; SEMARNAT 2010).

The natural pollinators of vanilla are also threatened: pesticide application and habitat modification have caused a reduction in populations of the bees that pollinate the vanilla, with a reported natural pollination of only 1% of the flowers (Childers and Cibes 1948). Faced with this situation, each vanilla flower must now be manually pollinated by the producers in order to form each fruit and obtain sufficient production for the harvest. This situation implies high labor costs for the crop (Soto-Arenas 1999; Richards 2001).

As with all orchids, vanilla requires mycorrhizal association for germination; however, due to the application of fungicides, the native fungi that integrated the natural rhizosphere of the vanilla have been reduced, diminishing the capacity for absorption of nutrients and causing almost null reproduction through seed germination (Soto-Arenas 2006).

We are now faced with a crop species of reduced genetic variability, problems of natural conservation, absence of pollinators and almost no natural germination. For this reason a program of germplasm conservation and genetic improvement of the vanilla is urgently required in order to produce varieties resistant to phytosanitary problems or drought and thus to provide solutions to the current problems faced by the cultivation of vanilla in the field.

Given the economic importance of vanilla and the problems described above, *in vitro* cultivation techniques have been developed with two main objectives: (1) to provide plants with genetic diversity and (2) to produce individuals free of pathogens for the establishment of new crops. Studies of asexual propagation have also been conducted, including tissue culture through various explants and sexual propagation through the micropropagation of seeds, and assaying the asymbiotic and symbiotic germination associated with specific mycorrhizae (Porras-Alfaro and Bayman 2007). *In vitro* propagation also offers an efficient method for propagating selected material for both commercial cultivation and reintroduction (Flanagan and Mosquera 2016).

This chapter summarizes these studies of *in vitro* propagation in the genus *Vanilla*, conducted in the orchidarium of the Centro de Investigaciones Tropicales of Universidad Veracruzana in Mexico. The studies address different micropropagation techniques such as the germination of seeds and immature embryos, tissue culture, generation of hybrids and maintenance of the *in vitro* germplasm bank. The results of these studies have defined and validated the methodological basis for initiating a program of genetic improvement of vanilla in Mexico.

12.1 Methods

12.1.1 *In vitro* Germination

In order to produce individuals with genetic variation, the vanilla species were sexually propagated by controlled manual pollination. The fruits were washed with liquid soap and placed in a 3% solution of commercial bleach (0.18 NaClO) for 15 min, followed by a bath in 96% ethanol for 10 min, and finally flame sterilized twice prior to sowing. The culture media used were Burgeff's N3F medium, supplemented with 12 mg L^{-1} of arginine and 18 mg L^{-1} of lysine, and Murashige and Skoog (MS) medium (1962), supplemented with 400 mg L^{-1} of glutamine and 80 mg L^{-1} of adenine sulfate. All of these media were supplemented with 20 g L^{-1} of D-Sucrose (Phytotechnology Labs S391). The pH was adjusted to 5.6 with 0.1 N of HCl or NaOH, 2 g L^{-1} of Gelrite were added (Phytotechnology Labs G469) and the media were sterilized in an autoclave at 120°C for 15 min and 15 psi of pressure. In order to achieve *in vitro* germination, the asymbiotic germination technique developed by Knudson (1922) was followed. Seeds of different species of the genus *Vanilla* were placed in a cultivation chamber under the following controlled conditions: photoperiod of 16 h light, temperature of 25°C and a light intensity of 164 fc.

12.1.2 Tissue Culture

In order to clonally propagate some selected plants, or increase the *in vitro* accessions of vanilla species belonging to wild populations, axillary tips and leaf and stem segments were used as explants (Lozano 2010; Montiel 2011). The MS medium was used with different concentrations of vegetal regulators, pH was adjusted to 5.6 and 2 g L^{-1} of Gelrite were added. The jars containing 25 mL of medium were sterilized in an autoclave at 120°C and a pressure of 15 psi for 15 minutes. The cultures were incubated at a temperature of 25 ± 2°C, with a photoperiod of 16 h light and a photonic flow density of 50 µMol $m^{-2} s^{-1}$.

12.2   Results and Discussion

12.2.1   Germination

In order to develop an appropriate protocol for the propagation of seeds or immature embryos of vanilla, it was considered necessary to conduct research to determine the state of maturity of the seed and the time necessary for germination, as well as the appropriate techniques of scarification and culture media.

12.2.2   Seed Maturity

In order to determine the optimal state of maturity of the seeds for germination of *Vanilla planifolia*, Menchaca (1999) utilized seeds that came from eight stages of maturity of the fruit, cultivated in four culture media (Knudson C, MS, Burgeff's N3F and Knop) and determined that the percentage of germination decreased with the hardening of the seed coat that occurs as the fruit matures. The highest percentages of germination (50%) were obtained from seeds at 154 to 200 days after manual pollination. This time of maturity of the fruit has since been shortened in the methodology, which now uses seeds of 44 days after pollination, obtaining germination of greater than 80% (9, 2011). Whitner (1955) states that seeds that have late dehiscence present strategies associated with animal dispersion, such as that of the aroma of dry vanilla fruits. Soto-Arenas (1999) states that the seeds can possibly present dispersion by bats, although the possible ingestion of seeds by rodents or even iguanas has been reported in the field. This animal dispersion and digestion could help to scarify the seeds with the stomach acids of the fauna that ingest them, enabling eventual mycorrhizal association and subsequent germination in nature.

12.2.3   Time for Germination

The different species and hybrids of vanilla presented different times for germination. In this sense, germination of *Vanilla planifolia* occurred at 100 to 130 days, while that of *V. insignis* occurred at 70 days and the hybrids of *Vanilla planifolia* × *Vanilla pompona* shortened the time for germination to 40 days.

12.2.4   Scarification

Vanilla seeds are very small and present an undifferentiated embryo and a very hard and waxy seed coat that also contains germination inhibitors (Bory *et al.* 2008). In order to soften this seed coat and facilitate germination, Lozano (2014) used a solution of 0.1N HCl to scarify mature seeds of *Vanilla inodora* that were subjected to different times of immersion in the acid, obtaining germination percentages of up to 100% in seeds submerged for 30 minutes and cultivated in MS medium (1962).

12.2.5   Tissue Culture

Montiel (2011) cultivated axillary tips of *Vanilla planifolia* in MS medium with different concentrations of BAP, from 0.5 to 2 mg L^{-1}. The treatment MS + 1.5 mg L^{-1} of BAP promoted greater growth of the sprouts, with an increase of 1.6 cm, as well as a greater number of leaves developed (mean value 1.22). On the other hand, the treatment 0 mg L^{-1} of BAP promoted the development of a greater number of roots with a mean of 1.03 new roots within a 60-day evaluation period. Lozano (2015) cultivated *in vitro* axillary tips of vanilla subjected to different treatments with vegetal regulators. These regulators were N6-benzylaminopurine (BAP), kinetin (Kin), meta-topolin (mT) and thidiazuron (TDZ). The Murashige and Skoog culture medium was used, supplemented with glycine at 2 mg L^{-1}, myo-inositol 100 mg L^{-1} and 20 g L^{-1} of sucrose. The variables of number of sprouts per explant and height of sprouts were evaluated. The explants were subcultured every 30 days, providing a second BAP treatment (2.22 µM) at 90 days in order to stimulate differentiation of the undefined sprouts. The best result was obtained with the treatment MS + BAP 8.88 µM after 120 days in cultivation, with an average of 6.1 new sprouts per explant and an average height of 5.59 cm for each sprout, demonstrating that the use of 8.88 µM of BAP increases the formation and subsequent development of adventitious sprouts. In the germplasm bank of the orchidarium, the methodology created by Lozano 2015 is currently used when the promotion of sprouts is desired, as well as carbon-poor media for maintenance of the accessions of the germplasm bank in the medium and long term.

12.2.6 Hybridization

Despite the fact that there are ten species within the genus (Soto-Arenas 1999), there is no program of hybridization and genetic improvement of vanilla in Mexico. The species that have been considered more interesting in terms of conducting a hybridization program with crop species are *Vanilla phaeantha, V. insignis, V. odorata* and *V. pompona.* The latter of these species is considered of particular utility for providing desirable characteristics due to the fact that it produces large, robust and strongly fragrant fruits for use in the manufacture of perfumes, is a vigorous and drought resistant plant and, unlike other species, can grow in granitic zones (both volcanic and calcareous), maintain a large number of fruits until harvest, produce large quantities of limonene and is naturally pollinated by bees of the genus *Eulaema* (Soto 1999), which would imply reduced input of labor in cultivation (Childers, 1959, cited by Soto-Arenas, 1999). Considering the above, crosses were performed between species that were found growing together in the vanilla crop in order to obtain new individuals with desirable characteristics. Menchaca (2011) conducted controlled crossing in order to obtain hybrids of the two Mexican vanilla species *V. planifolia* and *V. pompona.* Of the assayed crosses, the seeds from the interspecific cross *V. pompona* × *V. planifolia* presented the highest percentage of germination (85%), followed by the inverse cross *V. planifolia* × *V. pompona* (57.9%). The seeds produced by self-pollinated *V. pompona* produced very low germination values (10.8%), while germination was null in those obtained from *V. planifolia.* The most efficient culture medium of all of the treatments for germination of the different crosses was that of Murashige and Skoog, supplemented with 400 mg L^{-1} of glutamine and 80 mg L^{-1} of adenine sulfate. Morphological characterization of the plants revealed that the hybrid of *V. pompona* × *V. planifolia* presented higher values for the variables plant length and diameter, leaf length, width and number, and internodal length. The hybrid of *Vanilla planifolia* with *Vanilla insignis* was obtained following the same methodology.

Following pollination, it was observed that the fruits of *V. planifolia* that were pollinated by pollen from *V. pompona* increased in size by 17%, a phenomenon known as metaxenia, in which the pollen parent influences the development of the maternal tissues of the fruit. These results provide the possibility of obtaining, through cross-pollination between these two species, a larger fruit and different aromatic characteristics that could be subjected to further study.

12.2.7 *In vitro* Germplasm Bank

There have been efforts by different institutions to conserve the germplasm of vanilla. Globally, the collection of BRC Vatel CIRAD on the island of Reunion, France (Grisoni *et al.* 2007), the American Vanilla Collection in University of California, Davis, the Spice Board collection in Valikut, India and the Etablissement Vanille in Tahiti, French Polynesia (Lozano 2014). There are also *in vitro* collections reported in the Universidad Nacional de Costa Rica and in the germplasm bank of the CIAT in Palmira, Colombia. Of all these collections, that of the BRC Vatel is prominent for maintaining a collection of more than 250 accessions and 22 species of vanilla (Grisoni *et al.* 2007).

Currently, *Vanilla planifolia* is at great risk of disappearance in Mexico, for which reason it is important to safeguard the accessions of different species belonging to wild populations, considering that the challenge of their conservation is the maintenance of the genetic diversity that still exists both in the species and in its close relatives, since the probabilities of *in situ* conservation are very remote (Soto 2006). In order to increase the germplasm bank of the orchidarium of Universidad Veracruzana, Lozano (2014) conducted *in vitro* propagation of the following species of the genus *Vanilla*: *V. inodora*, *V. insignis*, *V. planifolia*, *V. pompona* and *V. tahitiensis*, plus two artificial hybrids created in the laboratory: *V. planifolia* × *V. pompona* and *V. planifolia* × *V. insignis*. At present, the germplasm bank of the Centro de Investigaciones Tropicales of the Universidad Veracruzana has 44 *in vitro* accessions of five species of the genus *Vanilla* and three hybrids (*V. planifolia* × *V. pompona*, *V. pompona* × *V. planifolia* and *V. planifolia* × *V. insignis*), as well as wild collections of *V. planifolia*.

The germplasm bank of the Universidad Veracruzana is registered with the Secretaría de Medio Ambiente y Recursos Naturales de México (SEMARNAT), under the key UMA_INV-viv-0129-Ver /11, and is in process of incorporation into the OSSSU Project – Orchid Seed Science/Stores for Sustainable Use.

12.2.8 Repatriation and Recovery of Mexican Species

Thanks to the convention of collaboration between the orchidarium of Universidad Veracruzana and CIRAD of the island of Reunión, France, the repatriation of four specimens of *Vanilla cribbiana* was achieved. This species had been reported in Mexico but has not been found in the field in that country for more than twenty years. At present, the specimens are safeguarded in the *in vitro* collection as well as under greenhouse conditions and have presented adequate growth for three years. This project of collaboration with CIRAD raises the possibility of recovering this genetic material for conservation and future crossing with the species *Vanilla planifolia*.

12.2.9   Method of *Ex vitro* Adaptation

One of the important steps in the methodology of *in vitro* propagation is that of cultivation of seedlings in the greenhouse. Ramos (2011) evaluated five substrate mixtures for acclimatization of *Vanilla planifolia* seedlings, observing that the best treatment was a mixture of volcanic rock, perlite, river sand and vegetal carbon in equal parts (23.5%) and *Sphagnum* peat at 0.5%, with a pH of 4.7, since the *Vanilla planifolia* seedlings of this treatment increased in size by 12% and maintained the highest number of leaves over a 90-day evaluation period. Following this step, the plants were fertilized with NPK 17:17:17, decreasing the frequency of application until development into adult plants. Similarly, it was observed that the application of biofertilizers stimulates growth in length of the hybrids *V. planifolia* x *V. pompona* (Menchaca *et al.* 2011). These results coincide with that observed by Baldy-Porras *et al.* (2014) in *Vanilla planifolia.*

12.2.10   Greenhouse Collection

At present, as well as the *in vitro* collection and in collaboration with small vanilla producers, the orchidarium maintains a collection of plants under greenhouse conditions where different accessions that originated in wild vanilla populations are conserved: *V. cribbiana, V. insignis, Vanilla odorata, Vanilla pompona* and *Vanilla tahitiensis*, as well as the hybrids of *Vanilla planifolia* × *Vanilla pompona*, and the backcross of the *V. planifolia* × *V. pompona* hybrid to *V planifolia*. The hybrids produced in the orchidarium have been monitored for eight years, evaluating morphological (leaf shape, stem diameter, number of internodes) and physiological (photosynthetic capacity, CO_2 assimilation, conductivity) variables, histological and stomatal characteristics, speed of growth, phytosanitary tests and tests of resistance to pathogens and viruses (in collaboration with Dr. Grisoni of CIRAD, France).

12.2.11   Social Linkage

Conservation of species is not an isolated task: in order to maintain a collection in the *in vitro* germplasm bank, it is very important to maintain links with small producers who conserve wild varieties of vanilla. These people can identify individual plants with prominent characteristics in the crop or provide new accessions, as was the case with the producer Oscar Mora of Coatepec, in Veracruz, Mexico, who created the hybrid of *Vanilla planifolia* and *Vanilla insignis* and donated the seeds to the germplasm bank. It is important to consider that the people who cultivate vanilla in the field are the most important actors in the conservation of these species, since they provide the genetic material for *in vitro* propagation and will ultimately be the users of the plants that they generate, providing valuable information for the production and development of these plants in the field.

In order to disseminate the use of natural vanilla, festivals have been held in which regional vanilla producers are put in contact with local consumers. In collaboration with Dr. Aracely Pérez Silva of the Tecnológico de Tuxtepec, Mexico, organoleptic tests were conducted on the extract from vanilla plants of different origins. In addition, workshops have been held in order to exchange experiences, with the aim of improving vanilla quality and drying processes. In these meetings, the producers suggest the species that they are interested in massively propagating *in vitro* and with which they have conducted pollination in their crops, obtaining crosses of selected plants as well as hybrids of the species that they conserve. In this sense, the orchidarium of Universidad Veracruzana has maintained links with small vanilla producers, providing training for cultivation in the communities of Veracruz and Oaxaca. Also, with support from the Consejo Nacional de Ciencia y Tecnología (CONACyT) and large commercial vanilla enterprises such as Gaya S.A. de C.V., training has been provided and a laboratory of *in vitro* propagation was established.

12.2.12 Human Resource Training and International Interaction

A germplasm bank is a long-term project, for which reason the training of new generations of human resources for the future monitoring of conservation activities is considered necessary.

In order to exchange experiences in terms of *in vitro* cultivation of vanilla, the researchers and students of the orchidarium have participated in academic placements in different laboratories worldwide: (1) In 2012, Centro de Cooperación Internacional en Investigaciones Agronómicas para el Desarrollo (CIRAD) on the island of Reunión, France, hosted R.M.A. Lozano; (2) In 2014, the Departamento de Ciencias Naturales Universidad Técnica Particular de Loja hosted Castelán-Culebro; (3) In 2016, the laboratory of *in vitro* propagation of the Universidad Nacional del Nordeste UNNE, in Corrientes, Argentina hosted Morales-Ruiz; and (4) In 2017, the Centro Internacional de Agricultura Tropical (CIAT) in Palmira, Colombia, hosted Moreno-Martínez.

Another aspect of the collaboration of the orchidarium has been through knowledge of indicators of sustainability of the vanilla crop. Thus, in 2017 Sánchez-Morales participated in an exchange in the ecoregion El Cerrado, in the tropical savannah of Brazil, in which a link was established with the project "Vainilla del Cerrado", led by Luiz Camargo and with the participation of Alex Atala. This project was focused on the agroforestry cultivation of the vanilla species found in the region, established by the Kalunga community in Brazil.

In addition to these research placements, training workshops have been held addressing the cultivation of vanilla in terms of aspects of sowing, pollination and development of products with added value in communities of Mexico, Colombia and Bolivia. This has been done with the aim of encouraging the small producers of the communities to exploit the species of their own particular country in a sustainable manner.

12.3 Conclusions

In summary, a validated methodology has been developed in the orchidarium of Universidad Veracruzana for protocols of *in vitro* propagation of different Mexican vanilla species, which have been obtained from seeds or immature embryos, as well as axillary tips. Hybrids of various vanilla species have been obtained and their development monitored from the *in vitro* phase up to the adult state. At present, pollen germination studies are being conducted, as well as those of indicators of sustainability in the vanilla crop, and the isolation and characterization of the mycorrhizae of wild populations.

Given the experience acquired and the results obtained, we propose a program of vanilla conservation and genetic improvement in the following phases:

1) The first phase is *in situ* and *ex situ* establishment of germplasm banks in order to safeguard the vanilla material that exists in the wild populations of *Vanilla planifolia*, as well as other species of the genus *Vanilla* that can provide different characteristics for genetic recombination with the crop species.
2) The second phase is identification of individuals, or selection of species, that provide desirable characteristics for *in vitro* reproduction by seeds generated through crossing of selected materials and through programs of hybridization, in order that a greater genetic diversity is expressed. Consideration should also be given to the use of mutagenic agents that can create diversity in the long-term, such as gamma rays or protoplast fusion in selected species (Reyes *et al.* 2006; Ortega *et al.* 2016).
3) In the third phase, it is considered necessary to return to clonal propagation, producing genetic lines with the potential for improvement and with homogeneous characteristics among the individuals that can be evaluated from initial growth to adult status, subjecting the plants produced to adverse situations and pathogen tests in order to evaluate their adaptability to climate change or resistance to diseases.
4) In the fourth phase, it is necessary to perform organoleptic evaluation and aromatic analysis of the fruits produced by the genetic lines in order to conduct commercial selection.
5) Finally, it is considered important to influence international norms, so that they permit the international trade of vanilla hybrids or new varieties that may be produced.

Incorporation of desirable traits to the vanilla crop could include higher percentages of natural pollination, which could reduce the labor costs and increase resistance to pests and diseases in order to reduce dependence on the application of fungicides and pesticides. It may also be possible to produce drought tolerant varieties that could enable a response to the effects of climate change, or the production of varieties that provide different aromatic alternatives and even medicinal properties as a means by which to diversify the product.

References

Baldí-Porras, Y., Paniagua-Vásquez, A., Azofeifa-Bolaños, J., Mora-Salas, J. and Azofeifa-Bolaños, M. (2014) Prueba de abonos foliares orgánicos en *Vanilla planifolia* producida in vitro durante las primeras etapas de crecimiento en invernadero. In: *Seminario Internacional de Vainilla*, eds. C.A. Fernandez, R.C. Solorzano, A.P. Vasquez and J.B.A. Bolanos. Instituto de Investigacion y Servicios Forestales, Universidad Nacional, Heredia, Costa Rica.

Bory, S., Lubinsky, P., Risterucci, A.M., Noyer, J.L., Grisoni, M., Duval, M. and Besee, P. (2008) Patterns of introduction and diversification of *Vanilla planifolia* (Orchidaceae) in Reunion Island (Indian Ocean). *American Journal of Botany* 95, 805–815.

Childers, N.F. and Cibes, H.R. (1948) *Vanilla culture in Puerto Rico*. Circular No 28, Federal Experiment Station in Puerto Rico of the United States Department of Agriculture, Washington DC.

Duval, M.F., Bory S., Andrzejewski, S., Grisoni, M., Besse, P., Causse, S., Charon, C., Dron, M., Odoux, E. and Wong, M. (2006) Diversité génétique des vanilliers dans leurs zones de dispersion secondaire. *Les Actes du BRG* 6, 181–196.

FAO (1995) *Conservación y utilización sostenible de los recursos fitogenéticos de América central y México*. Conferencia Técnica Internacional sobre los Recursos fitogenéticos. San José, Costa Rica.

Flanagan, N. and Mosquera-Espinosa, A.T. (2016) An integrated strategy for the conservation and sustainable use of native Vanilla species in Colombia. *Lankesteriana* 16(2), 201–218.

Grisoni, M., Besse, P., Bory, S., Duval, M.F. and Kahane, R. (2007) Towards an international plant collection to maintain and characterize the endangered genetic resources of Vanilla. *Acta Horticulturae* 760, 83–91.

Knudson, L. (1922) Nonsymbiotic germination of orchid seeds. *Botanical Gazette* 73, 1–25.

Lozano, R.M.A. (2010) *Propagación in vitro de Vanilla planifolia Andrews a partir de yemas axilares*. Tesis de Licenciatura. Facultad de Ciencias Agrícolas. Universidad Veracruzana.

Lozano, R.M.A. (2014) *Establecimiento del banco de germoplasma in vitro de vainillas mexicanas*. Tesis Maestría en Ecología Tropical. Universidad Veracruzana, México.

Lozano, R.M.A., Menchaca, G.R.A., Alanís, M.J.L. and Pech-Canché, J.M. (2015) Cultivo *in vitro* de yemas axilares de *Vanilla planifolia* Andrews con diferentes citocininas. *Revista Científica Biológico Agropecuaria Tuxpan* 4(6), 1153–1165.

Menchaca, G.R. (1999) *Germinación in vitro de Vainilla (Vanilla planifolia)*. Tesis de Licenciatura, Facultad de Biología Universidad Veracruzana, México.

Menchaca, G.R. (2011) *Obtención y caracterización morfológica de híbridos de Vanilla Planifolia G. Jackson in Andrews y Vanilla pompona Schiede*. Tesis de Doctorado en Ecología Tropical, Universidad Veracruzana, México.

Menchaca, G.R.A., Ramos, P.J., Moreno, M.D., Luna, R.M., Mata, R.M., Vázquez, G.L.M. and Lozano, R.M.A. (2011) Germinación *in vitro* de híbridos de *Vanilla planifolia y V. pompona*. *Revista Colombiana de Biotecnología* 13, 80–84.

Montiel, F.M.Y. (2011) *Reproducción in vitro de yemas axilares de Vanilla planifolia Andrews bajo diferentes concentraciones de 6-BAP*. Tesis Facultad de Ingenieria Agrohidráulica Benemérita Universidad Autónoma de Puebla.

Moreno, R.Y. (2006) *Cultivo in vitro de embriones inmaduros de vainilla (Vanilla planifolia)*. Tesis Facultad de Ciencias Agrícolas, Universidad Veracruzana.

Ortega Macareno, L.C., Iglesias Andreu, L.G., Beltrán H.J.D. and Ramírez Mosqueda, M.A. (2016) Aislamiento y fusión de protoplastos *Vanilla planifolia* Jacks. Ex Andrews y *Vanilla pompona* Schiede. *Revista de la Asociacion Colombiana de Ciencias Biologicas* 28, 16–24.

Porras-Alfaro, A. and Bayman, P. (2007) Mycorrhizal fungi of *Vanilla*: specificity, phylogeny and effects on seed germination and plant growth. *Mycologia* 99, 510–525.

Ramos, N.J. (2011) *Evaluación de sustratos y aclimatación de Vanilla planifolia*. Tesis Facultad de Ingenieria Agrohidráulica Benemérita Universidad Autónoma de Puebla.

Richards, A.J. (2001) Does low biodiversity resulting from modern agricultural practice affect crop pollination and yield? *Annals of Botany* 88, 165–172.

Reyes-López, D., Soto Castañeda, C., Huerta Lara, M., Avendaño, C.H., and Cadena Iñiguez, J. (2006) *Influencia de rayos gamma (60 co) en esquejes de vainilla (Vanilla planifolia Andrews.* Memorias Del XXII Congreso Nacional y II Internacional de Fitogenética.

SEMARNAT (2010) *Norma Oficial Mexicana (NOM-059-ECOL-2001) de Protección especial de especies nativas de México de Flora y Fauna silvestres.* Diario Oficial de la Federación, 31 de Dic.

Soto-Arenas, M.A. (1999) *Filogeografía y recursos genéticos de lasvainillas de México.* Instituto Chinoin AC. Informe final SNIB-CONABIO proyecto No. J101. México D.F.

Soto-Arenas, M.A. (2006) La vainilla: retos y perspectivas de su cultivo. *Biodiversitas* 66, 1–9.

Whithner, C.L. (1955) Ovule culture and growth of *Vanilla* seedlings. *American Orchid Society Bulletin* 24, 380–392.

13

Curing of Vanilla

Chaim Frenkel, Arvind S. Ranadive, Javier Tochihuitl Vázquez,
and Daphna Havkin-Frenkel

13.1 Introduction

The mature-green pod (fruit) of the vanilla species used in commerce, including *Vanilla planifolia* and *Vanilla tahitensis*, is subjected to a curing process as a means for developing the prized vanilla flavor. *V. planifolia*, a climbing orchid, is indigenous to Mexico and neighboring Mesoamerica regions. Recently, *V. tahitensis* has been shown to be a hybrid between *V. planifolia* (maternal parent) and *V. odorata* (paternal parent) (Lubinsky *et al.* 2008). The plants are cultivated commercially, mostly in various tropical regions, requiring 3 to 4 years to set the flower, and flowers once a year. The pod-like fruit, also termed the "vanilla bean", is allowed to develop for 8 to 10 months and is then harvested, usually at the mature-green stage, followed by a curing process as discussed in this chapter. Annual worldwide production of cured vanilla beans is around 2,000 tons (US Department of Commerce).

Vanilla, first introduced and cultivated in Europe in 1520 by the Spanish Conquistador Hernan Cortes would not set fruit, because fruit set in its native habitat is dependent on vanilla flower pollination by the Melipona bee, a local insect (Childers *et al.* 1959). The discovery by Edmond Albius that a vanilla flower could be hand-pollinated, around 300 years later, created an opportunity for the commercial cultivation of vanilla in alternative global regions, mostly around equatorial zones, such as Madagascar, Indonesia, Papua New Guinea (PNG), and India. On-the-vine growth and development of the vanilla pod is manifested by a rapid increase in tissue mass, followed by a stage of maturation required for the formation of precursor compounds that give rise to aroma and flavor constituents during on-the-vine senescence or by the off-the-vine curing process.

In commerce, cultivated vanilla beans are harvested when green and flavorless. To bring out the prized vanilla flavor, green beans are subjected to a curing process commonly lasting three to six months, depending on various curing protocols in different localities. The objective of the curing process is two-fold: i) Development of the vanilla flavor and ii) Creation of shelf-life for cured beans by drying. Cured dry beans can be stored, distributed, and used subsequently for an ethanolic-water extraction that

Handbook of Vanilla Science and Technology, Second Edition.
Edited by Daphna Havkin-Frenkel and Faith C. Belanger.
© 2019 John Wiley & Sons Ltd. Published 2019 by John Wiley & Sons Ltd.

renders the familiar vanilla extract, as well as usage in other vanilla products. Vanilla cultivation, biosynthesis of flavor constituents, and economic aspects are discussed extensively in other reviews (Ranadive 1994; Dignum *et al.* 2001a; Havkin-Frenkel and Dorn 1997; Rao and Ravishankar 2000). In this chapter, we focus on the curing process and outline various aspects that might influence the flavor quality of cured vanilla beans, including botany of the vanilla bean, the nature and purpose of the curing process, as well as effects of curing practices in various production regions (see Addendum).

13.2 Botany of the Vanilla Pod

13.2.1 Two Fruit Regions

The syncarpous fruit of *Vanilla planifolia*, that is, the fruit with fused ovarian carpels, develops from an inferior ovary that eventually splits open along three lines at maturity, thus forming a capsule. Apparently, two principal parts in the vanilla pod are important for flavor development in the course of the curing process (Figure 13.1):

i)   the fruit wall, containing a "green" region including the epidermis, ground and vascular tissues of the fruit wall, surrounding the cortex with lesser chlorophyll content and whitish appearance; and

ii) the "white" inner region composed of the three parietal placentae (not including seeds), and the three bands of glandular hair-like cells between them. The glandular hair cells might play a role in the biosynthesis of glucovanillin.

13.2.2 Fruit Components

The fruit wall containing the outer "green" region and the cortical region, with lesser chlorophyll content, comprise about 60 and 20% of the fruit weight, respectively. The inner "white" region, containing the placenta, hair cells, and the seed components comprise the balance. However, this weight ratio of the outer and inner portions appears to

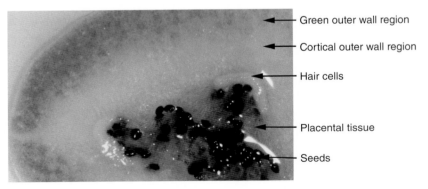

Green outer wall region

Cortical outer wall region

Hair cells

Placental tissue

Seeds

Figure 13.1 Cross-section (× 20) of freshly cut green vanilla bean. The figure shows the outer wall composed of a green wall region and the cortex, with lesser chlorophyll density. Also shown is the inner pod portion composed of placental tissue, hair cells, and seeds (dark bodies). Reproduced with permission from Havkin-Frenkel, D., French, J.C., Graft, N.M., Pak, F.E., Frenkel, C. and Joel, D.M. (2004) Interrelation of curing and botany in vanilla (*Vanilla planifolia*) bean. *Acta Hort. (ISHS)*, **629**, 93–102.

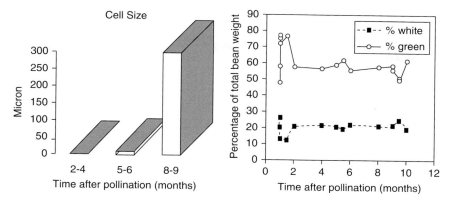

Figure 13.2 Time course change in cell size of vanilla pod during on-the-vine development (left) and associated changes in relative abundance of green and cortical white tissue in the outer cell wall tissue (right). Reproduced with permission from Havkin-Frenkel, D., French, J.C., Graft, N.M., Pak, F.E., Frenkel, C. and Joel, D.M. (2004) Interrelation of curing and botany in vanilla (*Vanilla planifolia*) bean. *Acta Hort. (ISHS)*, **629**, 93–102.

change during early and advanced stages of on-the-vine pod development, as cel size changes with pod development (Figure 13.2).

13.2.3 Fruit Anatomy

The epidermis, a cell layer enveloping the fruit, contains iso-diametric ground epidermal cells, which lack prominent chloroplasts. Each epidermal cell contains a rhomboidal crystal of calcium oxalate and is bounded by thickened, pitted cell walls. Stomata are widely spaced. In some varieties, dozens of extra floral nectaries occur on the fruit. In other varieties, these extra floral nectaries are entirely absent. The outer green fruit wall region contains a ring of about 15 vascular bundles. The vascular bundles are unbranched, and each contains a strand of xylem and phloem with a sclerotic bundle sheath. The xylem consists of annular to helical and reticulate elements. Tissue outside the ring of vascular bundles is composed of thin-walled parenchyma cells several times longer than wide. Each ground parenchyma cell in the cortex of the outer fruit wall contains chloroplasts and occasional rhomboidal calcium oxalate crystals. Needle-shaped crystal (raphide) "vessels" are abundant in the outer fruit wall and, when the fruit is cut, these cells release mucilage-containing raphides, which are highly irritating when coming into contact with skin. No attempt has been made to determine the development or structure of these large, complex cells, which are many times the length of ground epidermal cells and contain tightly packed bundles of raphides if undisturbed. Compared with the outer fruit wall region, the wall tissue inside the ring of vascular bundles contains larger cells with somewhat less abundant and smaller chloroplasts, and is much less green in freshly cut beans.

13.2.4 Pollination Initiates Ovary and Fruit Development

The inferior ovary of the non-pollinated vanilla flower has three weakly developed parietal placentae separated from each other by the smooth inner epidermis of the ovary.

Pollination triggers the placenta to begin extensive branching, followed by ovule development. Perhaps more important for the vanilla industry are unusual glandular hair cells that begin to develop quickly in the regions between the placentae. Each hair cell is unbranched and soon reaches a length of about 300 micrometers. Following pollination, large numbers of pollen tubes progress down the ovary moving in three groups, each located in a narrow pocket at one side of each of the three placentae, flanked by the hairs. The hairs become cemented together during their development, and later break down, releasing their contents into the surrounding locule. The developing hair cells have abundant endoplasmic reticulum, ribosomal structures, enlarged plastids containing lipid globules, and other features that are the hallmarks of metabolically active cells.

13.2.5 Mature Fruit

As the fruit develops, the inter-placental hairs develop thickened walls and a complex cytoplasm. Because of their size, number, and thick walls, the hairs are easily observed in transverse sections of vanilla pod, as three lustrous white bands. Many seeds become appressed into the hairs in the mature fruits. The three panels of hairs extend the full length of the fruit. The cells contain abundant lipids, which are released onto the locule and coat the seeds when the hairs senesce later in ripening. The hairs develop complex cell walls, which cement the hairs together in mature beans. Swamy (1947) suggested that vanillin is produced in these hairy cells. This suggestion has been confirmed by our work (Joel *et al.* 2003), showing that vanillin and related intermediates in the vanillin biosynthetic pathway accumulate in the inner white tissue of a developing vanilla pod, around the plancental hairs (Figure 13.3). This information may be important for understanding of the curing protocol, as outlined in Sections 13.3 and 13.4.

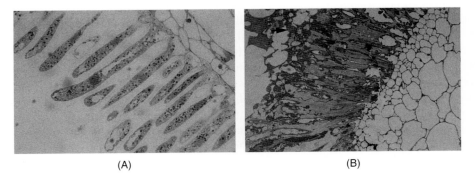

(A)   (B)

Figure 13.3 Magnified views (400 ×) of cross sections of green vanilla beans. (A) Cross section of a developing vanilla bean. The hair cells are distinct. (B) Cross section of an older bean showing the senescing inter-placental hairs (left) and white parenchyma cells of the fruit wall (right). The hair-like cells contain enzymes in the vanillin biosynthetic pathway. These cells release abundant lipid seen as globular bodies (arrow). The parenchyma cells, comprising the white cortical portion of the frit wall, contain degradative enzymes. Reproduced with permission from *Perfumer & Flavorist* magazine, Allured Business Media, Carol Stream, IL.

13.3 On-the-vine Curing Process in a Vanilla Pod

Naturally occurring on-the-vine senescence of a vanilla pod (Figure 13.4) might be a context for viewing cellular and metabolic changes occurring during off-the-vine (commercial) curing of vanilla bean, as discussed in Section 13.4. It is commonly observed that at the end of vanilla pod development and maturation, lasting around 9 to 10 months, the vanilla pod manifests de-greening and onset of yellowing. This change, made visible by chlorophyll degradation and, subsequently, unmasking of yellow carotenoid pigments, is a universal mark for the onset of ripening in fruit, including the vanilla pod. Another pronounced feature is the subsequent onset and progressive pod browning, stemming mostly from oxidative degradation of phenolic compounds. Yellowing and browning represent different and contrasting cellular states: The former marks an end point in pod development, where cellular processes are under genetic and tight metabolic control. Browning, in contrast, marks the loss of cellular organization and metabolic control and denotes the onset of degradative processes that have escaped cellular regulation. The latter may include degradation of vital cellular biopolymers and loss of membrane-driven compartmentalization of cellular constituents (Hopkins *et al.* 2007; Lim *et al.* 2007) and, importantly, activity of cell wall degrading enzymes, for example, protein degrading enzymes and enzymes that catalyze the hydrolytic cleavage of various glycosylated compounds, notably, glucovanillin. Further work might also reveal enzyme-catalyzed degradation of lipid and membrane-lipid and probably nucleic acids. Onset of pod yellowing and subsequent browning also represent a contrast from an energy perspective. Whereas yellowing and other ripening-related processes, representing organized cellular reactions, are predicated on free energy input, pod browning, which signifies destruction of cellular organization, is an entropy-driven process, entailing energy dispersion. Destruction of

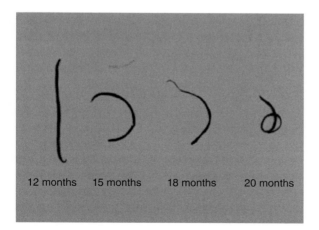

12 months   15 months   18 months   20 months

Figure 13.4 Vanilla bean undergoing on-the-vine senescence for 12, 15, 18, and 20 months after pollination. Continued water loss results in curling of the senescent vanilla pod. Reproduced with permission from Havkin-Frenkel, D., French, J.C., Graft, N.M., Pak, F.E., Frenkel, C. and Joel, D.M. (2004) Interrelation of curing and botany in vanilla (*Vanilla planifolia*) bean. *Acta Hort. (ISHS)*, **629**, 93–102.

cellular organization, particularly the loss of compartmentalization, is associated with unrestricted mobility and diffusion of matter between intra or inter-cellular compartments. For example, diffusion of vacuole-held constituents, such as organic acids, phenolic compounds, ions, or proteases onto the surrounding cytoplasm, as well as adjoining tissue regions, might lead to deleterious consequences. An example is cytoplasmic acidification and subsequent death, resulting from stress-induced leakage of organic acids from the cellular vacuole (Yoshida 1991, 1994). Pod browning is a salient manifestation of collapsed cellular organization, resulting, in part, from unrestricted and uncontrolled diffusion of harmful metabolites, unhindered enzyme-substrate interaction, as well as accessibility to ambient atmospheric oxygen. Whereas in viable plant tissue bio-membranes function as gas diffusion barriers (Grinberg *et al.* 1998), bio-membrane destruction in a browning pod results in removal of membrane hindrance to oxygen diffusion and, in turn, onset of enzymatic and non-enzymatic oxidative reactions and, moreover, formation of reactive oxygen species (ROS) arising, apparently, from lipid oxidation, as discussed in Section 13.5. Vanilla pod browning, a hallmark of on-the-vine senescence as well as off-the-vine bean curing, is an expression of collapsed biological order. In a senescing vanilla pod, enzyme-catalyzed hydrolytic cleavage and oxidation of cellular constituents might provide products useful in nutrition and protection of developing seeds.

Naturally occurring pod senescence and consequent browning may also be instigated by applied ethylene, as observed by Arana (1944). Our own work revealed that ethylene-induced browning is intensified when the gas is applied in oxygen. This is in accordance with the view that enhanced oxygen accessibility is stimulatory to the oxygen-dependent browning reaction (Figure 13.5). The treatment also stimulated bean-end splitting. Pod browning, signifying loss of cellular and tissue organizational integrity, instigated by on-the-vine senescence or by applied ethylene, is emulated by the off-the-vine curing process. It is worth noting that the Tahitian curing method (described in Section 13.9) is based on allowing on-the-vine pod browning and completion of the process without artificial killing after the bean has been harvested.

13.4 Off-the-vine Curing Process of Vanilla Beans

The commercial curing process, that is, off-the-vine induced destruction of cellular and tissue organization, creates conditions allowing the free flow and interfacing of previously compartmentalized cellular constituents, resulting in enzyme-substrate interactions as well as unrestricted access to atmospheric oxygen. These conditions launch the onset of hydrolytic and oxidative reactions that contribute to the formation of the prized vanilla flavor.

Cellular and tissue de-compartmentalization is obviously fundamental for accessing precursor metabolites for enzyme-catalyzed generation of flavor and aroma constituents. In whole green beans, phenolic compounds are restricted to the pod interior, as evidenced by catechin staining. In killed beans, the phenolic compounds have diffused from the inner portion of the pod and have populated the entire tissue, including the outer wall region (results not shown). Catechin staining indicated, moreover, that killing by freezing was more thorough and uniform than killing by dipping beans in hot water at 65 °C for 3 minutes (Havkin-Frenkel and Kourteva 2002). One important

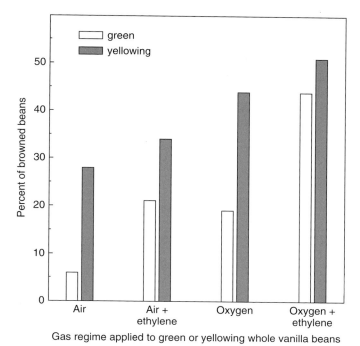

Figure 13.5 Mexican vanilla beans were harvested 7 months after pollination and sorted by background color (green and yellowing). Green or yellowing beans were held in 20-liter glass jars and ventilated at a rate of 200 ml/minute with different gas mixtures consisting of air, air + ethylene, oxygen, and oxygen + ethylene. The concentration of applied ethylene was 10 µl/liter gas. Reproduced with permission from *Perfumer & Flavorist* magazine, Allured Business Media, Carol Stream, IL.

consequence of killing and attendant de-compartmentalization is free diffusion of glu-covanillin, the vanillin precursor, from the bean interior to other regions in the vanilla pod. Unrestricted mobility of the compound creates conditions for contact with and hydrolytic cleavage of the compound by β-glucosidase-catalyzed action, observed mostly in the outer pericarp tissue (Arana 1943; Ranadive *et al.* 1983; Havkin-Frenkel and Kourteva 2002; Dignum *et al.* 2001a), and perhaps also with β-glucosidase found in the seeds (Jiang *et al.* 2000). Odoux *et al.* (2006) observed, in the same vein, that gluco-vanillin hydrolysis was low in green beans, even though β-glucosidase activity was sub-stantial, whereas in curing beans glucovanillin formation was robust although the enzyme activity was barely measurable. From these results they concluded that a state of de-compartmentalization defined the conditions for onset of glucovanillin hydroly-sis. It might be argued by extension, that these very conditions underscore the require-ment for contact between other glycosyl hydrolases and their respective precursor substrates that might contribute to the production of flavor and aroma constituents in the curing vanilla pod.

It has been suggested that curing-associated flavor formation might stem also from some synthetic activity and not merely from degradative processes. There is no hard evidence to support or refute this concept. We argue, however, that the probability of synthetic events occurring in cells and tissues experiencing organizational collapse is questionable because:

- Biosynthesis is often dependent on precise structural assembly of enzymes and proteins catalyzing biosynthetic pathways, mitochondrial ATP producing machinery for example (Lenaz and Genova 2009). These conditions are not expected to prevail in killed vanilla pods.
- Biosynthesis is generally dependent on free energy input and may not proceed in the killed pod, where metabolic machinery for energy production (ATP or reduced pyridine nucleotides) is not expected to survive.
- A strong proteolytic activity, unleashed by killing, can readily disrupt the molecular and structural integrity of enzymes required for biosynthetic processes in intact cells.

It has also been suggested that micro-organisms might contribute to an overall vanilla flavor (Ranadive 1994) and perhaps also to vanillin formation, because various microorganisms colonizing the vanilla pod during the curing process (Roling *et al.* 2001) manifest glycosidase activity and efficacy to convert ferulic acid to vanillic acid and related compounds (General *et al.* 2009). Additional studies might reveal whether hydrolytic release of vanillin is catalyzed exclusively by the bean endogenous enzyme or, alternatively, might also originate from colonizing micro-organisms.

13.4.1 Purpose of Curing

Following pollination and subsequent fruit set, the developing vanilla pod undergoes rapid growth for 3 months followed by growth cessation. Next, the fully grown vanilla pod enters a period of maturation, lasting several months. During on-the-vine bean development, lasting 8 to 10 months, flavor precursors accumulate, mostly in the placental tissue surrounding the seeds in the inner core of the bean (Figure 13.6). Premature harvesting of the pod, even at full size, results in formation of poor flavor upon subsequent curing.

However, when mature-green vanilla beans are harvested, they lack flavor. This phenomenon might stem from spatial separation of flavor precursors and corresponding enzymes that catalyze their breakdown to final flavor components. For example, glucovanillin, a vanillin precursor and β-glucosidase, which catalyzes the hydrolytic release of vanillin from glucovanillin, are apparently sequestered in different tissue regions in the vanilla pod. Thus, while glucovanillin is found mostly in the inner portion of the pod, estimation of β-glucosidase activity indicated that the enzyme reaction rate was roughly 10-fold higher in the outer fruit wall than in the inner pod region, including the placental tissue and the hair cells (see Section 13.5.3). This was also confirmed by histochemical staining of a cross-section of vanilla pod for the enzyme activity (results not shown). These data, indicating that β-glucosidase is localized mostly in the pod outer region, suggest that in intact tissues of green beans, the enzyme is spatially separated from glucovanillin and likewise, other glycosyl hydrolases might also be separated from their flavor precursors. The purpose of the curing process, then, is to create conditions for substrate-enzyme interaction and, thereby, onset of enzyme-catalyzed formation of vanillin or other flavor constituents, as well as onset of enzymatic and non-enzymatic oxidative reactions, by allowing contact with atmospheric oxygen. An additional objective is the drying of cured beans, as a preservation method for retaining the formed

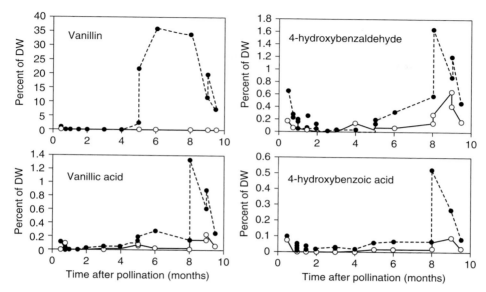

Figure 13.6 Time-course of change in the content of various metabolites in the green outer tissue (solid lines) and the inner white tissue (dashed lines) of a vanilla bean during pod development on the vine. Beans were harvested green at various stages of development. The various metabolites, present as glucosides, were hydrolyzed and the resulting aglycons determined, as described previously (Podstolski *et al.* 2002). Reproduced with permission from *Perfumer & Flavorist* magazine, Allured Business Media, Carol Stream, IL.

flavor compounds. Vanilla flavor contains around 250 identified constituents (Adedeji *et al.* 1993), chief among them is vanillin. Because the vanillin content in cured beans is a major criterion for bean quality, previous studies on the curing process focused on the production of vanillin from glucovanillin, and this topic will also receive special attention in the present chapter.

13.4.2 Traditional Methods of Curing

The curing process is comprised of four major stages including killing, sweating, drying, and conditioning.

13.4.2.1 Killing

The fundamental purpose of the killing stage is to bring about the cessation of the vanilla bean vegetative life and, furthermore, to disrupt cellular and tissue organization in the green bean, such that previously segregated enzymes and their corresponding substrates can come in contact and interact (Arana 1943; Theodose 1973). This reasoning is supported by the observation that disruption of green bean tissue by mechanical means, tissue maceration by chopping or grinding for instance, initiates a curing process (Towt 1952), including rapid degradation of glucovanillin to vanillin, apparently by β-glucosidase-glucovanillin interaction. Modern methods of killing, aimed at instigating cell and tissue disorganization, are based on the observation that killing by the ancient Mexican method consisted of wilting beans in the sun until they became brown, a manifestation of tissue death (Balls and Arana 1941a). Contemporary killing methods consist of:

- sun killing;
- oven killing;
- hot water killing;
- killing by scratching;
- killing by freezing (Childers *et al.* 1959).

The most practical and most commonly used killing methods are exposure of green beans to the sun, killing by oven heat, or hot water killing (Ranadive 1994). In sun-killing, a method originating from Mexico and practiced by the ancient Aztecs (Balls and Arana 1941a,b), beans are held under dark cloth and exposed to direct sun for several days, until the beans turn brown. In oven killing, the beans are tied in bundles and rolled in blankets and placed in an oven at 60 °C for 36 to 48 hours (Childers *et al.* 1959; Ranadive 1994). Hot water killing consists of placing the green beans in wire baskets and submerging them in hot water (60–70 °C) for several minutes. A variation of this method consists of repeated submersion for 10 seconds at a time at higher temperatures (80 °C) (Childers *et al.* 1959). Freezing, by dipping in liquid nitrogen or by holding the beans for a few hours in a freezer (0–80 °C), is yet another method of killing (Ansaldi *et al.* 1990). Our own experience and results by other studies (Dignum *et al.* 2001a, b) indicate, however, that it is preferable to quick-freeze and then store frozen beans at −70 °C or below, in order to preserve the viability of enzymes that upon subsequent thawing drive the curing process.

Jones and Vincente (1949a) evaluated various killing methods with respect to the quality of cured vanilla beans and found that killing by hot water scalding gave the best product, with freezing second, and scratching third. Although the various killing methods achieve the same objective, namely, disruption of tissue organization and creation of contact between enzymes and substrates, survival of different enzyme constellations and, subsequently, variation in flavor profile of cured beans might result from different killing methods. For example, the highest activity of β-glucosidase, peroxidase, polyphenol oxidase, and protease was found in heat-killed beans, whereas killing by freezing resulted in a different profile of enzyme activity (Ranadive *et al.* 1983; Dignum *et al.* 2001b, 2002b).

Severe killing conditions, excessive or prolonged heat for instance, may lead to a complete destruction of beneficial enzymes and, subsequently, arrest of enzymatic activity required for catalyzing formation of flavor compounds. We believe that rapid killing by heat or freezing is more beneficial for bean quality because these methods achieve the goal of killing, namely, tissue disorganization, while minimizing deleterious effects on the viability of beneficial enzymes.

13.4.2.2 Sweating

The killing stage is followed by "sweating", a condition consisting of high humidity and high temperature (45°–65 °C) for 7 to 10 days (Balls and Arana 1941a, 1942). During this period, the killed bean develops the characteristic vanilla flavor, aroma, and color. During sweating, high enough moisture content is retained as necessary for enzyme-catalyzed reactions. At the same time, enough moisture is allowed to escape in order to reduce water activity to a level that restricts activity and spoilage by micro-organisms.

Broadly speaking, enzyme activity during the sweating stage consists of hydrolytic and oxidative action. Apparently, some non-enzymatic oxidative reactions might also occur during this period. The use of elevated temperatures during this stage is to accelerate enzymatic and perhaps also non-enzymatic processes and is practical exploitation of heat stability of hydrolytic and oxidative enzymes stemming, in part, from the enzyme polypeptide side chain glycation with oligosaccharides (Nishi and Itoh 1992; Varki 1993). Traditionally, this process is carried out in Sweat Boxes, in a closed room but rarely in an oven (Ranadive 1994). High temperatures are also achieved by wrapping killed beans in various cloth materials, by densely stacking killed and warm beans in insulated containers and by re-warming with exposure to the sun for a few hours each day during the sweating period. In some instances, the sweating beans are dipped daily in hot water (Balls and Arana 1941a; Childers *et al.* 1959; Theodose 1973; Dignum *et al.* 2002b).

13.4.2.3 Drying and Conditioning

At the end of a sweating period, beans have attained a brown color and have developed most of the flavor and aroma characteristic of cured beans. However, at this stage, beans contain about 60 to 70% moisture and are, therefore, subject to spoilage by micro-organisms upon prolonged standing. Subsequently, beans that have completed the sweating period are dried to a moisture content of 25 to 30% of the bean weight (Ranadive 1994), a process that lends shelf-life to cured vanilla beans. Drying might also lead to the expulsion of volatile compounds, such as hexanal or other aldehydes and other compounds that impart "green" unripe notes to vanilla flavor. The most commonly used drying methods are sun and air-drying. These methods are occasionally supplemented by oven drying. Sun drying consists, traditionally, of spreading the beans on racks in the morning sun and transferring the sun dried beans to a shaded area in the afternoon. This protocol may be carried out daily for 3 months. Theodose (1973) divided the process into rapid and slow drying where, in the former, beans are held in the sun for a few hours every day and then wrapped in cloths and placed indoors. This process is repeated for 5 to 6 days until the beans become supple, a sign of sufficient drying. In slow drying, the beans are placed on shelves in a well-aerated room and are moved outside into the sun every 2 to 3 days. This method of drying may last one month. Other workers (Kamaruddin 1997; Ratobison *et al.* 1998) proposed using drying equipment based on solar energy. Because drying is the longest stage in the curing process, Theodose (1973) proposed combining traditional drying with hot air drying to shorten the drying period. Drying is the most difficult stage in the curing process. Uneven drying may result from varying bean size, differences in bean moisture content, and from variable environmental conditions, when outdoor drying methods are practiced. The latter may include weather conditions during sun drying or from variations in the relative humidity during sun or air-drying. The drying stage is apparently critical to the development of the full rich vanilla flavor, but prolonged drying may lead to loss of flavor and in vanillin content.

Bean appearance and suppleness are used by practitioners the trade as an index for moisture content. When beans are judged to have reached sufficient dryness, they are placed in wooden boxes and held for "conditioning" for an additional few months. This stage may be viewed as a continuation of the drying process where additional moisture and volatiles may be lost. However, this stage might also be accompanied by enzymatic and non-enzymatic oxidative processes that alter the vanilla flavor. Arana (1944) emphasized the probable importance of oxidative enzymes in general, and peroxidative enzymes in particular during conditioning, suggesting that vanillin or other phenolic compounds might be oxidized to quinones or other complex structures that might give rise to additional flavor notes. We show (below) that curing is also associated with lipid oxidation and, apparently, an additional origin for oxidant-induced flavor formation in a curing vanilla pod. The low rate of oxidative reactions might account for the prolonged conditioning period, lasting 5 to 6 months.

13.5 Activity of Hydrolytic Enzymes Occurring in a Curing Vanilla Pod

Senescence in plants is accompanied by extensive hydrolytic breakdown of cellular macromolecules, catalyzed by various hydrolytic enzymes (Rogers 2005; Hopkins *et al.* 2007). This process is illustrated, for example, by β-*D*-glucosidase-catalyzed hydrolytic cleavage of glucovanillin recorded in a senescing vanilla pod (Odoux *et al.* 2006). Given that the curing process is a mimic of senescence, it is reasonable to expect activity of a host of hydrolytic enzymes in killed vanilla beans, as outlined below.

13.5.1 Protease Activity

Killing and subsequent curing is associated with proteolytic activity in the vanilla pod. This conclusion is inferred from changes in the bean protein content, showing precipitous decline within 24 hours of killing, but a persistent level of protein content afterward (Figure 13.7). Wild-Altamirano (1969) showed that on-the-vine pod development is associated with a decline in protease activity, although the enzyme activity remains steady when beans have matured. Following killing the pod protease activity declined within 2 days to about 60 to 70% of the initial pre-killing level and remained steady afterward (Figure 13.8). Apparently proteases resist the severe

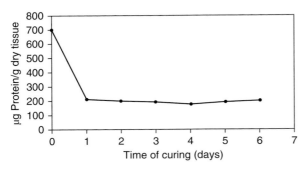

Figure 13.7 Time-course of change in total protein content in vanilla bean undergoing curing at 50 °C. Total proteins were extracted periodically from bean tissue and estimated as previously described (Ranadive *et al.* 1983). Reproduced with permission from *Perfumer & Flavorist* magazine, Allured Business Media, Carol Stream, IL.

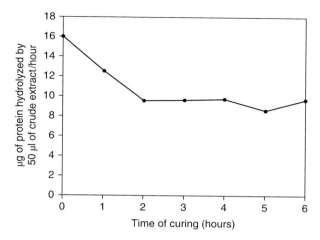

Figure 13.8 Time-course of change in proteolytic activity in vanilla bean undergoing curing at 50 °C. Fifty μl of crude extract, denoting unit of pod tissue, represents 14.3 mg fresh weight of curing vanilla bean tissue. Reproduced with permission from *Perfumer & Flavorist* magazine, Allured Business Media, Carol Stream, IL.

killing conditions, since proteolytic activity has been shown to survive extreme scalding, for example, 30 minutes at 80 °C. The same harsh conditions de-activated other enzymes, various glucosidases or phenylalanine ammonia lyase for instance (Dignum *et al.* 2002a).

Proteolytic activity in the curing vanilla pod may result from the release of cellular proteases, previously compartmentalized in the vacuole (Okamoto 2006; Muentz 2007) or perhaps from other cellular compartments. It is also likely that latent proteolytic activity is triggered by killing-induced protein denaturation, a process leading to surface exposure of the protein hydrophobic core and a mechanism for proteolytic targeting of denatured proteins (Bond and Butler 1987). Denaturation of cellular proteins may occur during killing by heat or freezing and, in addition, by previously compartmentalized cellular constituents that might be deleterious to correct folding and function of cellular proteins. Examples include cytosol acidification by organic acids or denaturation by phenolic compounds that have diffused out of the vacuole. Lipid peroxides formed in cured beans (Figure 13.9), and perhaps other oxidants, may also attack and denature cellular proteins (Bond and Butler 1987).

A marked decrease in protein content after a few hours of curing (Figure 13.7) suggests that the action of proteases diminished the level of enzymes and proteins. However, glycosyl hydrolases that catalyze the hydrolysis of glyco-conjugates, glucovanillin for example, may be temporarily spared from proteolytic degradation, because extracellular glycosyl hydrolases are glycoproteins. The latter, composed of a polypeptide glycated with oligosaccharide side chains (Trincone and Giordano 2006; Lopez-Casado *et al.* 2008), display resistance to proteolysis (Nishi and Itoh 1992; Varki 1993). This presumption is in keeping with the observation that β-glucosidase activity persists during the harsh curing conditions and is sufficient to carry out hydrolytic cleavage of glucovanillin to near completion (Figure 13.10), a conclusion verified also by Odoux *et al.* (2006).

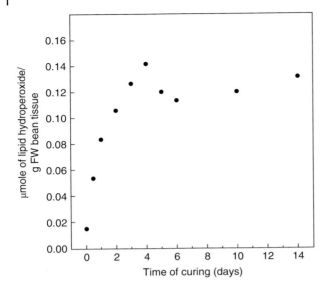

Figure 13.9 Change in the content of lipid hydroperoxides in bean tissue during curing at 50 °C. Tissue increments of vanilla bean were removed periodically during curing for the estimation of lipid hydroperoxide, as previously described (Eskin and Frenkel 1976). Reproduced with permission from Havkin-Frenkel, D., French, J.C., Graft, N.M., Pak, F.E., Frenkel, C. and Joel, D.M. (2004) Interrelation of curing and botany in vanilla (*Vanilla planifolia*) bean. *Acta Hort. (ISHS)*, **629**, 93–102.

13.5.2 Cell Wall Hydrolyzing Enzymes

Several studies observed that addition of commercial preparations of cell wall degrading enzymes accelerated the hydrolysis of glucovanillin to vanillin in curing vanilla pods (Mane and Zucca 1993; Brunerie 1998; Ruiz-Teran *et al.* 2001). These results suggest that glucovanillin or other glycosylated phenolic compounds trapped, apparently, in the wall matrix, might be released by the wall degradation and become accessible to their respective hydrolytic enzymes. We observed, accordingly, that addition of pure preparations of pectin-degrading enzymes to chopped green beans accelerated the conversion of glucovanillin to vanillin. However, at the end of the incubation period, the vanillin content was roughly the same in control beans (no enzyme added) as in enzyme-treated beans, the only difference being the rate of vanillin formation in control and treated beans (results not shown). These data suggest that activity of endogenous wall hydrolyzing enzymes is sufficient for the conversion of glucovanillin to vanillin, whereas applied enzymes merely accelerate the rate of the process. Additional studies may reveal whether the wall degradation is essential for accessing and subsequent conversion of glucovanillin to vanillin or whether cell wall dissolution might, alternatively, increase the extractability of released vanillin and other flavor constituents, as we previously found (unpublished data).

13.5.3 Glycosyl Hydrolases

Glucovanillin is a major glycosyl conjugate of vanillin, although trace amounts of other glycosyl conjugates of vanillin or other phenolic compounds containing mannose,

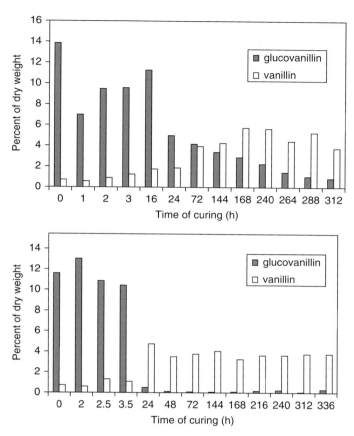

Figure 13.10 Time-course of change in the content of glucovanillin and in vanillin in whole vanilla beans (top) and chopped beans (bottom) undergoing curing at 50 °C. Reproduced with permission from *Perfumer & Flavorist* magazine, Allured Business Media, Carol Stream, IL.

galactose, and rhamnose are found in the developing vanilla pods (Leong *et al.* 1989a,b; Tokoro *et al.* 1990; Kanisawa *et al.* 1994; Pu *et al.* 1998; Dignum 2001a). According to Arana (1944), glucovanillin was first isolated from the vanilla bean in 1858 by Gobley, followed by a demonstration that the compound undergoes hydrolytic cleavage to vanillin and glucose during the curing process (Goris 1924). It is generally accepted that vanillin is formed by β-glucosidase-catalyzed hydrolytic cleavage of glucovanillin, although green vanilla beans contain other glycosyl hydrolases, including α- and β-glucosidase, α- and β-galactosidase, as well as α- and β-mannosidase (results not shown). Because of the importance of vanillin to vanilla flavor, β-glucosidase-catalyzed formation of vanillin is one of the most studied processes in a vanilla bean. The rate of glucovanillin conversion to vanillin and glucose may be measured by the rate of disappearance of glucovanillin and an accompanying accumulation of vanillin. Another approach is based on estimating the activity of β-glucosidase in bean tissue, assuming that the enzyme activity is an index of glucovanillin hydrolysis. Activity of β-glucosidase is measured traditionally with the use of *p*-nitrophenyl-β-glucopyranoside or with glucovanillin as substrates.

Temperature regimes during the killing and subsequent sweating stages appear to be critical to the activity of β-glucosidase (Marquez and Waliszewski 2008). Our studies revealed that temperature optima for enzymatic activity were 50 °C for β-glucosidase, 55 °C for α-galactosidase, and 60 °C for β-galactosidase. Activity of β-glucosidase and α- and β-galactosidase in curing vanilla beans held at 50 °C is substantial and measurable, whereas activity of other glycosyl hydrolases tends to be low(results not shown). Thermal-stability of glucosidases, arising from molecular features of the enzyme polypeptide chain, is discussed elsewhere (Sanz-Aparicio *et al.* 1998; Hrmova *et al.* 1999). Glycation of the polypeptide side chain with oligosaccharide is, apparently, another molecular feature conferring thermal stability on glycosyl hydrolases (Nishi and Itoh 1992; Varki 1993). Thermal tolerance of these enzymes is consistent with the empirical exploitation of elevated temperatures during the curing process, for the hydrolytic release of vanillin and perhaps other flavor components from glycol-conjugate precursors.

The conversion of glucovanillin to vanillin during the curing process is shown in Figure 13.11A. After 8 days of curing at 50 °C, the glucovanillin content decreased from an initial level of 14% to roughly6% on dry weight basis. During the same period, the vanillin content, liberated from glucovanillin, rose to approximately 6%. The content of the two compounds leveled off afterward. The hydrolytic release of vanillin appears to be accompanied also by the accumulation of vanillic acid, *p*-hydroxybenzaldehyde, and *p*-hydroxybenzoic acid (Figure 13.11B). An intriguing phenomenon is the accumulation of vanillin, whereas activity β-glucosidase, as well as other glycosyl hydrolases, declined during the same period. This occurrence casts doubt on the efficacy of the enzyme to catalyze hydrolysis of glucovanillin to vanillin. To explore this issue further, we examined the dependency of vanillin accumulation on enzymatic activity. Table 13.1 shows that application of either glucovanillin or β-glucosidase led to an increase in the vanillin content in curing fresh green beans and that vanillin content was increased further by the addition of both the substrate and the enzyme. However, when activity of

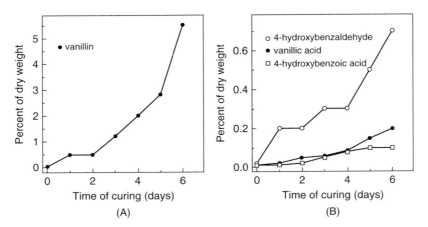

Figure 13.11 Time-course changes in the content of vanillin (A), vanillic acid, 4- hydroxybenzaldehyde and 4 hydroxybenzoic acid (B) in whole vanilla bean undergoing curing at 50 °C. Reproduced with permission from Havkin-Frenkel, D., French, J.C., Graft, N.M., Pak, F.E., Frenkel, C. and Joel, D.M. (2004) Interrelation of curing and botany in vanilla (*Vanilla planifolia*) bean. *Acta Hort. (ISHS)*, **629**, 93–102.

Table 13.1 Vanillin content in fresh and boiled whole vanilla beans supplemented with glucovanillin (GV) and β-glucosidase.

Hours of Curing	Tissue Condition	Compounds Added	Vanillin % of DW
0	fresh tissue	none	0.0
24			1.8
48			1.9
0	fresh tissue	glucovanillin	0.0
24			2.2
48			2.5
0	fresh tissue	β-glucosidase	0.0
24			4.2
48			3.2
0	fresh tissue	β-glucosidase + GV	0.0
24			6.7
48			6.5
0	boiled tissue	none	0.0
24			0.0
48			0.0
0	boiled tissue	glucovanillin	0.0
24			0.0
48			0.0
0	boiled tissue	β-glucosidase	0.0
24			4.5
48			4.5
0	boiled tissue	β-glucosidase + GV	0.0
24			6.8
48			6.7

endogenous β-glucosidase activity was abolished by tissue boiling, production of vanillin ceased altogether and could not be reconstituted, even by the addition of glucovanillin. Conversely, addition of β-glucosidase to the boiled tissue restored the process of hydrolytic release and accumulation of vanillin and was enhanced further by the addition of glucovanillin. Collectively, these results suggest that conversion of glucovanillin to vanillin is predicated on β-glucosidase-catalyzed action in the vanilla pod, a conclusion confirmed by Dignum *et al.* (2001b). Accordingly, controlled curing under laboratory conditions resulted in the disappearance of almost 95% of the glucovanillin with a potential yield of 5 to 7% vanillin on dry weight basis (Figure 13.10), suggesting sufficient enzymatic action to bring the hydrolytic release of vanillin to near completion. These results are supported by similar conclusions, suggesting that the level of β-glucosidase is not a limitation to the hydrolytic release of vanillin and merely determines the kinetics of the process (Odoux *et al.* 2006). Other studies (Odoux 2000) suggest, in contrast, that conversion of glucovanillin to vanillin during traditional curing in Reunion approached only 40% of the hydrolytic capacity of β-glucosidase. These results suggest that curing under traditional field conditions, yielding between 1.5 and 3% vanillin on dry weight basis, may not exploit the full potential of glycosyl hydrolases for vanillin release and accumulation. Alternatively, suboptimal levels of vanillin may reflect losses of the formed compound during the prolonged drying and conditioning stages (Arana 1943; Broderick 1956a,b; Odoux 2000).

Low activity of β-glucosidase in curing beans, resulting partially from heat de-activation (Marquez and Waliszewski 2008), may also stem from proteolytic destruction. Denaturation of the enzyme protein by phenolic compounds and perhaps by oxidants formed during the curing process might result in tagging of β-glucosidase for proteolytic degradation. These very conditions might be an impediment, however, for assessing the enzyme activity, because extraction and assay conditions may also lead to the enzyme denaturation and subsequent proteolytic degradation. For example, determination of enzyme-substrate affinity, measured by Km values in green pod extract, revealed that β-glucosidase affinity for natural or synthetic substrate was one order of magnitude lower than in other organisms (Dignum 2002b). These results suggest that low enzyme activity in a bean extract might reflect a dysfunctional state. Attempting to avert this problem, we found that protection of β-glucosidase from proteolytic degradation, using protease inhibitors in the extraction as well as the assay medium, resulted in increased enzyme-substrate affinity (lower Km) and increased reaction rate, compared to previously reported values (Dignum 2002b), and favorably comparable to those obtained by Hannum (1997). During curing, however, β-glucosidase and other glycosyl hydrolases, presumed to display resistance to proteolytic degradation (Nishi and Itoh 1992; Varki 1993), might persist at a level sufficient to carry out the hydrolytic release of vanillin or other glyco-conjugates in the vanilla pod. It is desirable to re-examine whether low β-glucosidase activity reflects the actual state of the enzyme protein in a curing bean or, alternatively, a dysfunctional state resulting from inappropriate extraction and assay protocols.

Substrate accessibility and subsequently enzyme-substrate interaction is yet another factor in the enzyme-catalyzed hydrolytic release of vanillin and perhaps other glyco-conjugates because glucovanillin, the vanillin parent compound and β-glucosidase that catalyzes the hydrolytic cleavage of glucovanillin, might reside in different regions of a vanilla pod. This view is the rationale and objective for killing, stated early on by Arana (1943) and confirmed by subsequent studies (Theodose 1973), namely, disorganization of vanilla pod tissue in order to establish contact between enzymes and their corresponding substrates, which are compartmentalized and separated in the green bean. This view is supported by studies showing that degradation of glucovanillin to vanillin, apparently by β-glucosidase as well as other flavor generating processes, is initiated by disruption of green bean tissue by mechanical means, tissue maceration by chopping or grinding, for instance (Towt 1952). Other studies indicate, similarly, that other killing methods lead to de-compartmentalization of enzymes and substrates and onset of flavor formation in curing vanilla bean (Odoux 2006). Assay for β-glucosidase activity, when protected against proteolytic degradation, revealed that the enzyme activity, expressed as μg product/hr/μg protein, was 75.2 in the green outer fruit tissue, 32.3 in the placental tissue, and 11.1 in the hair cells, respectively, suggesting enzyme localization mostly in the green outer region. Other studies (Odoux 2006) indicate that the enzyme is localized in the inner placental region. Knowledge of the enzyme localization in the vanilla pod is, therefore, contentious. While there is progress in the understanding of the site of synthesis and accumulation of glucovanillin, we do not have unequivocal knowledge on the localization of β-glucosidase in the vanilla pod. Molecular methods, for example, immuno-cytochemistry or β-glucosidase-green fluorescent protein fusion used for the enzyme visualization and localization in plants (Matsushima *et al.* 2003; Suzuki *et al.* 2006), might be used to elucidate the localization of the enzyme in the vanilla pod. Future studies might reveal whether glucovanillin and β-glucosidase are localized in different regions of the vanilla pod or, alternatively, in close proximity of the same tissue. This information may have a bearing on devising new killing and curing methods to optimize enzyme-substrate interaction and consequent flavor formation.

Accumulation of glucovanillin during on-the-vine development of a vanilla pod ensues during the fourth month after anthesis. It then rises sharply for the next 3 months and levels off during the last stages of pod development (Havkin-Frenkel *et al.* 1999). Formed glucovanillin may be sequestered mostly in the inner white placental tissue around the seeds (Figures 13.1 and 13.3). The distribution of glucovanillin along the longitudinal axis of green vanilla pods may also vary and was found to be as follows: 40% in the blossom end, 40% in the central portion, and 20% in the stem end, indicating uneven tissue distribution with respect to the substrate. This is in keeping with the observation by Childers *et al.* (1959) and by other studies, noting that vanillin crystals formed during curing appear mostly on the blossom end.

13.6 Activity of Oxidative Enzymes Occurring in a Curing Vanilla Pod

Killing of vanilla beans is associated with de-greening and onset of browning reactions, appearing to be a mimic of browning occurring during on-the-vine vanilla pod senescence (Figure 13.4). Browning is observed also during advanced senescence in other ripening fruit (Wilkinson 1970), as well as during stress or disease injury in plant tissues (Schwimmer 1972), arising mostly from oxidation of phenolic compounds (Broderick 1956b). Balls and Arana (1941a, 1942) observed that various killing methods, including chemical, mechanical, or heat stress, but not freezing stress, stimulated a temporary respiratory upsurge, suggesting that the killing-induced increase in oxygen consumption might contribute to onset of oxidative processes in the killed pod. This effect was simulated by applied ethylene, leading to a brief upsurge followed by a decline in CO_2 evolution in green beans, as the ethylene-treated pod continued to ripen (Balls and Arana 1941a, 1942). Furthermore, application of ethylene resulted eventually in pod browning (Arana 1944). Other studies, showing ethylene-induced H_2O_2 accumulation accompanying respiratory upsurge in plants (Chin and Frenkel 1976), suggest that ethylene-induced browning may stem from ethylene-induced oxidative stress, that is, oxygen consumption for the production and activity of reactive oxygen species (ROS). In keeping with this view, we observed that co-application of ethylene and oxygen amplified ethylene-induced browning (Figure 13.5). These data infer that bean browning, as occurring naturally during on-the-vine senescence or as induced by ethylene, may reflect onset of oxidative conditions in vanilla beans. Because various stress conditions are associated with the accumulation of H_2O_2 and other reactive oxygen species (Kocsy *et al.* 2001), it is a reasonable assumption that stress conditions, employed in vanilla bean killing, may also lead to the onset of oxidative conditions and in keeping with the results showing formation and accumulation of ROS in the curing bean (Figure 13.9).

Importantly, browning in a curing vanilla pod appears to be carried out by enzymatic activity, in agreement with the observation that browning was arrested by harsh killing conditions, prolonged or extreme heating, for example (Rabak 1916), apparently due to heat-denaturation of oxidative enzyme(s). A similar study showed arrest of browning in green beans using autoclaving at 120 °C and, furthermore, restoration of browning in autoclaved and non-browning beans upon the addition of oxidative enzymes of fungal origin (Jones and Vincente, 1949b). We also observed that while killing by freezing resulted in typical browning in mature green beans, excessive boiling in post-frozen bean resulted in the inhibition of browning (results not shown). Polyphenol oxidase (PPO) and peroxidase are two major enzyme systems that may catalyze browning processes in killed vanilla beans (Broderick 1956b). PPO, represented by a family of enzymes, utilizes molecular oxygen to catalyze the hydroxylation of monophenol to *O*-diphenol and subsequent removal of hydrogen atoms from *O*-diphenol to give *O*-quinone. The latter might spontaneously polymerize, resulting in the formation of dark oxidation products (Toscano *et al.* 2003). PPO-driven browning in vanilla pods may result mostly from the oxidation of tyrosine, caffeic, and chlorogenic acid, as well as other phenolic compounds (Schwimmer 1981). PPO-induced browning in killed vanilla pods, representing tissues in a state of stress, complies with the emergence of PPO activity in injured or stressed plants (Schwimmer 1981). Peroxidative enzymes, by comparison, utilize H_2O_2 and other hydroperoxides as well as molecular oxygen to catalyze the oxidation of various cellular substrates, including oxidation of aromatic compounds (Schwimmer 1981), oxidative bleaching of carotenoids (Ben-Aziz *et al.* 1971), discoloration of anthocyanins (Grommeck and Markakis 1964), or degradation of ascorbic acid (Blundstone *et al.* 1971). Because peroxidases catalyze the degradation of a wide array of cellular substrates, the role of the enzyme may be wider than just a contribution to browning. For example, peroxidase-driven peroxidation of unsaturated fatty acids, catalyzed by the heme group in the enzyme (Lilly and Sharp 1968), may lead to the utilization of lipids, apparently membrane-lipid, for the apoplastic production of H_2O_2 (Lindsay and Fry 2007; Kaerkoenen *et al.* 2009). Peroxidase-catalyzed production of peroxides might account for a marked increase in lipid hydroperoxides in killed vanilla beans (Figure 13.9), arising perhaps from the oxidation of abundant lipid bodies in placental hairs. Moreover, spontaneous propagation of formed lipid hydroperoxides might further amplify the oxidant effect of the enzyme.

Activities of PPO and peroxidase increase steadily during vanilla pod development (Wild-Altamirano 1969; Ranadive *et al.* 1983), although the enzyme activity is apparently latent and is not expressed in mature green beans. Activity of these enzymes appears to be unleashed by various killing methods, as occurs naturally in senescence or induced by ethylene. Several studies confirmed that activity of PPO remained high in vanilla beans following killing and during subsequent curing stages (Balls and Arana 1941a, 1942; Jones and Vincente, 1949c) or in tissue extracts of vanilla beans (Dignum

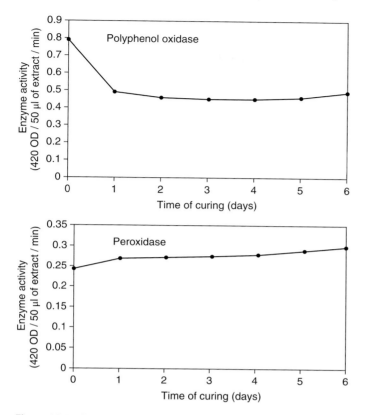

Figure 13.12 Time-course change in the activity of polyphenoloxidase (top) and peroxidase (bottom) in vanilla bean undergoing curing at 50 °C. Fifty μl of crude extract, denoting unit of pod tissue, represents 14.3 mg fresh weight of curing vanilla bean tissue. Reproduced with permission from *Perfumer & Flavorist* magazine, Allured Business Media, Carol Stream, IL.

2001a). Though Jiang *et al.* (2000) observed low PPO levels in cured beans, we found that PPO activity decreased during curing by approximately 50% but remained steady and substantial (Figure 13.12). Peroxidase activity, by comparison, increased with time of curing, suggesting that the enzyme protein resists proteolytic degradation occurring during the killing and subsequent curing stages (Figure 13.12). Persistence of peroxidases well into the conditioning stage (Broderick 1956b) may stem also from the enzyme heat-stability shown by continued peroxidative action, even after dipping green beans in 80 °C for 20 minutes (Dignum 2001b) or restoration of peroxidative activity after autoclaving vanilla pod tissue at 120 °C (Broderick 1956b), further indicating stability of peroxidative action, even under extreme conditions.

PPO and peroxidase-catalyzed oxidation of aromatic compounds is obviously important in pod browning. However, oxidative processes may also entail an oxidative degradation of other cellular compounds, lipid peroxidation for example (Figure 13.9). Arana (1944) emphasized the role of oxidative reactions, particularly quinone polymerization, in the formation of flavor notes during the prolonged conditioning phase. We suggest that the oxidation of lipids, which gives rise to volatile compounds including ketones, aldehydes, alcohols, or hydrocarbons (Niki 2008, 2009), might also give rise to flavor constituents found in cured vanilla beans (Adedeji *et al.* 1993), a view supported by recent studies (Dunphy and Bala 2009). This view underscores the need for further studies to understand the role and contribution of oxidant-induced reactions in the formation of aroma and flavor constituents during the curing process of a vanilla pod.

13.7 Vanilla Products

Cured vanilla beans are used for the extraction and the preparation of vanilla products. The four basic types of vanilla products are vanilla extract, by far the most used vanilla product, as well as vanilla oleoresin, vanilla absolute, and vanilla powder/sugar. Each form has its typical organoleptic, physical, and functional attributes tailored by the choice of beans used for the process and the processing conditions. In addition, the products in each category must meet the government regulations of the country where the products are manufactured or sold. Vanilla products are used in the food, dairy, confectionary, beverage, pharmaceutical, and fragrance industries.

13.8 Summary and Conclusions

The curing of green vanilla beans is intended to create the prized vanilla flavor and, in addition, lend shelf-life to cured beans. The process is predicated on the disruption of cellular organization and the consequent unleashing of the activity of hydrolytic and oxidative enzymes and, apparently, non-enzymatic processes that drive the formation of aroma and flavor constituents in cured vanilla bean. However, the curing process used in commerce, employing harsh temperature conditions to stimulate aroma and flavor formation, is a balancing act. Appropriate curing protocols, entailing controlled killing and a short sweating period around 50 °C, may be favorable to enzyme-catalyzed production of vanillin or other flavor constituents, but severe temperature conditions may arrest development of a full flavor complement, due to enzyme denaturation. Additional interference might stem from proteolytic activity or enzyme denaturation by de-compartmentalized phenolic compounds or other metabolites, which might lead to decay in activity of beneficial enzymes. Prolonged drying and conditioning may be yet another source of loss in formed flavor compounds.

The curing process is founded on the view that destruction of biological order in green vanilla beans unleashes enzymatic reactions by bringing the enzymes in contact with their respective substrates, which drives the formation of vanilla aroma and flavor. However, it is not entirely clear where these enzymes are localized in the cells or in the pod and, moreover, the conditions that allow their interaction with appropriate substrates. It is also desirable to understand whether sufficient activity of flavor-forming enzymes is spared from proteolytic degradation. Other issues regard the role of oxidative reactions in flavor formation. A growing body of studies might cast light on these problems and, moreover, reshape a working concept on the role and mechanism of the curing process in the formation of vanilla aroma and flavor.

13.9 Addendum: Commercial Curing Methods of Green Vanilla Bean

13.9.1 Traditional Methods

13.9.1.1 Mexican Curing Method

The two commonly used killing methods in Mexico are sun-killing and oven-killing (Childers and Cibes 1948; Theodose 1973). In the sun-killing method, green-mature beans are first sorted according to size and stage of maturity and defined as primes or seconds, or split-beans, based on size and appearance (Figure 13.13). Beans are then placed on dark woolen blankets and exposed to the sun for about 4 to 5 hours. When beans become warm to the touch, they are covered by the blanket edges and left in the sun until mid- to late afternoon. The blankets are rolled, taken indoors to sweat in mahogany boxes, which are lined and covered with mats and blankets. This process is repeated up to 6 to 8 times until all the beans have turned uniformly brown, evidence that the killing process has been completed. During this process, the beans lose moisture rapidly and become supple. This phase is followed by additional shorter "sunnings" and infrequent sweating for an additional 2 weeks. The beans are then placed indoors on racks and subjected to slow drying at ambient temperatures, lasting around 1 month, and are inspected regularly. Beans that have dried sufficiently are separated for the subsequent stage of conditioning. The entire process requires about 8 weeks.

In the oven-killing method, mature green beans are subjected to heat and high humidity in specially constructed rooms, called "calorifico", for about 36 to 48 hours. Around 500 to 1000 beans are piled on a jute cloth or a blanket, which is rolled and covered with matting or tied with ropes to form a "malleta". The malletas are then soaked with water and placed on shelves, lining the walls in the calorifico. The calorifico is heated with a wood-fired stove, with temperatures maintained at around 60° to 70°C, and is kept at very high humidity by pouring water on the floor. After 36 to 48 hours, the malletas are removed from the calorifico and placed in sweating boxes for an additional 24 hours to complete the killing process. Killed beans are then removed from the sweating boxes, inspected, and subjected to drying and conditioning as described Sections 13.3 and 13.4. Oven-killed Mexican vanilla beans are claimed to never "frost", that is, they do not

Figure 13.13 Mexican Curing method. Mature-green vanilla beans (left) are pressed by hand into a polyethylene-lined container and then sun-killed. Killed vanilla beans are then allowed to sweat (right). Reproduced with permission from *Perfumer & Flavorist* magazine, Allured Business Media, Carol Stream, IL.

become covered with vanillin crystals upon drying (Theodose 1973), perhaps because surface vanillin interacts with abundant ambient moisture and is volatilized (Frenkel and Havkin-Frenkel 2006).

13.9.1.2 The Bourbon Curing Method

The curing method commonly used by vanilla producers in the Indian Ocean basin is the Bourbon method, named after the former French colony of Réunion, previously known as Bourbon. Madagascar is presently the main producer of Bourbon-type beans, with Réunion and the Comoro Islands producing smaller but significant quantities. In the Bourbon method, also called scalding, beans are killed by hot water immersion. Beans are placed in perforated cylindrical baskets, which are then immersed in vats containing hot water, maintained at around 65 °C. Higher quality beans are scalded for about 2 to 3 minutes, while splits and beans deemed inferior, are scalded for 2 minutes or less. The metal vats, which are heated by a wood fire, have a capacity for scalding about 1.5 tons of vanilla beans in 4 to 5 hours. The scalded beans are quickly dried and while still very hot are wrapped in dark cloth or a blanket. The wrapped beans are then placed in sweating chests lined with cloth and other insulating materials. After about 24 hours of sweating, the beans are removed from the chests and dried in the hot sun for about 2 to 3 hours, rolled up in insulating cloths to retain as much heat as possible, and taken indoors to be replaced in the sweating boxes. The process is repeated for 6 to 8 days. The beans lose moisture quite rapidly and become very malleable. In the subsequent drying phase, lasting 2 to 3 months, the beans are allowed to dry slowly in properly ventilated rooms. During this time the beans are regularly sorted to remove pods that are adequately dried and ready for the next step of conditioning. For conditioning, the beans are placed for about 3 months in the air-tight chests or waxed paper lined metal containers. During this period beans are regularly inspected to ensure the development of the desired finished product. Satisfactorily cured beans are again graded according to size and quality and then bundled for shipment.

13.9.1.3 The Tahitian Curing Method

This method is markedly different in one important aspect from all other curing methods, as there is no artificial killing step involved. Mature green beans are allowed to reach on-the-vine senescence, characterized by yellowing and tip-browning of the beans, before harvesting for further curing. Next, beans are stacked in a cool environment for a few days to complete the browning process. The next stage consists of drying in the morning hours followed by stacking in piles for sweating for the remainder of the day, a process lasting for 15 to 20 days. The drying process is completed by holding cured beans in well aerated shade. In the final conditioning stage, beans are held in cases for 60 to 90 days.

13.9.1.4 Other Traditional Curing Methods

Scarification is used to kill beans in Guadeloupe. One to two millimeter deep scars are made lengthwise into the green vanilla pod, and scarred beans are then wrapped in a blanket and subjected to the hot sun. Sweating and slow drying are similar to the Mexican method (Arana 1945). Childers and Cibes (1948) describe a hot water killing method in Puerto Rico, adopted from the traditional Bourbon process, consisting of immersing green beans in hot water (around 80 °C) for 30 seconds. The process is repeated 3 times at 10-second intervals. Drained and blanket-wrapped beans are then placed in the sweating box overnight. Hot water treatment, at 70 °C and only 2 dippings, is repeated on the second day and again on the third day at 65 °C with only 1 dipping. Two hours sunning every day followed by sweating is repeated for 7 days, at the end of which the beans become ready for slow air or oven drying. In the "Guiana method" beans are killed in the ashes of a wood fire until they begin to shrivel, then wiped and rubbed with olive oil, and air dried (Purseglove *et al.* 1981). These curing methods are moot issues, however, because bean cultivation in these locations has ceased altogether or is of no commercial significance.

13.9.1.5 Indonesian Curing of Vanilla Bean

Growers in Java and other Indonesian islands carry out essentially the Bourbon curing method. However, in the past the practice lacked in consistency with respect to bean maturity and appropriate use of curing protocols, resulting in varied and often inferior products. This situation has been rectified by the proliferation of professionally-run curing houses, resulting in the production of good quality vanilla. About one-third of the total cured beans sold worldwide come from Indonesia.

13.9.2   Refinement of Traditional Curing Methods

Comparative studies on various commercial curing methods, carried out in Porto Rico by Arana (1944) and Jones and Vincente (1949a), led to the conclusion that all the various killing methods, namely, Mexican sun-killing, Mexican oven-killing, the Bourbon hot water scalding, Guadeloupe scratching, exposure to ethylene, and freezing gave a satisfactory product. However, the Bourbon scalding method was found to be preferential, based on ease of handling, resistance of killed beans to mold growth, finished product appearance, development of fine fragrance, vanillin content, and total phenol value. A related study in India (Muralidharan and Balagopal 1973) concluded that, although the Guiana curing process resulted in good flavor, US importers preferred the color and appearance of cured beans subjected to the Mexican or hot water killing method of vanilla beans. These studies indicate that flavor, although an important attribute, is judged in an overall context of other important features that contribute to quality. Arana (1944) found that sweating and drying of beans, carried out in an electric oven set at 45 °C, had lower incidence of mold growth. Moreover, drying was uniform and faster than sun drying, was less cumbersome and resulted in overall superior bean quality. Further studies revealed that sweating of beans at 38 °C produced a better product than accelerated drying at 45 °C (Rivera and Hageman 1951), suggesting that similar ambient temperature regimes occurring in tropical vanilla growing regions might be exploited for drying and conditioning of cured beans (Rivera and Hageman 1951; Broderick 1956a,b).

13.9.3   Novel Curing Methods

McCormick & Co., Baltimore, Maryland, was granted two US patents for curing vanilla by new and radically different curing methods. The first, a patent authored by Towt (1952), describes a protocol for an accelerated curing process, in which green beans are ground to a thick pulp of puree-like consistency and heated to about 48 ° to 54 °C in a tank, ventilated with forced air and with constant agitation for about 48 hours. After the pulp is cured, it is spread on trays and dried in the oven at 59° to 61 °C to a final moisture content of 20%. The dried pulp is then ground, packed, and shipped for use in various vanilla products. A second patent (Graves *et al.* 1958) deals with the curing of an aqueous extract from green beans. Mature green beans are chopped to fine consistency and agitated with water at room temperature for 3 minutes. Next, the pulp is filtered, followed by repeated washing with pure water and repeated filtration. The filtrates are combined and the extract concentrated under hypobaric condition at 29 °C and is then allowed to cure at 70 °C for 6 hours. After the curing is over, enough alcohol is added to obtain a single or multi-fold extract. Commercial enzymes could also be added to facilitate the curing process. According to the authors, the extracts produced by this process result in retention of vanilla flavor and odor to a far greater degree than conventional curing protocols.

Kaul (1967) describes a rapid curing method, in which whole or cut green vanilla beans are subjected to temperatures ranging from 35° to 60 °C and high humidity (80–100%), carried out in a closed system from 1 to 7 days. Cured beans are then dried at room or slightly higher temperature. This process claims to produce a more uniform curing without mold problems. In another rapid curing method, green vanilla beans are chopped into approximately one-half inch segments, and cured for about 70 to 78 hours in perforated trays within a closed tank maintained at about 140 °F (60 °C). Tissue exudates are returned to the curing beans and the mixture dried in a rotary drier at about 140 °F to a moisture content of about 35 to 40% by weight, transferred to a conditioner and dried with air at room temperature until moisture content is reduced to about 20 to 25% by weight (Karas *et al.* 1972). This method uses forced air for the removal of moisture much more rapidly than a method described by Kaul (1967). On the commercial scale, the entire curing and drying operation is said to be completed in 4 to 5 days.

Theodose (1973) described a curing method developed at the Antalaha Station in Madagascar, in which green beans are killed by scalding at 63 to 65 °C for 2 to 3 minutes, and allowed to sweat in closed chests for about 48 hours. The killed beans are cut into about 1 inch pieces and dried in a hot air drier at 65 °C for 3 hours each day for about 12 days. Each day, after the drying cycle is completed, the beans are placed in isothermal chests for accelerating enzymatic actions. After about 12 days, the beans are uniformly cured and have a moisture content of 20 to 25%. According to the US extract manufacturers, the quality of beans produced by this process is good and the vanillin content is higher than the conventionally produced Bourbon beans.

References

Adedeji, J., Hartman, T.G. and Ho, C-T. (1993) Flavor characterization of different varieties of vanilla beans. *Perfumer & Flavorist*, 18, 25–33.

Ansaldi, G.M., Marseille, G.G. and Aubagne, J.L.P. (1990) Process for obtaining natural vanilla flavor by treatment of green vanilla beans, and the flavor obtained. US Patent No. 4,956,192.

Arana, F.E. (1943) Action of β-glucosidase in the curing of vanilla. *Food Research*, 8, 343–351.

Arana, F.E. (1944) *Vanilla curing and its chemistry.* USDA Fed. Expt. Station, Mayaquez, Puerto Rico, Bulletin No. 42, Washington, DC, USDA.

Arana, F.E. (1945) *Vanilla Curing.* USDA Fed. Expt. Station, Mayaguez, Puerto Rico, Circular No. 25, Washington, DC, USDA.

Balls, A.K. and Arana, F.E. (1941a) The curing of vanilla. *Industrial and Engineering Chemistry*, 33, 1073–1075.

Balls, A.K. and Arana, F.E. (1941b) Determination and significance of phenols in vanilla extract. *Assoc. Off. Agr. Chem.*, J24, 507–512.

Balls A.K., and Arana F.E. (1942) Recent observations on the curing of vanilla beans in Puerto Rico. Proceedings of the 8th American Science Congress 1942, *Physics and Chemistry Science*, 7, 187–191.

Ben-Aziz A., Grosssman S., Ascarelli I., and Budouiski P. (1971) Carotene bleaching activities of lipoxygenase and heme proteins as studied by a direct spectrophotometric method. *Phytochemistry*, 10, 1445–1452.

Blundstone, H.A.W., Woodman, J.S. and Adams, J.B. (1971) Changes in vitamin C. In: *The Biochemistry of Fruits and their Products, Vol. II*, Hulme, A.C. (Ed.), Academic Press, London and New York, pp. 561–589.

Bond, J.S. and Butler, P.E. (1987) Intracellular proteases. *Annual Review Biochemistry*, 56, 333–364.

Broderick, J.J. (1956a) The science of vanilla curing. *Food Technology*, 10, 184–187.

Broderick, J.J. (1956b) A preliminary investigation of the quick curing of vanilla beans. *Food Technology*, 10, 188–189.

Brunerie, P.M. (1998) Process of the production on natural vanilla extracts by enzymatic processing of green vanilla pods, and extract thereby obtained. US Patent 5705205.

Childers, N.F. and Cibes, H.R. (1948) *Vanilla culture.* USDA Fed. Expt. Station, Mayaguez, Puerto Rico, Circular No.28, Washington, DC, USDA.

Childers, N.F., Cibes, H.R. and Hernandez-Medina, E. (1959) Vanilla – the orchid of commerce. In: *The Orchids: A Scientific Survey.* Withner, C.L. (Ed.), Robert E. Krieger Publishing Company, Malabar, Fl, pp. 477–508.

Chin, C-K. and Frenkel, C. (1976) Induction of upsurge in respiration and peroxide formation in potato tubers influenced by ethylene and oxygen. *Nature*, 264, 60.

Dignum, M.J.W., Kerler, J. and Verpoorte, R. (2001a) Vanilla production: technological, chemical, and biosynthetic aspects. *Food Res. Inter.*, 17, 199–219.

Dignum, M.J.W., Kerler, J. and Verpoorte, R. (2001b) β-Glucosidase and peroxidase stability in crude enzyme extracts from green beans of *Vanilla planifolia* Andrews. *Phytochemical Analysis*, 12, 174–179.

Dignum, M.J.W., Kerler, J. and Verpoorte, R. (2002a) Vanilla curing under laboratory conditions. *Food Chemistry*, 79, 165–171.

Dignum, M.J.W. (2002b) Biochemistry of the processing of vanilla beans, PhD Thesis, Leiden University, Leiden, Holland.

Dunphy, P., and Bala, K. (2009) Vanilla curing: the senescent decline of a ripe vanilla bean and the birth of vanillin. *Perfumer & Flavorist* 34, 34–40.

Eskin, N.A.M. and Frenkel, C. (1976) A simple and rapid method for assessing rancidity of oils based on the formation of hydroperoxides. *Journal of the American Oil Chemists' Society*, 53, 746–747.

Frenkel, C. and Havkin-Frenkel, D. (2006) The physics and chemistry of vanillin. *Perfumer & Flavorist*, 31, 28–36.

General, T., Mamatha, V., Divya, V. and Appaiah, K.A.A. (2009) Diversity of yeast with β-glycosidase activity in vanilla (*Vanilla planifolia*) plant. *Current Science*, 96, 1501–1505.

Goris, M.A. (1924) Sur la composition chimique des fruits verts de vanille et e mode de formation du parfum de la vanille. *Acad. des Sci. Colon Paris, Compt. Rend.*, 179, 70–72.

Graves, R.E., Hall, R.L. and Karas, A.J. (1958) Method of producing cured vanilla extract from green vanilla beans. US Patent 2,835,591.

Grinberg, O.Y., James, P.E. and Swartz, H.M. (1998) Are there significant gradients of pO_2 in cells? *Advances in Experimental Medicine and Biology*, 454, 415–423.

Grommeck, R. and Markakis, P. (1964) The effect of peroxidase on anthocyanin pigments. *Journal of Food Science*, 29, 53–57.

Hannum, T. (1997) Changes in vanillin and activity of β-glucosidase and oxidases during post-harvest processing of vanilla beans *Vanilla planifolia*. *Bulletin Teknologia dan Industri Pangan*, 8, 46–52.

Havkin-Frenkel, D. and Dorn, R. (1997) Vanilla. In: *Spices, flavor chemistry and antioxidant properties*. Risch, J. and Ho, C-T. (Eds), ACS Symposium, Series Vol. 660, American Chemical Society, Washington DC, pp. 29–40.

Havkin-Frenkel, D., Podstolski, A., Witkowska, E., Molecki, P. and Mikolajczyk, P. (1999) Vanillin biosynthetic pathways, an overview. In: *Plant Cell and Tissue Culture for the Production of Food Ingredients*. Fu, T.J., Singh, G. and Curtis W. R. (Eds), Kluwer Academic Press/Plenum Publishers, New York, pp 35–43.

Havkin-Frenkel, D. and Kourteva, G. (2002) *Biotechnological Production of Vanilla Flavor*. World Congress on Medicinal and Aromatic Plants, Budapest, Hungary.

Hopkins, M., Taylor, C., Liu, Z., et al. (2007) Regulation and execution of molecular disassembly and catabolism during senescence. *New Phytologist*, 175, 201–214.

Hrmova, M., Stewart, R.J., Varghese, J.N., Hoj, P.B. and Fincher, G.B. (1999) Three-dimensional structures, catalytic mechanisms and protein engineering of β-glucan hydrolases from barley. *Special Publication – Royal Society of Chemistry*, 246, 124–131.

Jiang, M., Pu, F., Xie, W.S., Hu, Y.Q. and Li, Y. (2000) Activity of three enzymes in vanilla capsule. *Acta Botanica Yunnanica*, 22, 187–190.

Joel D.M., French, J.C., Graf, N., Kourteva, G., Dixon, R.A. and Havkin-Frenkel, D. (2003) A hairy tissue produces vanillin. *Israel Journal of Plant Science*, 51, 157–159.

Jones, M.A. and Vincente, C.C. (1949a) Criteria for testing vanilla in relation to killing and curing methods. *Journal of Agricultural Research*, 78, 425–434.

Jones, M.A. and Vincente, C.C. (1949b) Inactivation of vacuum infiltration of vanilla enzyme systems. *Journal of Agricultural Research*, 78, 435–443.

Jones, M.A. and Vincente, C.C. (1949c) Quality of cured vanilla in relation to some natural factors. *Journal of Agricultural Research*, 78, 445–450.

Kaerkoenen, A., Warinowski, T., Teeri, T.H., Simola, L.K. and Fry, S.C. (2009) On the mechanism of apoplastic H_2O_2 production during lignin formation and elicitation in cultured spruce cells – peroxidases after elicitation. *Planta*, 230, 553–567.

Kamaruddin, A.M. (1997) Drying of vanilla pods using a greenhouse effect solar dryer. *Drying Technology*, 15, 685–669.

Kanisawa, T., Tokoro, K. and Kawahara, S. (1994) Flavor development in the beans of *Vanilla planifolia*. In: *Olfaction Taste, XI*. Kurihara, K., Suzuki, M. and Ogawa, H. (Eds), Proc. Int. Symp., Springer, Tokyo, pp. 268–270.

Karas, A.J., Hall, R.L. and Stahl, W.H. (1972) Vanilla bean drying and curing. US Patent 3,663,238.

Kaul, R.J. (1967) Curing of vanilla beans. US Patent 3,352,690.

Kocsy, G., Galiba, G. and Brunold, C. (2001) Role of glutathione in adaptation and signalling during chilling and cold acclimation in plants. *Physiologia Plantarum*, 113, 158–164.

Lenaz, G. and Genova, M.L. (2009) Structural and functional organization of the mitochondrial respiratory chain: A dynamic super-assembly. *International Journal of Biochemistry and Cell Biology*, 41, 1750–1772.

Leong, G., Archavlis, A. and Derbesy, M. (1989a) Research on the glucoside fraction of the vanilla bean. *Journal of Essential Oil Research*, 1, 33–41.

Leong, G., Uzio, R. and Derbesy, M. (1989b) Synthesis, identification and determination of glucosides present in green vanilla beans *Vanilla fragrans* Andrews. *Flavour and Fragrance Journal*, 4, 163–167.

Lilly, M.D. and Sharp, A.K. (1968) The kinetics of enzymes attached to water-insoluble polymers. *Chemical Engineer (London)*, 215, CE12–CE18.

Lim, PO, Kim, H.J. and Nam, H.G. (2007) Leaf senescence. *Annual Review of Plant Biology*, 58, 115–136.

Lindsay, S.E. and Fry, S.C. (2007) Redox and wall-restructuring. *Plant Cell Monographs*, 5, 159–190.

Lopez-Casado, G., Urbanowicz, B.R., Damasceno, C.M.B. and Rose, J.K.C. (2008) Plant glycosyl hydrolases and biofuels: a natural marriage. *Current Opinion in Plant Biology*, 11, 329–337.

Lubinsky, P., Cameron, K.M., Molina, M.C. et al. (2008) Neotropical roots of a Polynesian spice: the hybrid origin of Tahitian vanilla, *Vanilla tahitensis* (Orchidaceae). *American Journal of Botany*, 95, 1040–1047.

Mane, J. and Zucca, J. (1993) Process for production of natural vanilla flavor by treatment of vanilla pods and vanilla flavour so produced. French Patent Application PN FR 2691880A1.

Marquez, O. and Waliszewski, K.N. (2008) The effect of thermal treatment on β-glucosidase inactivation in vanilla bean (*Vanilla planifolia* Andrews). *International Journal of Food Science and Technology*, 43, 1993–1999.

Matsushima, R., Kondo, M., Nishimura, M. and Hara-Nishimura, I. (2003) A novel ER-derived compartment, the ER body, selectively accumulates a β-glucosidase with an ER-retention signal in Arabidopsis. *Plant Journal*, 33, 493–502.

Muentz, K. (2007) Protein dynamics and proteolysis in plant vacuoles. *Journal of Experimental Botany*, 58, 2391–2407.

Muralidharan, A. and Balagopal, C. (1973) Studies on curing of vanilla. *Indian Spices*, 10, 3–4.

Niki, E. (2008) Lipid peroxidation products as oxidative stress biomarkers. *BioFactors*, 34, 171–180.

Niki, E. (2009) Lipid peroxidation: physiological levels and dual biological effects. *Free Radical Biology and Medicine*, 47, 469–484.

Nishi, T. and Itoh, S. (1992) Qualitative improvement of therapeutic glycoproteins by glycotechnology. *Trends in Glycoscience and Glycotechnology*, 4, 336–344.

Odoux, E. (2000) Changes in vanillin and glucovanillin concentrations during the various stages of the process traditionally used for curing *Vanilla fragrans* beans in Réunion. *Fruits*, 55, 119–125.

Odoux, E. (2006) Glucosylated aroma precursors and glucosidase(s) in vanilla bean (*Vanilla planifolia* G. Jackson). *Fruits*, 61, 171–184.

Odoux, E., Escoute, J. and Verdeil, J-L. (2006) The relation between glucovanillin, β-*D*-glucosidase activity and cellular compartmentation during the senescence, freezing and traditional curing of vanilla beans. *Annals of Applied Biology*, 149, 43–52.

Okamoto, T. (2006) Transport of proteases to the vacuole: ER export bypassing Golgi? *Plant Cell Monographs*, 4, 125–139.

Podstolski, A., Havkin-Frenkel, D., Malinowski, J., Blount, J.W., Kourteva, G. and Dixon, R.A. (2002) Unusual 4-hydroxybenzaldehyde synthase activity from tissue cultures of the vanilla orchid Vanilla planifolia. *Phytochemistry*, 61, 611–620.

Pu, F., Zhang, S.J. and Jiang, M. (1998) Study on the components released from glucoside in vanilla bean of different curing stages. *Tianran Chanwu Yanjiu Yu Kaifa*, 10, 29–33.

Purseglove, J.W., Brown, E.G., Green, C.L. and Robbins, S.R.J. (1981) *Spices*. Vol. 2, Longman, London and New York, pp. 644–735.

Rabak, F. (1916) The effect of curing on the aromatic constituents of vanilla beans. *Industrial & Engineering Chemistry*, 8, 815–321.

Ranadive, A.S. (1994) Vanilla-cultivation, curing, chemistry, technology and commercial products. In: *Spices Herbs and Edible Fungi*, Charalambous, G. (Ed.), Developments in Food Science, Vol. 34, Elsevier Science Publishers BV, Amsterdam, The Netherlands, pp. 517–577.

Ranadive, A.S., Szkutnica, K., Guerrera, J.G. and Frenkel, C. (1983) Vanillin biosynthesis in vanilla bean. IX International Congress of Essential Oils. Singapore. Proceedings of the Congress, Book 147.

Rao, S.R. and Ravishankar, G.A. (2000) Vanilla flavour: production by conventional and biotechnological routes. *Journal of the Science of Food Agriculture*, 80, 289–304.

Ratobison, R., Zeghmati, B., Reddy, T.A. and Daguenet, M. (1998) Sizing of solar supplemented liquid and air heating systems for the treatment of vanilla. *Soar Energy*, 62, 131–138.

Rivera, J.G. and Hageman, R.H. (1951) *Vanilla curing*. USDA Fed. Expt. Station Report for 1950, Mayaguez, Puerto Rico, Report No. 30, Washington, DC, USDA.

Rogers, H.J. (2005) Cell death and organ development in plants. *Current Topics in Developmental Biology*, 71, 225–226.

Roling, W.F.M., Kerler, J., Braster, M., Apriyantono, A., Stam, H. and Van Verseveld, H.W. (2001) Microorganisms with a taste for vanilla: microbial ecology of traditional Indonesian vanilla curing. *Applied and Environmental Microbiology*, 67, 1995–2003.

Ruiz-Teran, F., Perez-Amador, I. and Lopez-Munguia, A. (2001) Enzymatic extraction and transformation of glucovanillin to vanillin from vanilla pod. *Journal of Agricultural and Food Chemistry*, 49, 5207–5209.

Sanz-Aparicio, J., Hermoso, J.A., Martinez-Ripoll, M., Gonzalez, B., Lopez-Camacho, C. and Polaina, J. (1998) Structural basis of increased resistance to thermal denaturation induced by single amino acid substitution in the sequence of β-glucosidase A from *Bacillus polymyxa*. *Proteins: Structure, Function, and Genetics*, 33, 567–576.

Schwimmer, S. (1972) Symposium: Biochemical control systems. Cell disruption and its consequences in food processing. *Journal of Food Science*, 37, 530–535.

Schwimmer, S. (1981) *Source Book of Food Enzymology*. AVI Publishing Comp, Westport, CT.

Suzuki, H., Takahashi, S., Watanabe, R. et al. (2006) An isoflavone conjugate-hydrolyzing β-glucosidase from the roots of soybean (*Glycine max*) seedlings: purification, gene cloning, phylogenetics and cellular localization. *Journal of Biological Chemistry*, 281, 30, 251–30, 259.

Swamy, B.G.L. (1947) On the life history of *Vanilla planifolia*. *Botanical Gazette*, 108, 449–456.

Theodose, R. (1973) Traditional methods of vanilla preparation and their improvement. *Tropical Science*, 15, 47–57.

Tokoro, K., Kawahara, S., Amano, A., Kanisawa, T. and Indo, M. (1990) Glucosides in vanilla beans and changes of their contents during maturation. In: *Flavour Science and Technology*. Bessièr, A.F. and Thomas, A. F. (Eds), John Wiley & Sons, Chichester, pp. 73–76.

Toscano, G., Colarieti, M.L. and Greco, G. (2003) Oxidative polymerization of phenols by a phenol oxidase from green olives. *Enzyme and Microbial Technology*, 33, 47–54.

Towt, L.V. (1952) Methods of dehydrating and curing vanilla fruit. US Patent 2,621,127.

Trincone, A. and Giordano, A. (2006) Glycosyl hydrolases and glycosyltransferases in the synthesis of oligosaccharides. *Current Organic Chemistry*, 10, 1163–1193.

Varki, A. (1993) Biological roles of oligosaccharides: all of the theories are correct. *Glycobiology*, 3, 97–130.

Wild-Altamirano, C. (1969) Enzymic activity during growth of vanilla fruit. I. Proteinase, glucosidase, peroxidase, and polyphenoloxidase. *Journal of Food Science*, 34, 235–238.

Wilkinson, B.G. (1970) *Physiological disorders of fruit after harvesting*. In: *The Biochemistry of Fruits and their Products, Vol. I.*, Hulme, A.C. (Ed.), Academic Press, London and New York, pp. 537–569.

Yoshida, S. (1991) Chilling-induced inactivation and its recovery of tonoplast H^+-ATPase in mung bean cell suspension cultures. *Plant Physiology*, 95, 456–460.

Yoshida, S. (1994) Low temperature-induced cytoplasmic acidosis in culture mung bean (*Vigna radiata* [L.] Wilczek) cells. *Plant Physiology*, 104, 1131–1138.

14

Fair Trade – The Future of Vanilla?
Richard J. Brownell Jr

14.1   The Crisis

In late 1999, the price of vanilla beans began to rise. Four months later, a Category 5 cyclone slammed directly into the center of the vanilla bean growing region of Madagascar. Initial reports claimed that 80% of the crop on the vines was destroyed. To make matters worse, beans from the prior year's crop awaiting export in rudimentary warehouses were damaged by the high winds and heavy rainfall as well. The price of vanilla beans quadrupled over night. And that was just the beginning. By the end of 2003, vanilla bean prices were fifteen times 1999 levels. What could have prevented this crisis? More importantly, what can prevent it from happening again? The answer may well be Fair Trade.

The roots of the crisis can be traced back almost a decade earlier. In 1994, under pressure from the World Bank, the government of Madagascar deregulated the pricing and allocation of vanilla beans. Throughout the balance of the decade prices for vanilla beans steadily fell. By contrast, worldwide demand for vanilla was growing robustly, fueled by three major food trends.

The first of these three trends was the growth of super premium ice cream, notably Haagen Dazs and Ben & Jerry's. The high fat content and indulgent platform of these products required an especially high flavor dose rate. The second major trend of this period was the shift toward natural, good for you foods, especially in Europe and the United States. What could be more natural than pure vanilla extract? Finally, food and beverage marketers looked to globalization for continued growth. Many of the world's most recognizable brands contain vanilla extract as an ingredient.

On the surface, it would seem inconsistent that vanilla bean prices would be falling, in a period of rising demand for vanilla extract. But, at the time of deregulation, there was a large surplus of vanilla beans. In the ensuing years, production shortfalls were met by the surplus stocks. Under proper conditions, vanilla beans can be stored for several years. By 1999, the surplus had been essentially exhausted. And many vanilla bean farmers, discouraged by years of falling prices had finally abandoned their vines and replaced them other, more lucrative crops.

Handbook of Vanilla Science and Technology, Second Edition.
Edited by Daphna Havkin-Frenkel and Faith C. Belanger.
© 2019 John Wiley & Sons Ltd. Published 2019 by John Wiley & Sons Ltd.

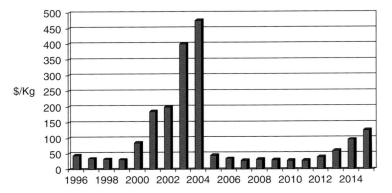

Figure 14.1 Historical pricing of Madagascar bourbon vanilla beans. *Source:* Virginia Dare Extract Company, Inc.

Once the deficit between production and demand became apparent, it was too late. It takes 3 to 4 years for new vines to produce vanilla beans. There would be no quick fix on the production side. Equally, vanilla extract consumption was not easily reduced. Reformulation to alternatives would be extremely costly for food and beverage manufacturers. There was also the risk that consumers would recognize and reject the reformulated products. Food and beverage companies initially tried to ride out the crisis. But eventually, the cost of vanilla extract became intolerable and they were forced to seek alternatives. By 2004, worldwide demand for vanilla beans was only half what it had been five years earlier.

When the crash finally came, it came with a vengeance. Vanilla bean prices dropped from over $500 per kilogram to $50 in twelve months (Figure 14.1). Even another Category 5 cyclone during this period only resulted in a brief pause in the downward spiral.

14.2   The Farmer

No one suffered more from the vanilla crisis than the farmers in Madagascar who grow the beans. Clearly there were windfall profits, but very little found its way back to the growers. New curing facilities and warehouses were built by the vanilla bean processors. They built offices for their managers and guest rooms for visiting bean buyers. Vanilla bean collectors traded in their motorcycles for SUVs. Some built luxurious homes near the coast and even air-conditioned hotels. A major highway was constructed connecting Anatalaha at the southern end of the vanilla bean growing region to Sambava to the north. By contrast, the farmers' lifestyle and average family income of $1.00 per day was virtually unchanged.

Vanilla cultivation is extremely labor-intensive. Vines are started from cuttings, which are gently tied to support trees. During the three years before the first flowering, the support trees must be regularly pruned to provide the correct amount of sunlight for the beans. Regular mulching at the base of the vines is required to provide the proper drainage and nutrients. The vines themselves must be trained to grow in loops so that the flowers and beans will be within the farmer's reach. Vanilla has no efficient natural pollinator, so each flower has to be pollinated by hand in order to produce a bean. The beans remain on the vines for ten months during which time in addition to the mulching, pruning and looping required, the beans must also be protected from thieves. Finally, when the beans are ready for harvest, they are picked individually, by hand, at the moment of optimal maturation.

Traditionally, vanilla bean collectors traveled into the remote villages where beans are grown to purchase beans from growers and transport them back to the curing facilities. The farmers were in a very weak position when it came to negotiating with the collectors for several reasons. Most importantly, green beans must begin the curing process within a week or so of harvest or they will quickly deteriorate and be worthless. If a farmer rejected the collector's offer, he risked losing everything. Secondly, the farmers seldom had the alternative of transporting the beans to the processor on their own. In some cases this would involve a journey of 50 kilometers on dirt roads barely passable under the best of conditions. And the farmers did not own motorcycles, much less SUVs. Lastly, communications in remote regions where vanilla is grown was largely nonexistent. Farmers in one village had little knowledge of what farmers in other areas were being paid for their beans. Often, they had to rely on the collectors themselves for market information.

Since the beginning of the crisis, the farmer's leverage has increased marginally. The introduction of cell phones increased access to market information and the going price for vanilla beans. Improved roads enabled some to transport beans to processing facilities on their own, bypassing the collectors. Perhaps more significantly, more and more farmers begin the curing process on their own. This adds value and provides the farmers more flexibility for selling the beans on their own terms.

However, the vanilla crisis left the farmers disadvantaged in other ways. Before, prices were low but at least they had an outlet for the beans as demand continually rose. After the crisis, they were left with low prices and low demand. Furthermore, there was now competition from new origins such as Uganda, India and Papua New Guinea (PNG) (Figure 14.2). At the conclusion of the crisis, PNG was producing more than 20% of the world's requirement for vanilla beans. Self-curing, which seemed to tip the scale in their

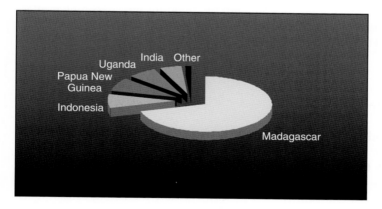

Figure 14.2 2009 Vanilla Bean Production by Country of Origin. *Source:* Virginia Dare Extract Company, Inc.

favor, often backfired as well. Without the expertise to cure properly, beans partially cured by the farmers often had off notes and were rejected by processors, leaving the farmers worse off than before.

14.3 Fast Forward

During the crisis, many new vines were planted in both traditional and new origins. By 2004, the average worldwide crop was approximately 2000 metric tons, while consumption had fallen to a little more than 1000 tons. Prices crashed in 2004 and have continued to drift lower ever since.

In early 2008, reports of extensive vine disease in Madagascar caused by the fungal pathogen *Fusarium* began to circulate. Met with skepticism at first, they still caused a shiver of concern throughout the industry. *Fusarium* is a fungus found in the soil all over the world. Historically, it has devastated many other crops including tomato, potato, pepper and eggplant. Prior out breaks of *Fusarium* in vanilla in India, China and Indonesia resulted in virtually the entire crop being infected and destroyed.

Fusarium is an opportunist organism. Healthy plants can often coexist with it. Favorable weather in 2008 seemed to keep the outbreak at bay. But, in the summer of 2009, the vines were vulnerable due to the stress of having just produced a bumper crop. Numerous crop surveys indicated that a growing percentage of the vines were infected. There is very little that can be done once a plant is infected other than to rip it out to try to prevent further spread of the disease.

In less than four years, worldwide demand for vanilla beans had essentially recovered to the pre-crisis level. In the fall of 2009, extremely light flowering in Madagascar was observed. Other producing countries, most notably India, Indonesia and Papua New Guinea were reporting significant declines in production. Many farmers have once again become discouraged by low bean prices and abandoned vanilla. An alarming percentage of vines in these countries have been removed and replaced with other crops, such rice, corn and cloves.

In 2009, vanilla bean exporters and dealers were still holding large inventories of beans from the 2008 crop. Furthermore, a bumper Madagascar crop estimated to be 2000 metric tons was being cured. So, even with the trend towards declining production and the bigger threat posed by *Fusarium*, there was likely to be a surplus of beans at least until 2011. However, eventually the boom–bust cycle is almost certain to reoccur.

14.4 Fair Trade – Background

In 2005, vanilla was added to the portfolio of Fair Trade Certified™ products. This may have been a development that could change the vanilla landscape forever. Fair Trade Certified Vanilla has the potential to eliminate the boom and bust cycles currently threatening the future of the vanilla industry itself.

Fair Trade initiatives began shortly after WWII, typically marketing artisan crafts from impoverished peoples in third world countries. The most common outlets for these products were World Shops, familiar to travelers in many airports around the world. The Fair Trade movement expanded in the 1960s, as dozens of Alternative Trading Organizations (ATOs) were established in Europe and North America. During this period, religious, student and various other groups embraced an anti-establishment sentiment. Fair Trade became a symbol of the exploitation and repression of the people, which needed to be changed.

The Fair Trade movement remained somewhat of an afterthought until three watershed events transformed it into a mainstream social engine for change. In the 1960s and 1970s, Fair Trade expanded its portfolio from simple artisan crafts to agricultural products. This exponentially increased the potential market size. Now, products that consumers around the world used on a daily basis became available with Fair Trade certification. These included coffee, tea, cocoa, rice and bananas. Today, the Fair Trade portfolio has added flowers, wine and even clothing, in addition to vanilla.

Even with the addition of agricultural products, for a long time the growth of Fair Trade was limited by its boutique-like distribution. In the late 1980s, an ATO in the Netherlands created the concept of a Fair Trade label. This label, awarded under the supervision of an independent organization, would allow Fair Trade products to be distributed through mainstream distribution channels, while still carrying the assurances of compliance with Fair Trade principles.

During the ensuing decade, Fair Trade grew exponentially, with coffee production being the most prominent example. The final watershed event for Fair Trade occurred in 1997, when Fair Trade Labeling Organizations (FLO) International was created. This organization ensured that inspection and certification of Fair Trade products were standardized throughout the world. This provided consistency and credibility to products carrying the Fair Trade label. In the United States, Transfair USA is the official FLO International representative.

14.4.1   Fair Trade Principles

The principles of Fair Trade are somewhat complex and far-reaching. However, the cornerstone is enabling small, disadvantaged producers the ability to compete in the global market place *and* receive a fair price for the goods and services they produce. A fair price reflects the true costs of sustainable production and a standard of living, which meets acceptable social and ethical norms.

The Fair Trade price provides the farmer a profit margin, which allows him to sustain production over the long term. This includes preservation of natural resources including clean water, soil, plants and animals. It also provides the capital required to reinvest in farm equipment, tools and supplies required for ongoing production.

Fair Trade also provides the farmer with the income necessary to maintain an acceptable standard of living. This includes adequate shelter, food, clothing, education and healthcare for the farmer and his family. The Fair Trade price includes a premium, which actually goes back to the cooperative or village, rather than the individual farmer. This money is used for a variety community improvements dependent upon their specific needs. Examples are school and medical facilities and supplies, bridges, wells and communications equipment such as cell phone towers and radios.

An acceptable standard of living applies not just to the farmer and his family, but to his employees as well. Fair Trade ensures that a safe working environment is maintained. Workers, particularly women and children must not be exploited in any way.

Unfortunately, from the coal mines in the United States a century ago to the diamond mines in South Africa today, exactly the opposite is often the case.

In summary, the principles of Fair Trade enables rural farmers in developing countries the ability to compete in the global marketplace. In many cases, without Fair Trade, they would otherwise have little contact outside their own villages, let alone the rest of the world.

14.4.2   Vanilla and Fair Trade

Vanilla is uniquely and ideally suited for Fair Trade. Virtually all vanilla grown throughout the world is grown by independent farmers in rural, sometimes even remote villages in developing countries. These farmers have limited access to their neighboring villages, let alone the global market place that Fair Trade offers.

Most farmers have limited means of transportation, perhaps a donkey or a bicycle. The roads are almost universally rudimentary. They are passable only with great discomfort and difficulty during the dry season and virtually impassable during the rainy season. Until the very recent spread of cell phone towers, farmers were basically isolated from the outside world. If Ben & Jerry's were launching a new super premium ice cream made with Bourbon Vanilla from Madagascar, the farmers who grew the vanilla beans would almost certainly be unaware of it.

Vanilla cultivation is uniquely compatible with the principles of Fair Trade. Vanilla grows best under a rain forest canopy. Native trees can be used to support the vines. As opposed to many other crops, there is no need for clear cutting virgin rain forests. Other vegetation can be used for mulching the roots of the vines.

In most origins, vanilla is grown without the use of chemical pesticides, fungicides or fertilizers. They are simply too expensive. Instead, the vines are kept healthy by proper cultivation methods, which are passed down from one generation to the next. These include adequate spacing of the vines, well-drained soil and not over pollinating the vines.

Virtually all vanilla grown in the world relies simply on a rainy climate for water. After the beans are harvested, the ideal curing technique involves almost exclusively solar energy. Alternate methods, such as oven drying are used only during long periods of cloudy weather. Some origins do use ovens and wood fires more extensively for curing, but the flavor profile is generally inferior to beans cured in the sun. Since vanilla cultivation and curing is extremely labor-intensive, keeping labor costs low is extremely important. In addition, because of the manual dexterity required, it has traditionally been done by women and sometimes children. In general, working conditions are safe and workers in the vanilla industry are treated fairly. They are often family members or neighbors. However, there is a clear pressure to keep wages low and possibility of exploitation does exist.

There is very little in the way of invested capital required to grow vanilla, so the profits from Fair Trade Vanilla can be channeled towards food, clothing, shelter and other aspects of an acceptable standard of living.

At the other end of the supply chain, are foods and beverages containing vanilla extract. Fair Trade is also a good fit for manufacturers and consumers of products containing vanilla. Vanilla is used in an astoundingly wide range of food and beverage products. In fact, vanilla is the most popular flavor in the world. It is the characterizing flavor in many products including ice creams, yogurts, cookies, shakes and lattes. But, vanilla is also used as a background note in many other products as a flavor enhancer. For example, vanilla provides creaminess and helps magnify the flavor of cola beverages. In another example, vanilla is used extensively as a masking agent for bitterness in chocolate.

Not only is vanilla used in a great range of products, but typically the usage rate is relatively small. On a pound for pound basis, vanilla may be relatively expensive. But, on a cost in use basis vanilla is usually a minor contributor to the overall cost of a finished food or beverage. So, the premium paid by manufacturers and consumers to support Fair Trade Vanilla is relatively small.

14.5   Commodity Cycles

In early 2000, the price of vanilla beans began to rise and by the end of 2003 had increased 15-fold. Worldwide consumption of vanilla beans dropped by 50% during the same period. Five years later, the price has fallen to near all-time lows and farmers around the world were abandoning and, in some cases, ripping out their vanilla vines.

Vanilla is particularly vulnerable to commodity cycles for several reasons. First, the global market is relatively small and lends itself to speculation with a relatively modest investment. Second, there is no futures market in play to offset current prices with expectations of future supply and demand. Third, vanilla is produced in a relatively small number of origins and is particularly influenced by events in Madagascar, which routinely accounts for approximately two thirds of the world's production.

Madagascar, located to the east of Africa, in the Indian Ocean lies directly in the path of some of the world's most intense cyclones (hurricanes). Additional vulnerability stems from the fact that vanilla grows in a relatively concentrated region of Madagascar. Military coups are also been relatively common in the Malagasy Republic. The last two Presidents have been forced to flee the country before their terms in office were completed.

Perhaps the greatest threat to vanilla in Madagascar is the spread of *Fusarium* wilt, which has been observed since early 2008. This disease has been largely responsible for the demise of vanilla in several other origins, most notably the island of Bali which was once a very significant producer. Hopefully, the impact of *Fusarium* in Madagascar will be mitigated by adherence to proper cultivation practices. However, its progress warrants close attention and is yet another example of the vulnerability of vanilla to external events and ultimately commodity cycles fueled by speculation.

Vanilla's vulnerability to commodity cycles is also affected by its natural growth cycle. When demand exceeds supply, farmers respond predictably by growing more vanilla. Unfortunately, however, newly planted vines typically do not produce significantly until the third year and don't reach full production until the fourth year. So, a severe supply shortage can only be remedied by a reduction in demand, at least for the first two or three years.

Fair Trade would help to offset the commodity cycles by preventing the price of Fair Trade Certified Vanilla beans to fall to a level that would cause farmers to abandon their vines. The extent of its impact would, of course, depend on the percentage of the overall crop that was produced and sold as Fair Trade. But, clearly the potential of Fair Trade to help stabilize the boom or bust nature of the vanilla market can be seen. And, this stability would benefit farmers, food and beverage manufacturers and consumers alike.

14.6 Issues

While the potential benefits of Fair Trade Vanilla are far reaching, there are still issues to be resolved in order to realize this potential including the following.

1) The price differential compared to conventional vanilla.
2) Vanilla quality is not well correlated with cost of production.
3) Limited availability.
4) Ensuring that farmers are really paid the FT price.
5) Consumer acceptance.

In 2009, FLO International revised the pricing structure for Fair Trade Vanilla, responding to suggestions made by industry experts. Several improvements were made, but the issues outlined above in large part remain unresolved. Let us examine these issues one at a time.

14.6.1 The Price Differential

The Fair Trade Certified price for cured vanilla beans from Madagascar is $47.50 per kg. This includes $6.50 per kg for the Fair Trade premium. As discussed above, the Fair Trade premium goes to the community as a whole. The balance of $41.00 per kg goes to the curer. By contrast, the current price for beans of identical quality, but without Fair Trade Certification, is roughly $20.00 per kg.

Is the true cost of sustainable production really more than double the current market cost? Notwithstanding the acknowledgement that the current market price has discouraged farmers to the point that many have turned to producing other crops instead, the answer is most likely no. Evidence of this excess can be observed by a related event that took place in the summer of 2009.

Concerned about declining production resulting from discouraged farmers abandoning vanilla, a group of exporters petitioned the Madagascar government to establish a minimum export price for vanilla beans. Some in the industry argued that the reason was really to offset poor decisions by exporters holding large inventories of beans purchased at prices well above the current market price. Regardless of their true intentions, the minimum price they recommended was $32.00 per kg.

This price was suggested by industry insiders who are extremely familiar with the costs of production. These are people who make their livelihood by buying, curing and exporting vanilla beans. They would certainly know what price level would provide farmers an adequate incentive to keep growing vanilla. In fact, one could argue that the recommended price was too high, designed to not only support ongoing production but to increase exporters' profit margins as well. Perhaps the Malgache government thought so too. A minimum export price of $27.00 per kg was signed into law by President Andry Rajoelina in the summer of 2009.

14.6.2   Vanilla Quality

As described above, growing and curing vanilla beans is extremely labor-intensive. The per capita annual income of a vanilla farmer is about $300. Vanilla beans from Madagascar are considered to be the best quality in the world, with the possible exception of those from Mexico. The beans are left on the vines for up to ten months. Then harvested one at a time to ensure each bean is at the peak of maturation.

By contrast, the per capita annual income of a vanilla farmer in Indonesia is closer to $1,000. However, Indonesian beans are often harvested months before maturity to prevent them from being stolen. This effectively lowers the cost of production, but it still likely exceeds the level in Madagascar. Many of the flavor precursors have not yet developed in early picked beans leaving the cured beans largely devoid of traditional vanilla flavor and aroma. So, the quality of the Indonesian beans at the time of harvest is decidedly inferior to those in Madagascar, even though the cost of production is substantially higher.

The curing process complicates the issue even further, because it is unrelated to the farmer's cost of production but has a major impact on the final quality of the beans. In Madagascar, curing is painstaking and lasts for several months. The end result is complex profile of both bold and delicate flavor and aroma. In Indonesia, curing is often done over a wood fire in a matter of hours.

So, there is the conundrum for Fair Trade pricing. At the farmer level, the country with the lower cost of green bean production produces the highest quality beans, while the country with the higher cost of production produces beans that taste like wood smoke.

14.6.3   Limited Availability

Since Fair Trade Vanilla was first established in 2005, India has been the only significant producer. This can be attributed to a number of factors. Perhaps the most important reason is the relatively high cost of production in India compared to other origins. Like their counterparts in Madagascar, Indian vanilla growers allow the beans to reach full maturity before harvest. Yet, the per capita income rate is roughly double that of Madagascar.

Indian growers often purchased cuttings to start their vanilla farms. By contrast, most Madagascar farmers had been growing vanilla on their farms for generations. They had ready sources of new cuttings literally in their own back yards. To offset higher costs of production, Indian growers quickly embraced the higher pricing afforded by Fair Trade Certification. As a result, a significant portion of Indian production was converted to Fair Trade.

Indian production of Fair Trade Certified vanilla beans probably averaged 20 to 30 metric tons per year in 2006–2008. Production from all other origins was probably just a few tons per year. Unfortunately, production during these years exceeded demand. Discouraged Indian farmers responded by cutting back. In 2009, global production of Fair Trade Certified vanilla beans is roughly one half the previous level; approximately 10 to 15 metric tons.

Large ice cream brands use many times that quantity of vanilla beans each year in the US alone. So do cola flavored carbonated soft drinks. There simply may not be enough

Fair Trade Certified vanilla beans to support a major new product introduction. And, because of the required record keeping, a significant increase in production doesn't happen overnight.

14.6.4 Ensuring that Farmers are Paid the FT Price

In Madagascar, there are reportedly 60,000 farmers growing vanilla beans. Most of these are located in remote locations, not easily accessible by rudimentary roads frequently impassable in the rainy season. The curing centers, by contrast are more centralized, typically in larger towns on the northeastern coast. Harvested beans begin to deteriorate in just a few days and will be completely spoiled if the curing process is not started within a week or so.

Traditionally, farmers sold their green beans to collectors who in turn transported and sold them to the curers. The collectors literally rode into the countryside on motorbikes with cash provided by the curers. Despite being advanced cash by the curers to finance bean purchases, the collectors were largely independent and uncontrollable. How much of the cash was actually paid to the farmers and how much remained in the pockets of the collectors was known only to them.

In many respects, the farmers were at the mercy of the collector. If they rejected the price offered by the collector, their beans would spoil. As a result, the farmers typically received very little for their beans, usually $1.00 to $2.00 per kg. More recently, farmers have responded by curing or partially curing their own beans. They hoped this would give them more leverage, because now they could store the beans while waiting for a more reasonable offer. However, unfortunately the farmers were not very adept at curing and the quality of their beans was poor, reducing their value. Even after curing, the farmers continued to receive minimal compensation for their beans.

The promise of Fair Trade is that the farmers will be paid a price that reflects the costs of sustainable production. FLO International has set the farm gate price of green vanilla beans at $5.60 per kg in Madagascar. However, can that promise be adequately enforced and/or verified among a highly fragmented, remote farmer network? Possibly, but it would require significant cultural and structural changes in an industry that has essentially operated the same way for generations.

14.6.5 Consumer Acceptance

In March of 1999, the Dow Jones Industrial average of 30 large cap stocks crossed 10,000 for the first time. In October 2009, the average once again moved above 10,000. The last decade has not been kind to the American investor.

The fate of the American consumer has been no better. By the fall of 2009, unemployment was just shy of 10%. Taking into account unemployed workers who had given up trying to find a new job, the actual rate was estimated to be 17%. American consumers had stopped spending. Those who were still working were trying to rebuild their retirement savings accounts. Those without jobs have little to spend or save.

How can Fair Trade Certified products, with significantly higher prices than their conventional counterparts, be expected to gain consumer acceptance in the face of such difficult economic times? One way is to educate consumers about the benefits of Fair Trade. Most consumers know very little about Fair Trade beyond the concept of paying a "fair" price to the farmer. Even fewer can identify the Fair Trade Certified label or even know that one exists.

Starbucks published an "Annual Corporate Social Responsibility Report" in 2007 addressed to "Stakeholders" not stockholders. It goes into great detail on all aspects of sustainability including those related to Fair Trade. Unilever, the producer of Ben & Jerry's brand Fair Trade Certified vanilla ice cream, published a similar brochure in 2008 titled "Sustainable Development – An Overview". Among Unilever's commitments to sustainability is the goal to source all tea for Lipton tea bags from Rainforest Alliance Certified™ farms by 2015. These are just two examples of the efforts that many major food and beverage corporations are making to educate their customers on the critical need for and benefits of a commitment to sustainability and Fair Trade.

Another important element required for gaining consumer acceptance of Fair Trade products is credibility. Most consumers want assurance that higher prices paid for Fair Trade Certified products are truly going to the farmer. There is skepticism among consumers that this is in fact the case. One way to build credibility is to establish a direct connection between the consumer and the farmer. For example, Dole Food Company sells its bananas with a farm code on the label, identifying the farm where the bananas were grown. The consumer can visit the "Planet Dole" website, enter the code and take a virtual tour of the farm, meeting the farmers and learning more about their products and their lives. Perhaps the same could be done with vanilla.

14.7 Conclusions

The vanilla industry is highly dependent on impoverished farmers in developing countries. When vanilla beans are in short supply and prices rise, they plant more. When there is a surplus of beans and prices fall, they get discouraged and abandon their vines. This creates a boom and bust cycle, which ultimately has a negative impact on worldwide demand for natural vanilla and threatens the survival of the industry.

Fair Trade would provide farmers a sustainable price for vanilla beans in good times and bad. It would stabilize the market by keeping production in line with demand and accommodating continued growth. Certainly the size, culture and structure of the vanilla market present formidable obstacles to implementing Fair Trade on a widespread basis. But, can we afford not to try?

14.7.1   Update 2017 – Fair Trade Vanilla: Today

Despite its potential, Fair Trade Certified vanilla has seen limited growth since it was added to the Fair Trade portfolio more than a decade ago. Among the issues discussed above, are three primary reasons Fair Trade vanilla has not yet fulfilled its promise.

First and foremost, it has been a victim of the commodity cycle described above. After almost a decade of low and stable pricing, vanilla bean prices began to rise in 2012. The large surplus that had been accumulated following the last crisis was largely depleted. Low prices had discouraged production from all origins other than Madagascar, which was now approaching 80% of the world's production of vanilla beans. In years when the country produced a small crop, it was not enough to meet global requirements. As a result, prices began to climb.

In April 2017, the market price for cured Madagascar vanilla beans was roughly $550 per kg – more than 20 times what it was in 2009. By comparison, the Fair Trade Certified minimum price was set at $47.50 per kg. While Fair Trade Certification still provides a premium over conventional beans, at these price levels there is little incentive for farmers to invest the time and effort to obtain it – especially if there is little demand for it.

The second major cause of the stagnant growth of Fair Trade vanilla is that it has not resonated with consumers, especially at these price levels, as well competing social and ecological platforms. For example, consumption of Organic Certified foods and beverages has grown more than 10% per annum over the last 5 years. Total organic sales in the US of $43.3 billion in 2015 accounted for almost 5% of total food and beverage sales. In many European countries, the growth rate is even higher.

Today's socially conscious consumers are most concerned about health, safety and environmental sustainability. Fair Trade is perceived to be based on a relatively narrow platform – fair compensation to growers. In fact, Fair Trade certification does include a broad spectrum of social and environmental sustainability practices, but this may not have been adequately communicated to consumers.

Finally, Fair Trade vanilla has historically had difficulty providing traceability back to the grower. Consumers who are willing to pay a premium for Fair Trade certified foods and beverages want that connection. However, the vanilla supply chain in Madagascar has historically been extremely complex, with multiple levels and roughly 80,000 individual growers.

The vast majority of these growers live in remote villages, which are extremely difficult for Fair Trade certifiers to visit. Even those that are more accessible typically lack the training for record keeping required for certification. And, each grower produces such a small quantity of vanilla beans that many growers must participate in order to meet the needs of major food and beverage brands. Grower cooperatives have been established, but the knowledge required to properly cure the beans is a critical requirement that takes time and resources to develop.

Sourcing of most vanilla beans continues to follow the traditional supply chain model. Collectors buy directly from growers and transport beans in bulk to other collectors and/or processors. The processors typically keep a log of their purchases but these records are already several steps removed from the growers.

14.7.2   Update 2017 – Fair Trade Vanilla: The Future

The compelling reasons for Fair Trade vanilla are just as important today as they were a decade ago. In fact, the standard of living for rural villages in Madagascar continues to decline. Vanilla remains the key component for providing food, shelter, health care and education for rural farmers and their families.

At the same time, the vanilla supply chain is slowly changing. Flavor companies convert vanilla beans to extracts and supply them to food manufacturers. In recent years, they have increasingly bypassed the traditional supply chain and become directly involved with vanilla bean collection and curing. This transition provides an opportunity to strengthen the connection between the grower and the consumer.

This consolidation can also help the industry manage the commodity cycle, as flavor companies tend to have a longer economic time horizon, than traditional players in the vanilla supply chain. This alone should alleviate some of the historical causes of wide swings in vanilla bean prices. Further, once the initial hurdles associated with Fair Trade certification, including education and training, are cleared the costs of ongoing certification are relatively minor. Thus, the Fair Trade premium would theoretically be lower, especially during periods when bean prices are high.

Although Fair Trade vanilla has not grown significantly over the past decade, several consumer product companies have committed to Fair Trade Certified Vanilla and their products continue to enjoy a loyal following in the marketplace. However, until the economic volatility is greatly diminished, long-term growth of the vanilla industry as a whole will be challenged and the promise of Fair Trade vanilla may go unrealized for the vast majority of vanilla growers. Ultimately, as the vanilla industry continues to consolidate, the implementation, acceptance and benefits of Fair Trade vanilla should become more widely achievable and attainable.

Part II

Authentication and Flavor Analysis

15

Quality Control of Vanilla Beans and Extracts

Arvind S. Ranadive

15.1 Introduction

Consistent quality is the hallmark of a good product. If the raw ingredients used in the manufacture of a product are consistent, if the manufacturing process is sound, and if the handlers or operators are properly trained, there is no reason why a consistent quality product cannot be made at all times. Quality can be defined as a sum of desirable attributes of a product requested or developed by a customer or consumer, and quality control (QC) is maintenance of the defined attributes designed to insure adequate quality in the manufactured products. Maintenance of quality is important not only for the manufacturer's reputation but also for profitability. Manufacturers whose mantra is "do it right the first time" save substantial costs in reworks and lost products. In the flavor industry, QC refers to two important attributes of the product, namely sensory quality, which deals with organoleptic evaluation and analytical quality, which deals with physical and chemical properties. A high quality commercial vanilla product relies on both quality control of the product itself, as well as the vanilla beans from which it is made. Since vanilla is an expensive product, adulteration is an ongoing challenge and determination of authenticity of vanilla products becomes an important activity. This chapter is divided into three sections:

i)   Quality control of vanilla beans
ii) Quality control of commercial vanilla products
iii) Determination of authenticity of vanilla extracts

15.2 Quality Control of Vanilla Beans

Commercial vanilla products are produced from the cured, dried, and conditioned pods of fully mature fruit of the orchid genus *Vanilla*. Of the 110 known species of vanilla, only *Vanilla planifolia* Andrews, *Vanilla tahitensis* Moore, and *Vanilla pompona* Shiede are of commercial importance today. In the United States, vanilla beans have a

Handbook of Vanilla Science and Technology, Second Edition.
Edited by Daphna Havkin-Frenkel and Faith C. Belanger.

standard of identity and are defined in the *US Code of Federal Regulations for Vanilla, 21 CFR 169.3*, as follows:

> Vanilla beans are defined as the properly cured and dried fruit pods of *Vanilla planifolia* Andrews and *Vanilla tahitensis* Moore. Unit weight of vanilla beans refers to 283.85 g of moisture free beans or 13.33 oz of beans with moisture content of 25%.

It is important to emphasize that beans of *V. pompona* are not covered under the standard of identity for use in the United States. However, *V. pompona* is used in Central America to some extent.

The parameters used to determine the quality of vanilla beans are:

i) Grade.
ii) Aroma.
iii) Moisture content.
iv) Vanillin content.
v) Microbial Limits.

15.2.1   Grading of Vanilla Beans

Vanilla is graded before it is sold in the market. Grading depends on the color, size, blemishes, and sheen on the cured and conditioned beans. Each producing country has its own grading system but the basic grading in all vanilla bean producing countries is "Whole beans", "Split beans", and "Cuts" (Figure 15.1). The price of vanilla beans is dependent on its grade category.

Figure 15.1 Basic grading of vanilla beans.

15.2.1.1 Vanilla Grading in Mexico

Grading is based on the following categories:

- *Extras*: Thick beans with dark brown to black color with oily sheen and exceptional appearance.
- *Superiors*: Similar to extras but not as thick and distinguished.
- *Good Superior*.
- *Good*: Slightly paler color than "Good Superior" and less luster, more drier than superiors.
- *Medium*.
- *Ordinary*: Poor color and appearance, can have surface blemishes.
- *Cuts*: Longs, shorts, superior cuts, ordinary cuts.

15.2.1.2 Vanilla Grading in Madagascar

- *Black beans*: Dark chocolate brown, supple beans with oily luster, more than 30% moisture.
- *TK vanilla*: TK(or TeKa) is a Malagasy term for vanilla beans whose quality falls between black beans and Red Fox beans, moisture in these beans is 25 to 30%.
- *Red Fox*: European quality, slightly thinner than TK vanilla with reddish chocolate brown color, 25% moisture.
- *American Quality*: Red Fox quality with 22 to 25% moisture and slightly more blemishes;
- *Cuts*: Short, cut, split beans with substandard aroma and color.

15.2.1.3 Vanilla Grading in Indonesia

Beans are graded into six categories, as shown in Table 15.1:

15.2.1.4 Vanilla Grading in Uganda

Vanilla beans in Uganda are classified into two grades, A and B. Both grades are marketable:

- *Grade A1*: Whole, healthy, supple, dark brown or reddish shiny beans. More than 12 cm long and maximum moisture content of 30%.
- *Grade A2*: Split vanilla with same characteristics as A1 beans.

Table 15.1 Six categories of bean grading.

Type	Vanillin (%)	Moisture (%)
Gourmet	1.6	35
Grade 1	1.6	25
Grade 2	1.4	20
Grade 3	1.0	17
Cuts	0.5	15
EP Cuts	<0.2	12

- *Grade B*: Whole oven dry, short, broken or cut beans. Poor quality beans with irregular shapes, poor color but with fair vanilla aroma, maximum moisture 25%.

15.2.1.5 Vanilla Grading in Tahiti

- *Category Extra*: 16 cm and longer beans.
- *Category 1*: 14 and 15 cm beans.
- *Category 2*: 13 cm and under beans.

Often 13 cm beans are included in Category 1. Maximum moisture in the beans is 44%.

In addition to the grading systems described above, the following specific terms are often used to describe vanilla beans:

- *Blistered beans*: Vanilla beans with blisters. The skin is discolored and easily peels off. This is caused by poor curing of rot afflicted immature beans.
- *Cuts*: Cured broken and cut vanilla beans.
- *Frosted beans*: Cured beans with vanillin crystals naturally formed on the surface (Figure 15.2).
- *Immature beans*: Beans harvested before they are mature. These beans have very low or no vanillin depending upon the degree of maturity (harvest time before maturity).
- *Moldy beans*: The beans are affected by fungal contamination and have a moldy smell, and/or have white or gray mycelium on the bean.
- *Oxidized beans*: Beans with dark spots and a metallic smell. This is caused by beans coming in contact with rusty iron during curing and storage.
- *Phenolic beans*: Beans with a creosote smell. The main reason for this to occur is the storage of harvested green beans a long time before they are killed for curing.

Figure 15.2 Frosted vanilla beans.

Anaerobic storage of cured beans in vacuum packaging is also likely to lead to phenolic odors.

- *Stinky beans*: Beans with offensive smells like smoke, phenolics, kerosene, etc.
- *Split beans*: Beans that are split along their length. This happens when beans are left on the vines well past their maturity. These beans lose some of their resins during curing.

15.2.2   Aroma of Vanilla Beans

The aroma of fully matured and well cured clean vanilla beans can be described with the following terminology:

- Barn yard.
- Floral.
- Hay like.
- Prune/Raisin.
- Resinous.
- Rummy.
- Smokey.
- Sweet.
- Spicy.
- Tobacco-like.
- Vanillin.
- Woody.

The qualitative variations in the aroma and flavor of vanilla of different species and geographic origins are quite distinctive. Various types of beans (Figures 15.3, 15.4 and 15.5) can be described as follows:

- Bourbon vanilla (*V. planifolia*): The term used collectively for beans from Madagascar, Réunion, Comoro Islands, and the Seychelles. The aroma is sweet, creamy, rich, full bodied, tobacco-like, somewhat woody and animal, deep balsamic, and has sweet spicy back notes.
- Mexican vanilla (*V. planifolia*): The aroma of beans from Mexico is sharp, slightly pungent, sweet, and spicy, but lacks body as compared to Bourbon vanilla.
- Indonesian vanilla (*V. planifolia*): The aroma of beans from the Indonesian Islands is less sweet and creamy than Bourbon vanilla. It lacks bouquet, but has a strong woody and slightly smoky character and a freshly sharpened pencil note.
- Indian Vanilla (*V. planifolia*): The aroma of beans from India is full bodied but less sweet and creamy than Bourbon vanilla. It lacks balsamic notes but has slight pungent sour notes.
- Tahitian vanilla (*V. tahitensis*): The aroma of beans from Tahiti has distinct perfumed, flowery, fragrant, and anisic notes. It has a rather shallow and less intense vanilla character.
- PNG vanilla (*V. tahitensis, V. hapape*, or possibly *V. politi):* The aroma of beans from Papua New Guinea is weakly flowery and perfumed, and possesses anisic notes. Its flavor and aroma characters are weak in general.
- Guadeloupe vanilla (*V. pompona*): The aroma of beans from Central America has perfumed, floral, and sweet aromatics. Extracts made from these beans lack body.

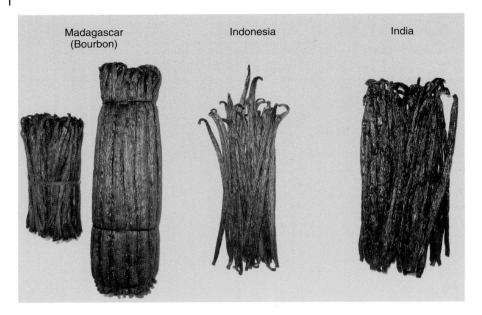

Figure 15.3 Vanilla beans from Madagascar, Indonesia and India.

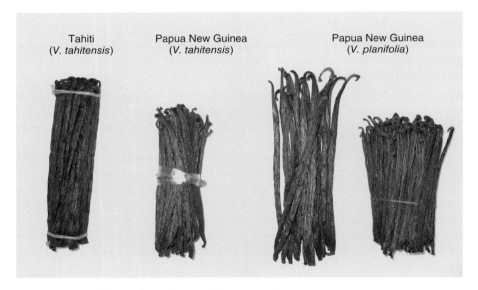

Figure 15.4 Vanilla beans from Tahiti, and Papua New Guinea.

A number of volatile aroma chemicals have been identified in the three vanilla species. A qualitative comparison of some of the major volatiles is presented in Table 15.2. An extensive characterization of the hundreds of volatile compounds present in vanilla beans is presented in chapter 11 by Toth et al. in this book. Characteristically 95% or more of the volatile components are present at concentrations below 10 parts per million.

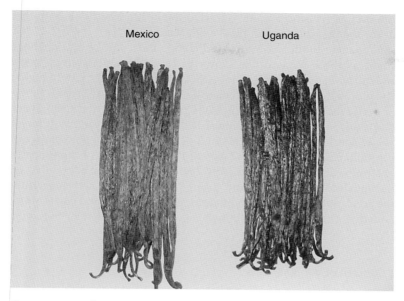

Mexico   Uganda

Figure 15.5 Vanilla beans from Mexico and Uganda.

Although all chemicals that might be responsible for characterizing different species are not identified, anisic alcohol, anisic aldehyde, anisic ethers, anisic acid esters, and *p*-hydroxybenzoic acid, which may be responsible for perfumed, floral aromas, are found to be present in relative abundance in *V. tahitensis*. They are non-detectable or are present in trace quantities in *V. planifolia*, which lacks the floral notes. Adedeji *et al.* (1993) and Tabacchi *et al.* (1978) have confirmed the presence of *p*-anisyl alcohol, *m*-anisaldehyde, *m*-anisate, and anisic acid in Tahitian and Guadaloupe vanillas. Adedeji *et al.* (1993) identified acetic acid, 3-methyl-2-pentanone, methyl pyruvate, anisyl

Table 15.2 Characterizing volatiles in three vanilla species. The number of + signs indicates relative abundance; − sign indicates absence; +/− sign indicates traces to absence (Tabacchi *et al.* 1978, Lhuguenot 1978, Oliver 1973).

	V. planifolia	*V. tahitensis*	*V. pompona*
Vanillin	+ + + + +	+ + + +	+ + +
Anisyl alcohol	−	+ + +	+
Anisic acid	+/−	+ + +	+
Anisic aldehyde	+/−	+ +	+
p-Hydroxybenzoic acid	+	+ + +	+ +
p-Hydroxybenzaldehyde	+ +	+ +	+
Protocatechaldehyde	+	+ +	+
Vanillic acid	+ +	+ +	+
Protocatechuic acid	+/−	+/−	+

acetate, acetylfuran, methyl furoate, 5-methyl-2(3H)-furanne, and half a dozen other minor components for the first time as Tahitian vanilla volatiles.

Although we can associate floral notes to anisyl compounds, smoky and phenolic notes to guaiacol and *p*-cresol, tobacco-like character to nicotine derivatives, sweet notes to vanillin and its derivatives, or balsamic notes and resinous character to higher molecular weight diketones, far more work needs to be done to correlate aroma chemicals identified in a particular vanilla bean type to its peculiar aroma and taste. Black (2005) has identified many known aroma chemicals in cured vanilla beans of different geographic origins and species. His work seems to indicate how the preponderance of certain chemicals in each case determines the overall flavor character of those beans. Black's work shows strong correlation between butter notes and the presence of diacetyl and acetoin, between guaiacol and *p*-cresol and strong phenolic notes, between pentanal, hexanal, and the green notes, between limonene, nonanal, decanal, and citrus notes, between vanillin, methyl cinnamate, t-caryophyllene, and typical vanilla, balsamic, woody notes.

15.2.3 Moisture Content of Vanilla Beans

The moisture content of commercial vanilla beans varies from 10% for poor quality lower grade beans to 35% for gourmet beans. The extraction grade beans contain from 20 to 25% moisture, the maximum limit set by the US standard of identity. Drier beans are less aromatic and less appealing than supple, shiny, high moisture beans. However, under poor handling and storage conditions, the high moisture beans can easily develop mold.

Moisture content in vanilla beans can be determined by using the following methods:

- Hot Oven Method.
- Vacuum Oven Method.
- Infrared Heating Method.
- Chemical Methods.
  - Distillation with toluene.
  - Karl Fischer method.

15.2.4 Vanillin Content

Vanilla derives its flavor character and flavoring strength from one of its most important components, vanillin. Almost one-third of the flavor strength of a vanilla product is attributed to its vanillin content. Naturally, one of the criteria for determining the vanilla bean price is its vanillin content. Beans with higher vanillin content fetch a higher price.

The vanillin content of cured vanilla beans of different geographic origins varies considerably, from trace quantities to almost 3% by weight. However, this is the function more of cultivation and curing practices than of the geographic location. Beans harvested at full maturity and cured under properly controlled conditions yield highest vanillin. Fully mature beans, if cured poorly, can produce very low vanillin. On the other hand, immature or under matured beans will produce low or no vanillin, even if curing is carried out meticulously. Gassenmeier *et al.* (2008) note that vanilla beans produced using alternative vanilla-processing methods (as opposed to traditional curing method) can produce beans with higher vanillin content but can be poor in overall aroma characteristics.

Vanillin content of different species of beans is significantly different. Beans of *V. planifolia* can yield 2 to 2.5% vanillin under optimum conditions, whereas beans of *V. tahitensis* normally will produce no more than 1.2 to 1.5% vanillin, even under optimum conditions.

Standardized methods for vanillin quantification from beans are:

15.2.4.1 Vanilla Bean Extraction

Weigh 50 g of dry cured beans. Cut into quarter inch length pieces, place in a 500 ml Erlenmeyer (conical) flask, add 250 ml of 35% aqueous ethanol, and close the flask with a stopper. Place in a hot water bath maintained at 55 to 60 °C for at least 8 to 12 hours, shaking the flask once every hour. Then decant the extract and collect it in a 500 ml volumetric flask.

Add 250 ml fresh solution of 35% aqueous ethanol to the Erlenmeyer flask containing extracted beans, close with the stopper and place in the hot water bath for another 8 to 12 hours, shaking the flask occasionally. At the end of this period, decant the extract and add to the volumetric flask containing the first extract. Bring the volume of the combined extract to 500 ml with the addition of 35% aqueous ethanol, if required. Shake well. This is equivalent to one-fold vanilla extract.

15.2.4.2 Vanillin Determination

UV (Spectrophotometric Method)   Standardization of 1 ppm vanillin. This is done every 3 to 6 months.

Dissolve 0.1000 g vanillin in 3 ml 190 proof ethyl alcohol in a 100 ml volumetric flask and bring to volume with distilled water.

Prepare 0.1 N sodium hydroxide (NaOH) solution by dissolving 4 g NaOH pellets in distilled water and bringing the volume to 1 liter in a 1-liter volumetric flask.

Pipette 3 ml vanillin solution into a 1-liter flask, add 2 ml of 0.1 N NaOH and bring to volume with distilled water.

Prepare a blank by adding 2 ml 0.1 N NaOH in a 100 ml volumetric flask and bring to volume with distilled water.

Determine absorbance of vanillin solution at 270, 348, and 380 nm using the blank solution to zero the machine at each wavelength.

Calculate absorbance of 1 ppm vanillin as follows:

$$\text{Absorbance}(\text{corrected}) = \frac{A_{348} - (0.29 A_{270} + 0.71 A_{380})}{3}$$

A_{348} is the peak of absorption of vanillin. Background absorption from polyphenolic and other interfering compounds in the sample is corrected for by subtracting a percentage of the absorbance at 270 and 380 nm.

Standard Vanillin Factor (SVF) = Absorbance (corrected) × 10. The SVF should be in the range of 1.5 to 1.54.

15.2.4.3   Vanillin Determination in Vanilla Extracts and Other Vanilla Products

See Table 15.3 for using appropriate amounts for analysis for different folds of extracts and for using appropriate multiplier for calculating vanillin. The following procedure is given for one-fold vanilla extract.

Pipet 5 ml vanilla extract in a 100 ml volumetric flask and bring to volume with distilled water. (Solution A)

Pipet 2 ml Solution A into another 100 ml volumetric flask and add 2 ml 0.1 N NaOH solution and bring to volume with distilled water. (Solution B)

Pipet 2 ml Solution A into another 100 ml volumetric flask and bring to volume with distilled water. (Solution C)

Determine the absorbance of Solution B at 348 nm using a spectrophotometer and using Solution C as a reference solution to zero the machine.

Vanillin content in the extract is calculated as:

$$\frac{A_{348}\left(\text{Multiplier from the table}\right)}{\text{Standard Vanillin Factor}}$$

15.2.4.4   HPLC Method

This method gives a more accurate estimate of vanillin content of a vanilla product. Vanilla extracts (and other vanilla products) show lower values for vanillin when analyzed using the HPLC method than when the UV spectrophotometric method is used. This is because in HPLC analysis, interference from other vanilla components, such as *p*-hydroxybenzaldehyde and *p*-hydroxybenzoic acid, is eliminated. Typical system parameters are as follows:

i)   The liquid chromatograph instrument consists of a solvent delivery system, sample injector valve or auto sampler and a light absorbance detector with UV detection set at 254 nm;
ii)   Column: stainless steel with dimensions of 100 × 4.6 mm to 250 × 4.6 mm packed with 5 to 10 μm particle size C8 stationary phase;
iii) Solvents: LC grade methanol, LC grade water, 95% ethanol and glacial acetic acid;
iv)   Mobile phase is methanol-acidified water (10 + 90). Water is acidified by adding 10 mL glacial acetic acid per 800 mL water;
v)   Sample filters used are alcohol compatible 0.45 μm nylon 66 membrane.

Table 15.3 Sample preparation for vanillin determination.

Vanilla Product	To make solution A Use	To make solution B Use	0.1N NaOH Use	Multiplier
1-Fold Extract	5 ml	2 ml	2 ml	1
2-Fold Extract	5 ml	2 ml	2 ml	1
3-Fold Extract	5vml	1 ml	2 ml	2
4-Fold Extract	5 ml	1 ml	2 ml	2
5-Fold Extract	2 ml	1 ml	2 ml	5
10-Fold Extract	1 ml	1 ml	2 ml	10
Oleoresin	0.1 g (wt)	2 ml	2 ml	50

15.2.5   Microbial Contaminant Limits

To carry out the analysis for microbial contamination, beans are taken from many different boxes of a shipment of beans, so that a representative sample is obtained. A weighed sample of beans is then suspended in sterile distilled water and shaken on a shaker for at least 1 hour. The beans are then removed and the extraction liquid is diluted serially with sterile distilled water for appropriate microbial load for culture and testing.

To insure the microbial safety, beans are tested for contaminants, as shown in Table 15.4:

MPN stands for Most Probable Number and is a standard parameter of reporting *E. coli* analysis.

Contaminant levels exceeding the recommended limits indicate improper handling and unhygienic conditions during curing and storage.

15.3   Quality Control of Commercial Vanilla Products

15.3.1   Definition of Vanilla Products

The US Code of Federal Regulations for Vanilla, 21 CFR 169.175–169.182, defines commercial vanilla products.

15.3.1.1   Vanilla Extracts

According to the US Food and Drug Administration regulations, Vanilla extract is the solution in aqueous ethyl alcohol of the sapid and odorous principles extractible from vanilla beans. Ethyl alcohol content of such an extract is not less than 35% by volume, and the extractible matter of one or more units of vanilla constituent. A unit of vanilla constituent is 13.35 oz of beans containing not more than 25% moisture per gallon of finished extract. This amounts to the extractible matter of not less than 10.0125 oz (FDA Standard) of beans on the moisture-free basis. FDA will not object to the production of vanilla extract through the use of 10 oz of vanilla beans on a dry weight basis, per gallon of vanilla extract. This means that the weight of beans to manufacture each gallon of vanilla extract can vary, depending on the moisture content of the beans. Vanilla extract may contain one or more of the following ingredients:

- Glycerin.
- Propylene glycol (usually no more than 2%).

Table 15.4 Recommended limits for contaminants.

Test	Recommended Limits, Counts/g
Total Plate Count	1,000–10,000
Yeast	<10
Mold	<1
E. coli 3 Tube MPN	<3/g
Salmonella	Negative

- Sugar (including invert sugar).
- Dextrose.
- Corn syrup (or corn syrup solids).

Vanilla extracts are available in 1- to 10-fold strengths. Almost all household extracts are single-fold strength. Two-fold and higher strength extracts are preferred by the industrial users such as ice cream, chocolate, candy, bakery, beverage, and other food manufacturers.

15.3.1.2   Vanilla Flavoring
This is similar to vanilla extract but contains less than 35% ethyl alcohol by volume.

15.3.1.3   Vanilla-Vanillin Extract and Flavoring
These are synthetic vanillin fortified extracts and flavorings. One ounce of vanillin per fold of extract per gallon can be legally added. For example, 1 gallon of vanilla-vanillin 2-fold extract is made by adding 1 oz of vanillin to 1 gallon of 1-fold vanilla extract. It is assumed that 1 oz of pure vanillin has a flavoring strength of 1 gallon of 1-fold vanilla extract.

15.3.1.4   Concentrated Vanilla Extract and Flavoring
It is usually not practical to prepare extracts of more than two-fold strength by the straightforward extraction although, with proper equipment, it is possible to obtain higher strength extracts. Vanilla extracts or flavorings of three-fold or higher strengths are prepared by removing part of the solvent by vacuum distillation. The recovered alcohol is then added back to bring the alcohol level to a minimum of 35% by volume in conformation with the CFR. Because of the standard of identity, these products cannot be made by adding vanilla oleoresin to vanilla extract. The latter product needs to be declared as "concentrated vanilla extract × fold made from oleoresin." The quality of concentrated vanilla extract is superior to vanilla concentrate made from oleoresin. Concentrated extracts are available in 3- to 10-fold strengths.

15.3.1.5   Vanilla Oleoresin
This solvent-free viscous concentrate is composed of 50% aqueous ethanol soluble vanilla solids. Vanilla oleoresins are standardized to 4, 6, and 8 oz strengths with glycerin or propylene glycol. Four ounces by volume of a 4 oz strength vanilla oleoresin when dissolved in sufficient quantity of 35% aqueous ethanol will yield 1 gallon of a 1-fold extract. Similarly, 6 oz strength oleoresin requires 6 oz and 8 oz strength oleoresin requires 8 oz to produce 1 gallon of 1-fold extracts, respectively. It follows then that 4 oz oleoresin is designated 32-fold strength. In the process of manufacturing oleoresin, there is loss of some volatile aromatics during distillation of the solvent. A small amount of polymerization of various vanilla constituents also occurs during concentration. Because of this, oleoresins do not have the full flavor of percolated vanilla extracts. They are often used for producing compounded and vanilla-vanillin type products.

15.3.1.6   Vanilla Absolute
This, the most concentrated form of vanilla, is available as a hydrocarbon solvent (such as hexane) extracted product or as a supercritical liquid CO_2 extracted product. The two products are quite different in their aroma character, color, and solubility in various

carrier solvents. Absolutes are useful in products where intense vanilla aroma is desirable.

15.3.1.7 Vanilla Powder And Vanilla-Vanillin Powder

Vanilla powder is available as a spray dried or plated product. Starch, gum, or a mixture of starch and gum are used as carriers for spray dried products. Sugar or maltodextrins are used for plated vanilla powders. The powders are available in one- to five-fold strength, although higher strength powders can be prepared.

Vanilla-vanillin powder is a synthetic vanillin fortified product. This is prepared by adding 1 oz of synthetic vanillin per unit of vanilla constituent, for example, a 2-fold vanilla-vanillin powder can be obtained by adding 1 oz of vanillin to vanilla powder containing one unit of vanilla.

15.3.1.8 Vanilla Tincture for Perfumery

This is prepared by macerating vanilla beans with specially denatured alcohol. The finished tincture contains 90% alcohol and is soluble in perfume oils.

15.3.2 Vanilla Extract Quality Parameters

Quality of a vanilla extract can be judged using following parameters:

15.3.2.1 Appearance: Color and Clarity

Pure single-fold vanilla extract is a pale to dark amber brown clear liquid. Some vanillas have a reddish hue. The color depends on the quality of beans, extraction method used and, to some extent, on the geographic origin of the beans. Clarity of the extract is related to the resin and wax content of the beans. Usually beans with higher resin contents produce cloudier products. Clarity of the product can be achieved through optimization of extraction methods.

15.3.2.2 Flavor

Flavor quality of vanilla extract depends, to a large extent, on the vanilla beans and the extraction method used for its manufacture. In turn, the flavor quality of beans depends on vanilla genotype, geographic region, environment (climate, soil, etc.), harvest maturity of beans, and overall curing technique. Freshly percolated vanilla extract has a slightly harsh alcohol bite, unless tempered with the addition of sugar syrup or glycerin. The product mellows as it ages in a tank over a period of time. Stored under proper conditions (60–65 °F) and away from light, vanilla develops pleasant, sweet, somewhat complex and mellow notes.

The following terminology is commonly used to describe vanilla flavor:

- Anisic (Tahitian).
- Spicy, Acidic (Indian).
- Balsamic (Bourbon).
- Caramelized.
- Chocolate.
- Creamy.
- Floral.

- Hay-like.
- Prune/Raisin.
- Resinous.
- Smokey (Indonesian).
- Sweet.
- Tea-like.
- Vanillin.
- Rummy.
- Woody.

Hoffman *et al.* (2005) have combined gas chromatography, high pressure liquid chromatography, and sensory testing methods in their quest to distinguish "good" vanilla extracts from low quality vanilla extracts. They report that esters of ethyl acetate, ethyl hexanoate, and ethyl octanoate as having pleasant fruity aromas and are very effective at very low concentrations. Acetaldehyde ethyl acetal was associated with a nutty aroma, whereas ethyl nonanoate aroma was similar to cognac and reminiscent of rummy resinous notes in good quality vanilla. Guaiacol, vanillic acid, and *p*-hydroxybenzoic acid contribute to a smoky odor and faintly phenolic taste. Vanillyl ethyl ether and 4-(ethoxy methyl) phenol are described as having sweet vanilla-like flavors, with creamy coconut undertones.

15.3.2.3 Soluble Solids Content

Typical single-fold vanilla extract made without any additives has soluble solids content of about 1.8 to 2.2% on the weight basis. Soluble solids are composed of sugars, non-sugar carbohydrates, amino acids and proteins, phenolic compounds, and minerals. Lower values for soluble solids in extracts can arise from incomplete extraction of beans or use of lower than required amount of beans.

15.3.2.4 Vanillin Content

This principal flavoring component of vanilla is present in the single-fold extract at levels ranging from 0.2 g per 100 mL (0.2%) for a good quality extract to less than 0.02 g per 100 mL (0.02%) for inferior quality product. Vanillin content of an extract depends on the quality of beans used. Higher quality and higher grade beans produce extracts with higher vanillin content and vice versa. It should be remembered that high vanillin is not the only defining character of high quality of vanilla extract. Vanilla extracts made from Indian vanilla beans tend to yield vanillin content in the range of 0.2 to 0.25% per fold, although Gassenmeier *et al.* (2008) have found lower values for the 2006/2007 crop of Indian beans. In the last few years, beans from Madagascar have declined in their quality consistency and are averaging less than 0.18% vanillin content per fold. It is generally observed that Indonesian and PNG (Papua New Guinea) bean-based extracts are less consistent in their vanillin content and average less than 0.14%. Besides, a number of blended vanilla products are available, where vanillin content varies from 0.075% to 0.15%. Routinely, vanillin content of vanilla extract is determined using either the UV method (AOAC) or a high-performance liquid chromatography (HPLC) method (Guarino and Brown 1985; Ranadive 1993; Wallace 1983). Vanillin content of vanilla extracts is determined using the procedures described in earlier sections. Of the two methods used, UV (spectrophotometric) and HPLC, the latter gives more accurate but typically 5 to 8% lower values than the values obtained using UV method. Gas chromatography techniques have also been used for vanillin determination.

15.3.2.5 Organic Acids – (Wichmann) Lead Number

Vanilla extracts show at least 20 organic acid peaks in gas chromatographic analysis (Fitelson and Bowden 1968). Of the eight major peaks, only malic acid is positively identified; citric acid, isocitric acid, and succinic acid are suspected to be among other major acids.

Quantitatively, the organic acids content of vanilla extracts is expressed as "lead number". Lead number is determined by precipitating lead salts of organic acids with the addition of lead acetate solution to the vanilla extract. Unreacted excess lead acetate is back-titrated to determine the amount of precipitated lead. Often, the lead number is used to determine the purity as well as the strength of vanilla extract. The lead number also indicates whether the extract has been prepared with the correct amount of beans and it used to be the standard method for such a determination. Typically, Bourbon extracts have lower lead numbers, in the range of 0.65 to 0.85 per fold of extract, compared to a 1.1 to 1.4 per fold range for Indonesian extracts. It is not uncommon to see the lead numbers falling out of either side of these ranges for both types of authentic extracts. For practical reasons, it is assumed that a pure single-fold vanilla extract should have a minimum lead number of 0.7. The procedure to measure lead number is described in the publication *Official Methods of Analysis of AOAC International*, 18th edition (2005), section 36.2.1.

This method of ascertaining the purity of vanilla extract is easy to circumvent by the addition of foreign acids to enhance the lead number. However, if used in conjunction with the gas chromatography method of Fitelson and Bowden (1968), it can be very useful in detecting adulteration of vanilla with foreign organic acids. The volatile trimethylsilylated (TMS) derivatives of vanilla organic acids show a typical pattern of 8 major peaks upon separation by gas liquid chromatography. The TMS pattern for extracts made from beans of different geographic areas varies in peak heights as well as certain peak ratios. A detailed procedure to determine the organic acids pattern in vanilla is described in the publication Official *Methods of Analysis of AOAC International*, 18th edition (2005), Section 36.2.22.

Many vanilla companies have stopped using the lead number method, because of environmental considerations regarding lead disposal.

15.3.2.6 Resin Content

Single strength Bourbon extracts contain slightly more than 0.1% resins. Similar strength Indonesian extracts contain 0.04 to 0.06% resins. These pleasant taste producing polysubstituted polymerized phenols are determined quantitatively and their authenticity confirmed by a paper chromatography method, as described in AOAC procedure (2005).

15.3.2.7 Microbial Limits

Chances of microbial contamination coming from the use of vanilla products are very small. But it is a concern to most food manufacturers and is of particular concern to the manufacturers of cultured dairy products. Most of vanilla buyers have adapted microbial contaminant limits (Table 15.5):

Table 15.5 Recommended limits for microbial contaminants.

Test	Recommended Limits, Counts/g
Total Plate Count	1000–10,000
Yeast	<10
Mold	<10
E. coli 3 Tube MPN	<3/g
Salmonella	Negative
Thermophilic Aerobic Spores	<10
Mesophilic Aerobic Spores	<10

15.4 Determination of Authenticity of Vanilla Extracts

Since vanilla is an expensive product and the cost of vanilla beans can fluctuate significantly over a relatively short period of time, vanilla products are tempting targets for sophisticated adulteration schemes. It is therefore a continuing challenge to verify authenticity of vanilla extracts and natural vanilla products. Since addition of synthetic or semi-synthetic vanillin is the simplest and easiest method to reduce the product cost, vanillin is the main target for authenticity determination.

In order to distinguish between a pure vanilla extract and one to which synthetic vanillin has been added, Martin *et al.* (1977) developed an identification ratios method. They determined the absolute levels of vanillin, potassium, nitrogen, and phosphate in the authentic extracts made from beans from Madagascar, Mexico, Indonesia, Comoros, and Tahiti. Next, they determined the vanillin/K, vanillin/N, and vanillin/PO_4 ratios for each type of product (Table 15.6) and established that these ratios were different for different type of beans but fairly constant for authentic extracts made from same type of beans. Their work further established that when these ratios, for a particular vanilla extract, were considerably higher than the authentic vanilla extracts, it was a strong evidence of added vanillin. On the other hand, low absolute values for vanillin, K, N, and PO_4 in a vanilla extract were an indication of inadequate amounts of beans used.

15.4.1 Guidelines for Determination of Authenticity

Authenticity of vanilla products is now accepted if it conforms to criteria defined by: i) the US standard of identity for vanilla, and ii) IOFI (International Organization of the

Table 15.6 Ratios of vanillin and N, PO_4, and K in authentic vanilla extracts.

Vanillin/N*	11(M), 2.2 (J), 8 (MX), 7.5 (T)
Vanillin/PO_4*	11.8 (M), 1.6(J), 8 (MX), 7.5 (T)
Vanillin/K*	1.5 (M), 0.25 (J), 1.45 (MX), 0.8 (T)
Vanillin/*p*-hydroxybenzaldehyde	12-15 (M,J,MX) 7.9 (T), 27(I)

* These ratios are not yet available for Indian, Ugandan, or PNG beans.
M – Madagascar, J – Java (Indonesia), MX – Mexico, T – Tahiti, I – India

Flavor Industry) guidelines, which has adapted the French government's document specifications titled DGCCRF (Directorate General for Competition, Consumption, and Fraud Repression).

The definition for the US standard of identity for vanilla has been given before. This requirement basically defines pure vanilla as a product made from only two species of vanilla, *V. planifolia* and *V. tahitensis*, using the specified amount of vanilla beans and containing at least 35% ethyl alcohol by volume.

IOFI Guidelines N 1271 has adopted three criteria to determine authenticity of natural vanilla products.

15.4.1.1 Evaluation of the Ratios Between Specific Components

There are four major components in vanilla, which are used as marker compounds to determine quality and authenticity of vanilla products. The specific components are *p*-hydroxybenzoic acid (*p*-HB acid), *p*-hydroxybenzaldehyde (*p*-HBAld), vanillic acid, and vanillin. These four components are easily separated and calculated by the HPLC method of analysis. For authentic vanilla extracts, these ratios are fixed within a certain range. The revised guidelines by DGCCRF – Notification note – 2003-61, and adopted by IOFI for these ratios are given in Table 15.7. However, the ratios for vanillin/*p*-HBAld (R1) and vanillic acid/*p*-HBAld(R2) for Indian vanilla beans were found to be higher than the DGCCRF guideline by John and Jamin (2004) and Gassenmeier *et al.* (2008).

15.4.1.2 Isotope-ratios Mass Spectrometry

This guideline for authenticity verification involves the determination of the ^{13}C deviation of each of the four important vanilla aromatics, vanillin, *p*-hydroxylbenzaldehyde (*p*-HB), *p*-hydroxybenzoic acid (*p*-HB acid), and vanillic acid.

DGCCRF- Notification note – 2003-61 has determined acceptable limits for ^{13}C deviation values for four main components of vanilla. These values are given in Table 15.8. Excellent information on ^{13}C deviations is available in Scharrer and Mosandl (2002) and John and Jamin (2004) (Table 15.8).

15.4.1.3 Site-specific Quantitative Deuterium NmR

This method depends on determining the authenticity of pure vanilla by determining the four site-specific natural isotope ratios (D/H) of its vanillin molecule (Figure 15.6) by NMR (Remaud *et al.* 1997).

Table 15.7 DGCCRF Guideline values.

Ratios between flavor components	Revised ranges
R1 = Vanillin/*p*-HBAld	10–20
R2 = Vanillic acid/*p*-HBAld	0.53–1.5
R3 = *p*-HB acid/*p*-HBAld	0.15–0.35
R4 = Vanillin/Vanillic acid	12–29
R5 = Vanillin/*p*-HB acid	40–110

Table 15.8 ^{13}C Deviation values.

Vanillin	−21.2%
p-Hydroxybenzaldehyde	−19.2%
Vanillic Acid	−24.0%
p-Hydroxybenzoic acid	−23.0%

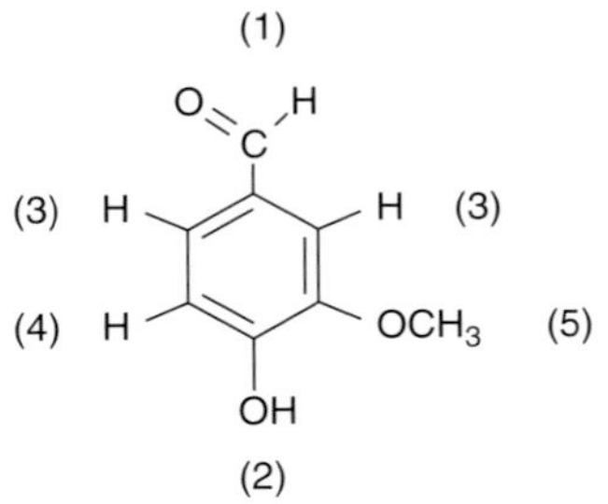

Figure 15.6 Specific sites on vanillin molecule for natural isotope ratios. Vanillin from vanilla beans of many origins (49 samples) yielded the following values: D/H ratios: 1–130.8 SD ±3.1, −157.3±3, 4–196.4 SD ± 2.5, 5–126.6 SD ± 1.7.

15.4.2 Other Methods to Determine Authenticity

15.4.2.1 Stable Isotope Ratio Analysis (SIRA)

The structure and flavoring properties of vanillin are the same, whether derived from vanilla beans or prepared synthetically from lignin, eugenol, and guaiacol or petroleum by-products. The ^{13}C/^{12}C ratio for natural vanillin from vanilla beans, however, is different from vanillin produced synthetically from lignin, eugenol, or guaiacol, all of which are also plant materials. In plants, the isotopic fractionation of carbon takes place during photosynthesis via CO_2 fixation. Plants fix CO_2 by one of three metabolic pathways: Calvin synthesis (C$_3$ plants), Hatch-Slack pathway (C$_4$ plants), or Crassulacean acid metabolic (CAM) pathway, each of which produce natural products with different ^{13}C/^{12}C ratios. In the CAM plants, both the Calvin cycle as well as Hatch-Slack pathway modes are operational. In these plants, CO_2 is absorbed at night and combines with phosphoenolpyruvate to form 4-carbon organic acids (malic and isocitric acids). During daylight, C$_4$-acid breaks down releasing CO_2 to be utilized by the Calvin cycle. The CAM pathway operates in vanilla and enriches vanillin with the ^{13}C isotope. Bricout (1974), Bricout et al. (1981), and Hoffman and Salb (1979) applied the radio carbon technique to determine the δ^{13}C values (generally reported as the stable isotope ratio analysis, or SIRA) of vanillin from different sources and used the information to establish the origin of vanillin in vanilla extracts. The δ^{13}C results are reported in parts per thousand relative to the fossil Pee Dee (South Carolina) Belemnite Standard (PDB). They are computed as follows:

$$\delta^{13}C_{sample} = \frac{\left(^{13}C/^{12}C\right)_{sample}}{\left(^{13}C/^{12}C\right)_{PDB}} - 1 \times 1000$$

The $\delta\,^{13}C$ values for vanillin from different sources are given in Table 15.9.

It is apparent that the addition of synthetic vanillin to vanilla extract can be detected by SIRA. However, a means of circumventing SIRA of vanillin was discovered, which involved addition of ^{13}C-enriched vanillin to lignin-derived vanillin. This was done by replacing the methyl group of lignin-derived vanillin with methyl-^{13}C, which made the synthetic vanillin appear natural to routine SIRA. Bricout et al. (1981) and Krueger and Krueger (1983) developed new tests to detect this addition of [methyl-^{13}C]-vanillin to lignin-derived vanillin. Their technique involves removal of methyl carbon from the vanillin molecule prior to performing SIRA on the resulting dihydrobenzaldehyde (DHB) or CH_3I. This analysis reveals whether lignin vanillin is altered with methyl-^{13}C and makes it even more difficult to adulterate pure vanilla extracts with synthetic vanillin. This is illustrated in Table 15.10.

Besides the methyl group, there are seven other molecular positions in vanillin, in which the ^{13}C label could be placed without detection by any of the above mentioned tests. Out of concern that some day an economical method could be developed for producing one of these, [carbonyl-^{13}C]-vanillin, Krueger and Krueger (1985) have developed a method for detecting such an adulteration. Their method involves oxidation of vanillin to vanillic acid in the first step. This is followed by decarboxylation of vanillic acid with bromine, and performing SIRA on the resulting CO_2 from the carbonyl carbon. Vanillin carbonyl SIRA values for lignin and natural vanillin are given in Table 15.11

Table 15.9 Carbon SIRA of vanillin from various sources (from Hoffman and Salb 1979).

Source of Vanillin	$\delta\,^{13}C_{total}$
Vanilla (Madagaxscar)	−20.4
Vanilla (Indonesia)	−18.7
Vanilla (Mexico)	−20.3
Vanilla (Tahiti)	−16.8
Lignin	−27.0
Eugenol (clove oil)	−30.8
Guaiacol	−32.7

Table 15.10 Analysis of [methyl -^{13}C]-Vanillin (from Krueger and Krueger 1985).

Sample	$\delta\,^{13}C_{total}$	$\delta\,^{13}C_{DHB}$	$\delta\,^{13}C_{methyl}$
Lignin vanillin	−27.0	−26.7	
Altered lignin vanillin	−20.0	−26.0	
Lignin vanillin	−27.3		−28.4
Altered lignin vanillin	−20.6		+25.8

Table 15.11 Vanillin carbonyl SIRA on various vanillin sources (from Krueger and Krueger, 1985).

Sample	$\delta^{13}C_{carbonyl}$	$\delta^{13}C_{total}$	Remarks
Lignin	-37.7 ± 1.4	-27.22	Known lignin vanillin
Altered lignin	$+17.1$	-19.94	Lignin vanillin with added [carbonyl ^{13}C]-vanillin
Bourbon	-25.7	-21.4	Known natural vanillin
Bourbon	-27.7		Known natural vanillin
Bourbon	-23.2		Known natural vanillin
Bourbon	-29.9		Known natural vanillin
Bourbon	-24.7		Known natural vanillin
Bourbon	-24.8		Known natural vanillin
Commercial	-24.5		Probably natural
Commercial	$+67.1$	-15.6	Adulterated carbonyl-^{13}C

15.4.2.2 SNIF-NMR Technique

A new technique known as SNIF-NMR (site-specific natural isotope fractionation by nuclear magnetic resonance), was developed in 1988 by Gerard Martin at the University of Nantes, France. The natural replacement of hydrogen by its heavy isotope deuterium, in certain natural molecules, is measured by this technique.

The technique has been shown to be very useful in determining the authenticity of pure vanilla by determining the four site-specific natural isotope ratios (D/H) of its vanillin (Figure 15.6) by NMR (Maubert *et al.* 1988). Depending upon the source of vanillin, natural from vanilla beans, natural made by fermentation, or synthetic from lignin or guaiacol, these (D/H) ratios will vary due to the fact that physical, chemical, and biological processes enrich some materials with deuterium and deplete others. Rain water in the tropics, where vanilla grows, is relatively rich in deuterium (160 ppm). This deuterium concentration gradually decreases as we move away from the equator towards the North and South Poles. The rain (snow) water near the South Pole, for example, is only 90 ppm. The plants absorb rain water containing different concentrations of deuterium, depending on the region where they are growing. Altitude, distance from the sea, and level of rainfall of the growing area also affect the distribution of deuterium in local water and hence the site-specific ratios. All these factors and plant biochemical processes are responsible for particular level of deuterium enrichment at specific sites on the vanillin molecule. Vanillin molecules from different geographic origins and sources will have different site specific natural (D/H) ratios. Thus measurement of (D/H) ratios on all four sites on the vanillin molecule can precisely reveal its origin.

John and Jamin (2004) reported that the SNIF-NMR pattern of vanillin and *p*-HB for vanilla beans of Indian origin "showed significant differences from beans from other geographical sources." They surmise this could be due to effect of latitude, altitude, distance from the sea, rainfall, etc. on the Indian vanilla growing areas. They, however, conclude that SNIF-NMR data can discriminate the vanillin and *p*-hydroxybenzaldehyde of the Indian beans from synthetic or semi-synthetic sources.

15.5   Summary

The factors that define quality of vanilla beans and vanilla extracts have been reviewed in this chapter. Grading of vanilla beans based on appearance, aroma, taste, and vanillin content was explained. Aroma compounds, besides vanillin, in vanilla and their role in vanilla quality was described. Vanilla quality control parameters and their methods of analysis were presented. Values for the acceptable ratios of important marker compounds were presented. Methods for determining authenticity of vanilla products were described and the recommended authenticity values for ^{13}C deviation (SIRA) and SNIF-NMR analysis were given. Information in this chapter should help vanilla growers, processors, suppliers, and vanilla products manufacturers to understand and maintain quality of vanilla. To vanilla consumers, big and small, the information can help in choosing the appropriate and best quality products in their applications.

Acknowledgment

I want to thank my wife, Sunanda Ranadive, for reading and providing valuable comments and suggestions during the preparation of this manuscript.

References

Adedeji, J., Hartman, T.G. and Ho, C.T. (1993) Flavor characterization of different varieties of vanilla beans. *Perfumer & Flavorist*, 18, 25.

Black, M. (2005) Vanilla bean volatile analysis, origin and species. Paper presented at Vanilla 2005 conference, Veracruz, Mexico.

Bricout, J. (1974) Analysis of stable isotopes of carbon in quality control of vanillas. *J. Assoc. Offic. Agric. Chemists*, 57, 713.

Bricout, J., Kozlet, J., Derbesy, M. and Beccat, B. (1981) New possibilities with analysis of stable isotopes of carbon in quality control of vanillas. *Ann. Fals. Exp. Chim.*, 74, 691.

Fitelson, J. and Bowden, G.L. (1968) Determination of organic acids in vanilla extracts by GLC separation of trimethylsilylated derivatives. *J. Assoc. Offic. Agric. Chemists*, 51, 1224.

Gassenmeier, K., Riesen, B. and Magyar, B. (2008) Commercial quality and analytical parameters of cured vanilla beans (*Vanilla planifolia*) from different origins from the 2006–2007 crop. *Flavour Fragr. J.*, 23: 194–201.

Guarino, P.A. and Brown, S.M. (1985) Liquid chromatographic determination of vanillin and related flavor compounds in vanilla extract: Cooperative Study. *J. Assoc. Offic. Anal. Chem.*, 68, 1198.

Hoffman, P.G. and Salb, M. (1979) Isolation and stable isotope ratio analysis of vanillin. *J. Agric. Food Chem.*, 27, 352.

Hoffman, P., Harmon, A., Ford, P. *et al.* (2005) Analytical approaches to vanilla quality and authentication. In: Vanilla: First International Congress. Allured Publishing, Carol Street, IL pp. 41–49.

John, T. V. and Jamin, E. (2004) Chemical investigation and authenticity of Indian vanilla beans. *J. Agric. Food Chem.*, 52, 7644.

Krueger, D.A. and Krueger, H.W. (1983) Carbon isotopes in vanillin and the detection of falsified "natural" vanillin. *J. Agric. Food Chem.*, 31, 1265.

Krueger, D.A. and Krueger, H.W. (1985) Detection of fraudulent vanillin labeled with ^{13}C in the carbonyl carbon. *J. Agric. Food Chem.*, 33, 323.

Lhuguenot, J.C. (1978) A propos de quelques particularités des--extraits de vanille Tahiti. *Ann. Fals. Exp. Chim.*, 71, 115.

Martin, G.E., Ethridge, M.W. and Kaiser, F.E. (1977) Determining the authenticity of vanilla extracts. *J. Food Sci.*, 42, 1580.

Maubert, C., Guérin, Mabon, F. and Martin, G.J. (1988) Détermination de l'origine de la vanilline par analyse multidimensionnelle du fractionnement isotopique naturel spécifique de l'hydrogéne. *Analysis*, 16, 434.

Oliver, R. (1973) Methods for the study of the aromatic components of vanilla extracts. *Proceedings of the Conference on Spices, London,* 1972.

Official Methods of Analysis of the AOAC International, 18th edition (2005) Sections 36.2.06–36.2.07.

Ranadive, A.S. (1993) Vanillin and related flavor compounds in vanilla extracts made from beans of various global origins. *J. Agric. Food Chem.*, 40, 1922–1924.

Remaud, G., Martin, Y.L., Martin, G.G. and Martin G.J. (1997) Detection of sophisticated adulterations of natural vanilla flavors and extracts: application of the NIF-NMR method to vanillin and *p*-hydroxybenzaldehyde. *J. Agric Food Chem.*, 45, 859.

Scharrer, A. and Mosandl, A. (2002) Progress in the authenticity assessment of Vanilla. *Dtsch. Lebensm.–Rundsch.*, 98, 117.

Tabacchi, R., Nicollier, G. and Garnero, J. (1978) A propos de quelques particularités des extraits de vanille Tahiti. *Ann. Fals. Exp. Chim.*, 71, 109.

Wallace, E.M. (1983) HPLC of vanillin and related phenols. *J. Chromat. Sci.*, 21, 139.

16

Flavor, Quality, and Authentication

Patrick G. Hoffman and Charles M. Zapf

16.1   Introduction

The wonder of vanilla has existed ever since the Aztecs concocted the royal drink, chocolatl, made with cocoa, corn, honey, and vanilla. The flavor was brought to Europe by Cortez. For several hundred years the only source of vanilla beans was Mexico, but this changed in the nineteenth century when the methods to not only propagate the vine but also to cause the flowers to produce beans through hand pollination were discovered. Because of the popularity of the product and the demand, vanilla plantations were started around the world. It was at this time that the science basis of vanilla became possible, starting with Gobley's (1858) discovery of the major flavor ingredient, vanillin (4-hydroxy-3-methoxybenzaldehyde). The investigation into the flavor of this unique product accelerated. A significant step forward occurred in 1976, when Klimes and Lamparsky utilized unique gas chromatographic techniques in order to identify 169 vanilla constituents. While this was an important step forward, little effort was made to determine the flavor significance of these compounds. This was typical of most of natural products research at the time, where identification of the unique constituents in various products overshadowed the interest in their importance to flavor impact.

Vanilla is an extremely unique product and not only because it is the only orchid which produces an edible product. This tropical vine has limited production and is found only in countries within 15 to 20 degrees above and below the equator. The two main cultivated species are *Vanilla planifolia* and *V. tahitensis*. In the United States, these are the only species permitted to be used in food products. The vanilla plant requires years in order to mature to fruit bearing. The plant then requires 6 to 9 months to produce vanilla beans, which requires a unique hand pollination of the flower, which is open for only a few hours. Once the mature green beans are harvested, processing and curing the beans is an intensive hands-on process designed to produce a flavorful product. Several rapid, controlled curing processes have been developed, but most curing is still manual. During the mid-1900s, many beans were stored for up to 3 years, continuing the development of more flavor constituents. However, this later storage stage was virtually discontinued during government unrest in a major producer, Madagascar, during the 1970s and 1980s. All the stored stocks were sold. Now most

Handbook of Vanilla Science and Technology, Second Edition.
Edited by Daphna Havkin-Frenkel and Faith C. Belanger.
© 2019 John Wiley & Sons Ltd. Published 2019 by John Wiley & Sons Ltd.

commercial vanilla beans are annually produced and sold. A very detailed description of vanilla horticulture and curing is found in Gillette and Hoffman (2000).

The cured beans can be sold as they are, but the vast majority is used to produce vanilla extract. Extraction is also an intensive process requiring multiple aqueous alcoholic washes, which can be achieved by a variety of methods. The one-fold extract (1×) can be sold or concentrated to multifold 2×, 3 1/3×, 5×, 10×, 20×, and even 26×. The latter is an oleoresin with little to no residual solvent. The vanilla extract fold is regulated in the United States by the US Code of Federal Regulations (21CFR169.175). This regulation, considered the "standard of identity", defines a "unit" of vanilla beans as 13.35 ounces of beans of not more than 25% moisture in one gallon of aqueous ethanol at not less than 35%. The standard describes folds, labeling, and other products. More detail on the manufacture of vanilla, as well as the regulation of this product, can be found in Gillette and Hoffman (2000) and in the chapter in this book by Ranadive.

Because of the limited origins and the intensive and complex production sequence from plantation to store shelf, there are many financial incentives to short-cut this process at many points. In addition, one of the major producing countries, Indonesia, has an intensive population, in excess of 230 million in a country slightly smaller than Texas. These factors encouraged pilfering of beans maturing on the vine and early harvesting by many farmers in order to prevent this loss. These early picked beans, even if cured correctly, lack many of the necessary precursors for the production of the unique flavor constituents, as well as reduced amounts of others. Also, the proper, substantial, and costly curing process can also be quickened or shortened, again producing a less flavorful and lower quality product. Furthermore, the high cost of quality beans encourages extension and dilution of these extracts. And finally, while fully imitation products, if properly labeled, are perfectly acceptable and in some instances a close approximation of the "real thing", there are those unscrupulous suppliers that sell these as "pure product".

Since the discovery and use of this product, it has been extensively analyzed in order to understand the origin of this unique and desirable flavor. In addition, it has been analyzed in order to assure that the vanilla's quality is maintained and to assure the consumer that the product is indeed "real" vanilla. The early wet chemistry techniques used to identify the major flavor constituents, as well as to affirm authenticity, included painstaking extraction, isolation, and purification, resin determination, paper and thin layer chromatography, or other methods (Gillette and Hoffman, 2000 and references therein). As new analytical techniques were developed, so did the application of these methods to understanding the flavor of vanilla as well as to detecting adulterated products. These methods range from the chromatographic techniques such as gas chromatography (GC) and high performance liquid chromatography (HPLC), to the most recent nuclear magnetic resonance (NMR) and accelerator mass spectrometer (AMS) instrumental techniques. Also, new sophisticated isolation techniques beyond solvent/solvent extraction were developed and applied to vanilla. Supercritical CO_2 extraction, solid phase micro extraction (SPME), and stir-bar sorptive extraction (SBSE) are examples.

16.2   Vanilla Flavor Analyses

A significant review of vanilla flavor research is found in the 1991 *Encyclopedia of Food Science & Technology*, which covers vanilla flavor analyses up to that time fairly extensively.

In 1992 and 1993, researchers at Rutgers University used direct thermal desorption-gas chromatography (DTD-GC) and thermal desorption-gas chromatography-mass spectrometry (DTD-GC-MS) methodologies and identified 60 flavor compounds among 10 vanilla bean samples of different origin (Hartman et al. 1992; Adedeji et al. 1993). Several of the compounds were identified in vanilla for the first time. The phenols identified in Bourbon vanilla were suggested to be used as flavor quality indicators. These investigations identified a wide variety of constituents, which contribute to the "characteristic" flavor and aroma, mostly aldehydes, ketones, acids, alcohols, esters, ethers, and long and short chain hydrocarbons.

Taylor (1993), from the Natural Resources Institute, developed an HPLC procedure capable of separating 40 constituents of vanilla extracts. Various method conditions were investigated and retention times and wavelength absorbencies for the constituents were reported. However, this effort focused on the quantitation of the four main vanilla constituents for which regression equations had greater than 0.99 correlation coefficients and a small comparison was made between HPLC and the ISO 5565:1982 (withdrawn in 1999 and replaced) spectrophometric method.

Hydroxymethylfurfural (HMF) was identified as a minor constituent of vanilla extracts after isolation by liquid-liquid extraction and column chromatography, followed by a thin layer chromatographic method. Typical HMF concentrations ranged from 0.2 to 34 mg/100 mL using a rapid test amendable to any laboratory (Kiridena et al. 1994). German researchers (Ehlers et al. 1994) compared V. tahitensis and V. planifolia extracts by HPLC analyses. The Tahitian beans had low levels of vanillin and vanillic acid and relatively high quantities of p-hydroxybenzaldehyde. These beans also had significant levels of anisic acid and anisyl alcohol imparting a unique flavor to this product. In addition, they indicated that piperonal was probably not a natural constituent of vanilla, as had been thought previously.

Gas chromatography-olfactometry (GC-O) was applied by US researchers (Yang and Peppard 1996) "to bridge the gap between instrument analysis and sensory perception." They used an "odor intensity technique" (OSME) of GC-O to explore the origin of the "beany notes" in vanilla extract. Perez-Silva et al. (2006) used GC-MS and GC-O to identify 65 vanilla volatiles in a pentane/ether extract of vanilla beans. Sensory analysis showed this extract to be the most representative of vanilla beans. GC-O revealed 26 odor active compounds and that guaiacol, 4-methylguaiacol, acetovanillin, and vanillyl alcohol, although found in much lower concentrations then vanillin, are as intense.

Ramaroson-Raonizafinimanana et al. (1999), from France, analyzed the neutral lipids from V. fragrans (an older name for V. planifolia) and V. tahitensis, which resulted in the identification of a new product family, long chain γ-pyrones. Three γ-pyrones were identified and their variability studied in relation to bean maturity. They proposed that the pyrones may have anti-inflammatory and potential anti-platelet effects. The next year this group again isolated and identified an additional product family, β-dicarbonyl compounds (Ramaroson-Raonizafinimanana et al. 2000). The most abundant was nervonoylacetone, representing 74.5% of this fraction. These constituents like the γ-pyrones, were studied in relation to bean maturity. The β-dicarbonyl family was found in the beans of the aromatic species V. fragrans and V. tahitensis, but not in V. madagascariensis, which produces non-aromatic beans.

Australian researchers (Sostaric et al. 2000) developed a solid-phase micro-extraction (SPME) method coupled with GC-MS and used it to profile vanilla extracts. An evaluation of several fibers and analytical conditions was undertaken. Poly (acrylate) was identified as the best fiber. Their analysis was able to distinguish among three vanilla extracts, a Bourbon, Tahitian, and an Indonesian. The extract of Tahitian vanilla was unique in the presence of *p*-methoxybenzaldehyde (anisaldehyde) and an unidentified component. Bourbon and Indonesian extracts could be distinguished by the quantities of hexanoic acid, 5-propenyl-1, 3-benzodioxole, and ethyl nonanoate. They proposed the use of this method for identifying adulterated products as well as indicating that this method had the potential to be used to identify the type of vanilla extract or flavoring used to flavor foods.

Chinese researchers investigated supercritical CO_2 (SCO_2) flavor component isolation from vanilla beans (Fu *et al.* 2002a,b). In the first study they used SCO_2 at two different pressures, 35 and 30 MPa. In the second study they compared the SCO_2 isolates to those from a 95% ethanol extract. With NIST107LTB and NIST21LIB databases, 28 and 32 constituents were identified in the first study, and 28 and 31 in the second, respectively.

Da Costa and Pantini (2006) investigated Tahitian vanilla with International Flavor & Fragrances' Generessence® program. Imported beans were analyzed by GC/MS and dynamic headspace directly, and after methylene chloride extraction, and with sorptive stir-bar extraction (Twister™). These analyses identified 276 compounds. Sixteen were targeted by this program as being of interest, including several not previously reported in vanilla. Table 16.1 contains the newer components discovered and Figure 16.1 the structures of the novel anisyl esters. Joulian *et al.* (2007) used a variety of techniques, including stir-bar solvent extraction (SBSE), GC-MS, and LC/MS/MS, in order to confirm that the *p*-methoxy aromatic compounds, including anisaldehyde, are the major aroma contributors to characteristic Tahitian vanilla. Most interestingly, after careful analysis, heliotropin (piperonal) was shown not to occur naturally in this product. This confirmed what Ehlers *et al.* (1994) had reported.

An Indian research group (Sharma *et al.* 2007) developed a reverse-phase HPLC (RP-HPLC) method, which separated 10 phenolic compounds efficiently and could be used for analytical research as well as for quality assurance. The compounds were 4-hydroxybenzyl

Table 16.1 Components of interest in Tahitian vanilla. (Reproduced from Da Costa and Pantini 2006, with permission from Elsevier.)

Component	Concentration ppt[a]	Component	Concentration ppt
Anisyl alcohol	225.8	Caffeine	0.1
Anisic acid	87.4	Theobromine	0.1
Anisaldehyde	25.0	α-Ionone	0.4
Dianisyl ether	3.1	β-Ionone	0.4
Anisyl ethyl ether	15.0	Dihydroactinidiolide	0.2
Anisyl methyl ether	0.8	Vitispirane	0.3
Anisyl anisate	6.6	Anisyl 4-hydroxybenzoate	7.4
Anisyl *trans*-cinnamate	0.5	Anisyl *cis*-cinnamate	0.2

a) parts per thousand in the liquid extract as determined by GC analysis.

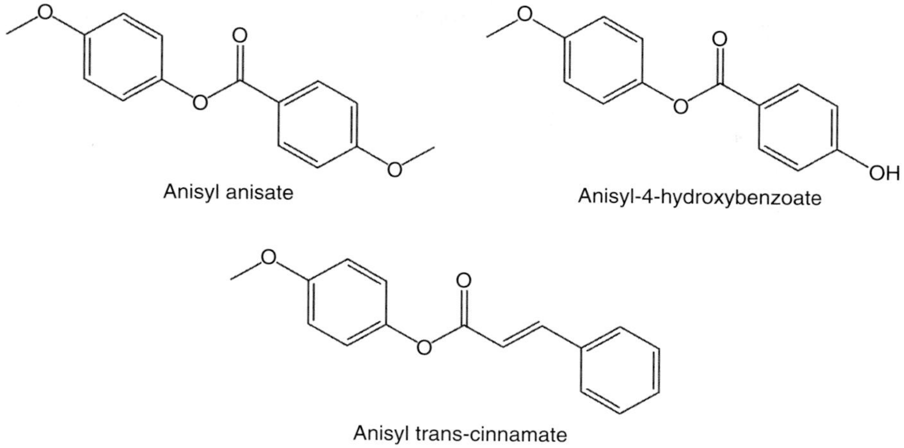

Figure 16.1 Novel compounds identified in Tahitian Vanilla. (Reproduced from Da Costa and Pantini, 2006, with permission from Elsevier.)

alcohol, vanillyl alcohol, 3,4-dihydroxybenzaldehyde, 4-hydroxybenzoic acid, vanillic acid, 4-hydroxybenzaldehyde, vanillin, *p*-coumaric acid, ferulic acid, and piperonal.

FDA researchers (de Jager *et al.* 2007) developed a liquid chromatography mass spectrometer (LC-MS) method to determine coumarin, vanillin, and ethyl vanillin in vanilla extract products. Both UV and MS signals were used to quantify the data. Electrospray ionization (ESI) for MS detection was utilized with selected ion monitoring (SIM). Limits of detection (LOD), accuracy, and precision were determined for each compound. Twenty-four commercial vanilla extracts were analyzed, 9 were domestic and 15 were imported. It was determined that the adulteration of Mexican vanilla was not widespread, as was reported in 1988. Further investigation of these vanilla products with HS-SPME and GC-MS methods at CFSAN(FDA) was also undertaken. A survey of these 24 vanilla products revealed that not only could the method be used for analysis of coumarin, vanillin, and ethyl vanillin, but also could screen for 18 other vanilla flavor constituents. This data, when compared with LC-ESI-MS, demonstrated that these two conceptually different methods could be confirmatory of each other.

Sharma *et al.* (2007) developed an RP-HPTLC method for the quantitative determination of vanillin and related phenols. The method also involves an accelerated solvent extraction (ASE). Silica gel coated aluminum plates were used with a methanol/water/isopropanol/acetic acid mobile phase. This was a simple, fast, sensitive method useful for analytical research and for on-line analyses of vanilla extracts for QC. A Chinese group (Zhou *et al.* 2007) investigated the use of different solvents during the ultrasonic extraction of vanilla beans. They identified acetone as the best solvent. The extract was separated with SPE and the analysis by GC-MS showed a detection limit of vanillin of 0.04 mg/L and that of *p*-hydroxybenzaldehyde of 0.06 mg/L.

A German research group (Schwarz and Hoffman 2009) investigated non-volatile taste contributing constituents in vanilla. Specifically, they were attempting to detect components beyond vanillin and divanillin that contribute to the "mouth-feel" of this product. A Madagascar vanilla extract was extensively extracted by partitioning with pentane and then ethyl acetate. Further fractionation of the solvent free ethyl acetate fraction by gel permeation chromatography (sephadex LH-20) with varying concentrations of aqueous methanol produced nine sub-fractions. Several targeted fractions were further purified by HPLC and nine "active" constituents were isolated, including vanillin and divanillin. All were fully characterized and their sensory impact determined with taste dilution analysis (TD) with the half-tongue test. The constituent with the most mouth-feel was 4′, 6′ – dihydroxy – 3′, 5′ – dimethoxy – (1, 1′ – biphenyl) – 3 – carboxaldehyde. These researchers proposed that quantitative monitoring of these compounds could be used in order to tailor the curing process to enhance these orosensory active molecules.

16.3   Biochemistry and Genetic Research on Vanilla

During the 1990s, research into the biochemistry of vanilla flavor was undertaken. In 1989, a French group (Leong *et al.* 1989) synthesized the glycosides of the four major vanilla flavor phenolics. The analytical characteristics (m.p., RF, UV, HPLC, and ^{13}C-NMR) of each of these glucosides were determined. These synthetic glucosides were then used to confirm the presence and quantity of these glycosides in green vanilla beans from Comoros, Madagascar, Reunion, and Indonesia. They also showed the absence of free aldehydes and acids in green uncured vanilla beans.

Ranadive (1992) demonstrated that the major vanilla flavor constituents were present as glycosides prior to curing. Using reverse phase HPLC, he determined the amount of vanillin, *p*-hydroxybenzaldehyde, vanillic acid, and *p*-hydroxybenzoic acid and also identified vanillyl alcohol and *p*-hydroxybenzyl alcohol. He investigated vanilla beans from seven growing regions and showed that added glucosidase increased the amount of these flavor phenolics when added to heat treated enzyme-deactivated green Jamaican beans.

Brodelius (1994) also used reverse-phase HPLC to study the formation of phenolic flavor compounds in developing vanilla fruits following hand pollination. He found that vanillin formation began at 16 weeks and maximized at 26 weeks after pollination. The other three major vanilla phenolics were also detected.

In the late 1990s, three studies were undertaken by Chinese researchers. Pu *et al.* (1998a) demonstrated that the treatment of green beans with the enzyme *β*-glucosidase under optimized conditions improved the hydrolysis of the glycoside and increased vanillin content, as well as the content of the other phenolic flavor compounds also tied up as glycosides. The second study (Pu *et al.* 1998b) investigated *β*-glucosidase activity at different stages of green bean maturity. This information could be used for improving the quality of vanilla beans as well as the curing process. The third study (Jiang *et al.* 1999) investigated vanilla enzymes. *β*-glucosidase had greater activity in the seeds than the pod. Peroxidase had high activity in the pod and little in the seeds. Jiang *et al.* (2005) extended this research by investigating changes in *β*-glucosidase and peroxidase activity during different processing conditions, which included different deactivation of enzyme and roasting processes.

Early in the twenty-first century, a Dutch group (Dignum *et al.* 2002) used a laboratory model of curing to investigate the effect of the curing processes on the activities of the variety of enzymes found in vanilla beans. *β*-Glucosidase, peroxidase, protease, and phenylalanine ammonia-lyase were studied. They later investigated the source of the key minor flavor constituents in vanilla (Dignum *et al.* 2004). They were curious from where p-cresol, creosol, guaiacol, and 2-phenylethanol were derived. Were they derived from glucosides or were they formed during the curing process? Glucosides of p-cresol and creosol were detected as minor compounds in green beans, but glucosides of guaiacol and 2-phenylethanol were not detected. In addition, they found glucosides of vanillin, vanillyl alcohol, vanillic acid, p-hydroxybenzaldehyde, and two tartrates. With all these glucosides, these researchers investigated the β-glycosidase enzyme action on these compounds. This was an intensive investigation of enzyme activity on a series of natural and synthetic glycosides.

The quantitative conversion of glucovanillin to vanillin during the curing process was investigated by the Symrise group (Gatfield et al. 2007), which reported losses up to 50% of the vanillin during the bean curing process. Analysis indicated approximately 9 g glucovanillin per 100 g of dry substance initially. During the curing, vanillin was dimerized to divanillin accounting for some of the vanillin loss.

An interesting and intensive genetic investigation of the origin of Tahitian vanilla was reported by researchers in 2008 (Lubinsky et al.). They collected a variety of vanilla species from natural and cultivated populations. From their genetic analysis they concluded that *V. planifolia* and *V. odorata* were the parental species of the interspecific hybrid species V. tahitensis and that there has been little evolutionary divergence since its probable hybridization during the Late Classic (1350–1500) in Mesoamerica.

Waliszewski *et al.* (2009) extensively studied the activity of the vanilla enzyme polyphenol oxidase (PPO), which was considered to be a factor in the browning of vanilla beans during curing. The stability of the extracted and purified enzyme was studied under a variety of conditions, including pH and temperature. Also, the inhibitory effects of several chemicals were investigated. The optimum temperature was 37 °C and the pH, 3.0 and 3.4, depending on substrate. These researchers proposed that because vanilla PPO utilizes mono- and diphenols in order to catalyze oxidation, that the enzyme may decrease the vanillin content in the cured bean.

French (La Reunion), Dutch and Malaysian scientists (Palama et al. 2009) applied ^{1}H-NMR and LC-MS in the study of the metabolic variation of developing *V. planifolia* pods. This metabolomic analysis was performed on the green pods at several developmental stages. Analysis of the data was performed with multivariate data analysis. Generally, younger pods contained more glucose, malic acid, and homocitric acid while older pods contained more sucrose, glucovanillin, vanillin, p-hydroxybenzaldehyde glucoside, and p-hydroxybenzaldehyde. They suggested their results may serve as a base to clarify the biosynthetic pathway for vanillin.

16.4 Vanilla Quality and Authentication Analyses

In order to assure the consumer that the product purchased is indeed vanilla, governments worldwide have tried to define what constitutes a product called vanilla. As discussed earlier, in the United States, vanilla products are defined by a "standard of identity" found in 21CFR169.175. Similar laws occur throughout the world in legal statements such as that found in New South Wales, Australia: "the quantity of soluble substances in their natural proportions that are extracted by aqueous Alcoholic solution..." (Archer 1989). These regulations require a means to police compliance and many such methods have been developed over the last half century.

The early means to detect the formulation or extension of vanilla products with other botanical extracts and juices used paper and thin layer chromatography, which could separate and identify the various adulterating botanicals such as licorice root, cascara, yarrow, cherry, and prune extracts. In addition, the analysis of the vanilla organic acids by ion-exchange chromatography, as well as the analysis of amino acids, helped to validate vanilla products. Interestingly, also used during this period was the questionable lead number (Pb#), which supposedly established the vanilla bean content with a lead precipitation. However, not only was this method dangerous to the analyst as well as the environment, but it was also notoriously inaccurate.

Botanical dilution and substitution could be detected by these early methods but the most common "extender" was the addition of synthetic vanillin as discussed earlier. Vanillin was identified by Gobley in 1858 as the main flavor constituent of vanilla. The synthetic form of vanillin quickly became the compound used to enhance poor quality products and/or produce a purely imitation product. This form of adulteration was difficult if not impossible to detect by the techniques available at that time. The analysis and quantitation of amino acids (Stahl *et al.* 1962) was also added to the variety of testing methods. In the mid-1970s, a variety of combined and instrumental techniques were investigated. Martin *et al.* at ATF (1977) investigated the use of constituent ratios in order to detect not only synthetic vanillin addition but also the strength or dilution of the extract. These ratios involved vanillin, potassium, inorganic phosphates, and nitrogen.

While all these methods helped in thwarting adulteration, clever adulterers were still circumventing all these efforts. In the mid- to late 1970s and into the 1980s, two techniques were intensely investigated focusing primarily on detecting the addition of non-bean derived vanillin to the extract. These include high pressure liquid chromatography (HPLC) and methods based on the isotopic composition of the vanillin found in the vanilla product.

16.4.1 Liquid Chromatographic Methods

Ever since the proposal that the vanilla HPLC constituent ratios could be used to verify the origin of vanilla in the early 1980s, a lot of effort has been expended on the application of this concept. Initially, the vanillin/*p*-hydroxybenzaldehyde ratio was proposed but later the use of the ratios between all of the four major vanilla phenolics, including vanillin/vanillic acid and vanillin/*p*-hydroxybenzoic acid as well as other constituent ratios were incorporated. Researchers from around the world investigated a variety of HPLC eluents and columns as well as a variety of HPLC conditions (Figure 16.2). Most used reverse phase columns and methanol, acetonitrile and water, in a variety of combinations, as the eluent and a variety of acids as modifiers. However, it appeared that if the analysis was effective in the separation and quantification of these constituents, the various ratios were reasonably consistent, no matter the geographic origin of the vanilla. It was proposed in the 1980s and 1990s that these ratios should be used as a means to authenticate vanilla and eventually these ratios began to be incorporated into the national regulations of various countries, primarily in Europe. The ratios were eventually promulgated into EU regulations for vanilla. While regulations make these absolutes, it soon became clear that there were natural vanilla exceptions that fell outside the established ranges. Because this is a natural product, natural constituent variability is

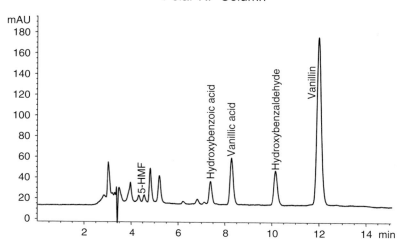

Figure 16.2 Typical HPLC profile of a vanilla extract from the author's laboratory (courtesy of Alan Harmon).

inevitable, so the use of rigid ratio limits as a regulatory standard may be asking too much of Nature. However, the ratio of vanillin/*p*-hydroxybenzaldehyde has shown the most consistency with fewer natural outliers. The usefulness of this ratio is primarily due to the greater concentrations of these two vanilla constituents, which results in lower statistical error. A better use of these HPLC ratios, especially the vanillin/*p*-hydroxybenzaldehyde ratio, would be their use as an initial check. When the vanilla product falls beyond the established ratio limit, it should be analyzed by another more definitive and more expensive method, such as SIRA, NMR, etc.

HPLC methods are based primarily on separating four key vanilloid type constituents. These constituents are often summed and compared to the common UV result for total "vanillin", expressed as g/dL and/or to establish ratios in order to authenticate beans or extracts. The four major non-volatile constituents separated and quantified are vanillin, *p*-hydroxybenzaldehyde, *p*-hydroxybenzoic acid, and vanillic acid (Figure 16.3). Other constituents and non-vanilla adulterants are also often separated and identified, for example, coumarin, ethyl vanillin, piperonal, etc. These constituents and others may provide further information as to the quality and source of the vanilla and can be helpful to the analyst for assessing the overall age, purity, quality, or sensory properties of vanilla products (Kahan *et al.* 1997; Taylor 1993; Hoffman *et al.* 2005). The ratios used for authentication are fairly consistent. The most significant ratio is the vanillin to *p*-hydroxybenzaldehyde ratio, which falls within the 10 to 20 range for authentic vanilla products. These four constituents in vanilla products are often measured to resolve many requirements. In pursuit of this information, many HPLC method variations have been developed but most involve aqueous methanol as the eluant, several include acetonitrile and most include an acid modifier and use a variety of reverse phase columns. Many investigators have contributed to the number of HPLC methods (Sinha *et al.* 2008).

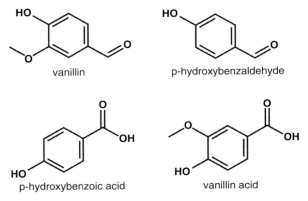

vanillin   p-hydroxybenzaldehyde

p-hydroxybenzoic acid   vanillin acid

Figure 16.3 Structures of the four vanilloid compounds used for calculation of component ratios, typically from HPLC analysis.

Among the many published methods, some might be adopted for use by quality control chemists or researchers, if applicable to their needs for either routine or investigative studies. Clear methods and simple procedures are preferred in the quality lab; commercial labs may use a number of methods to fulfill their purpose for customer studies and the researcher may apply methods which maximize the chemical information or provide accurate analysis of a key analyte.

A method with potential for routine analysis was published by TTB (Tax, Trade Bureau – US government, then BATF) researchers (Jagerdeo *et al.* 2000). They prescribed a simple HPLC method with good sensitivity and linearity for the desired components, vanillin, etc., to be employed for product surveillance. Several gradient and isocratic profiles were studied but were not disclosed. Waliszewski *et al.* (2007) developed a very rapid HPLC method for determining the vanillin content in vanilla extracts. Through a very intensive evaluation of columns, mobile phases, and detection wavelengths, a method was developed that could be completed in approximately 2 minutes using a Nucleosil C18 column, 60:40 methanol:water mobile phase, and detection at 231 nm with a gradient elution, as compared to the 7 to 36 minutes for other known methods.

An extension of AOAC Official Method 990.25, for the HPLC determination of vanillin, vanillic acid, *p*-hydroxybenzaldehyde, and *p*-hydroxybenzoic acid in fortified and imitation vanilla flavors, was developed in order to include ethyl vanillin. From this activity, Method 990.25 was modified (Kahan *et al.* 1997) to include the determination of ethyl vanillin in vanilla products using a C₈ column, methanol-acidified water (10 + 90), and UV detection at 254 nm.

Chemists continue to develop methods to measure the chemicals expected in natural vanilla extracts, and also try to improve methods to detect other added flavor chemicals and even banned substances. An example is the more sophisticated LC-MS method developed by de Jager *et al.* (2007) at CFSAN-FDA (Center for Food Safety and Nutrition, US Food & Drug Administration). Specifically ethyl vanillin and coumarin were measured in 24 vanilla extracts obtained principally from the United States and Mexico. With this optimized method, which was compared to the routine UV detection method, the authors claimed a more accurate assessment and concluded that the samples were devoid of contamination. This is in contrast to an earlier study where nearly 68% of the Mexican samples contained coumarin (Thompson and Hoffmann 1988).

Jurgens (1981) used the ratios of key vanilloid constituents, notably vanillin/*p*-hydroxybenazaldehyde, calculated from the HPLC analysis, to verify the authenticity of vanilla extracts. Several papers have appeared that reviewed the value of these ratios and have presented results that have either challenged or verified the utility of the measured constituents and their calculated ratios. Ehlers *et al.* (1999) reported on an investigation of the effects of extraction solvents, time of extraction, as well as extraction equipment. They noted that the quantity of vanillin did vary considerably with extraction conditions. Interestingly, the ratios of the components were not affected. However, because regulations especially in France require specific vanillin content as well as the correct ratios, they suggested that the French regulation should be revised. German researchers (Scharrer and Mosandl 2001) analyzed a variety of vanilla bean samples of different origins and years of harvest with typical aqueous ethanol and diethyl ether extractions for the four vanilloid components, which were analyzed by HPLC and GC. Ethyl vanillin and veratraldehyde were used as internal standards when developing the two methods. Complete constituent tables were included for the GC and HPLC results. From these analyses they too believed that the ratios currently being used were too restrictive.

French researchers did an extensive study of the 2000 vanilla crop (Charvet and Derbesy 2001). They analyzed 265 vanilla bean samples using HPLC, including those from Madagascar and Comores. The data they obtained on the four key vanilla components were compared to data for other samples collected over the previous five years. They noted the decreasing content of the two aldehydes, vanillin and *p*-hydroxybenzaldehyde, while the content of the two acids seemed nearly constant. A statistical assessment was not presented beyond the charts used to depict the trends.

Further investigation into the applicability of the vanilloid ratios was extended to those vanilla components measured in dairy products (Littmann-Nienstedt and Ehlers 2005). Changes can occur in the ratios when vanilla is used in dairy products. Milk enzymes stoichiometrically convert vanillin and *p*-hydroxybenzaldehyde to their respective acids. Therefore, we need to take these potential conversions into account when calculating the vanilla constituent ratios in dairy products. The authors presented an adjusted formula that could be useful for these products.

An evaluation of 93 samples of vanilla beans of different grades from the 5 producing countries was reported by Givaudan experts (Gassenmeier *et al.* 2008). The samples were extracted with ethanol in a Soxhlet unit, and analyzed by HPLC. Sensory analyses of the extracts found little correlation between "good vanilla" and vanillin content. Their HPLC analyses revealed a significant lack of compliance of the "ratios" (Table 16.2) with the accepted and regulated values notably for the case of Madagascar, even when the approved extraction conditions were used. The research discusses each bean source critically. They suggest that the strict enforcement of these ratios has lead to products produced to meet these parameters but that may be less flavorful. This confirms earlier research findings that vanillin explains less than 12% of the organoleptic variability (Hoffman *et al.* 2005). "A parameter that focuses on the purity of an extract or the amount of vanilla beans per solvent seems to be more appropriate," according to these researchers. This would be very similar to the US Standard of Identity for vanilla (21CFR169.175). However, these researchers acknowledge that the vanillin/*p*-hydroxybenzaldhyde ratio is still useful to detect the addition of vanillin to vanilla extracts, perhaps as an initial investigation.

Table 16.2 Ratios for vanillin: *p*-hydroxybenzaldehyde from various sources compared to the IOFI expected values. (I.O.F.I. International Organization of the Flavor Industry. Information letter No. 775, France: Vanilla Flavor 25.01.1998.)

Source	Sample Preparation	Number of samples/type	Vanillin/*p*-hydroxybenzaldehyde HPLC-RATIO
Ranadive, 1992	Ethanol/Water	7 various	7.9–25.7
Lampprecht *et al.*, 1994	Ethanol/Water NF	6 Bourbon	11.4–24.4
Charvet and Derbesy, 2001	AFNOR NF ISO 5565	93 Madagascar, 29 Comores	15.1–20.0
Charvet and Derbesy, 2001	AFNOR NF ISO 5565	265 various	16.2–16.4
Scharrer and Mosandl, 2001	Ethanol/Water (AOAC)	15 various	10–28
Gassenmeier *et al.*, 2008	Ethanol (French)	85 various	10–27.5
Expected IOFI			10–20

In addition to HPLC methods, thin layer chromatography (TLC) is often investigated and useful. A thin layer variation of the liquid chromatographic method was developed by Belay and Poole (1993), which uses automated multiple development (AMD-TLC). With this technique, multiple samples can be analyzed simultaneously providing greater throughput while requiring minimal sample preparation by the analyst. This technique was capable of separating and quantifying the same vanilla flavor constituents as the common HPLC methods, and could be used to verify the authentication of products based on the constituent ratios.

A Russian group (Gerasimov *et al.* 2003) also developed a TLC procedure, which could be used to authenticate vanilla flavor in products. The methodical work optimized the chromatographic conditions specifically to separate vanillin and ethyl vanillin, as well as enabling the analysis of other chemicals commonly added to vanilla flavorings.

16.4.2 Isotopic Techniques

The isotopic methods used for vanilla authentication involve a variety of analytes, ^{14}C, ^{13}C, ^{12}C, H, D, ^{16}O, ^{18}O, ^{3}H, as well as a number of techniques, liquid scintillation (LS), stable isotope ratio analysis (SIRA), gas proportional counting (GPC), nuclear magnetic resonance (NMR) including site-specific nuclear isotopic fractionation (SNIF-NMR) and accelerator mass spectrometry (AMS).

16.4.3 Radiometric and Stable Isotope Ratio Analysis

Researchers (Culp and Noakes 1992) at the University of Georgia's Center for Applied Isotope Studies (CAIS) investigated the application of ^{14}C, $^{12}C/^{13}C$ and D/H isotopic ratios to establish the naturalness of a variety of 11 flavors, including vanilla. The targeted flavor compound was isolated and then the stable isotope ratio determined by an isotope ratio mass spectrometer (IRMS). They reported that D/H ratios were better at differentiating the source of these flavor compounds. For example, vanillin from guaiacol had a D/H value at −20‰, while pulp vanillin was at −186‰, and vanillin from vanilla beans had a D/H value of −77‰. They advocated the use of a combination of all isotopic measurements in order to enhance the interpretation and define the flavor materials' origin.

A coupled technique, gas chromatography–isotope ratio mass spectrometry (GC-IRMS) was described by Fayet *et al.* (1996). The technique allows direct analyses of constituents, such as vanillin, in complex mixtures by the application of gas chromatographic separation and direct injection into the isotope ratio mass spectrometer. In addition, a German group (Kaunzinger *et al.* 1997) used coupled GC-C-IRMS as well as the gas chromatographic quantification of vanilla constituents to develop an integrated authenticity evaluation. The developed profile could be used for vanilla quality assurance. As they stated, "Nevertheless, our investigations clearly define the 4-hydroxybenzaldehyde/vanillin concentration ratios in connection with their $\delta^{13}C$ values as characteristic parameters of genuine vanilla."

In the mid-1990s, further efforts at the CAIS involved the investigation of tritium (^3H) analysis (Neary *et al.* 1997). While it was demonstrated that vanilla bean vanillin and petroleum-derived vanillin could be differentiated with a gas proportional counter (GPC), the technique combusts vanillin and then electrolytically concentrates the tritium, but required inordinate amounts of material, 5–10 grams or more of vanillin.

Culp *et al.* (1998) at CAIS further investigated the application of isotopic analysis for the detection of vanilla adulteration. They applied accelerator mass spectrometry (AMS) and gas chromatography/combustion/isotope ratio mass spectrometry (GC/C/IRMS) in order to authenticate flavors, with vanilla as one example. The vanilla extract was derivatized with bis (trimethylsilyl) triflouroacetamide (BSTFA). They were able to obtain the $\delta^{13}C$ values of the major as well as ancillary compounds in the extract, *p*-hydroxybenzaldehyde, *p*-hydroxybenzoic acid, vanillin, vanillic acid, *o*-vanillin, and syringaldehyde. They proposed the use of these $\delta^{13}C$ values in order to generate a fingerprint pattern that could be used to authenticate these flavors. Also, because this GC/IRMS analysis typically measures in nanograms to picograms, flavors in finished products could be analyzed for the detection of adulterated products. This demonstrated the application of AMS for ^{14}C analysis and they hoped to reduce the amount of vanillin required so as to interface the AMS directly to the gas chromatography.

The French regulatory group, at Laboratoire DGCCRF (Fayet *et al.* 1999), used GC/IRMS to investigate the isotopic deviation of vanillin, vanillic acid, *p*-hydroxybenzaldehyde, and *p*-hydroxybenzoic acid. Twenty-two samples of vanilla beans from Madagascar, Indonesia, and Reunion were analyzed. The analytical data for the first three constituents were consistent with that previously reported. The stable isotope ratio analysis (SIRA) for *p*-hydroxybenzoic acid was first reported here and was suggested as a criterion for authenticity of vanilla.

Expanding on the work of Fayet *et al.* (1999), German researchers (Scharrer and Mosandl 2002) optimized the use of gas chromatography combustion-isotope ratio mass spectrometry (GC-C-IRMS) on the difficult-to-analyze phenolics, vanillin, and *p*-hydroxybenzaldehyde. In addition, analyte extraction efficiency was also investigated. They determined the optimum combustion temperature by moving the tip of the capillary column inside the heater and demonstrated that complete component extraction was not necessary in order to obtain a consistent $\delta^{13}C/^{12}C$ SIRA. They analyzed 15 samples of *V. planifolia* and 4 of *V. tahitensis*. The $\delta^{13}C_{v\text{-pdb}}$ was determined on vanillin and *p*-hydroxybenzaldehyde and demonstrated its reliability in differentiating vanillin, ex-vanilla, from vanillin of synthetic or biotechnological origin.

Over the years, adulteration techniques that could circumvent isotopic detection have been developed, one being syntheses of isotopically altered ingredients. Researchers at Eurofins, a company specializing in isotopic analyses, demonstrated a method to detect this form of adulteration (Bensaid *et al.* 2002). They described a technique in which vanillin is deformylated and the product, guaiacol, analyzed by IRMS. This ^{13}C information can be used in order to improve the authentication potential of C-IRMS. In addition, the ^{18}O of the guaiacol is a better authentication tool then ^{18}O of vanillin, which has exchangeable oxygen in the formyl group. Further collaborative studies were suggested in order to confirm that consistent results could be obtained between laboratories.

Most recently, Culp (2009) at the CAIS presented a two decade review of isotopic data on flavors, including vanilla. He demonstrated the need to have an in-depth understanding of the changing atmospheric ^{14}C radiocarbon levels, as these are changing because of the atmospheric nuclear testing in the 1950s and 1960s and its subsequent cessation. He discussed data gathered on over 3,800 flavor samples during these two decades. Data included levels of radiocarbon (^{14}C) and the stable isotopes of $^{13}C/^{12}C$ and D/H. This data can be used to authenticate flavor materials. Also, due to the yearly predictable decay of atmospheric ^{14}C, the year of origin of the material can possibly be established.

16.4.4 Nuclear Magnetic Resonance (NMR)

Martin *et al.* (1993) described the NMR authentication technique that they had developed, designated site-specific natural isotope fractionation (SNIF™). A variety of applications for different products, including vanilla, were detailed. Nuclear magnetic resonance (NMR) determines the deuterium (^2H) concentration at each of the five isotomers on the vanillin molecule. They also point out that unlike other analytical methods, SNIF-NMR can detect ^{13}C enriched vanillin but it does require long spectrometer time. A review by Martin (1998) then described the use of deuterium SNIF-NMR for the authentication of a variety of flavors, including vanilla, coupled with additional application of both deuterium and ^{13}C NMR. This permitted greater classification of the vanillin origins than that allowed by SNIF-NMR alone.

Jamin *et al.* (2007) reported on a collaborative study of site-specific isotope ratios of deuterium/hydrogen (D/H) in vanillin analyzed by deuterium-NMR (2H-NMR) spectrometry. Nine laboratories participated and the results supported the recommendation that the method be adopted as a First Action Official Method by AOAC International.

16.4.5 Isotopic Techniques Summary

Reported here are just a few examples of these isotopic methods, which can be used to authenticate the bean-derived origin of vanillin. There are many other reports of the applications of these techniques. Those reported here demonstrate the extent to which researchers are going in order to protect this valued product. Also indicated are the continued extension and development of these isotopic methods.

16.4.6 Integrated and Miscellaneous Methodologies

German researchers (Kaunzinger *et al.* 1997) used gas chromatographic profiling with quantitation for vanillic acid, 4-hydroxybenzaldehyde, anisic alcohol, anisic acid, and 4-hydroxybenzoic acid with subsequent $\delta^{13}C$ isotopic analysis using IRMS, to map both the quantity and isotopic ratios for these components to cross-reference the authenticity of several vanilla extracts. They proposed that a database of these joint measures would give a more accurate assessment of authentic vanilla and acknowledged the intensive work needed to complete this goal.

Another approach combining analytical technologies to ensure authentic natural vanilla was described in the work of John and Jamin (2004). Because the HPLC ratios appeared to be changing, these researchers compared the HPLC ratios, stable isotope ratio analyses (SIRA), and the SNIF-NMR data on Indian vanilla. They report that for these products, many of the HPLC ratios were inconsistent with the French regulation, especially the vanillin to *p*-hydroxybenzaldehyde ratio, which was consistently high. The SIRA values were also slightly different than stipulated by regulation and they suggested a change to a minimum limit of −21.5‰ for vanillin and −19.2‰ for *p*-hydroxybenzaldehyde. With respect to SNIF-NMR, differences in D/H values were noted, and the discrimination power of this technique to detect adulterated products was demonstrated.

Calabretti *et al.* (2005) reported on investigating an explicit experimental design methodology in order to determine the best parameters to be used for solid phase micro-extraction (SPME) analysis of vanilla. These researchers used a patented "cogrounds", solid state activation (SSA) mixed with vanilla extract. They focused the analyses on volatiles "characterizing vanilla", myrcene, methoxybenzaldehyde, ethylbenzoate, ethyl-vanillin, vanillin, coumarin, *p*-hydroxybenzaldehyde, and vanillin. Confusingly, there was no explanation as to why ethylvanillin and coumarin, which do not occur naturally, were targeted. However, this very detailed and statistically based experiment identified the 2 cm length DVB/CARB/PDMS fiber as the optimum to be used for further investigation.

Italian researchers (Bettazzi *et al.* 2006) developed an interesting disposable electro-chemical sensor, which can be used for the determination of vanillin in natural extract concentrates and in the finished products yogurt and compote. This vanillin method was comparable to HPLC. The electrochemical behavior of other compounds: vanillic acid, *p*-hydroxybenzaldehyde, and *p*-hydroxybenzoic acid was also investigated.

As noted earlier, the extraction of vanilla using organic solvents can affect the measurement of the chemistry and indeed the efficiency too. Mexican researchers (Valdez-Flores and Canizares-Macias 2007) developed a method for efficient vanillin extraction with ultrasound assisted extraction (USAE). This extraction with the continuous flow dilution was shown to be more efficient and quicker than the more common soxhlet and maceration extraction. Mexican researchers (Longares-Patron and Canizares-Macias 2006) developed a microwave based extraction process for vanillin and *p*-hydroxybenzaldehyde and obtained the ratios with UV analyses at 348 and 329 nm, respectively, with Vierordt's method. A statistical optimization of the extraction conditions was investigated, which permitted the claim of faster extraction and increased vanillin and *p*-hydroxybenzaldehyde concentrations.

Sinha *et al.* (2008) published an interesting review of vanilla, which included an extensive review of biological activity of this product. Antimicrobial, antioxidant, and a variety of disease resistant, amelioration effects were described, as well as insect repellent activity. Extraction methods were described, both conventional percolation and soaking as well as non-conventional methods including SCO_2. A variety of analytical techniques used to investigate vanilla were also discussed, including TLC, HPLC, GC, and electrophoresis. They also discussed the production of vanillin and the methods to detect the adulteration of this unique product.

An interesting use of capillary electromigration-microchips was investigated by Avila *et al.* (2007). They applied this electrochemical detection technique to five flavor constituents: vanillyl alcohol, ethyl maltol, maltol, ethyl vanillin, and vanillin. When analyzing actual samples they focused on vanillin and ethyl vanillin, as the latter would immediately reveal adulteration, but all constituents were analyzed.

While the application of either a GC or HPLC procedure in order to determine vanilla quality is a desirable goal, Hoffman *et al.* (2005) have shown a more comprehensive analytical approach using both techniques may lead to a procedure more applicable to determining the quality of vanilla extract and its correlation to sensory measurements. In this approach, the authors used a combination of GC analysis, using the SPME sampling technique and a stir bar sorptive extraction (SBSE) sampling for reversed phase HPLC analysis. Gas chromatography analysis separated greater than 100 constituents. Ten of the volatiles, which occur in most vanilla extracts, were targeted. Fifty-five multifold vanilla extracts of beans from Madagascar, Indonesia, Comores, Mexico, Tonga, Uganda, and Tahiti, as well as extracts of blended origins, were analyzed by GC for volatile constituents and by HPLC for the four noted vanilla-phenolic compounds. These 55 extracts were also extensively evaluated by a sensory group of highly trained descriptive panelists, defining the vanilla's flavor and aroma characteristics. Table 16.3 summarizes the statistical influence of several key compounds in relation to the sensory parameters described by the panelists. A total quality vanilla (TQV) assessment factor established a means to quantify the sum of the quality attributes in vanilla. Factor analysis was used to evaluate the analytical chemical data from 14 chemical constituents (10 by GC and 4 by HPLC) in each of the 55 vanilla samples. Factor analysis could explain 75% of the variability with 3 factors and 4 factors could describe 86% of the variability. Discriminate analysis was also used to predict sensory quality with selected chromatographic components. The best resulting formulae correctly predicted 92% of the 55 vanilla extracts as acceptable compared to the sensory panel. A validation model was successfully performed with a separate set of 11 vanilla extracts.

Perez-Silva *et al.* (2006) provided further evidence as to the importance of chemicals other than vanillin to the complexity of the aroma of vanilla extract. They compared the aroma of several solvent extracts to that of vanilla powder prepared from ground beans from the Tuxtepec region of Mexico with sensory panels. The extracts were analyzed by GC-FID and GC-MS. They concluded that the pentane/ether extract correlated well with the panel evaluation for several chemicals at the ppm level. Guaiacol, 4-methylguaiacol, acetovanillone, and vanillyl alcohol had a major contribution to the aroma, even though they were present at levels 1,000 times lower than vanillin. Such studies verify the contribution to the aroma of components that are typically measured

Table 16.3 Statistical factor analysis of combined GC and HPLC data for sensory analysis. Bolded compounds contribute positively to the statistical factor.

	Factor Structure (correlations)			
	Factor 1 Age-related Compounds	Factor 2 Rummy Resinous	Factor 3 Vanillin Flavor	Factor 4 Smokey Phenolic
Total Variability explained = 86%	45%	19%	12%	10%
4-Hydroxybenzoic acid	−0.01226	−0.05422	0.16541	**0.75172**
Vanillic acid	0.28421	0.26838	0.60821	**0.70989**
4-Hydroxybenzaldehyde	0.56672	0.42593	**0.93793**	0.28968
Vanillin	0.50772	0.39799	**0.92226**	0.25951
Ethyl acetate	**0.97844**	0.34279	0.56633	0.06165
Acetaldehyde diethyl acetal	0.32885	**0.73236**	0.59983	−0.12226
Ethyl Hexanoate	**0.96116**	0.11165	0.37054	−0.0442
Guaiacol	0.0023	0.5506	0.01162	**0.80818**
Hexanal diethyl acetal	**0.97141**	0.23478	0.47905	0.03083
Ethyl Octanoate	0.64018	0.51133	0.54316	−0.03314
Nonanoic acid	0.58698	0.64114	0.36559	**0.67612**
Ethyl Nonanoate	**0.97227**	0.38472	0.5459	0.15428
4-Ethoxymethyl phenol	0.40717	**0.92058**	0.49555	0.13943
4-Ethoxy-3-methoxybenzaldehyde	0.14387	**0.89371**	0.17131	0.40043

for vanilla quality and authenticity assessment, but also indicate that other, sometimes minor components, contribute to the sensory properties (Table 16.4). Thus, while chemists seek to measure components that attest to "quality" and authenticity, they also seek to measure components that validate the aroma and flavor quality of vanilla extracts.

16.5 Conclusion

Vanilla continues to be the world's most favored flavoring ingredient, with broad appeal and application. Over the last several decades, sophisticated research continued to unlock the secrets of this complex flavor. It is well established that the main constituent, vanillin, while playing a significant role, contributes less than one-third of the overall flavor/aroma impact. Researchers, such as Perez-Silva et al. (2006), with GC-MS and GC-O, were able to identify 26 odor active constituents, many at parts per million concentrations but still providing a sensory impact. Many other researchers, actively applying state of the art technologies, identified many constituents as well as compounds for

Table 16.4 Aroma-active compounds detected by GC-O analysis of a representative aroma extract from cured vanilla beans. (Reproduced from Perez-Silva *et al.* (2006), with permission from Elsevier.)

Compounds	ppm	Odor quality	Intensity[a]
Phenols			
Guaiacol	9.3	Chemical, sweet spicy	+++
4-Methylguaiacol	3.8	Sweet, woody	+++
p-Cresol	2.6	Balsamic, woody, spicy	++
4-Vinylguaiacol	1.2	Chemical, phenolic	+
4-Vinylphenol	1.8	Sweet, woody	++
Vanillin	19118	Vanilla, sweet	+++
Acetovanillone	13.7	Vanilla, sweet, honey	+++
Vanillyl alcohol	83.8	Vanilla-like	+++
p-Hydroxybenzaldehyde	873	Vanilla-like, biscuit	++
p-Hydroxybenzyl alcohol	65.1	Vanilla-like, sweet	++
Aliphatic acids			
Acetic acid	124	Sour, vinegar	++
Isobutyric acid	1.7	Buttery	++
Butyric acid	<1	Buttery, oily	+
Isovaleric	3.8	Buttery, oily	++
Valeric acid	1.5	Cheese	+++
Alcohols			
2,3-Butanediol (isomer 2)	8.0	Floral, oily	+
Anisyl alcohol	2.4	Herbal	++
Aldehyde			
2-Heptenal	2.1	Green, oily	+
(*E*)-2-decenal	1.8	Herb-like, floral	++
(*E*,*Z*)-2,4-decadienal	1.4	Herb-like, fresh	++
(*E*,*E*)-2,4-decadienal	1.2	Fatty, wood	++
Esters			
Methyl salicylate	<1	Chalk	+++
Methyl cinnamate	1.1	Sweet	++
Ethyl linolenate	13.5	Sweet	++
Ketone			
3-Hydroxy-2-butanone	14.6	Buttery	+
Unknown[b]	6.2	vanilla-like, chemical	+++

a) (+) Weak, (++) Medium, (+++) Strong.
b) Mass fragmentation (91(90), 74(37), 69(34), 89(25)) and RI (2528).

the first time in vanilla. However, as researchers in the past, they were more interested in identifying new compounds and not in determining or reporting their contribution to the flavor of vanilla. These analytical data do contribute to the overall understanding of the constituents in vanilla and therefore can be used to maintain and improve the quality of this unique product.

Like the flavor analytical research, the biochemical research progressed significantly during these same 20 years. Significant research identified the flavor precursors for the major flavor constituents of vanilla, the phenolic glycosides, and demonstrated the formation of these flavor constituents through the action of β-glucosidase during curing. Vanilla browning during curing was also attributed to the enzyme polyphenol oxidase, which also decreases the vanillin content. In addition, genetic research demonstrated the relationships among the *Vanilla* spp. Also researchers were able to develop hybrids, which were resistant to *Fusarium*. All of these investigations can contribute to vanilla quality.

As technologies advanced flavor research and biotechnology, unfortunately they were also used by unscrupulous suppliers to take advantage of the limited supply of this costly product and circumvent existing authentication methods. In response, the legitimate flavor industry, regulators, and academics applied sophisticated technologies to thwart these dishonest suppliers. Liquid chromatographic techniques, including HPLC and HP-TLC, were enhanced and applied to vanilla authentication. The key vanilla constituents were separated and quantified by researchers around the world with a variety of liquid chromatographic techniques. Ratios of these constituents were determined and shown to be relatively consistent, regardless of researcher or origin of the vanilla. Unfortunately, these ratios were promulgated into national regulations but, because of the natural constituents involved, there were legitimate products that fell outside the limits set in regulations. However, the vanillin/p-hydroxybenzaldehyde ratio is the most consistent and should be used as an initial analysis, which if necessary, could be followed by the more accurate isotopic analysis.

Like all technologies, the isotopic methods evolved dramatically during this time. Radiocarbon analyses (^{14}C), stable isotope ratio analysis (SIRA), nuclear magnetic resonance analysis (NMR), accelerator mass spectrometry AMS, and even gas proportional counting (GPC), were all updated, refined, and applied to authenticating vanilla products. These techniques were also coupled with other technologies in order to enhance the detection of adulterated products, for example GC-C-IRMS and methods linking HPLC and SIRA.

Finally, recent research has linked minor flavor constituents to the overall vanilla flavor. Additional research in this area needs to be undertaken and reported.

References

Adedeji, J., Hartman, T. and Ho, C. (1993) Flavor characterization of different varieties of vanilla beans. *Perfumer & Flavorist*, 18, 25–33.

Archer, A. (1989) Analysis of vanilla essences by high-performance liquid chromatography. *Journal of Chromatography*, 462, 461–466.

Avila, M., Gonzalez, M.C., Zougagh, M., Escarpa, A. and Rios, A. (2007) Rapid sample screening method for authenticity controlling vanilla flavors using a CE microchip approach with electrochemical detection. *Electrophoresis*, 28, 4233–4239.

Belay, M.T. and Poole, C.F. (1993) Determination of vanillin and related flavor compounds in natural vanilla extracts and vanilla-flavored foods by thin layer chromatography and automated multiple development. *Chromatographia*, 37, 365–373.

Bensaid, F.F., Wietzerbin, K. and Martin, G. (2002) Authentication of natural vanilla flavorings: isotopic characterization using degradation of vanillin into guaiacol. *Journal of Agricultural and Food Chemistry*, 50, 6271–6275.

Bettazzi, F., Palchetti, I., Sisalli, S. and Mascini, M. (2006) A disposable electrochemical sensor for vanillin detection. *Analytica Chimica Acta*, 555, 134–138.

Brodelius, P.E. (1994) Phenylpropanoid metabolism in *Vanilla planifolia* Andr. (V) High performance liquid chromatographic analysis of phenolic glycosides and aglycons in developing fruits. *Phytochemical Analysis*, 5, 27–31.

Calabretti, A., Campisi, B., Procida, G., Vesnaver, R., and Gabrielli, L. (2005) Analysis of volatile compounds of *Vanilla planifolia* essential oil by using headspace soild-phase microextraction. *Proceedings of the 2nd Central European Meeting [and] 5th Croatian Congress of Food Technologists, Biotechnologists and Nutritionists*, Opatija, Croatia, October 17–20, 2004, pp. 249–257.

Charvet, A.S. and Derbesy, M. (2001) Vanilla beans analysis. Samples analyzed between January 1 and December 31, 2000. *Annales des Falsifications de l'Expertise Chimique et Toxicologique*, 94, 79–84.

Culp, R.A. (2009) Two decades of flavor analysis, trends revealed by radiocarbon (^{14}C) and stable isotope analysis. *12th International Flavor Conference*, Skiathos, Greece

Culp, R.A. and Noakes, J.E. (1992) Determination of synthetic components in flavors by deuterium/hydrogen isotopic ratios. *Journal of Agricultural and Food Chemistry*, 40, 1892–1897.

Culp, R.A., Legato, J.M. and Otero, E. (1998) Carbon isotope composition of selected flavoring compounds for the determination of natural origin by gas chromatography/ isotope ratio mass spectrometer. *ACS Symposium Series 705* (Flavor Analysis). American Chemical Society, Washington, DC, pp. 260–287.

Da Costa, N.C. and Pantini, M. (2006) The analysis of volatiles in Tahitian vanilla (*Vanilla tahitensis*) including novel compounds. *Developments in Food Science*, 43, 161–164.

de Jager, L.S., Perfetti, G.A. and Diachenko, G.W. (2007) Determination of coumarin, vanillin, and ethyl vanillin in vanilla extract products: liquid chromatography mass spectrometry method development and validation studies. *Journal of Chromatography A*, 1145, 83–88.

Dignum, M.J.W., Heijden, R.V.D., Kerler, J., Winkel, C. and Verpoorte, R. (2004) Identification of glucosides in green beans of *Vanilla planifolia* Andrews and kinetics of vanilla β-glucosidase. *Food Chemistry*, 85, 199–205.

Dignum, M.J.W., Kerler, J. and Verpoorte, R. (2002) Vanilla curing under laboratory conditions. *Food Chemistry*, 79, 165–171.

Ehlers, D., Schafer, P., Doliva, H. and Kirchhoff, J. (1999) Vanillin content and component ratios of vanilla beans. The influence of extraction conditions on the detection yields. *Deutsche Lebensmittel-Rundschau*, 95, 123–129.

Ehlers, D., Pfister, M. and Bartholomae, S. (1994) Analysis of Tahiti vanilla by high-performance liquid chromatography. *Zeitschrift fuer Lebensmittel-Untersuchung und –Forschung*, 199, 38–42.

Fayet, B., Derbesy, M. and Guerere, M. (1996) Analytical validation of isotopic analysis of vanilla products. *Analysis*, 24, 398–400.

Fayet, B., Saltron, F., Tisse, C. and Guerere, M. (1999) Contribution to the isotopic characterization of vanilla beans. *Annales des Falsifications de l'Expertise Chimique et Toxicologique*, 92, 11–16.

Fu, S., Huang, M., Zhou, J. and Li, S. (2002a) Determination of chemical constituents of vanilla by supercritical CO_2 fluid extraction. *Huaxue Yanjiu Yu Yingyong*, 14, 208–210.

Fu, S., Huang, M., Zhou, J. and Li, S. (2002b) Study on components of vanilla extract by different extracting technology. *Shipin Kexue*, 23, 109–112.

Gassenmeier, K., Riesen, B. and Magyar, B. (2008) Commercial quality and analytical parameters of cured vanilla beans (*Vanilla planifolia*) from different origins from the 2006–2007 crop. *Flavour and Fragrance Journal*, 23, 194–201.

Gatfield, I.L., Hilmer, J-M., Weber, B., Reiß, I., and Bertram, H-J. (2007) Chemical and biochemical changes occurring during the traditional Madagascan vanilla curing process. Flavor Biotechnology Recent Highlights in Flavor Chemistry & Biology, [Proceedings of the Wartburg Symposium on Flavor Chemistry and Biology], 8th, Eisenach, Germany, Feb. 27-Mar. 2, 2007, 395–398.

Gerasimov, A.V., Gornova, N.V. and Rudometova, N.V. (2003) Determination of vanillin and ethylvanillin in vanilla flavorings by planar (thin-layer) chromatography. *Journal of Analytical Chemistry*, 58, 677–684.

Gillette, M. and Hoffman, P. (2000) Vanilla extract. In: *Encyclopedia of Food Science and Technology*. 2nd Edn, Francis, F.J. (Ed.), John Wiley & Sons, New York, pp. 2383–2399.

Gobley, N.T. (1858) Recherches sur le principe odorant de la vanille. *Journal de Pharmacie et de Chimie*, 34, 401–405.

Hartman, T.G., Karmas, K., Chen, J., Shevade, A., Deagro, M. and Hwang, Hi. (1992) Determination of vanillin, other phenolic compounds, and flavors in vanilla beans by direct thermal desorption-gas chromatography and gas-chromatography-mass spectrometric analysis. *ACS Symposium Series* 506 (Phenolic Compounds in Food Their Effects on Health), American Chemical Society, Washington, DC, pp. 60–76.

Hoffman, P., Harmon, A., Ford, P. *et al.* (2005) Analytical approaches to vanilla quality and authentication. *Vanilla, 1st International Congress*. Princeton, NJ, November 11–12, 2003, Allured Publishing Corp., Carol Street, IL, pp. 41–49.

Jagerdeo, E., Passetti, E. and Dugar, S. (2000) Liquid chromatographic determination of vanillin and related aromatic compounds. *Journal of the AOAC International*, 83, 237–240.

Jamin, E., Martin, F. and Martin, G.G. (2007) Determination of site-specific (Deuterium/ Hydrogen) ratios in vanillin by ^{2}H-nuclear magnetic resonance spectrometry: collaborative study. *Journal of AOAC International*, 90, 187–195.

Jiang, M., Liu, T., Yang, Z., Zhou, B. and Hu, Y. (2005) The changes activity of two endogenous enzymes in differently processed *Vanilla planifolia* capsule. *Yunnan Zhiwu Yanjiu*, 27, 310–314.

Jiang, M., Pu, F., Xie, W.S., Hu, Y.Q. and Li, Y. (1999) Study on the activity of enzymes from vanilla bean. *Zengkan, Proceedings for '99 China's Symposium on Technology Development and Application of Perfume and Essenc*. Jingxi Huagong Bianjibu, pp. 372–374.

John, T.V. and Jamin, E. (2004) Chemical investigation and authenticity of Indian vanilla beans. *Journal of Agricultural and Food Chemistry*, 52, 7644–7650.

Joulian, D., Raymond, L., Jerome, M., Jean-Claude, B. and Hughes, B. (2007) Heliotropin, heliotrope odor and tahitian vanilla flavor: the end of saga? *Natural Product Communication*, 2, 305–308.

Jurgens, U. (1981) The vanillin/*p*-hydroxybenzaldehyde ratio in bourbon vanilla. *Lebensmittelchem. Gerichtl. Chem.*, 35, 97.

Kahan, S., Krueger, D.A. and Berger, R. (1997) Liquid chromatographic method for determination of vanillin and ethyl vanillin in imitation vanilla extract (modification of AOAC Official Method 990.25): collaborative study. *Journal of AOAC International*, 80, 564–570.

Kaunzinger, A., Juchelka, D. and Mosandl, A. (1997) Progress in the authenticity assessment of vanilla. 1. Initiation of authenticity profiles. *Journal of Agricultural and Food Chemistry*, 45, 1752–1757.

Kiridena, W., Poole, S.K. and Poole, C.F. (1994) Identification of 5-(hydroxymethyl)-2-furfural in vanilla extracts. *Journal of Planar Chromatography – Modern TLC*, 7, 273–277.

Klimes, I. and Lamparskey, D. (1976) Vanilla volatiles – A comprehensive analysis. *International Flavours Food Additives*, 7, 272–291.

Lamprecht, G., Pichlmayere, F. and Schmid, E.R. (1994) Determination of the authenticity of vanilla extracts by stable istotope ratio analysis and component analysis by HPLC. *Journal of Agricultural and Food Chemistry*, 42, 1722–1727.

Leong, G., Uzio, R. and Derbesy, M. (1989) Synthesis, identification and determination of glucosides present in green vanilla beans (*Vanilla fragrans* Andrews). *Flavour and Fragrance Journal*, 4, 163–167.

Littmann-Nienstedt, S. and Ehlers, D. (2005) Evaluation method for products with vanilla or vanilla flavorings by using a modified ratio equation. *Deutsche Lebensmittel-Rundschau*, 101, 182–187.

Longares-Patron, A. and Canizares-Macias, M.P. (2006) Focused microwaves-assisted extraction and simultaneous spectrophotometric determination of vanillin and *p*-hydroxybenzaldehyde from *Vanilla fragans*. *Talanta*, 69, 882–887.

Lubinsky, P., Cameron, K.M., Molina, M.C., Wong, M., Lepers-Andrzejewski, S., Gomez-Pompa, A., and Kim, S-C. (2008) Neotropical roots of a Polynesian spice: the hybrid origin of Tahitian vanilla, *Vanilla tahitensis* (Orchidaceae). *American Journal of Botany* 95, 1040–1047.

Martin, G., Remaud, G. and Martin, G.J. (1993) Isotopic methods for control of natural flavours authenticity. *Flavour and Fragrance Journal*, 8, 97–107.

Martin G., Ethridge, M.W., Kaiser, F.E. (1977) Determining the authenticity of vanilla extracts. *Journal of Food Science*, 42, 1580–1586.

Martin, G.J. (1998) Recent advances in site-specific natural isotope fractionation studied by nuclear magnetic resonance. *Isotopes in Environmental and Health Studies*, 34, 233–243.

Neary, M.P., Spaulding, J.D., Noakes, J.E. and Culp R.A. (1997) Tritium analysis of burn-derived water from natural and petroleum-derived products. *Journal of Agricultural and Food Chemistry*, 45, 2153–2157.

Palama, T.L., Khatib, A., Choi, Y.H., Payet, B., Fock, I., Verpoorte, R. and Kodja, H. (2009) Metabolic changes in different developmental stages of *Vanilla planifolia* pods. *Journal of Agricultural and Food Chemistry* 57, 7651–7658.

Perez-Silva, A., Odoux, E., Brat, P. *et al.* (2006) GC-MS and GC-olfactometry analysis of aroma compounds in a representative organic aroma extract from cured vanilla (*Vanilla planifolia* G. Jackson) beans. *Food Chemistry*, 99, 728–735.

Pu, F., Jiang, M., Zhang, Z., Zhang, J. and Kong, F. (1998a) Study on vanilla curing by enzyme treatment method. *Yunnan Zhiwu Yanjiu*, 20, 355–361.

Pu, F., Zhang, J., Zhang, Z. and Jiang, M. (1998b) Study on the components released from glycosides in vanilla beans at different curing stages. *Tianran Chanwu Yanjiu Yu Kaifa*, 10, 29–33.

Ramaroson-Raonizafinimanana, B., Gaydou, E.M. and Bombarda, I. (1999) Long-chain γ-pyrones in epicuticular wax of two vanilla bean species: *V. fragrans* and *V. tahitensis*. *Journal of Agricultural and Food Chemistry*, 47, 3202–3205.

Ramaroson-Raonizafinimanana, B., Gaydou, E.M. and Bombarda, I. (2000) Long-chain aliphatic β-diketones from epicuticular wax of vanilla bean species. *Synthesis of nervonoylacetone. Journal of Agricultural and Food Chemistry*, 48, 4739–4743.

Ranadive, A.S. (1992) Vanillin and related flavor compounds in vanilla extracts made from beans of various global origins. *Journal of Agricultural and Food Chemistry*, 40, 1922–1924.

Scharrer, A. and Mosandl, A. (2001) Reinvestigation of vanillin contents and component ratios of vanilla extracts using high-performance liquid chromatography and gas chromatography. *Deutsche Lebensmittel-Rundschau*, 97, 449–456.

Scharrer, A. and Mosandl, A. (2002) Progress in the authenticity assessment of vanilla. 2. $\delta^{13}C_{V\text{-}PDB}$ correlations and methodical optimizations. *Deutsche Lebensmittel-Rundschau*, 98, 117–121.

Schwarz, B., and Hofmann, T. (2009) Identification of novel orosensory active molecules in cured vanilla beans (*Vanilla planifolia*). *Journal of Agricultural and Food Chemistry* 57, 3729–3737.

Sharma, U., Kumar, S., Nandini, G., Ajai, P., Kumar, V. and Sinha, A.K. (2007) RP-HPTLC densitometric determination and validation of vanillin and related phenolic compounds in accelerated solvent extract of *Vanilla planifolia. Journal of Separation Science*, 30, 3174–3180.

Sinha, A.K., Sharma, U.K. and Sharma, N. (2008) A comprehensive review on vanilla flavor: Extraction, isolation and quantification of vanillin and others constituents. *International Journal of Food Sciences and Nutrition*, 59, 299–326.

Sostaric, T., Boyce, M.C. and Spickett, E.E. (2000) Analysis of the volatile components in vanilla extracts and flavorings by solid-phase microextraction and gas chromatography. *Journal of Agricultural and Food Chemistry*, 48, 5802–5807.

Stahl, W.H., Voelker, W.A. and Sullivan, J.H. (1962) Analysis of vanilla extracts. IV. Amino acid determination. *Journal of the Association of Official Agricultural Chemists*, 45, 108–113.

Taylor, S. (1993) Improved determination of vanillin and related phenolic components in vanilla (*Vanilla fragrans* (Salissb.) Ames by high-performance liquid hromatography. *Flavour and Fragrance Journal*, 8, 281–287.

Thompson, R.D. and Hoffmann, T.J. (1988) Determination of coumarin as an adulterant in vanilla flavoring products by high-performance liquid chromatography. *Journal of Chromatography*, 438, 369–382.

Valdez-Flores, C. and Canizares-Macias, M.P. (2007) On-line dilution and detection of vanillin in vanilla extracts obtained by ultrasound. *Food Chemistry*, 105, 1201–1208.

Waliszewski, K.N., Ofelia, M. and Pardio, V.T. (2009) Quantification and characterization of polyphenol oxidase from vanilla bean. *Food Chemistry*, 117, 196–203.

Waliszewski, K.N., Pardio, V.T. and Ovando, S.L. (2007) A simple and rapid HPLC technique for vanillin determination in alcohol extract. *Food Chemistry*, 101, 1059–1062.

Yang, X. and Peppard, T. (1996) Aroma evaluation of natural products by gas chromatography-olfactometry. Book of Abstracts, 212th ACS National Meeting, Orlando, FL, August 25–29 (1996), AGFD-131. Publisher: American Chemical Society, Washington, DC.

Zhou, B., Zhang, C., Ren, H. and Zhang, C. (2007) Analysis of vanillin in vanilla bean by GC-MS. *Xiangliao Xiangjing Huazhuangpin*, 4, 1–4.

17

Volatile Compounds in Vanilla

Stephen Toth, Keun Joong Lee, Daphna Havkin-Frenkel, Faith C. Belanger, and Thomas G. Hartman

Vanilla flavor is often described by the layperson as being "plain" and unremarkable. However, specialists involved in the vanilla industry and flavorists or perfumers who appreciate the unique qualities vanilla brings to their creations know that vanilla flavor is extremely complex and far from "plain". Just like a fine wine, vanilla flavor is an orchestra of individual nuances that combine to form the unique bouquet we know as vanilla. Although the compound 3-methoxy-4-hydroxybenzaldehyde (vanillin) is characterizing for vanilla, it alone does not constitute vanilla flavor. Indeed, sensory evaluation of pure synthetic vanillin or the similar compound, 3-ethoxy-4-hydroxybenzaldehyde (ethyl vanillin), which are often used for synthetic imitation vanilla renditions, both are found to possess a lack-luster vanilla-like odor with a somewhat "chemical-like" nuance. The characterizing notes of vanillin must be accompanied by other compounds to produce the familiar flavor and aroma we know as vanilla. Indeed, in some cases, extracts with high amounts of vanillin will not taste as good as other extracts with lower vanillin content but containing the other flavor components. Flavor chemists call these additional compounds vanilla fortifiers or synergists. The following is a lexicon of aroma and flavor descriptors that sensory evaluators often use to grade vanilla.

17.1 Lexicon of Vanilla Aroma/Flavor Descriptors

Acidic, Anisic, Aromatic, Balsamic, Barnyard, Caramelized, Chocolate, Creamy, Earthy, Floral, Fruity, Hay-like, Moldy, Musty, Phenolic, Prune, Pungent, Raisin, Resinous, Rummy, Smokey, Sour, Spicy, Sweet, Tea-like, Tobacco-like, Vanillin, Vinegar, Woody

As can be seen from this list of descriptors, vanilla flavor is not "plain" but rather complex and intriguing. Volatile and semi-volatile compounds other than vanillin that are found in vanilla extracts are responsible for many of these sensory attributes and a compilation of these many compounds is the focus of this chapter.

Vanilla renders different flavor and aroma profiles, depending on geographic growing region, cultivation methods, curing, extraction, or storage conditions. The flavor and aroma qualities undergo a constant transition as the initially flavorless green vanilla

Handbook of Vanilla Science and Technology, Second Edition.
Edited by Daphna Havkin-Frenkel and Faith C. Belanger.
© 2019 John Wiley & Sons Ltd. Published 2019 by John Wiley & Sons Ltd.

beans are cured and used directly or extracted with water and ethanol to produce vanilla extract. Reactions occurring in the bean curing process are largely enzymatic and involve conversion of nonvolatile, glycosidically-bound phenolic precursor molecules such as glucovanillin into the free forms (Walton *et al.* 2003; Havkin-Frenkel *et al.* 2004, 2005; Korthou and Verpoorte 2007). In addition, many compounds present in cured vanilla beans are further chemically transformed into different compounds as they react with ethanol in the vanilla extraction process. Pungent acids can be converted into ethyl esters with more pleasant floral and fruity aromas. Esters originally present in beans can convert into acids or alcohols. Many esters originally present as methyl esters in beans can trans-esterify into corresponding ethyl esters in the extract process. Aldehydes can be converted into more delicate acetals, which have potential to revert back when exposed to higher water concentration in downstream applications. Finally, when vanilla is used in different foods, it renders different flavor profiles. Vanillin is an aromatic aldehyde and it readily reacts with amino groups of proteins and amino acids to form resonance-stabilized, Schiff-base compounds that often attenuate its flavor impact. Frozen desert manufacturers that tried to lower calories by substituting textured proteins for milk fat learned this the hard way. Similar reaction pathways produce colored compounds that limit vanillin use in perfumery. Vanilla is actually a versatile and dynamic flavoring substance, the potential of which has still not been fully realized.

Vanilla flavor begins with the vanilla bean. The vanilla bean is the sole edible fruit of the orchid family. Out of 110 known *Vanilla* species only two, *Vanilla planifolia* Andrews and *Vanilla tahitensis* Moore, are allowed to be used in foods. The estimated world production of vanilla beans is around 2,000 tons per year (Havkin-Frenkel *et al.* 2005). Most of the growing areas, notably Madagascar, cultivate *V. planifolia*, while Tahiti, Papua New Guinea (PNG), and limited parts of Indonesia produce mainly *V. tahitensis*. Mesoamerica is known as the origin of *V. planifolia* but the origin of *V. tahitensis* was only recently been determined and appears to be a hybrid of *V. planifolia* and *V. odorata* (Lubinsky *et al.* 2008). Novel compounds described for the first time in this chapter appear to support this conclusion.

Vanilla was introduced to Europe by the Spanish Conquistadores in 1520 but commercial production of vanilla started about 300 hundred years later with the discovery of hand pollination of the vanilla flower. In the wild, insects carry out the pollination of vanilla flowers (Childers *et al.* 1959). In commerce, vanilla is cultivated in tropical regions and is propagated by cuttings. The plant requires 3 to 4 years to set the flower, and afterward flowers once a year. The pod-like fruit (vanilla bean) is allowed to develop for 8 to 10 months before harvesting.

Vanilla beans are harvested green and are initially flavorless. The green beans are subjected to a curing process for 3 to 6 months or longer, depending on various curing protocols in different production regions. The objective of the curing process is to develop the prized vanilla flavor and to dry the cured beans to prevent microbial growth during transport and storage. Vanilla cultivation, vanillin biosynthesis, and economic aspects are discussed extensively in other reviews (Havkin-Frenkel *et al.* 2004, 2005). These reports provide a wealth of information on the botany of the vanilla bean, including detailed descriptions of special papillary cells that are believed to be the site of vanillin biosynthesis. This information is vital to the understanding of the curing process and for further improvement of this process. The curing process is described in detail in Chapter 6 by Frenkel *et al.*

In the United States, vanilla is the only flavor that has its own standard of identity, which defines how it is produced and labeled. For more details, see section 21 CFR 169.175 of the US Code of Federal Regulations. *V. planifolia* and *V. tahitensis* are the only two *Vanilla* species allowed to be used in food. The beans need to be properly cured, with no more than 25% moisture content. A one-fold vanilla extract is defined as the total of flavor and odor principles of 13 oz of beans per gallon of water and alcohol. Alcohol content should be at a minimum of 35% by volume. Additional ingredients also permitted are glycerin, propylene glycol, sugar, and corn syrup. Vanilla flavors are vanilla extracts that contain less than 35% alcohol.

The authors' research group has been collecting and analyzing vanilla, chiefly in the form of cured vanilla beans produced in different geographic regions, for over 20 years. Other investigators have focused their analytical efforts on vanilla extract rather than cured beans. The list of volatile and semi-volatile compounds detected in vanilla and/or vanilla beans is at least 494 and growing. Back in 1993, investigators in the authors' lab at Rutgers University reported on volatile compounds in vanilla beans grown in Madagascar, Tahiti, Indonesia, Mexico, Tonga, Costa Rica, and Jamaica (Adedeji *et al.* 1993; Adedeji 1993). Since this time, vanilla cultivation has spread to some new regions, including India, Papua New Guinea, and Uganda. Our lab has more recently conducted studies of vanilla beans produced in the 2004 and 2005 growing seasons from a wide range of geographic regions. Some of the compounds detected in these analyses are reported in this chapter for the first time.

This chapter is a compilation of volatile/semi-volatile compounds detected in vanilla, as reported in the published literature. Table 17.1 is a list of volatile and semi-volatile compounds reported to be naturally present in vanilla. For the purposes of this chapter, vanilla is defined as vanilla beans and/or extracts prepared thereof. Compounds are listed in alphabetical order and are grouped by chemical class. For each entry, the individual compounds are listed using International Union of Pure and Applied Chemistry (IUPAC) nomenclature, common name or synonym, Chemical Abstracts Services Registry Number if available (CAS), and their reported concentrations (if available) in vanilla beans/vanilla extract from different vanilla species and geographic origins. The table also provides references to the original investigators. Please note that many volatile compounds occurring in vanilla and listed in the table do not contribute significantly to flavor or aroma. For instance, many of the compounds (i.e. hydrocarbons, long chain fatty acids, etc.) are odorless and have no flavor or aroma value. Entries highlighted in bold font in Table 17.1 are volatile compounds reported by the authors in this chapter for the first time.

We recently analyzed some cured, wild vanilla beans collected from a Peruvian rainforest. The beans are of unknown species but morphologically they resembled *V. odorata*. Interestingly, the beans were found to contain a novel series of anisyl alcohol esters that have not been described previously as occurring in vanilla, or any other natural products for that matter. In addition to anisyl formate, anisyl acetate, anisaldehyde, and anisic acid, which have been reported previously in vanilla, the compounds anisyl acrylate, anisyl salicylate, anisyl anisate, anisyl vanillate, anisyl protocatechuate, anisyl myristate, anisyl pentadecanoate, anisyl palmitate, anisyl linoleate, anisyl oleate, anisyl linoleate, and

Table 17.1 Volatile compounds detected in vanilla beans from different species and from different geographic growing.

IUPAC Name	Synonyms	CAS #	Bourbon	Tahitian	Balli	Java	Mexican	Tonga
ACIDS								
(9Z)-Octadec-9-enoic acid	oleic acid	112-80-1	13[f], 110-356[h]	280[h]			16[b], 203[h]	239[h]
2,2-dimethyl-3-butenoic acid		10276-09-2				48[a]		
2,3-dihydroxypropyl ester 9,12-octadecanoic acid								
2-ethoxy-carbonylbenzoic acid	mono ethyl phthalate							
2-heptenoic acid	hept-2-enoic acid	18999-28-5					2[b]	
2-hydroxy-2-(3-hydroxy-4-methoxyphenyl)acetic acid	isovanillyl-mandelic acid	3695-24-7	3[h]					
2-hydroxybenzoic acid	salicylic acid	69-72-7						
2-hydroxy-propanoic acid	lactic acid	50-21-5						
2-methyl-1-(1,1-dimethyl)-2-methylpropanoic acid			32[a]					
2-methyl-propanoic acid	isobutyric acid	79-31-2					2[b]	
2-propenoic acid	**acrylic acid**	79-10-7	22[h]					
3,4-dimethoxy-benzoic acid	**veratric acid**	93-07-2						
3-methylbutanoic acid	isovaleric acid	503-74-2				62[a]	4[b]	34[h]
4-hydroxy-3-methoxybenzoic acid	vanillic acid	121-34-6	110-586[a], 440[f], 246-825[h]	112[a], 186[h,t]	520-869[a]	270[a]	994[a], 1315[b], 887[h]	439[a], 472[h]
4-hydroxybenzoic acid	p-hydroxybenzoic acid	99-96-7	6[a], 28[f]	136[a], 2478[h,t],ab	194[a]		218[a], 255[b]	49[a]
4-methoxybenzoic acid	p-anisic acid	100-09-4	69-267[a], 49[f]	6471[a], 87370[c], 4129[h,s,x]			10[b], 22[h]	

Costa Rican	Jamaican	V. fragrans origin Unknown	V. pompona origin unknown	V. pompona from Madagascar	Unidentified from Madagascar	Indian	Comoros	Ugandan	Indonesian	Hawaiian	PNG	Wild Type from Peru
				852^h	63^h	$156-834^h$	529^h	253^h	3639^h	291^h	$71-2288^h$	1033^h
						47^h						
		g										
						189^h						
		g l				108^h					$30-2971^h$	
		g										
		o										
						$6-30^h$						11^h
									52^h			
	35^a	g o			40^h	$47-138^h$	86^h		9^h	9^h	$30-441^h$	
787^a	830^a	l o t y ae	t	1011^h	915^h	$440-1963^h$	1077^h	1158^h	390^h	577^h	$244-726^h$	1360^h
310^a	47^a	l t ab	t								30^h	121^h
		sx		39^h		79^h				33^h	$23-7945^h$	1488^h

(Continued)

Table 17.1 (Continued)

IUPAC Name	Synonyms	CAS #	Bourbon	Tahitian	Balli	Java	Mexican	Tonga
4-methyl-2 oxovaleric acid	**isopropyl pyruvic acid**	816-66-0		97[h]			3[h]	
4-methyl-3-cyclohexene-1-carboxylic acid, methyl ester						36[a]		
4-oxopentanoic acid	levulinic acid	123-76-2	14[a], 120-126[h]	460[h]	52[a]			2[a], 209[h]
4-phenoxybenzoic acid		2215-77-2		25[a]				
9,12-octadecadienoic acid	linoleic acid	60-33-3	59-509[a], 135[f], 117-411[h]	190[a], 509[h]	367-561[a]	66[a]	70[a], 225[b], 408[h]	54[a], 451[h]
9-hexadecanoic acid	palmitoleic acid	373-49-9	38[h]	26[h]			6[b], 8[h]	24[h]
a,d-dimethyltetronic acid								
acetic acid	ethanoic acid	64-19-7	1130-2348[a], 151[f], '1078-1347[h]	1515[a], 2844[h]	937-1153[a]	955[a]	361[a], 124[b], 683[h]	135[a], 1820[h]
benzene propanoic acid	hydro-cinammic acid	501-52-0					4[b]	
benzoic acid	dracylic acid	65-85-0	38[f]				3[b]	
butanoic acid	butyric acid	107-92-6					<1[b]	
cinnamic acid	3-phenyl-2-propenoic acid	140-10-3	114[a]		52-342[a]		3[b]	
cyclohexaneacetic acid	**cyclohexyl-acetic acid**	5292-21-7	13[h]					
decanoic acid	capric acid	334-48-5						
dodecanoic acid	lauric acid	143-07-7	6-20[h]		583[a]	74[a]	2[b]	8[h]
formic acid	methanoic acid	64-18-6	130[a], 62[f], '314-342[h]	471[h]			1904[a], 98[h]	72[a], 319[h]
heptadecanoic acid	margaric acid	506-12-7	8-27[a], 30[f], 27[h]	20[h]	4[a]	35[a]	110[a], 6[b], 7[h]	18[a], 10[h]

Costa Rican	Jamaican	V. fragrans origin Unknown	V. pompona origin unknown	V. pompona from Madagascar	Unidentified from Madagascar	Indian	Comoros	Ugandan	Indonesian	Hawaiian	PNG	Wild Type from Peru
									112^h		130^h	174^h
		o		282^h	148^h	$54-213^h$	184^h	128^h	173^h	131^h	$85-239^h$	
											7^h	
142^a	1421^a			3805^h	7072^h	$211-942^h$	857^h	586^h	7600^h	252^h	$89-14484^h$	2609^h
					300^h	$47-269^h$			65^h		95^h	61^h
				177^h						17^h		
1587^a	1613^a	$g\ o$		1647^h	1135^h	$849-1855^h$	1343^h	1713^h	2178^h	365^h	$1488-2417^h$	2667^h
		$o\ ae$										
		g										22^h
		s										
										79^h		
		$g\ p$										
		g			69^h	$21-42^h$	73^h	3^h	80^h	36^h	$11-77^h$	
168^a	185^a	g		507^h	293^h	$168-550^h$	730^h	599^h	289^h	198^h	$340-1280^h$	635^h
	35^a			39^h	51^h	$47-224^h$	37^h	62^h	124^h	11^h	$63-300^h$	54^h

(Continued)

Table 17.1 (Continued)

IUPAC Name	Synonyms	CAS #	Bourbon	Tahitian	Balli	Java	Mexican	Tonga
heptanoic acid	enanthic acid	111-14-8					2^b	
hexadecanoic acid	palmitic acid	57-10-3	$23\text{-}54^a$, 258^f, $279\text{-}754^h$	383^a	$60\text{-}516^a$	28^a	127^a, 127^b, 475^h	18^a, 568^h
hexanoic acid	caproic acid	142-62-1	23^h				102^a, $<1^b$, 39^h	
hydroxy acetic acid	glycolic acid	79-14-1						
m-anisic acid, methyl ester				76^h				
methoxyacetic acid	methoxyethanoic acid	625-45-6						
nonanoic acid	pelargonic acid	112-05-0	95^a, 31^f				130^a, 16^b	
octadecanoic acid	stearic acid	57-11-4	$42\text{-}587^a$, 39^f, $41\text{-}99^h$	15^a, 108^h	$11\text{-}80^a$	95^a	67^a, 14^b, 47^h	60^a, 81^h
octanoic acid	caprylic acid	124-07-2				157^a	6^b	
pentadecanoic acid	pentadecylic acid	1002-84-2	$38\text{-}254^a$, 24^f, $12\text{-}29^h$	34^a, 33^h	$69\text{-}89^a$	192^a	43^a, 13^b, 11^h	37^a, 17^h
pentanoic acid	valeric acid	109-52-4					2^b	
propanoic acid	ethanecarboxylic acid	79-09-4	$18\text{-}199^a$	67^a			2^b	6^h
spirohexane-1-carboxylic acid, ethyl ester				26^a				34^a
tetradecanoic acid	myristic acid	544-63-8	$22\text{-}343^a$, 12^f, $45\text{-}51^h$	29^a, 80^h	$102\text{-}336^a$	66^a	42^a, 12^b, 16^h	15^a, 20^h
ALCOHOLS								
(2E)-3,7-dimethylocta-2,6-dien-1-ol	nerol	106-25-2						

Costa Rican	Jamaican	V. fragrans origin Unknown	V. pompona origin unknown	V. pompona from Madagascar	Unidentified from Madagascar	Indian	Comoros	Ugandan	Indonesian	Hawaiian	PNG	Wild Type from Peru
		I									60^h	
	30^a			859^h	2054^h	658-2020^h	1314^h	790^h	4240^h	323^h	350-5795^h	1659^h
		g i										
		g				34^h	6^h				11-56^h	
		g							11^h			
		i										
17^a	97^a			421^h	544^h	108-474^h	191^h	98^h	885^h	58^h	34-1094^h	1758^h
		g i p										
8^a	22^a			59^h	143^h	51-178^h	101^h	45^h	243^h	10^h	12-385^h	92^h
						6-68^h	30^h	8^h			3-11^h	
20^a	78^a	*g*		59^h	269^h	58-240^h	467^h	44^h	191^h	19^h	25-293^h	193^h
		g										

(Continued)

Table 17.1 (Continued)

IUPAC Name	Synonyms	CAS #	Bourbon	Tahitian	Balli	Java	Mexican	Tonga
(4-methoxyphenyl) methanol	anisyl alcohol	105-13-5	24^a, 9^f, 25-54^h	1165^a, 225830^e, $9515^{h,\,i,}$ $_{s,\,v,\,x,\,ac}$			71^a, 2^b, 7^h	17^a, 49^h
(7,7-dimethyl-4-bicyclohept-3-enyl) methanol	myrtenol	6712-78-3						
(E,7R,11R)-3,7,11,15-tetramethylhexadec-2-en-1-ol	phytol	150-86-7	6^h				7^h	
1-(4-methoxy-phenyl)-2-methyl-3-buten-1-ol				15^h				
1,2,3,4-butanetetraol	erythritol	149-32-6				43^a		
1,2,3-propanetrio	glycerol, glycerine	56-81-5	978^a					
1,2-cyclohexanediol		931-17-9						
1,2-dihydroxybenzene	catechol, pyrocatechol	120-80-9	29^f					
1,2-dimethoxybenzene	veratrol	91-16-7					18^h	
1,2-propandediol	propylene glycol	57-55-6					$<1^b$	
1,3-butanediol	b-butylene glycol	107-88-0	219-326^a, 47^f, 55-65^h		238-334^a	44^a	301^a, 95^h	570^h
1,3-cyclohexanediol, trans-	**hexahydroresor cinol**	504-01-8	84^h	110^h				
1,4-butanediol	tetramethylene glycol	110-63-4	89-259^a				226^a	
1-methoxy-2-(4-methoxyphenyl) methylbenzene				245^a				
1-octen-3-ol	octenol, vinyl hexanol	3391-86-4					$<1^b$	
1-phenyl-1,2-butanediol			130^a		63^a			
2-(4-methyl-cyclohex-3-en-1-yl) propan-2-ol	alpha-terpineol	98-55-5						

Costa Rican	Jamaican	V. fragrans origin Unknown	V. pompona origin unknown	V. pompona from Madagascar	Unidentified from Madagascar	Indian	Comoros	Ugandan	Indonesian	Hawaiian	PNG	Wild Type from Peru
		gm	*su*	410[h]	70[h]	31-128[h]	129[h]	40[h]	42[h]		38-4387[h]	9011[h]
		g				66[h]	56[h]				15[h]	
						6-16[h]		3[h]			16[h]	74[h]
40[a]												
				42[h]						11[h]		
	227[a]					133-2895[h]	37[h]	279[h]			258[h]	
		gi										
		h										

(Continued)

Table 17.1 (Continued)

IUPAC Name	Synonyms	CAS #	Bourbon	Tahitian	Balli	Java	Mexican	Tonga
2,2,4-trimethyl-3-penten-1-ol								64[a]
2,2-dimethylpentan-1-ol	**neoheptanol**	2370-12-9	103[h]					
2,3-butanediol	dimethylene glycol, 2,3-butyleneglycol	24347-58-8			951[a]		17[b] 174[h]	
2,4-dimethyl-1-heptanol							20[a]	
2,6-dimethoxy-4-methylphenol	4-methyl syringol	6638-05-7						
2,6-dimethyl-4-ethyl-4-heptanol								43[a]
2-acetoxy-1-propanol				8[h]				
2-butyne-1,4-diol						17[a]		
2-cis-9-octadecenyl-oxyethanol			184[a]	28[a]				
2-ethylcyclobutanol		35301-43-0	14[h]					
2-methoxy-3-methylbutane					1091[a]			
2-methoxy-4-(2-propenyl)-phenol	eugenol	97-53-0						
2-methoxy-4-methylphenol	creosol	93-51-6					4[b]	
2-methoxy-4-prop-2-enylphenol	**engenol, caryophyllic acid**	97-53-0						
2-methoxy-p-cresol	2-Hydroxy-5-methylanisole	93-51-6	3[h]				12[h]	10[h]
2-methoxyphenol	guaiacol, methylcatechol	9009-62-5	19[f], 220-322[h]				9[b], 3[h]	77[h]
2-methyl-butan-1-ol		137-32-6						
2-octen-4-ol	butyl propenyl carbinol	20125-81-9	25[f]					
2-phenylethanol	phenylethyl alcohol	60-12-8					1[b]	

Costa Rican	Jamaican	V. fragrans origin Unknown	V. pompona origin unknown	V. pompona from Madagascar	Unidentified from Madagascar	Indian	Comoros	Ugandan	Indonesian	Hawaiian	PNG	Wild Type from Peru
				11[h]	6[h]						76[h]	
	g									3[h]	177[h]	
									90[h]		155[h]	
						6[h]						
				45[h]								
15[a]	*i*											
	g i											
									12[h]			
				18[h]		4-83[h]		8[h]	13[h]	4[h]	4-6[h]	6[h]
	g i p			207[h]	126[h]	11-17[h]	131[h]		14[h]		7-59[h]	27[h]
	g											
	gi											

(Continued)

Table 17.1 (Continued)

IUPAC Name	Synonyms	CAS #	Bourbon	Tahitian	Balli	Java	Mexican	Tonga
3-(hydroxymethyl) phenol		620-24-6	20[a]		30[a]	202[a]		
3,4-Dimethoxybenzyl alcohol	veratryl alcohol	93-03-8						
3,7,11-trimethyl-1,6,10-dodecatrien-3-ol	nerolidol	40716-66-3			110[a]			
3,7-Dimethyl-2,6-octadien-1-ol	geraniol	106-24-1						
3,7-dimethyl-6-octen-1-ol	citronellol, dihydrogeraniol	106-22-9						
3-7-dimethylocta-1,6-dien-3-ol	linalyl alcohol, linalool	78-70-6						
3-methyl-2-buten-1-ol	prenol	556-82-1						
3-methyl-2-propylpentan-1-ol			47[a]	102[a]				7[a]
3-methyl-butan-1-ol	isopentanol	6423-06-9						
3-methylhexan-2-ol		2313-65-7	178[h]		14[a]			171[a]
3-methylpentan-1-ol		589-35-5						
3-methylphenol	cresol	108-39-4						
3-phenyl-2-propen-1-ol	cinnamyl alcohol	104-54-1					<1[b]	
3-phenylpropan-1-ol	benzene propanol	122-97-4					<1[b]	
4-(3-hydroxy-1-propenyl)-2-methoxyphenol	coniferyl alcohol	458-35-5						
4-(hydroxymethyl)-2-methoxyphenol	vanillic alcohol	498-00-0		j			84[b]	12 [(8)]
4-(hydroxymethyl) phenol	p-hydroxy benzyl alcohol	623-05-2	29[f], 15[h]	j			65[b]	4[h]
4,5-octanediol		22607-10-9						

Costa Rican	Jamaican	V. fragrans origin Unknown	V. pompona origin unknown	V. pompona from Madagascar	Unidentified from Madagascar	Indian	Comoros	Ugandan	Indonesian	Hawaiian	PNG	Wild Type from Peru
						45^h		82^h				
												371^h
		g										
		g										
		g										
		g										
	243^a											
		g										
											831^h	
	68^a	g										
									34^h			
		l										
		j ad af	j		$11\text{-}2676^h$		78^h				16^h	23^h
		j n p af	j	62^h	$13\text{-}74^h$						9^h	
										22^h		

(Continued)

Table 17.1 (Continued)

IUPAC Name	Synonyms	CAS #	Bourbon	Tahitian	Balli	Java	Mexican	Tonga
4-butoxy-1-butene								
4-ethenyl-2-methoxyphenol	4-Hydroxy-3-methoxystyrene	7786-61-0	318[h]				1[b], 33[h]	
4-ethenylphenol	4-vinyl phenol	2628-17-3					2[b]	
4-ethenylphenol	p-vinyl phenol	2628-17-3						
4-Ethyl-1,3-benzenediol	4-ethylresorcinol	2896-60-8	48[a]		156[a]			
4-ethyl-2-methoxyphenol	p-ethyl guaiacol	2785-89-9	33[f]					
4-Methoxybenzene propanol		5406-18-8						
4-methyl-1-propan-2-ylcyclohex-3-en-1-ol	terpinen-4-ol	562-74-3						
4-methylphenol	p-cresol	106-44-5				27[a]	3[b]	
9,12-octadecadien-1-ol	linoleyl alcohol	506-43-4		64[a]				
9-octadecen-1-ol	oleyl alcohol	143-28-2		28[a]				
benzene-1,4-diol	hydroquinone, quinol	123-31-9			143[a]			
but-2-yne-1,4-diol		110-65-6	10[h]					
docosan-1-ol	behenyl alcohol	661-19-8	44[a]		201[a]			
dodecan-1-ol	lauryl alcohol	112-53-8						
dodecylcyclohexanol					33[a]			
heptacosan-1-ol		2004-39-9	140-1353[h]	216[h]			223[h]	641[h]
heptadecan-1-ol	heptyl alcohol	111-70-6						
heptan-2-ol	amyl methyl carbinol	543-49-7			612[h]	40[a]		
hexacosan-1ol		506-52-5	222[h]	50[h]			186[h]	

Costa Rican	Jamaican	V. fragrans origin Unknown	V. pompona origin unknown	V. pompona from Madagascar	Unidentified from Madagascar	Indian	Comoros	Ugandan	Indonesian	Hawaiian	PNG	Wild Type from Peru
	138[a]											
		g		577[h]					290[h]			
		g										
173[a]								20[h]		4[h]	7[h]	
		g			4[h]						103[h]	
		g										
		g i p										
											64[h]	
		g										
		g	180[h]	1034[h]	1391–1582[h]		781[h]		1483[h]	1705[h]	486–1063[h]	203[h]
		g										
			36[h]		8–299[h]			249[h]		164[h]	94[h]	88[h]

(Continued)

Table 17.1 (Continued)

IUPAC Name	Synonyms	CAS #	Bourbon	Tahitian	Balli	Java	Mexican	Tonga
hexadecan-1-ol	cetyl alcohol, palmityl alcohol	36653-82-4	38^f					
hexan-1-ol	hexyl alcohol	111-27-3						
hexan-2-ol		626-93-7	$175\text{-}629^h$					867^h
hydroxytricosane			$14\text{-}31^a$					
methyl-2-vinylethyl carbinol							6^h	
nonan-2-ol		628-99-9						
nonan-3-o1		624-51-1	277^a					107^h
o-catechol			$85\text{-}2224^a$	27^a	1966^a	60^a		
octacosan-1-ol	montanyl alcohol	557-61-9	1289^h	427^h			2224^h	556^h
octan-1-ol	octyl alcohol, capryl alcohol	111-87-5					1^b	
pentan-1-ol		71-41-0						
pentan-2-ol	amyl alcohol	6032-29-7	95^f					
phenol	phenyl alcohol, benzenol	108-95-2	$362\text{-}553^a, 28^f, 26\text{-}161^h$	$143^a, 23^h$	263^a	201^a	$190^a, 2^b$	14^h
phenylmethanol	benzyl alcohol	100-51-6					3^b	
tetradecan-1-ol	myristyl alcohol	112-72-1	13^f					
ALDEHYDES								
(2E,4E)-deca-2,4-dienal		80-56-8	$209^a, 25^f$	34^a		83^a	$1^b, 168^h$	
2,2-dimethylpent-4-enal		5497-67-6				97^a		
2,6,6-trimethylcyclohexene-1-carbaldehyde	beta-cyclocitral	52844-21-0						
2-hydroxy-2-methylpropanal		20818-81-9	13^a					

Costa Rican	Jamaican	V. fragrans origin Unknown	V. pompona origin unknown	V. pompona from Madagascar	Unidentified from Madagascar	Indian	Comoros	Ugandan	Indonesian	Hawaiian	PNG	Wild Type from Peru
		g										
					5[h]	179-849[h]		1023[h]	22[h]		4-18[h]	17[h]
		g										
109[a]								51[h]				
	133[a]											
						22-1732[h]	966[h]	7[h]		703[h]	7[h]	244[h]
		g										
		g										
		g										
95[a]	210[a]	g		17[h]	5[h]	3-32[h]	54[h]		27[h]		6-311[h]	8[h]
		g										
	112[a]							126[h]	127[h]			
		g										

(Continued)

Table 17.1 (Continued)

IUPAC Name	Synonyms	CAS #	Bourbon	Tahitian	Balli	Java	Mexican	Tonga
2-hydroxybenzaldehyde	salicylic aldehyde	90-02-8						
2-methylbut-2-enal	2-methyl-2-butenal	6038-09-1					140[a]	
2-oxopropanal	pyruvaldehyde	78-98-8	44[a]					
3,3-dimethylhexanal								
3,4-dihydroxybenzaldehyde	protocatechualdehyde	139-85-5		23[h]				
3,5-Dimethyoxy-4-hydroxybenzaldehyde								
3-methoxybenzaldehyde	m-anisaldehyde	591-31-1		954[a]				
3-methylbutanal	Isovaleraldehyde	590-86-3	37-48[a], 27[f]	32[a], 13[h]	18-370[a]	459[a]	150[a]	36[a], 37[h]
3-methylpentanal	3-methyl valeraldehyde	15877-57-3	24-378[a], 15-242[h]	368[a], 6[h]	27-222[a]	566[a]	104[a], 2[h]	44[a], 13[h]
4-hydroxy-3,5-dimethoxybenzaldehyde	syringic aldehyde	134-96-3	26-457[a], 10[f], 25[h]		113[a]			12[h]
4-hydroxy-3-methoxybenzaldehyde	vanillin	121-33-5	15896-17465[a], 19400[f], 16674-22435[h]	6797[a,g], 5798[h]	15227-20054[a]	3413[a]	9296[a], 19118[b], 22757[h]	12193[a], 10429[h]
4-hydroxybenzaldehyde		123-08-0	790-950[a], 1040[f], 1396-1414[h]	1387[a], 812[h v aa]	772-1114[a]	55[a]	635[a], 873[b], 1549[h]	288[a], 467[h]
4-methoxybenzaldehyde	anisaldehyde	123-11-5	12[h]	25030[c], 329[h i x ac]				
acetaldehyde	ethanal	75-07-0						
benzaldehyde		100-52-7						
dec-2-enal	2-decenal	3913-81-3					2[b]	
decanal		112-31-2	4-5[h]					

Costa Rican	Jamaican	V. fragrans origin Unknown	V. pompona origin unknown	V. pompona from Madagascar	Unidentified from Madagascar	Indian	Comoros	Ugandan	Indonesian	Hawaiian	PNG	Wild Type from Peru
		g										
227[a]				26[h]							31-148[h]	
						24[h]						
45[a]	6[a]					19[h]			123[h]	15[h]	48-87[h]	
81[a]	36[a]					76-181[h]		157[h]	98[h]		67-321[h]	308[h]
					24[h]	29-61[h]	45[h]	23[h]	8[h]	8[h]	42[h]	24[h]
17613[a]	12271[a]	g	g	23252[h]	26713[h]	12178-28593[h]	29171[h]	11901[h]	1439[h]	22186[h]	9024-16630[h]	4393[h]
765[a]	604[a]	g l m r v aa		3533[h]	1891[h]	380-1103[h]	1564[h]	659[h]	94[h]	897[h]	460-1265[h]	364[h]
		g i o									20-51[h]	134[h]
		o										3[h]
		g o y										3[h]
				55[h]				75[h]		24[h]		12[h]

(Continued)

Table 17.1 (Continued)

IUPAC Name	Synonyms	CAS #	Bourbon	Tahitian	Balli	Java	Mexican	Tonga
hept-2-enal	2-heptenal	29381-66-6					2[b]	
hexacosanal		26627-85-0	417[h]	32[h]			677[h]	166[h]
hexanal	hexanaldehdye	66-25-1	277[a]		389[a]		206[a]	724[a]
nonanal	nonanaldehyde	124-19-6	1901[a], 53[f], 26-52[h]		110[a]			
octacosanal		22725-64-0	61[h]					
octadec-9-enal	octadecenyl aldehyde	5090-41-5			54[a]			
pentanal	valeraldehyde	110-62-3						
ALKANES								
1,3,5-trimethyl-cyclohexane		1839-63-0					11[h]	
10-methyleicosane		54833-23-7	25[f]					
11-decylheneicosane		55320-06-4	23[a]	60[a]	14-44[a]	33[a]	38[a]	
2,2,4,6,6-pentamethylheptane	isododecane	31807-55-3	16[a]	23[a]				
2,2,4-trimethylpentane	**isooctane**	540-84-1		136[h]				
2,3,3,4-tetramethylpentane		16747-38-9				62[a]		
2,6,10,15-tetramethyl heptadecane				12[a]		1587[a]		
2,7,10-triethyldodecane						21[a]		
2-cyclohexyleicosane		4443-56-5	10[a]					
3-ethyl-2-methylpentane		68333-81-3					56[a]	
3-ethyl-3-methylpentane		1067-08-9					32[a]	
4-ethylheptane		2216-32-2						22[a]

Costa Rican	Jamaican	V. fragrans origin Unknown	V. pompona origin unknown	V. pompona from Madagascar	Unidentified from Madagascar	Indian	Comoros	Ugandan	Indonesian	Hawaiian	PNG	Wild Type from Peru
				87^h	432^h	$108-484^h$	267^h	511^h	877^h	300^h	$148-825^h$	
	356^a											
				142^h				24^h			$81-151^h$	16^h
					2049^h	$4-18019^h$					$13-2101^h$	
		g										
				23^h								
								24^h				
							283^h					

(Continued)

Table 17.1 (Continued)

IUPAC Name	Synonyms	CAS #	Bourbon	Tahitian	Balli	Java	Mexican	Tonga
4-ethyltetradecane		55045-14-2	33[a]					
6-ethyl-2-methyldecane		62108-21-8				25[a]		
decane		73138-29-1						
docosane		629-97-0	7-11[a], 22[f], 31[h]	36[a], 111[h]	14[a]	687[a]	67[a], 29[h]	34[a], 7[h]
dodecane		94094-93-6						
heneicosane		629-94-7	23[f]					
hentriacontane		630-04-6		359[h]				
heptacosane		593-49-7	38-204[h]	26[h]			162[h]	79[h]
heptadecane		629-78-7						
hexacosane		630-01-3	24[f], 16-90[h]	18[h]			49[h]	30[h]
hexadecane		544-76-3						
hexane		92112-69-1				263[a]		
hexatricontane		630-06-8	78-114[a]	126[a]	32[a]	17[a]	15[a]	
icosane	eicosane	112-95-8	39[f]					
nonacosane		630-03-5	25-203[h]	79[h]			188[h]	197[h]
nonane		111-84-2						
octacosane		630-02-4			37[a]			
pentacosane		629-99-2	98-404[h]	91[h]			20[b], 235[h]	158[h]
pentadecane		629-62-9						

Costa Rican	Jamaican	V. fragrans origin Unknown	V. pompona origin unknown	V. pompona from Madagascar	Unidentified from Madagascar	Indian	Comoros	Ugandan	Indonesian	Hawaiian	PNG	Wild Type from Peru
		g										
64[a]	25[a]	g		34[h]	321[h]	8-136[h]	79[h]	40[h]	569[h]	126[h]	15-908[h]	
		g										
					746[h]	78-4637[h]	517[h]	56[h]	2215[h]		372-452[h]	451[h]
				123[h]	279[h]	49-327[h]	226[h]	265[h]	331[h]	177[h]	60-293[h]	66[h]
		g										
				46[h]	272[h]	18-138[h]	156[h]	150[h]	160[h]	73[h]	17-205[h]	36[h]
		g i										
432[a]												
26[a]	43[a]											
		g										
				364[h]	278[h]	79-431[h]	236[h]	126[h]	933[h]	246[h]	43-362[h]	120[h]
		g										
											8[h]	
				118[h]	364[h]	80-599[h]	459[h]	446[h]	739[h]	290[h]	93-917[h]	158[h]
		g										

(Continued)

Table 17.1 (Continued)

IUPAC Name	Synonyms	CAS #	Bourbon	Tahitian	Balli	Java	Mexican	Tonga
pentatriacontane		630-07-9	47-108[a]	34[a]	13-170[a]	242[a]		43[a]
tetracosane		646-31-1	13[f], 42-198[h]	30[h]			91[h]	61[h]
tetradecane		629-59-4						
tricosane		638-67-5	89[h] 101-585[a], 26[f], 314[h]	325[a], 57[h]	38-191[a]	249[a]	54[a], 16[b], 172[h]	29[a], 119[h]
undecane		1120-21-4						
x-decane (branched)								
ALKENES								
(1R,4aS,8aS)-7-methyl-4-methylidene-1-propan-2-yl-2,3,4a,5,6,8ahexahydro-1H-naphthalene	gamma-cadinene	39029-41-9						
(1S,4aS,8aR)-1-isopropyl-4,7-dimethyl-1,2,4a,5,6,8ahexahydro-naphthalene	alpha-muurolene	10208-80-7						
1,2,3-trimethoxy-5-methylbenzene	**3,4,5-trimethoxytoluene**	6443-69-2						
1,2,3-trimethylbenzene		526-73-8						
1,2-dimethoxybenzene	veratrole	91-16-7						
1,3-dimethylbenzene	xylene	1330-20-7						
1-cyclohexen-1-ylethanone	**1-acetylcyclohexene**	932-66-1	9[h]					
1-ethyl-4-methylbenzene	p-ethyl toluene	622-96-8						
1-methyl-4-(6-methylhept-5-en-2-yl)benzene	alpha-curcumene	644-30-4						

Costa Rican	Jamaican	V. fragrans origin Unknown	V. pompona origin unknown	V. pompona from Madagascar	Unidentified from Madagascar	Indian	Comoros	Ugandan	Indonesian	Hawaiian	PNG	Wild Type from Peru
23[a]												
				55[h]	161[h]	42-279[h]	211[h]	222[h]	338[h]	122[h]	40-470[h]	85[h]
		g										
36[a]	224[a]			80[h]	233[h]	442[h] 104-541[h]	352[h]	351[h]	611[h]	212[h]	140[h] 74-821[h]	174[h]
		g										
		g										
		g										
		g										
											107[h]	
		g										
		g										
		g										
					15[h]							
		g										
		g										

(Continued)

Table 17.1 (Continued)

IUPAC Name	Synonyms	CAS #	Bourbon	Tahitian	Balli	Java	Mexican	Tonga
1-methyl-4-(6-methylhepta-1,5-dien-2-yl) cyclohexene	beta-bisabolene	495-61-4						
1-methyl-4-prop-1-en-2-ylcyclohexene	limonene	5989-27-5						
1-methyl-4-propan-2-ylbenzene	cymene	99-87-6						
1-methyl-4-propan-2-ylcyclohexa-1,3-diene	alpha-terpinene	99-86-5						
1-Tricosene		18835-32-0	73[h]					25[h]
2,6,10,15,19,23-hexamethyltetracosa-2,6,10,14,18,22-hexaene	squalene	111-02-4	217[a], 514-592[h]	22[a], 363[h]	24-83[a]	487[a]	392[h]	148[h]
3-eicosene		74685-33-9			195[a]	453[a]		
3-methylidene-6-propan-2-ylcyclohexene	beta-phellandrene	555-10-2						
4,7,7-trimethylbicyclo[3.1.1] hept-3-ene	alpha-pinene	80-56-8						
5-eicosene		74685-30-6	16-42[a]			16[a]		49[a]
7,7-dimethyl-4-methylidenebicyclo[3.1.1]heptane	beta-pinene	127-91-3						
7-methyl-3-methylideneocta-1,6-diene	myrcene	123-35-3						
9-tricosene		27519-02-4	17-29[a]		23-83[a]	2345[a]	18[a]	38[a]
benzene	benzol	71-43-2						
docos-1-ene	docosene	1599-67-3	14-44[h]	23[h]			14[h]	16[h]
ethenylbenzene	styrene	100-42-5						
ethyl benzene	ethylbenzol, phenylethane	100-41-4						

Costa Rican	Jamaican	V. fragrans origin Unknown	V. pompona origin unknown	V. pompona from Madagascar	Unidentified from Madagascar	Indian	Comoros	Ugandan	Indonesian	Hawaiian	PNG	Wild Type from Peru
		g										
		g										
		g										
		g										
				20[h]	117[h]	39-115[h]	74[h]	71[h]	168[h]	55[h]	29-333[h]	62[h]
				466[h]	1325[h]	146-1841[h]	1057[h]	156[h]	1136[h]	255[h]	133-496[h]	1350[h]
		g										
		g										
30[a]	16[a]											
		g										
		g										
25[a]	86[a]					46[h]						
		g										
						7-83[h]	20[h]	4[h]			19[h]	48[h]
		g										
		g										

(Continued)

Table 17.1 (Continued)

IUPAC Name	Synonyms	CAS #	Bourbon	Tahitian	Balli	Java	Mexican	Tonga
heptacos-1-ene	heptacosene		12-42[h]	35[h]			36[h]	52[h]
hexacos-1-ene	hexacosene	18835-33-1	32[h]				21[h]	3[h]
hexadec-1-ene	**hexadecene**	629-73-2		92[a]				
methylbenzene	toluene	108-88-3						
napthalene	naphthalin, antimite	91-20-3						
non-4-ene		68526-55-6						11[a]
nonacos-1-ene			79-423[h]	211[h]			621[h]	194[h]
pentacos-1-ene			39[f], 24-180[h]				142[h]	69[h]
propylbenzene	1-phenylpropane, isocumene	103-65-1						
tricosene			38[f]					
x-dodecene								
x-eicosene								
x-tetradecene								
ESTERS								
(1,7,7-trimethyl-6-bicyclo[2.2.1]heptanyl) acetate	bornyl acetate	76-49-3	[d]					
(4-formyl-2-methoxyphenyl) acetate	acetovanillin	881-68-5	27[f]					
(4-methoxyphenyl) methyl (E)-3-phenylprop-2-enoate	anisyl transcinnamate			530[c]				
(4-methoxyphenyl) methyl (Z)-3-phenylprop-2-enoate	anisyl ciscinnamate			150[c]				
(4-methoxyphenyl) methyl acetate	anisyl acetate	104-21-2		233[a,i]				
(4-methoxyphenyl) methyl anisate	anisyl anisate			6640[c]				

Costa Rican	Jamaican	V. fragrans origin Unknown	V. pompona origin unknown	V. pompona from Madagascar	Unidentified from Madagascar	Indian	Comoros	Ugandan	Indonesian	Hawaiian	PNG	Wild Type from Peru
					735^h	$15\text{-}221^h$		73^h	1413^h	56^h	$70\text{-}1467^h$	
					49^h	$13\text{-}60^h$	42^h		121^h	13^h	$15\text{-}129^h$	19^h
		g										
		$g\ q$										
				177^h	1052^h	$103\text{-}811^h$	371^h	539^h	1436^h	354^h	$170\text{-}996^h$	262^h
					202^h	$83\text{-}278^h$	172^h	227^h	464^h	222^h	$48\text{-}327^h$	78^h
		g										
		g										
		g										
		g										
		q									45^h	27^h

(Continued)

Table 17.1 (Continued)

IUPAC Name	Synonyms	CAS #	Bourbon	Tahitian	Balli	Java	Mexican	Tonga
(4-methoxyphenyl) methyl formate	anisyl formate	122-91-8	*d*	25[a]			2[b]	
(4-methoxyphenyl) methyl hexadecanoate	**anisyl stearate**			472[h]				
(4-methoxyphenyl) methyl hexanoate	**p-anisyl hexanoate**	6624-60-8						
(4-methoxyphenyl) methyl hydroxybenzoate	anisyl 4-hydroxybenzoate			7400[c]				
(4-methoxyphenyl) methyl octadec-9-enoate	**anisyl oleate**			2720[h]				
(4-methoxyphenyl) methyl octadecanoate	anisyl petadecanoate							
(4-methoxyphenyl) methyl petadecanoate	**anisyl palmitate**			160[h]				
(4-methoxyphenyl) methyl tetradecanoate	anisyl tetradecanoate							
(4-propan-2-ylphenyl) acetate	**4-isopropylphenyl acetate**							
[(1R,4S,6R)-1,7,7-trimethyl-6-bicyclo[2.2.1]heptanyl] acetate	isobornyl acetate	125-12-2	*d*					
[(2S)-2,3-dihydroxypropyl] (9Z,12Z)-octadeca-9,12-dienoate	2,3-dihydroxypropyl linoleate	2277-28-3					62[a]	
[(E)-3-phenylprop-2-enyl] (E)-3-phenylprop-2-enoate	cinnamyl cinnamate	122-69-0						
[(E)-3-phenylprop-2-enyl] benzoate	cinnamyl benzoate	5320-75-2						
2-(2-hydroxybenzoyl) oxybenzoic acid	**salsalate**	552-94-3		186[h]				

Costa Rican	Jamaican	V. fragrans origin Unknown	V. pompona origin unknown	V. pompona from Madagascar	Unidentified from Madagascar	Indian	Comoros	Ugandan	Indonesian	Hawaiian	PNG	Wild Type from Peru
												50^h
			117^h								107-204^h	248^h
											9^h	
											53-178^h	937^h
												18^h
			46^h								37^h	512^h
												34^h
											34^h	
		q										
		q										
											20^h	

(Continued)

Table 17.1 (Continued)

IUPAC Name	Synonyms	CAS #	Bourbon	Tahitian	Balli	Java	Mexican	Tonga
2-(4-methylcyclohex-3-en-1-yl) propan-2-yl acetate	alpha-terpinyl acetate	80-26-2	[d]					
2,3-dihydroxypropyl acetate	**glycerolmono-acetate**	93713-40-7	75-2204[a], 287-444[h]	27[a]			100[a], 251[h]	266[h]
2-acetyloxyethyl acetate	ethylene diacetate	111-55-7						
2-ethylhexyl (E)-3-(4-methoxyphenyl) prop-2-enoate	**2-Ethylhexyl-4-methoxy-cinnamate**	5466-77-3						
2-ethyl-trans-bicyclohexyl methyl 14-methylpenta-decanoate			118[a]		7[a]	34[a]		
2-hydroxyethyl acetate	ethylene glycol acetate	65071-98-9						29[a]
2-methylpropyl pentanoate	isobutyl valerate	10588-10-0						
2-pentanal propanoate								7[a]
2-phenethyl formate		104-62-1	[d]					
2-phenylethylacetate		103-45-7						
3,7-dimethyloct-6-enyl 2-methylpropanoate	citronellyl isobutyrate	97-89-2	[d]					
3,7-dimethylocta-1,6-dien-3-yl acetate	linalyl acetate	115-95-7	[d]					
3-hydroxypropyl (Z)-octadec-9-enoate	3-hydroxypropyl oleate				54[a]			
3-hydroxypropyl prop-2-enoate	**2-hydroxypropyl acrylate**	25584-83-2						
3-methylbutyl 2-hydroxybenzoate	isoamyl salicylate	87-20-7	[d]					
4-hexen-1-ol acetate							33[a]	
7-methyl-4-octanol acetate			7[a]		28[a]			

Costa Rican	Jamaican	V. fragrans origin Unknown	V. pompona origin unknown	V. pompona from Madagascar	Unidentified from Madagascar	Indian	Comoros	Ugandan	Indonesian	Hawaiian	PNG	Wild Type from Peru
	251[a]			480[h]		138-711[h]	391[h]	408[h]	50[h]	149[h]	109-125[h]	
						8-71[h]		15[h]			15-73[h]	
						5[h]						
44[a]												
		g										
		g										
												218[h]
	52[a]											

(Continued)

Table 17.1 (Continued)

IUPAC Name	Synonyms	CAS #	Bourbon	Tahitian	Balli	Java	Mexican	Tonga
bis(2-ethylhexyl) benzene-1,2-dicarboxylate	**dioctyl phthalate**	117-81-7						
bis(2-ethylhexyl) hexanedioate	dioctyl adipate	70147-21-6	20-774[a]				9[a]	
bis(2-methylpropyl) benzene-1,2-dicarboxylate	diisobutyl phthalate	84-69-5	62[a]	1090[a]				
bis(6-methylheptyl) benzene-1,2-dicarboxylate	isooctylphthalate	27554-26-3	14-111[a], 18-88[h]	36[a], 41[h]	227[a]	28[a]	19[a], 41[h]	76[h]
butyl hexanoate	butyl caproate	626-82-4	[d]					
butyl pentanoate	butyl valerate	591-68-4						
dibutyl benzene-1,2-dicarboxylate	dibutyl phthalate	84-74-2						
diethyl benzene-1,2-dicarboxylate	diethyl phthalate	84-66-2						
diethyl benzoate								
dipropyl benzene-1,2-dicarboxylate	dipropyl phthalate	53-59-8						
ethenyl formate	vinyl formate	692-45-5			3095[a]			17[a]
ethyl 2-(4-hydroxy-3-methoxyphenyl) acetate	**ethyl homovanillate**	60563-13-5	4[h]					
ethyl 2-hydroxy-2-methylbutanoate	ethyl-2-methyl butyrate	77-70-3						
ethyl 2-hydroxybenzoate	ethyl salicylate	118-61-6	[d]					
ethyl 2-hydroxypropanoate	ethyl lactate	97-64-3						
ethyl 2-methoxyacetate	ethyl methoxyacetate	3938-96-3						
ethyl 4-hydroxy-3-methoxybenzoate	ethyl vanillate	617-05-0						15[h]

Costa Rican	Jamaican	V. fragrans origin Unknown	V. pompona origin unknown	V. pompona from Madagascar	Unidentified from Madagascar	Indian	Comoros	Ugandan	Indonesian	Hawaiian	PNG	Wild Type from Peru
						71[h]						
	11[a]			47[h]	69[h]	59-216[h]	77[h]	957[h]	48[h]	43[h]	51-704[h]	63[h]
		g										
		g										
		g					114[h]					177[h]
		g										
		g										
						249[h]		28[h]			31-36[h]	
		g										
		g										
		g				192[h]					32[h]	

(Continued)

Table 17.1 (Continued)

IUPAC Name	Synonyms	CAS #	Bourbon	Tahitian	Balli	Java	Mexican	Tonga
ethyl 4-oxopentanoate	ethyl levulinate	539-88-8						
ethyl acetate		141-78-6						
ethyl hexadecanoate	ethyl palmitate	628-97-7						
ethyl hexanoate	ethyl caproate	123-66-0						
ethyloctadeca-9,12,15-trienoate	ethyl linolenate	1191-41-9					14[b]	
hexyl 2-hydroxybenzoate	hexyl salicylate	6259-76-3	d					
hexyl acetate	methamyl acetate	142-92-7	178[h]	12[h]			21[h]	
hexyl acetate		142-92-7						
hexyl butanoate		2639-63-6	d					
methoxymethyl acetate		4382-76-7					218[a]	
methyl (9Z,12Z)-octadeca-9,12-dienoate	methyl linoleate	112-63-0	21[f]					
methyl 10,13-octadecadienoate			16-70[a]	27[a]	39[a]	179[a]	70[a]	
methyl 11-octadecenoate						136[a]		
methyl 2-hydroxyacetate	methyl glycolate	96-35-5						
methyl 2-hydroxybenzoate	methyl salicylate	119-36-8					<1[b]	
methyl 2-hydroxypropanoate	methyl lactate	547-64-8						
methyl 2-oxopropanoate	methyl pyruvate	600-22-6	11[a]	66[a]	134-369[a]		57[a]	
methyl 2-phenylacetate		101-41-7						

Costa Rican	Jamaican	V. fragrans origin Unknown	V. pompona origin unknown	V. pompona from Madagascar	Unidentified from Madagascar	Indian	Comoros	Ugandan	Indonesian	Hawaiian	PNG	Wild Type from Peru
		g										
				455[h]	74[h]		32[h]		50[h]			
		g		31[h]		8-72[h]						
		g										
						17-274[h]	440[h]	145[h]			515[h]	
		g										
10[a]	39[a]											
		g										
		g										
		g										
120[a]	252[a]										199[h]	
		g										

(Continued)

Table 17.1 (Continued)

IUPAC Name	Synonyms	CAS #	Bourbon	Tahitian	Balli	Java	Mexican	Tonga
methyl 3-methoxybenzoate	methyl m-anisate	5368-81-0		218[a]				
methyl 4-(2-hydroxyethoxy) benzoate								
methyl 4-(hydroxymethyl) benzoate		6908-41-4	36[f]					
methyl 4-hydroxy-3-methoxybenzoate	methyl vanillate	3943-74-6	15[h]					
methyl 4-hydroxybenzoate		99-76-3	13-77[a]		275[a]			
methyl 4-methoxybenzoate	methyl anisate	121-98-2		i				
methyl acetate	Tereton	79-20-9	24[a], 84[f],15[h]					
methyl acrylate	**methyl prop-2-enoate**	96-33-3	99-103[h]	110[h]			69[h]	13[h]
methyl benzoate		93-58-3						
methyl dodecanoate	methyl laurate	111-82-0						
methyl heptadecanoate	methyl margarate	1731-92-6						
methyl heptanoate		106-73-0						
methyl hexadecanoate	methyl palmitate	112-39-0	13[f], 4[h]					23[h]
methyl hexanoate	methyl caproate	106-70-7						
methyl icosanoate	methyl arachidate	1120-28-1					6[a]	
methyl nonanoate		1731-84-6						
methyl pentadecanoate		7132-64-1						
methyl pentanoate	methyl valerate	624-24-8						
methyl tetradecanoate	methyl myristate	124-10-7						

Costa Rican	Jamaican	V. fragrans origin Unknown	V. pompona origin unknown	V. pompona from Madagascar	Unidentified from Madagascar	Indian	Comoros	Ugandan	Indonesian	Hawaiian	PNG	Wild Type from Peru
											30[h]	
		g		17[h]	8[h]	5-120[h]	18[h]	136[h]		16[h]	4[h]	119[h]
10[a]	65[a]											
								21[h]			85[h]	
				131[h]	128[h]	58-464[h]		269[h]	171[h]	5[h]	41-346[h]	200[h]
		g										
		g										
		g										
		g										
		g, i			13[h]	5-30[h]	20[h]		25[h]		36[h]	8[h]
		g										
		g										
		g										
		g										

(Continued)

Table 17.1 (Continued)

IUPAC Name	Synonyms	CAS #	Bourbon	Tahitian	Balli	Java	Mexican	Tonga
methyl-2-(4-hydroyphenoxy)benzoate								
methyl-3-phenylprop-2-enoate	methyl-transcinnamate	1754-62-7	27-33[h]	32[a], 24[h,i]			1[b]	4[h]
methyl-4-hydroxybenzoate			19[h]					
methyl-8-methyldecanoate			21[a]				21[a]	6[a]
pentyl 2-hydroxybenzoate	pentyl salicylate	2050-08-0	d					
pentyl acetate	n-amyl acetate	628-63-7						
phenylmethyl (E)-3-phenylprop-2-enoate	benzyl cinnamate	103-41-3						
phenylmethyl acetate	benzyl acetate	140-11-4						
phenylmethyl benzoate	benzyl benzoate	120-51-4						
phenylmethyl butanoate	benzyl butyrate	103-37-7						
phenylmethyl formate	benzyl formate	104-57-4						
prop-2-enyl octadecanoate	allyl stearate	6289-31-2			77[a]			
propan-2-yl acetate	isopropyl acetate	108-21-4			53[a]			
propan-2-yl pentanoate	isopropyl valerate	18362-97-5						
propyl 4-hydroxybenzoate	propyl paraben	94-13-3	83[a], 24[h]		134[a]			
propyl pentanoate	propyl valerate	141-06-0						
ETHERS								
1,2-dimethoxyethane	ethylene glycol dimethyl ether	110-71-4	264[a]					
1-methoxyhexane	**methyl hexyl ether**	4747-07-3	7[h]					

Costa Rican	Jamaican	V. fragrans origin Unknown	V. pompona origin unknown	V. pompona from Madagascar	Unidentified from Madagascar	Indian	Comoros	Ugandan	Indonesian	Hawaiian	PNG	Wild Type from Peru
												52^h
	$g\ i\ q\ s$			18^h	15^h	$20\text{-}28^h$	51^h	14^h				52^h
						120^h		108^h		36^h	728^h	
	g											
	q											
	g											
	q											
	q											
	g											
				8^h								
	g											
				30^h								
	g											
				36^h								

(Continued)

Table 17.1 (Continued)

IUPAC Name	Synonyms	CAS #	Bourbon	Tahitian	Balli	Java	Mexican	Tonga
1-methoxypropane	**methyl propyl ether**	557-17-5	15[h]					
1-propoxypropane	propyl ether	111-43-3		74[h]				
2-ethoxypropane	isopropyl ethyl ether	625-54-7	40[a]		115[a]		30[h]	
2-methoxypropane	methyl isopropyl ether	598-53-8	275-388[a]					
3-ethenoxyprop-1-ene	vinyl allyl ether	3917-15-5	3[h]					10[a]
3-methoxypentane	1-Ethylpropyl methyl ether	36839-67-5			4[a]			
4-(ethoxymethyl)-2-methoxyphenol	vanillyl ethyl ether	13184-86-6						
4-(ethoxymethyl)phenol	p-hydroxybenzyl ethyl ether	57726-26-8						
4-(methoxy)-2-methoxyphenol	vanillyl methyl ether						<1[b]	
anisole	methoxybenzene	100-66-3						
phenoxybenzene	diphenyl ether	101-84-8						
	anisyl ethyl ether				15040[c]			
	anisyl methyl ether				800[c]			
	dianisyl ether				3130[c]			
	isopentyl methyl ether				76[h]		4[h]	
	methyl cyclobutyl ether							
	p-cresyl isopropyl ether							
	p-hydroxybenzyl-methyl ether			17[f]				
KETONES								
(E)-4-(2,6,6-trimethylcyclohex-2-en-1-yl)but-3-en-2-one	alpha-ionone	9066-80-2			410[c]			

Costa Rican	Jamaican	V. fragrans origin Unknown	V. pompona origin unknown	V. pompona from Madagascar	Unidentified from Madagascar	Indian	Comoros	Ugandan	Indonesian	Hawaiian	PNG	Wild Type from Peru
321[a]												
219[a]						20[h]					107[h]	
		e i										
		e g										
		e aa										
		g										
		g										
						133[h]						
		g										
		eg										

(Continued)

Table 17.1 (Continued)

IUPAC Name	Synonyms	CAS #	Bourbon	Tahitian	Balli	Java	Mexican	Tonga
(E)-4-(2,6,6-trimethyl-cyclohexen-1-yl)but-3-en-2-one	beta-ionone	79-77-6		1000^c				
(E)-oct-3-en-2-one		1669-44-9						
1-(2,4,6-trimethylphenyl)ethanone	2,4, 6-trimethyl-acetophenone		$42^a, 69^f$			47^a		52^a
1-(2,4-dihydroxyphenyl)ethanone	2,4-dihydroxy-acetophenone	89-84-9	$4-22^h$		$80-233^a$	21^a		
1-(2-hydroxy-5-methylphenyl)ethanone	o-acetyl-p-cresol	1450-72-2	$24^f, 221^h$	87^h				26^h
1-(4-hydroxy-3-methoxyphenyl)ethanone	acetovanillone	498-02-2	$35-82^a, 37^f,$ $12-23^h$			70^a	$309^a,$ $14^b, 6^h$	$8^a, 9^h$
1-(4-hydroxy-3-methoxyphenyl)propan-2-one	vanillyl methyl ketone	2503-46-0	2^h				167^a	32^h
13-methyl-oxacyclo-tetradecane-2-11-dione			27^a					
1-hydroxyheptan-2-one								
1-hydroxypentan-2-one								
1-hydroxypropan-2-one	hydroxy acetone, pyruvic alcohol	116-09-6	$63-216^h$		587^a	546^a	53^h	8^h
1-methoxypropan-2-one	methoxyacetone	5878-19-3	$1391^a, 5^h$	$23^a, 13^h$	$44-177^a$	6^a		15^a
1-phenylethanone	acetophenone	98-86-2						
2,3,3,4-tetramethyl-cyclobutan-1-one								
2,4-dimethylpentan-3-one	**diisopropyl ketone**	565-80-0						
2-butan-2-ylcyclopentan-1-one			34^h				12^h	
2-hydroxy-3-methylcyclopent-2-en-1-one		80-71-7						3^a

Costa Rican	Jamaican	V. fragrans origin Unknown	V. pompona origin unknown	V. pompona from Madagascar	Unidentified from Madagascar	Indian	Comoros	Ugandan	Indonesian	Hawaiian	PNG	Wild Type from Peru
		g										
						3[h]						6[h]
					177[h]	91-511[h]	68[h]	165[h]			129-953[h]	134[h]
95[a]	59[a]				40[h]	8-226[h]	37[h]	38[h]	20[h]	35[h]	25-52[h]	44[h]
					10[h]	31-690[h]	58[h]				60[h]	
		g										
		g										
				349[h]	29[h]	114-351[h]	436[h]	462[h]	114[h]		66-397[h]	73[h]
1[a]	146[a]											
		o										
						4[h]						
											57[h]	
								66[h]			113-124[h]	

(Continued)

Table 17.1 (Continued)

IUPAC Name	Synonyms	CAS #	Bourbon	Tahitian	Balli	Java	Mexican	Tonga
2-methylpentan-2-one			117-130^a	21^a	8-332^a	5^a	361^a	
3-hydroxy-1-(4-hydroxy-3-methoxyphenyl)propan-1-one	**b-hydroxypropiovanillone**							
3-hydroxybutan-2-one	2-acetoin	513-86-0					15^b	10^h
3-methyl-3-decen-5-one								49^a
3-methylbut-3-en-2-one		814-78-8						
3-methylbutan-2-one	isopropyl methyl ketone	563-80-4	21^a			67^a		
3-methylcyclohex-2-en-1-one		1193-18-6	34-154^a	77^a	102^a	40^a		49^a
3-methylcyclohexan-1-one		625-96-7	68^a					
3-methylcyclopentan-1-one		6672-30-6	159-500^a, 207^h, 25^f	10^a	106-132^a	117^a	181^a	225^a
3-methylpentan-2-one	**sec-butyl-methyl ketone**	565-61-7						
3-Oxabicyclo[3.2.0]heptane-2,4-dione		7687-27-6					62^h	
3-penten-2-one	methyl propenyl ketone	625-33-2						
4-(4-hydroxy-3-methoxyphenyl)butan-2-one	zingerone, vanillyl acetone	122-48-5	46^f					9^h
4-(4-hydroxyphenyl)-2-butanone	raspberry ketone	5471-51-2	31^f					
4,4-dimethyl-2-oxethanone					79^a			
4-acetyl-2-hydroxy-5-methylbenzene								
4-butoxy-3-methylbutan-2-one							149^a	
4-methylene-2-oxethanone							95^a	

Costa Rican	Jamaican	V. fragrans origin Unknown	V. pompona origin unknown	V. pompona from Madagascar	Unidentified from Madagascar	Indian	Comoros	Ugandan	Indonesian	Hawaiian	PNG	Wild Type from Peru
45[a]												
						18-280[h]					27[h]	
		g				147-215[h]	21[h]	73[h]			28[h]	23[h]
252[a]												
									65[h]			
90[a]	139[a]											
211[a]	34[a]					10[h]						
						9-19[h]					50[h]	27[h]
									194[h]	183[h]		
		g										
					19[h]	72[h]					58[h]	31[h]
246[a]												
						4[h]						

(Continued)

Table 17.1 (Continued)

IUPAC Name	Synonyms	CAS #	Bourbon	Tahitian	Balli	Java	Mexican	Tonga
5-hydroxy-2,3-dimethyl-2-cyclopentene-1-one								
5-hydroxyheptan-2-one								
6,10,14-trimethyl-pentadecan-2-one	hexahydrofarnesyl acetone	502-69-2	23^f				40^a	6^h
butan-2-one	methyl ethyl ketone	78-93-3			35^a			
butane-2,3-dione	diacetyl	431-03-8	6-22^h	25^h				19^a, 2^h
cycloheptanone	suberone	502-42-1	35^a				6^h	
cyclohexane-1,3-dione	dihydro-resorcinol	504-02-9	69^a					
cyclohexane-1,4-dione	tetrahydro-quinone	637-88-7					10^a	
cyclohexanone		108-94-1	24^h	413^h			194^h	
cyclopent-4-ene-1,3-dione		930-60-9	16-49^h	10^h				
decan-2-one	octyl methyl ketone	693-54-9						
di(phenyl)methanone	**diphenyl ketone**	119-61-9						
hentriacontene-2,4-dione			966^h	481^h			621^h	
heptacosane-2,4-dione			3696-5110^h					
heptacosene-2,4-dione			6272^h	15897^h			55475^h	11073^h
heptan-2-one	amyl methyl ketone	29308-56-3						
heptan-4-one	dipropyl ketone	123-19-3						
heptane-2,4,6-trione	diacetyl acetone	626-53-9	57^f					
hex-5-en-2-ol		626-94-8	58^a					
hexan-2-one	propyl acetone	591-78-6						
hexane-2,3-dione		3848-24-6						

Costa Rican	Jamaican	*V. fragrans origin Unknown*	*V. pompona origin unknown*	*V. pompona from Madagascar*	*Unidentified from Madagascar*	Indian	Comoros	Ugandan	Indonesian	Hawaiian	PNG	Wild Type from Peru
						98[h]						
	g											
	g					9-28[h]			56[h]		21-72[h]	
	oaa				7[h]	2-31[h]		150[h]	23[h]		7-25[h]	5[h]
								65[h]		2[h]	55[h]	
								58[h]				
				147[h]	103[h]	36-201[h]	97[h]		125[h]	8[h]	61-458[h]	195[h]
				453[h]	51[h]	22-50[h]	157[h]	137[h]	272[h]	5[h]	21-318[h]	
	g											
										78[h]		
					511[h]	593-1574[h]	608[h]	1119[h]	1040[h]	873[h]	296-970[h]	154[h]
				4053[h]	17124[h]	8479-19721[h]	13970[h]	21844[h]		22878[h]	5387-14038[h]	6186[h]
	g											
110[a]												
	g											
									438[h]			249[h]

(Continued)

Table 17.1 (Continued)

IUPAC Name	Synonyms	CAS #	Bourbon	Tahitian	Balli	Java	Mexican	Tonga
hexane-2,4-dione		3002-24-2	162-236[a]		9-137[a]	56[a]		
nonacosene-2,4-dione			145[h]	6154[h]			15434[h]	2572[h]
nonan-2-one	heptyl methyl ketone	821-55-6						
octa-4,6-dien-3-one								
octan-2-one	hexyl methyl ketone	27457-18-7						
pentacosane-2,4-dione			107[h]				197[h]	36[h]
pentacosene-2,4-dione								
pentan-2-one	**ethyl acetone**	27157-48-8	26[h]					
pentane-2,3-dione	acetyl propionyl	600-14-6						
HETEROCYCLIC								
1-(1H-pyrrol-2-yl)ethanone	2-acetyl pyrrole	1072-83-9						
1,3,7-trimethylpurine-2,6-dione	caffeine	95789-13-2		12[c]				
1,3-benzodioxole-5-carbaldehyde	heliotropine, piperonal	120-57-0		k w z aa ac				
1,4-dimethylpiperazine	**Lupetazine**	106-58-1						
1-furan-2-ylethanone	acetyl furan, 2-furl methyl ketone	80145-44-4	42-152[a], 4-45[h]	138[a], 22[h]	75[a]			33[a]
1-furan-2-ylpropan-1-one	2-propionylfuran	3194-15-8						20[a]
1H-pyrrole-2,5-dione, ethyl-4-methyl							2[b]	
2-(hydroxymethyl)-5-hydroxy-4H-pyran-4-one					80-156[a]		169[a]	8[a]
2,2,4,5-tetramethyl-1,3-dioxolane								
2,20-bi-1,3-dioxolane					263[a]			

Costa Rican	Jamaican	V. fragrans origin Unknown	V. pompona origin unknown	V. pompona from Madagascar	Unidentified from Madagascar	Indian	Comoros	Ugandan	Indonesian	Hawaiian	PNG	Wild Type from Peru
	149[a]					22[h]		79[h]		2[h]		
				846[h]	4801[h]	4663-24524[h]	5822[h]	5869[h]	5257[h]	6002[h]	426-2086[h]	2366[h]
		g										
		g										
		g										
						75-2503[h]		110[h]	15902[h]	18[h]	31[h]	
								2107[h]				
						39-106[h]			89[h]		8-257[h]	272[h]
					19[h]	4-26[h]	33[h]	30[h]	18[h]		20-34[h]	584[h]
		g										
		g	wz									
												1411[h]
		g		62[h]	16[h]	39-81[h]	25[h]				5[h]	
202[a]												
159[a]												
										45[h]		

(Continued)

Table 17.1 (Continued)

IUPAC Name	Synonyms	CAS #	Bourbon	Tahitian	Balli	Java	Mexican	Tonga
2,3-dihydro-1-benzofuran	coumaran	496-16-2	69-125[a], 114[f], 21-128[h]	95[a], 21[h]	63-286[a]		366[a]	113[a], 7[h]
2,3-dihydro-2,5-dimethylfuran			57[a]	110[a]	21[a]		256[a]	
2,5-dimethyl furfural								
2,6,6-trimethyl-10-methylidene-1-oxaspiro[4.5]dec-8-ene	vitispirane	65416-59-3		280[c]				
2,6-dimethyl-3(2H)-benzofuranone				12[a]			17[a]	
2,6-dimethyl-4-pyranone			10-72[a]	216[a]				
2-butyltetra-hydrofuran			40[a]		40-78[a]	254[a]		
2-ethyl-1,3-dioxolane							3[a]	
2-furancarboxylic acid methyl ester			14[f]					
2-hydroxy-5-methyl furan								
2-pentylfuran	2-amyl furan	64079-01-2					39[a]	
2-propylfuran		4229-91-8	216-263[a]					
3,4-dimethylfuran-2,5-dione	dimethylmaleic anhydride	766-39-2	150-737[h]	1604[h]			493[h]	75[h]
3,5-dihydroxy-6-methyl-2,3-dihydro-4H-pyran-4-one			14-3268[a]	3601[a]	84-4076[a]	1780[a]	3886[a]	1416[a]
3,5-dimethyl-2,4-(3H,5H)-furandione			202-419[a]	14[a]	42-375[a]	200[a]		
3,7-dimethylpurine-2,6-dione	theobromine	83-67-0		40[c]				
3H-pyran-2,6-dione	**glutaconic anhydride**		18[h]	10[h]				3[h]
3-hydroxy-2-methylpyran-4-one	maltol	118-71-8	84-176[a], 42-66[h]	54[h]	23[a]	415[a]	10[h]	12[a], 200[h]
3-hydroxy-4,4-dimethyl-dihydro-2(3H)-furanone						66[a]		

Costa Rican	Jamaican	V. fragrans origin Unknown	V. pompona origin unknown	V. pompona from Madagascar	Unidentified from Madagascar	Indian	Comoros	Ugandan	Indonesian	Hawaiian	PNG	Wild Type from Peru
195[a]	52[a]			89[h]	26[h]	19-111[h]	106[h]	104[h]	138[h]		96-131[h]	
195[a]				312[h]			14[h]					
		o										
		ag										
18[a]												
							11[h]			99[h]		
		g										
	50[a]	g				3[h]						
520[a]	368[a]											
				1142[h]	533[h]	499-975[h]	569[h]	707[h]	1331[h]	565[h]	685-1216[h]	
2350[a]	2476[a]											
	171[a]											
				53[h]	12[h]	9-32[h]	73[h]			10[h]	24-100[h]	236[h]
78[a]	119[a]			42[h]	50[h]	56-186[h]	74[h]	77[h]	27[h]	11[h]	17-61[h]	76[h]

(Continued)

Table 17.1 (Continued)

IUPAC Name	Synonyms	CAS #	Bourbon	Tahitian	Balli	Java	Mexican	Tonga
3-hydroxy-4,4-dimethyloxolan-2-one	pantolactone	79-50-5	51[f]				1[b]	
4-(4,5-dimethyl-1,3-dioxolan-2-yl)-2-methoxyphenol	vanillin-2,3-butyleneglycol acetal	63253-24-7	10[f]					
4,4,7a-trimethyl-6,7-dihydro-5H-1-benzofuran-2-one	dihydro-actinidiolide	19432-05-4		230[c]				
4-hydroxy-2,5-dimethylfuran-3-one	furaneol, strawberry furanone	3658-77-3	36-99[h]	86[h]			11[h]	10[h]
4-methyl-2H-furan-5-one		22122-36-7	20[h]	15[h]				
5-(hydroxymethyl)furan-2-carbaldehyde	hydroxymethyl-furfural	76330-16-0	460[f], 1511-6707[h]	6878[h]			596[h]	3287[h]
5-(hydroxymethyl)-tetrahydro-2-furanol								
5,6-dihydro-2H-pyran-2-carboxaldehyde					102[a]			
5-ethylfuran-2-carbaldehyde	**5-ethylfurfural**	23074-10-4						
5-hydroxy-2-(hydroxymethyl)pyran-4-one	**kojic acid**	67-99-2	5-18[h]	12[h]			6[h]	63[h]
5-isopropyldihydro-3(2H) furandione						20[a]		
5-methyl-3H-furan-2-one		591-12-8	22-272[a]	143[a]	100-360[a]	66[a]	179[a]	114[a]
5-methylfuran-2-carbaldehyde	5-methyl-2-furfural	620-02-0	534-646[a], 24[f], 21-105[h]	217[a], 26[h]	430-606[a]	421[a]	36[a], 3[h]	19[a], 93[h]
5-methyl-tetrahydro-2-furan-methanol			340[a]		79-748[a]			
5-pentyloxolan-2-one	coconut aldehyde	104-61-0						
6-undecyloxan-2-one	hexadecalactone	7370-44-7					45[a]	
dihydro-5-isopropyl-3(2H)-furanone			23[h]					
dihydroxydi-hydromaltol			335[f]					

Costa Rican	Jamaican	V. fragrans origin Unknown	V. pompona origin unknown	V. pompona from Madagascar	Unidentified from Madagascar	Indian	Comoros	Ugandan	Indonesian	Hawaiian	PNG	Wild Type from Peru
		g										
		g										
				259[h]	133[h]	20-44[h]	218[h]	59[h]	217[h]	9[h]	28-330[h]	15[h]
								49[h]				
				10448[h]	14506[h]	737-6361[h]	12826[h]	4170[h]	6220[h]	6174[h]	893-6980[h]	404[h]
						10[h]						
						25[h]						
				139[h]	105[h]	56-211[h]	194[h]	211[h]			74-92[h]	
						17-49[h]	30[h]	95[h]		14[h]	17-82[h]	
230[a]	80[a]											
406[a]	134[a]	g		229[h]	147[h]	27-485[h]	56[h]	68[h]	319[h]	12[h]	49-106[h]	14[h]
		g										

(Continued)

Table 17.1 (Continued)

IUPAC Name	Synonyms	CAS #	Bourbon	Tahitian	Balli	Java	Mexican	Tonga
furan-2,5-dione	maleic anhydride	24937-72-2	4[a]					
furan-2-carbaldehyde	furfural	98-01-1	560-672[a], 54[f], 131-645[h]	838[a], 598[h]	449-1176[a]	560[a]	1039[a], <1[b], 63[h]	74[a], 147[h]
furan-2-ylmethanol	furfuryl alcohol, 2-furancarbinol	98-00-0	214-625[a], 15[f], 225-354[h]	362[a], 41[h]	230-347[a]	268[a]	468[a], 45[h]	73[a], 174[h]
furan-2-ylmethyl acetate	furfuryl acetate	623-17-6						
furan-2-ylmethyl benzoate	furfuryl benzoate	34171-46-5						
furan-3-ylmethanol	**3-furan-methanol**	4412-91-3						
furfuryl hydroxy methyl ketone								
hydroxydi-hydromaltol			2912-4805[h]	4556[h]			2131[h]	3836[h]
hydroxymaltol			12[f]					
methyl 5-oxotetrahydro-2-furancarboxylate			31[h]				25[h]	
methyl furan-2-carboxylate	2-furan carboxylic acid methyl ester	611-13-2	12-69[a]	19[a], 104[h]			369[a], 9[h]	
methyl pyridine-3-carboxylate	methyl nicotinate	93-60-7						
oxolan-2-one	gamma-butyrolactone	96-48-0	3-20[h]				<1[b], 6[h]	
oxolan-2-ylmethyl acetate	tetrahydro-2-furfuryl acetate	637-64-9				49[a]		
pyran-4-one		29943-42-8	130[a]	219[a], 8[h]				
tetrahydro-4,4,6,6,-tetramethyl-2Hpyran-2-one						60[a]	52[h]	
tetrahydroxy-methylfurfuryl alcohol			17[f]					
thiophene	thiofuran	110-02-1						
trans-2,3-dimethyl-tetrahydro-2-furanol			4[a]			628[a]		

Costa Rican	Jamaican	V. fragrans origin Unknown	V. pompona origin unknown	V. pompona from Madagascar	Unidentified from Madagascar	Indian	Comoros	Ugandan	Indonesian	Hawaiian	PNG	Wild Type from Peru
					18^h							
882^a	542^a	gio		1632^h	1122^h	128-4588^h	1600^h	790^h		599^h	157-860^h	785^h
664^a	406^a	g		469^h	165^h	92-7845^h	339^h	340^h	228^h	8^h	214-590^h	92^h
		q										
		q										
						1-12^h						
		g										
				5569^h	2320^h	1411-3487^h	3466^h	2308^h	3489^h	1738^h	2383-3838^h	786^h
							173^h					
				10^h		51^h	25^h	78^h		67^h		
		g										
		g		210^h	20^h	51-100^h	44^h	146^h	60^h		50-147^h	66^h
		g										

(Continued)

a) Adedeji *et al.* 1993
b) Perez-Silva *et al.* 2006
c) DaCosta and Pantini 2006
d) Werkhoff and Guntert 1996
e) Galetto and Hoffman 1978
f) Hartman *et al.* 1992
g) Klimes and Lamparsky 1976
h) Lee 2006
i) Shiota and Itoga 1975
j) Prat and Subitte 1969
k) Lhugenot *et al.* 1971
l) Anwar 1963
m) Bohnsack 1965
n) Bohnsack and Seibert 1965
o) Bohnsack 1967
p) Bohnsack 1971a
q) Bohnsack 1971b

r) Chovin, *et al.* 1954
s) Gnadinger 1925
t) Stoll and Prat 1960
u) Simony 1953
v) Bonnet 1968
w) Pritzer and Jungkunz 1928
x) Walbaum 1909
y) Tiermann and Haarmann 1876
z) Busse 1900
aa) Kleinert 1963
ab) Morison-Smith 1964
ac) Cowley 1973
ad) Goris 1924
ae) Goris 1947
af) Chevalier *et al.* 1972
ag) Schulte-Elte *et al.* 1978

approximately 10 unidentified anisyl alcohol esters were found. We believe that this data further supports the hypothesis that *V. tahitensis* appears to be a hybrid of *V. planifolia* and *V. odorata*, based on the nuclear and plastid DNA sequence analysis recently reported by Lubinsky *et al.* (2008). It is well known that *V. tahitensis* beans have a floral/perfume nuance imparted by high concentrations of anisaldehyde, anisyl alcohol, anisic acid, and anisyl esters that are found at much lower concentration or not at all in *V. planifolia*. In contrast, the putative *V. odorata* beans are very high in anisyl alcohol derivatives and it stands to reason that interspecific hybridization of *V. planifolia* and *V. odorata* may be responsible for the genetic basis of anisyl traits in *V. tahitensis*.

In addition to the compounds listed in Table 17.1, which are assumed to be naturally occurring in vanilla, analyses often reveal synthetic compounds as common contaminants and/or adulterants. Many of these compounds are packaging-borne migrants. For instance, vanilla beans are often bundled and shrink-wrapped in polyvinyl chloride (PVC) films. Synthetic plasticizers, stabilizers, and antioxidants present in these films migrate into the beans and are commonly detected. Plasticizers such as diethylphthalate (DEP), dibutylphthalate (DBP), di-2-ethylhexylphthalate (DEHP), butyl, benzyl phthalate (BBP), diisononylphthalate (DINP), diisodecylphthalate (DIDP), dibutyladipate (DBA), dioctyladipate (DOA), tributylphosphate (TBP), tributyl, acetyl citrate (Citroflex A), and many others are commonly found. Synthetic hindered phenol-type antioxidants such as butylated hydroxyl toluene (BHT) are ubiquitous in vanilla beans that we have analyzed.

The high cost of vanilla has historically led to adulteration by unscrupulous merchants eager to gain an economic advantage. The practice is widespread enough to have generated elaborate analysis methods to detect adulteration (i.e. isotope ratio mass spectrometry, stable isotope fractionation, nuclear magnetic resonance spectroscopy, etc.) and has been addressed by special committees of the American Chemical Society (ACS), industry groups such as the US Flavor and Extracts Manufacturers Association (FEMA), and many others worldwide. Most of these methods are focused on detection of adulteration based on "boosting" vanillin concentration in vanilla beans or extract by adding synthetic vanillin. However, our group has also observed a less obvious method of adulteration that involves spraying vanilla beans with compounds to increase their weight. Compounds such as glycerol, 1,3-butylene glycol, and others are naturally occurring in vanilla beans to some extent but can be sprayed on beans to increase their weight and thus the price paid per unit weight basis. Beans with unusually high concentrations of these type glycols/polyols should be considered suspect.

We trust that the list of volatile compounds offered in this chapter will be useful to investigators studying quality attributes of vanilla, for fingerprinting vanilla bean chemistry from different species or geographic regions of production, as an aid for detecting contaminants/adulterants and a guide for flavorists seeking vanilla fortifiers or synergists for the production of synthetic renditions of vanilla flavor.

References

Adedeji, J. (1993) Flavor characterization of different varieties of vanilla beans by direct thermal desorptiongas chromatography and gas chromatography-mass spectrometry. PhD Thesis, Rutgers University.

Adedeji, J., Hartman, T.G. and Ho, C-T. (1993) Flavor characterization of different varieties of vanilla beans. *Perfumer & Flavorist*, 18, 25–33.

Anwar, M.H. (1963) Paper chromatography of monohydroxyphenols in vanilla extract. *Analyical Chemistry*, 35, 1974–1976.

Bohnsack, H. (1965) Uber die Inhaltstoffe der Bourbon-Vanilleschote (1 Teil) und Vanillin – butylenglycolid. *Riechstoffe, Aromen, Koeperflegemittel*, 15, 284–287.

Bohnsack, H. (1967) Uber die Inhaltstoffe der Bourbon-Vanilleschote (4 Teil). *Riechstoffe, Aromen, Koeperflegemittel*, 17, 133–136.

Bohnsack, H. (1971a) Uber die Inhaltstoffe der Bourbon-Vanilleschote (5 Teil). *Riechstoffe, Aromen, Koeperflegemittel*, 21, 125–128.

Bohnsack, H. (1971b) Uber die Inhaltstoffe der Bourbon-Vanilleschote (6 Teil): Neutraloel. *Riechstoffe, Aromen, Koeperflegemittel*, 21, 163–166.

Bohnsack, H. and Seibert, W. (1965) Uber Inhaltstoffe der Bourbon-Vanilleschote (2 Teil). *p*-Hydroxybenzyl-alkohol und *p*-hydroxybenzylmethyl ether. *Riechstoffe, Aromen, Koeperflegemittel*, 15, 321–324.

Bonnet, G. (1968) Separation chromatographique en phase gazeuse et sur couche mince de quelques composes a saveur vanillee. *Annales des Falsifications et de l'Expertise Chimique*, 61, 360–371.

Busse, W. (1900) Ueber die Bildung des Vanillins und der Vanillefrucht. *Zeits, Untersuch, Nahr, Genussmtl.*, 3, 21–25.

Chevalier, M., Pray, Y. and Navellier, P. (1972) Caracterisation par chromatographie en couche mince des constituents aromatiques de la vanille dans les glaces et crème glaces. *Annales des Falsifications et de l'Expertise Chimique*, 65, 12–16.

Childers, N.F., Cibes, H.R. and Hernandez-Medina, E. (1959) Vanilla-the orchid of commerce. In: *The Orchids. A Scientific Survey*, Withner, C.L. (Ed.), Robert E. Krieger Publishing Company, Malabar, FL, pp. 477–508.

Chovin, P., Stoll, S. and Bouteville, Y. (1954) Un compasant naturel de la Vanille: L'aldehyde p-hydroxybenzoique. Extraction, caracterisation. *Annales des Falsifications et des Fraudes.*, 47, 187–191.

Cowley, E. (1973) Vanilla and its uses. Proceedings of the Conference on Spices, London, April 1972, Tropical Products Institute, London, pp. 79–84.

DaCosta, N. and Pantini, M. (2006) The analysis of volatiles in Tahitian vanilla (*tahitensis*). *Developments in Food Science*, 43, 161–164.

Galetto, W. and Hoffman, P.G. (1978) Some benzyl ethers present in the extract of vanilla. *Journal of Agricultural and Food Chemistry*, 26, 195–197.

Gnadinger, C.B. (1925) Identification of sources of vanilla extracts. *Journal of Industrial and Engineering Chemistry*, 17, 303–304.

Goris, M.A. (1924) Sur la composition chimique des fruits verts de vanille et la mode formation du parfum de la vanille. *Compt. Rend.*, 179, 70–72.

Goris, M.A. (1947) Formation du parfum de la vanille. *Ind. Parfumerie*, 2, 4–11.

Hannum, T. (1997) Changes in vanillin and activity of β-glucosidase and oxidases during post harvest processing of vanilla beans *Vanilla planifolia*. *Bulletin Teknologia dan Industri Pangan*, 8, 46–52.

Hartman, T.G., Karmas, K., Chen, J., Shevade, A., Deagro, M. and Hwang, H-I. (1992) Determination of vanillin, other phenolic compounds, and flavors in vanilla beans by direct thermal desorption-gas chromatography and gas chromatography-mass spectrometry analysis. In: *Phenolic Compounds in Food and Health, ACS Symposium Series 506*, Ho, C.-T., Lee, C.Y. and Huang, M.-T. (Eds), American Chemical Society, Washington, DC, pp. 60–76.

Havkin-Frenkel, D., French, J.C., Graft, N.M., Pak, F.E., Frenkel, C. and Joel, D.M. (2004) Interrelation of curing and botany in vanilla (*Vanilla planifolia*) bean. *Acta Horticulturae*, 629, 93–102.

Havkin-Frenkel, D., French, J., Pak, F. and Frenkel, C. (2005) Inside vanilla. *Perfumer&Flavorist*, 30, 2–17.

Kleinert, J. (1963) Vanille und Vanillin. *Gordian*, 53, 809–814, 840–854, 892–898.

Klimes, I. and Lamparsky, D. (1976) Vanilla volatiles – A comprehensive analysis. *International Flavours and Food Additives*, November/December 272–291.

Korthou, H. and Verpoorte, R. (2007) Vanilla. In: *Flavours and Fragrances*. Berger, R.G. (Ed.), Springer, Berlin, pp. 203–217.

Lee, K.J. (2006) Flavor characterization of bean produced in different geographical regions. MS Thesis, Rutgers University.

Lhugenot, C. Maume, B.F. and Baron, C. (1971) Gas chromatographic and mass spectrometric study of vanilla related compounds. *Chromatographia*, 4, 204–208.

Lubinsky, P., Cameron, K.M., Molina, M.C. *et al.* (2008) Neotropical roots of a Polynesian spice: the hybrid origin of Tahitian vanilla, *Vanilla tahitensis* (Orchidaceae). *American Journal of Botany*, 95, 1040–1047.

Morison-Smith, D. (1964) Determination of compounds related to vanillin in vanilla extracts. *Journal of Association of Official Agricultural Chemists*, 47, 808–815.

Perez-Silva, A., Odoux, E., Bart, P. and Ribeyre, F. (2006) GC-MS and GC-olfactometry analysis of aroma compounds in a representative organic aroma extract from cured Vanilla (*Vanilla planifolia*) beans. *Food Chemistry*, 99, 728–735.

Prat, Y. and Subitt, J. (1969) Contribution a l'etude des constituants aromatiques de la vanille. *Annales des Falsifications et de l'Expertise Chimique*, 62, 225–230.

Pritzer, J. and Jungkunz, R. (1928) Analytisches uber Vanillin und Vanillinzucker. *Chemiker-Zeitung*, 52, 537.

Schulte-Elte, K.H., Gautschi, F., Renold, W. *et al.* (1978) Vitispiranes, important constituents of vanilla aroma. *Helvetica Chimica Acta*, 61, 1125–1133.

Shiota, H. and Itoga, K. (1975) The study of the aromatic components of vanilla beans (*Vanilla planifolia* Andrews and *V. tahitensis* Moore). *Koryo (Japan)*, 113, 65–71.

Simony, R. (1953) Etude Morphologique, Histologique et Chimique des Vanilles. PhD Thesis, Univ of Strasbourg.

Stoll, S. and Prat, Y. (1960) Etude de la degradation de la vanilline, de l'ethylvaniline et de l'aldehyde p-hydroxybenzoic, par autoxidation. *Annales des Falsifications et de l'Expertise Chimique*, 53, 316–317.

Tiermann, F. and Haarmann, W. (1876) Ueber Bestandtheile der naturlichen Vanille. *Chem. Ber.*, 9, 1287–1292.

Walbaum, H. (1909) Das Vorkommen von Anis-alkohol und Anisaldehyd en den Fruchten der Tahitivanille. *Wallachs Festschrift*, 1, 649–653.

Walton, N.J., Mayer, M.J. and Narbad, A. (2003) Vanillin. *Phytochemistry*, 63, 505–515.

Werkhoff, P., and Guntert, M. (1996) Identification of some ester compounds in Bourbon Vanilla beans. *Lebensm-Wiss u-Technol*, 30, 429–431.

18

A Comprehensive Study of Composition and Evaluation of Vanilla Extracts in US Retail Stores

Daphna Havkin-Frenkel, Faith C. Belanger, Debra Y.J. Booth, Kathryn E. Galasso, Francis P. Tangel, and Carlos Javier Hernández Gayosso

18.1   History

Vanilla is the most popular and most widely used flavor. Vanilla was introduced to Europe by the Spanish conquistadors who brought vanilla beans from Mexico in the sixteenth century and vanilla flavor has been gaining in popularity ever since. Vanilla is the most popular and loved flavor, apparently because it evokes emotions of comfort and familiarity (Havkin-Frenkel 2009). Vanilla beans contain over 300 individual chemicals (see Chapter 11 by Toth *et al.*), which create a unique pleasant experience in a wide variety of applications.

In the United States, vanilla has been sold in retail stores for over 100 years. Vanilla is part of our everyday life. It can be found in ice cream, candies, baking products, drinks, and food flavorings. The sale of vanilla in retail stores declined in the 1980s and 1990s, because women joining the work force preferred ready-made baked goods. In the last few years there has been a shift in trends towards gourmet, natural, organic, and health consciousness and vanilla has found its original place restored. With additional explosions of new dishes and new health and culinology trends (Havkin-Frenkel 2009) we can now find vanilla in every type of store, such as supermarkets, drug stores, art stores, department stores, convenience stores, dollar stores, fresh produce and flower stores, restaurants and even clothing stores. There are hundreds of sites on the internet selling vanilla products.

18.2   Uses of Vanilla in the Industry

The desirable flavor attributes of vanilla make it one of the most common ingredients used in the global marketplace, whether as a primary flavor, as a component of another flavor, or for its desirable aroma qualities.

In the industry, vanilla can be used in many forms. However, in retail stores or household uses, the use of vanilla is divided between vanilla beans and vanilla extract. In the United States, most households will use vanilla extract. In the industry, vanilla products

Handbook of Vanilla Science and Technology, Second Edition.
Edited by Daphna Havkin-Frenkel and Faith C. Belanger.

are defined by the FDA standard of identity (discussed further in Section 18.2.3). However, in the retail market, many vanilla extracts are poorly defined (as discussed in Section 18.4). The industrial uses of vanilla can be summarized as follows:

18.2.1   Household Products

Vanilla, or synthetic vanillin, can be used as secondary aroma modifiers in everyday household products such as cleansers, laundry detergent, and dish washing liquid to help impart a pleasant aroma to the product. Vanilla can also be used in adhesives and glue items, such as in book bindings, to mitigate the undesirable aroma inherent in these products.

18.2.2   Dairy Products

Perhaps the most widely used market segment for vanilla is in dairy applications, such as ice cream, yogurt, flavored milk, coffee creamers, and the like.

18.2.3   Ice Cream (Frozen Dairy Products)

There are three categories of vanilla ice cream, as defined by the FDA Standard of Identity. Vanilla ice cream Category I contains only vanilla extract. Vanilla ice cream Category II contains vanilla made up of 1 oz of synthetic vanillin per 1 gallon of 1-fold vanilla extract. Vanilla ice cream Category III contains synthetic ingredients.

In ice cream, vanilla is used in a variety of ways. In some countries, regulatory conditions dictate whether natural, reinforced, or artificial vanilla is used. Single strength bourbon vanilla is used, generally at a use level of 4 to 6 oz per 5 gallons of mix. Higher strength vanilla, usually 2- to 4-fold and vanilla-vanillin (as defined by the FDA Standard of Identity), can be used to provide a scale of economy while providing a pleasant vanilla flavor. Products containing an artificial vanilla must be labeled as such.

Generally, bourbon or mixtures of bourbon and Indonesian beans are used in ice cream products. The vanilla used in ice cream must have a clean flavor and be low in smoky notes. Vanilla pieces (ground vanilla beans) can be used in ice cream products to impart a visual effect.

Generally, the lower the fat content of the frozen dairy product, the higher the level of vanilla needed. The same is true for sugar content; the lower the sugar, the more vanilla needed to impart a desirable flavor. In some instances, the spent vanilla beans, that is, the beans that have been previously extracted, can be added to ice cream products to enhance the visual appeal of the product. Since these spent beans have little flavor value, vanilla must also be used to impart flavor.

18.2.4   Yogurt

Yogurt, being acidic, requires a more intense vanilla profile than ice cream. For this reason, Indonesian vanilla, sometimes blended with bourbon vanilla, is used. The smoky, woody notes of the Indonesian beans help balance and mask the acidic taste of the yogurt, while allowing the delicate vanilla notes to come through.

18.2.5 Puddings

A wide array of vanillas can be used in pudding products, depending on the pudding composition, and how the pudding is processed. Generally for heat processed puddings, blends of Indonesian and bourbon vanilla are used. Higher than one-fold vanilla extracts can be used or powdered vanilla flavors made from highly concentrated vanilla extracts (oleoresins) can also be used. Some flans and delicate pudding products can use bourbon vanilla extracts.

18.2.6 Chocolate

Vanilla, or synthetic vanillin, is an essential ingredient in chocolate and chocolate products. The vanillin in vanilla smoothes out the sharp, dry flavor characteristics of the cocoa in chocolate. Oil soluble vanilla, vanilla oleoresin, and vanillin are the vanilla products commonly used in chocolate bars or chocolate coatings. Usage levels can vary between 0.01 and 0.1%, depending upon the level of cocoa and sweetness in the product. Any product containing chocolate or cocoa such as baked goods, syrups, chocolate milk, chocolate drink mixes, and chocolate puddings generally contain vanilla or vanillin to balance out the flavor.

18.2.7 Confections

In addition to vanilla's widespread use in chocolate, vanilla is used in non-chocolate confections as well. Vanilla is used as a primary flavor or flavor modifier in many confections such as hard candy, toffee (taffy), fruit chews, creams, gummies, jelly beans, tablets, candy coated pieces, and breath mints. In coated candy, vanillin is added to impart a pleasant flavor to the outside coating when placed in the mouth.

18.2.8 Baked Goods

Vanilla is widely used in the baking industry, either as a primary flavor or as a secondary flavor, in chocolate cupcakes, for example. The composition of the product and the baking temperature play a vital role in the type and level of vanilla needed. Generally, the lower the fat content of the product, the more vanilla required, the higher the protein, the more vanilla needed, and the higher the baking temperature, the more vanilla needed. In baked goods, spray dried flavors or encapsulated vanilla flavors are preferred over liquid due to the steaming off phenomenon, which occurs during the baking process. If liquid vanilla must be used, it is preferable to pre-blend the vanilla with the fat, if possible, and to incorporate the vanilla as late in the mixing phase as possible to minimize evaporative flashing off from the vanilla. In addition, lower baking temperatures help retain the vanilla in the product.

18.2.9 Beverages

In beverage applications, vanilla is used as a primary flavor or as a flavor toner. Vanilla smoothes out and balances the flavor profile in carbonated beverages. Vanilla soda (carbonated beverages) exists in the marketplace where a good bourbon vanilla or a blend of bourbon and Indonesian vanilla is used as the primary flavor. Root beer soda and cream soda use vanilla as a major flavor modifier. Just about all sodas, including colas, contain vanilla or synthetic vanillin or ethyl vanillin. In ready-to-drink beverages, such as ready-to-drink coffee products, vanilla plays a significant role in smoothing the harsh bitter notes, which develop during retorting. In functional beverages, vanilla helps mask undesirable off notes commonly associated with these types of products. In chocolate beverages, vanilla smoothes out the drying cocoa notes while enhancing the desirable flavor characteristics of the cocoa. Alcoholic beverages, including creme cordials, liqueurs, and specialty items, use vanilla or vanillin as a primary flavor or secondary flavor to smooth out the overall flavor profile of the product.

18.2.10 Pet Products

Vanilla flavors can be used to enhance the consumer (purchaser) appeal of the aroma of pet food by helping to mitigate off odors associated with some pet food products. Vanilla flavors can be used to mask the undesirable breath aroma in ruminants.

18.2.11 Pharmaceutical Products

Vanilla flavors are commonly used to help mask bitter, unpleasant flavors normally associated with the actives found in cough and cold preparations. The addition of vanillin to a flavor used in a cough preparation will help mask the undesirable off notes caused by the medicinal components present.

18.2.12 Oral Care

Vanilla flavors are used in conjunction with other flavors to impart a pleasant aftertaste in toothpaste and mouth rinses. Since vanillin can discolor over time, using vanillin or ethyl vanillin in a white base will result in a slight browning over time.

18.2.13 Perfume

Vanilla can be used in perfumes, candles, and potpourris to impart pleasing "feel good" aromatic qualities. Bourbon vanilla beans can be cut and placed in potpourris to impart a pleasant aroma. Extractives of bourbon vanilla can impart a delicate nuance to fine fragrances. Generally vanilla absolutes with high vanillin content, and little color value, are used in fine fragrances. Vanilla can also be used to balance or "tie in" the individual components of a fragrance.

18.2.14 Toys

Vanilla derived compounds can be safely used in children's toys to mask the unpleasant notes associated in rubber and plastic items.

Overall, the desirable flavor and aromatic aspects of vanilla, whether in its natural or reconstructed form, make it one of the most widely used flavor and aroma items in the world.

18.3   Major US Vanilla Companies

Vanilla extract has been sold commercially in the United States for over a century. Many companies have been producing vanilla extract since the introduction of vanilla to the United States, and are still major producers of vanilla extract today. The oldest company on record is Watkins Natural Gourmet, founded in 1868. Vanilla was added to the company's inventory in 1895. McCormick was started in 1889 and natural flavorings were some of their first products. David Michael & Company was established in 1896 and started producing vanilla sugar in the late 1890s. Nielsen-Massey Vanillas Inc. and Virginia Dare have been producing vanilla products since 1907 and 1923, respectively. Rodelle Laboratories was established in 1936.

Today, the major companies that produce vanilla extract are David Michael & Company, Elan Vanilla Company, Firmenich, Givaudan, International Flavors & Fragrances, McCormick, Nielsen-Massey Vanillas Inc., Symrise GmbH & Co., and Virginia Dare.

In addition, there are numerous small- to medium-sized companies that sell vanilla extract in the retail markets. These companies either produce their own products or buy vanilla products from the larger companies and repackage the products using their own label.

18.4   Introduction to the Study

The purpose of this study was to evaluate the composition of products sold in retail stores as "vanilla extract". Labeling of vanilla extract is regulated by the FDA Standard of Identity Title 21 Part 169 (http://www.access.gpo.gov/nara/cfr/waisidx_00/21cfr169_00. html). Accordingly, a package that carries the name "vanilla extract" has to be at least one-fold vanilla extract (see Chapter 9 by Ranadive). The ingredients have to be listed from the highest to the lowest amount. The ingredients for a basic vanilla extract should appear in the following order: water, alcohol, vanilla bean extractives. If other permitted optional ingredients (sugar, glycerin, propylene glycol, corn syrup, dextrose) are added, they should follow the same consistent order. Vanilla flavor is exactly like vanilla extract but with less than 35% alcohol.

We collected, from all over the United States, bottles that carry the label "Vanilla Extract", sold in various retail stores, including supermarkets, specialty gourmet stores, craft stores, and department stores. We analyzed a large sample of retail vanilla extracts for the four major components commonly associated with vanilla extract. From these analyses it is clear that many of the companies selling vanilla extract in retail stores do not adhere to the standard of identity or to their own labeling specifications (the label does not accurately describe the contents). The study also revealed that large companies that produce and assign their own name to the label do adhere to the standard of identity and provide consistent products. Private labels, supermarket brands, home-made products, and club stores do not have consistent products and do not always adhere to the standard of identity.

18.5   Materials and Methods

We collected, from retail stores all over the country, bottles labeled as "Vanilla Extract". Two samples labeled as "Artificial Vanilla Flavor" or "Imitation Vanilla Extract" were also included. A total of 65 samples were used in this study. Each sample received a code number, to protect the manufacturer's name. We assumed that all the vanilla extracts we purchased were one-fold, unless otherwise noted.

For HPLC analysis of the vanillin, vanillic acid, *p*-hydroxybenzaldehyde, and *p*-hydroxybenzoic acid contents, the extracts were diluted with 50% ethanol to ratios of 1:10 and 1:100. The diluted samples were analyzed as previously described (Havkin-Frenkel *et al.* 1996). The flavor components were identified and quantified using standard chemicals from Sigma (St Louis). The quantification was based on a standard curve.

For analysis of total phenols, the samples were diluted 1:10 and 1:20 with 50% ethanol, and 20 µL of the diluted samples was used for the assay of total phenols as described by Booth *et al.* (2008). The amount of total phenols was expressed as µg chlorogenic acid equivalents mL^{-1} vanilla extract.

18.6 Results and Discussion

18.6.1 Labeling of Retail Vanilla Extracts

The retail vanilla extracts analyzed were packaged in sizes from 1 to 32 oz, in plastic, glass, or metal containers. The bottles were clear, opaque, or dark. Some of the bottles were sold in boxes. Table 18.1 describes the information stated on the bottles or boxes. The information includes product name, amount of extract, type of package, list of ingredients, and certification.

Most labels carried the name "Pure Vanilla Extract" that which conforms to the FDA standard of identity, which does not require specification of the origin or species of vanilla beans used for the preparation of the extract. However, when a label states "Pure Bourbon Vanilla", "Pure Tahitian Vanilla", or "Pure Madagascar Vanilla Extract", customers expect the product to adhere to what is stated on the label. Our HPLC analysis (described below) questions the authenticity of some of the listed samples.

There was a wide variation in the order of listing ingredients, although common practice in food labeling follows the convention of listing ingredients from the highest to the lowest amounts present in the product. Table 18.1 indicates some samples (2, 3, 4, 5, 12, 15, 19, 20, 33, 34, 36, 38, 52, 54, 56, 60, and 64), which chose to start the list of ingredients with vanilla beans, rather than listing water, the main ingredient, as the first on the list. If we choose to specify the type (premium quality for instance), origin (Mexico, Madagascar or Tahiti, for example), or species of vanilla bean (*V. planifolia* or *V. tahitensis*), it could be part of the label name but should appear in the ingredient list in the order of amount.

Many credible vanilla extracts list ingredients such as "vanilla bean extractives in water" at the top of the ingredient list. The present authors believe that this statement is confusing because the label conveys the impression that product content is mostly vanilla extractive rather than water. Another source of vagueness is confusion of bean origin and bean species. For example, "Tahitian bean" denotes a geographic origin, although the term is commonly used to imply that the species used is *Vanilla tahitensis*. Farmers in Papua New Guinea (PNG) grow both *Vanilla tahitensis* and *Vanilla planifolia*. However, some vanilla extracts carrying the name pure Tahitian beans in fact originate from PNG (samples 24 and 33, personal communication).

Table 18.1 Characteristics of the retail vanilla extracts analyzed in this study.

Sample Number	Product Name	Amount	Type of Package	List of Ingredients	Certification
1	Vanilla Pure Extract	2 oz/59 mL	Plastic	Water. Alcohol, Extractives of Vanilla Beans	None
2	Pure Vanilla Extract	2 oz/59 mL	Glass	Vanilla Bean Extractives in Water, 35% Alcohol and Corn Syrup	Kosher
3	Pure Vanilla Extract	2 oz/59mL	Glass	Vanilla Bean Extractives in Water, 35% Alcohol and Corn Syrup	Kosher
4	Pure Vanilla Extract	1 oz/29 mL	Glass	Vanilla Bean Extractives in Water, 35% Alcohol and Corn Syrup	Kosher
5	Pure Vanilla Extract	2 oz/59 mL	Glass	Vanilla Bean Extractives in Water, 35% Alcohol and Corn Syrup	Kosher
6	Pure Vanilla Extract	1 oz/29 mL	Plastic	Water, Alcohol, Corn Syrup, and Vanilla Bean Extractives	Kosher
7	Pure Vanilla Extract	2 oz/59 mL	Glass	Water, Alcohol, and Bourbon Vanilla Bean Extractive	Kosher
8	Pure Vanilla Extract	2 oz/59 mL	Plastic	Water, Alcohol, Extractives of Vanilla Beans	None
9	Pure Vanilla Extract	2 oz/59 mL	Plastic	Water, Alcohol, Corn Syrup, and Vanilla Bean Extractives	Kosher
10	Pure Vanilla Extract	2 oz/59 mL	Glass	Water, Alcohol, Corn Syrup, and Vanilla Bean Extractives	Kosher
11	Pure Vanilla Extract	1 oz/29 mL	Glass	Water and Alcohol, Extract of Vanilla Beans, Sugar	None
12	Pure Vanilla Extract	2 oz/59 mL	Glass	Vanilla Bean Extractives in Water, 35% Alcohol and Corn Syrup	Kosher
13	Pure Vanilla Extract	1 oz/29 mL	Plastic	Water, Alcohol, Extractives of Vanilla Beans	Kosher
14	Pure Vanilla Extract	1 oz/59 mL	Glass	Water, Alcohol (35%), Corn Syrup, and Vanilla Bean Extractives	Kosher
15	Vanilla Extract	2 oz/59 mL	Plastic	Vanilla Bean, Water, Alcohol	None
16	Vanilla Extract	4 oz/118 mL	Plastic	Water, Alcohol, Vanilla Bean Extractive	None

(Continued)

Table 18.1 (Continued)

Sample Number	Product Name	Amount	Type of Package	List of Ingredients	Certification
17	Pure Vanilla Flavor	4 oz/118 mL	Plastic	Water, Glycerine, Sugar Cane, Vanilla Bean Extractives	Kosher
18	Planifolia Vanilla Extract	4 oz/120 mL	Plastic	Water, Alcohol, Planifolia Vanilla Beans	None
19	Pure Bourbon Vanilla Extract	4 oz/120 mL	Plastic	Bourbon Vanilla Bean Extractives in Water, Alcohol 35%	Kosher
20	Natural Bourbon Vanilla	4 oz/120 mL	Plastic	Madagascar Bourbon Vanilla Bean Extractives and other natural flavors in alcohol, water, and sugar	Kosher
21	Pure Vanilla Extract	1 oz/59 mL	Plastic	Water, Alcohol, Extractives of Vanilla Beans	None
22	Pure Vanilla Extract	7 oz/207 mL	Plastic	Water, Alcohol (35%), Corn Syrup, and Vanilla Bean Extractives	None
23	Pure Bourbon Vanilla Extract	32 oz/946 mL	Plastic	Water, Alcohol, Vanilla Bean Extractive	Kosher
24	Pure Tahitian Vanilla Extract	32 oz/946 mL	Plastic	Water, Alcohol, Vanilla Bean Extractive	Kosher
25	Pure Vanilla Extract from Papantla	30 mL	Glass	Water, Alcohol, Extractives of Vanilla Beans	Kosher
26	Vanilla Extract Pure	2 oz/59 mL	Glass	Water, Alcohol, Extractives of Vanilla Beans	None
27	Vanilla	2 oz/59 mL	Glass	Water, Alcohol, Extractives of Vanilla Beans	None
28	Pure Vanilla Extract	2 oz/59 mL	Glass	Water, Alcohol, Vegetable Glycerine, and Vanilla Bean Extractives	Kosher
29	Pure Vanilla Extract	2 oz/59 mL	Glass	Water, Alcohol, Vanilla Extractives	None
30	Vanilla Extract	2 oz/59 mL	Glass	Water, Organic Alcohol, Organic Vanilla Bean Extractives	Kosher
31	Pure Vanilla Extract	2 oz/59 mL	Glass	Water, Alcohol, Vanilla Extractives	None
32	Pure Vanilla Extract	2 oz/59 mL	Glass	Water, Alcohol, Vanilla Bean Extractives	None
33	True Tahiti Vanilla	2 oz/59 mL	Glass	Tahitian Vanilla Beans, Alcohol, Water, Sugar	None

Table 18.1 (Continued)

Sample Number	Product Name	Amount	Type of Package	List of Ingredients	Certification
34	PNG Tahitensis Vanilla	2 oz/59 mL	Glass	Papua New Guinea Vanilla Beans, Alcohol, Water, Sugar	None
35	Madagascar Premium Vanilla	4 oz/118 mL	Glass	Water, Alcohol, Vanilla Bean Extractives	Kosher
36	Premium Quality Pure Vanilla Extract	4 oz/118 mL	Glass	Vanilla Bean Extractives in 35% Alcohol	Kosher
37	Madagascar Bourbon Premium Vanilla Extract	4 oz/118 mL	Glass	Water, Alcohol, Sugar, Vanilla Bean Extractives	Kosher, Gluten Free
38	Pure Vanilla Extract	8 oz/236 mL	Glass	Bourbon and Tahitian Vanilla Bean Extractives in Purified Water and Alcohol (40%)	Kosher
39	Pure Madagascar Vanilla Extract	8 oz	Glass	Water, Alcohol, Sugar, Vanilla Bean	None
40	Mexican Vanilla Extract	250 mL	Glass	Water, Alcohol, and Vanilla Beans Extract	None
41	Vanilla	59 mL	Glass	Water, Alcohol, Vanilla Beans, Sugar	None
42	Double Intensity Pure Vanilla Extract	100 mL	Glass	Water, Alcohol, Vanilla oleoresin from Vanilla planifolia	None
43	Vanilla	20 mL	Glass	N/A	Kosher
44	Organic Vanilla Extract	59 mL	Glass	Water, Organic Alcohol, Organic Vanilla Bean Extractives	USDA Organic
45	Vanilla Pure Extract	118 mL	Plastic	Water, Alcohol, Vanilla Extractives	Kosher
46	Artificial Vanilla Flavor	118 mL	Plastic	Water, Alcohol, Caramel color, Vanillin and Artificial Flavor	Kosher
47	Vanilla	20 mL	Glass	N/A	Kosher
48	Imitation Vanilla Extract	59 mL	Plastic	Water, Sugar, Caramel Color, Vanillin, Ethyl Vanillin Artificial Flavor	Kosher
49	Pure Vanilla Extract	480 mL	Plastic	Water, Alcohol, Corn Syrup, and Vanilla Bean Extractives	Kosher
50	Pure Vanilla Extract	473 mL	Plastic	Water, Alcohol, Vanilla Bean Extractives, and Cane Sugar	None

(Continued)

Table 18.1 (Continued)

Sample Number	Product Name	Amount	Type of Package	List of Ingredients	Certification
51	Vanilla Extract	4 oz/118 mL	Plastic	Water, Alcohol, Vanilla Bean Extractives	None
52	Pure Vanilla Extract	2 oz/59 mL	Plastic	Vanilla Bean Extractives in Water, Alcohol (35%), and Corn Syrup	Kosher
53	Madagascar Bourbon Pure Vanilla Extract	9 oz/267 mL	Glass	Water. Ethyl Alcohol (35%), and Vanilla Bean Extractives	None
54	Pure Madagascar Bourbon Vanilla Extract	4 oz/118 mL	Glass	Extractive Matter of Vanilla Beans in 35% Alcohol	None
55	Pure Vanilla Extract	2 oz/59 mL	Plastic	Water, Alcohol, Vanilla Bean Extractives	None
56	Pure Vanilla Extract	2 oz/59 mL	Plastic	Vanilla Bean Extractives in Water, Alcohol (35%), Corn Syrup	Kosher
57	Pure vanilla Extract	2 oz/59 mL	Plastic	Water, Alcohol, Vanilla Bean Extractives, and Sugar	None
58	Pure Vanilla Extract	4 oz/118 mL	Plastic	Water, Alcohol, Corn Syrup, and Vanilla Bean Extractives	Kosher
59	Premium Quality Madagascar Bourbon Pure Vanilla Extract	2 oz/59 mL	Glass	Pure vanilla extract[water, alcohol (35%), extractive of Madagascar Bourbon vanilla beans], glucose	Kosher
60	Pure Vanilla Extract	4 oz/118 mL	Plastic	Vanilla Bean Extractives in Water, Alcohol (35%), and Sugar	Kosher
61	Vanilla Extract Pure	4 oz/118 mL	Glass	Water, Alcohol, Vanilla Bean Extractives	Kosher
62	Pure Vanilla Extract	4 oz/118 mL	Glass	Water, Alcohol, Vanilla Bean Extractives	None
63	Organic Vanilla Extract	2 oz/59 mL	Glass	Water, Alcohol, Vanilla Bean Extractives	Kosher, USDA Organic
64	Vanilla Extract	3.38 oz/100mL	metal	Madagascar Bourbon Vanilla, water, ethanol (40%), glycerin	Kosher
65	Pure Vanilla Extract	4 oz/118 mL	Glass	Water, Ethyl Alcohol (35%), vanilla bean extractives	Kosher

Another issue is plastic vs. glass or metal containers. Traditionally, vanilla extract, which contains at least 35% ethanol, is stored in dark glass bottles to protect it from light-induced darkening of phenolics and avoid reaction with the alcohol during long storage. During short handling periods, common in industry and for restaurant use, it is appropriate to hold vanilla extract in plastic containers, to avoid problems with breakable glass. Many supermarkets also concerned with the breakability of glass keep vanilla extract in glass containers, but protect the glass bottles by placing them in protective boxes.

Some manufacturers specify Kosher certification (Jewish law) or Organic certification. These certifications designate yet another level of inspection and add to the product credibility. These certificates are not issued to cottage industry-made vanilla products.

18.6.2 Flavor Components in the Retail Vanilla Extracts

Although vanilla extract contains more than 300 compounds (Toth *et al.*, Chapter 9), the most studied and commonly associated with vanilla extract quality are vanillin, vanillic acid, *p*-hydroxybenzaldehyde, and *p*-hydroxybenzoic acid. A label carrying the statement "Vanilla Extract" must contain at least one-fold vanilla. An average one-fold bourbon vanilla extract usually contains 0.1 to 0.2% vanillin, 0.01 to 0.05% vanillic acid, 0.01 to 0.03% *p*-hydroxybenzaldehyde, and 0.01 to 0.02% *p*-hydroxybenzoic acid. These values provide the standard acceptable ratios of these compounds in a one-fold vanilla extract. The concentrations of these compounds in vanilla beans have been the subject of debate, because vanilla grown in different regions does show variation in the compound ratios (Gassenmeier *et al.* 2008). However, for the purpose of this study, these values provide a useful guideline.

Figures 18.1, 18.2, 18.3 and 18.4 present the content of each compound, as a percent of the extract, in each of the tested samples. Figure 18.1 indicates a range of 0.01 to 0.8%

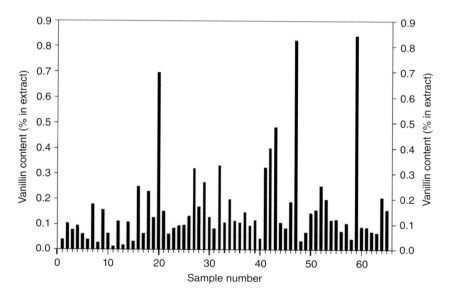

Figure 18.1 Vanillin content in samples 1 to 65.

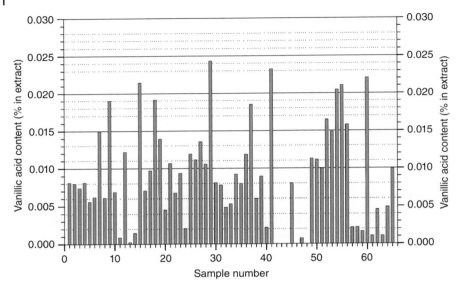

Figure 18.2 Vanillic acid content in samples 1 to 65.

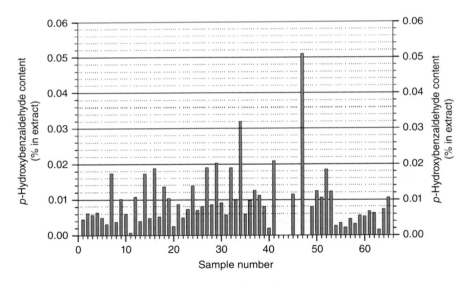

Figure 18.3 *p*-Hydroxybenzaldehyde content in samples 1 to 65.

vanillin in the tested extracts. The vanillin content is 0.1% or above in most samples. Samples showing vanillin content of 0.3% and above (0.7, 0.5, 0.8% in samples number 20, 43, and 47 respectively, for example) raise doubts about the authenticity of these extract preparations, suggesting supplementation with exogenous vanillin. Sample 20 lists "other natural flavors" (WONF) in the ingredients list. In the authors' opinion, there is no provision in the standard of identity that allows for addition of other natural

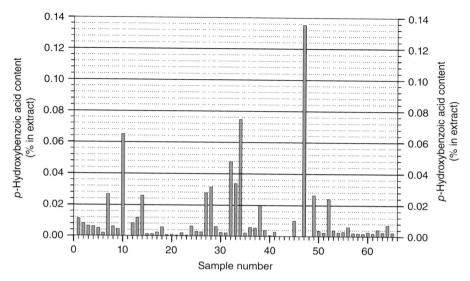

Figure 18.4 p-Hydroxybenzoic acid content in samples 1 to 65.

flavors to vanilla extract. By labeling the product "Natural Bourbon Vanilla", rather than "Vanilla Extract", the producer may not be subject to the standard of identity. Legally this may be a gray area and is certainly deceiving the consumers. Samples 46 and 48 were labeled as synthetic and were included in the analysis for comparison with those labeled as vanilla extract. These samples use synthetic vanillin or ethyl vanillin as the main flavor component. Ethyl vanillin was not analyzed in this study.

The concentration of vanillic acid in the samples ranged from zero (no detectable vanillic acid) to around 0.025%, although there was considerable variation (Figure 18.2). However, most samples were below 0.01% vanillic acid. The concentration of p-hydroxy-benzaldehyde in the samples ranged from zero to around 0.05%, with most samples containing below 0.01% (Figure 18.3). The concentration of p-hydroxybenzoic acid in the samples ranged from zero to around 0.13%, with most samples containing below 0.01% (Figure 18.4). Vanilla extract is often a blend of vanilla grown in different regions and we might expect variation in the tested compounds. However, an extract lacking some of these other flavor compounds suggests either a low amount, or even complete absence, of vanilla matter in the extract.

A graph illustrating the relative amounts of the four compounds in five of the samples is shown in Figure 18.5. The samples were chosen to represent the range in concentration of the flavor compounds observed. Sample #19 carries the name "Pure Bourbon Vanilla Extract". Sample #12 is "Pure Vanilla Extract". Sample #42 is named "Intensified Double Pure Vanilla Extract". Sample #61 is "Pure Vanilla Extract" and Sample #65 is "Pure Vanilla Extract Madagascar Bourbon".

Sample #19 is a private label of a well-respected supermarket chain and the label states that the product is "Pure Bourbon Vanilla Extract". Based on common under-standing, Bourbon vanilla indicates beans from Madagascar, Reunion, and the Comoros islands, although in some opinions Bourbon denotes the beans of V. planifolia species grown elsewhere. Moreover, the expression "Bourbon" is also used to indicate quality

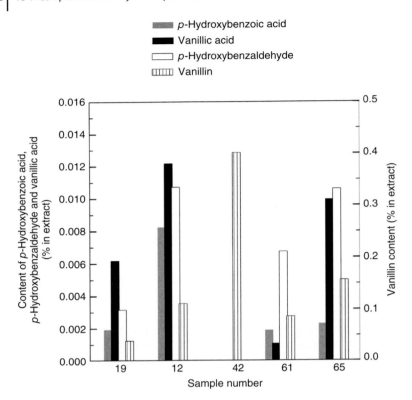

Figure 18.5 A comparison in the composition of samples representing vanilla extracts found in the retail market. Data used in the graph is from Table 18.1 and Figures 18.1 through 18.4. The percent p-hydroxybenzoic acid, vanillic acid, and p-hydroxybenzaldehyde contents are on the left axis and the percent vanillin content is on the right axis.

and hence is not relevant for our analysis. This sample is apparently a blend of some Bourbon and Indonesia-grown beans based on organoleptic characteristics.

Sample #12 is a blend of good quality vanilla beans with corn syrup added for flavor enhancement. Sample #42 was obtained from a gourmet retail store and is packaged in an appealing, distinguished bottle that carries a premium price. The label states that it is a double intensity vanilla extract made from vanilla oleoresin, from beans originating from Veracruz, Mexico. There is no requirement to state the origin of the vanilla beans. However, when the name states a place of origin, Mexico for example, the extract should be only from Mexican beans. The exceptionally high level of vanillin, the lack of the other flavor components associated with vanilla extracts (Figure 18.5), and the lack of sediment in the bottle, strongly suggest there is actually no vanilla oleoresin in the product. The organoleptic properties also suggest this product is actually synthetic vanillin.

Sample #61 is an industrial sample of a blend made up from Madagascar- and Indonesia-grown beans. This sample is a good reference for the content and ratio of constituents in a blended vanilla extract.

Sample #65 is 100% Madagascar Bourbon vanilla extract. It is the top of the line and sells in gourmet stores. It is a good reference for comparing against other vanilla extracts.

18.6.3 Total Phenol Content of the Retail Vanilla Extracts

Total phenol is a non-specific assay to measure the total phenols in an extract. In vanilla extract, total phenol is a good quality indicator and can be related to the strength of the extract or to its fold. According to the standard of identity, vanilla extract must be at least one-fold. If we are certain that the vanilla extract does not contain any added synthetic chemicals, such as synthetic vanillin or ethyl vanillin, the total phenol assay is a very simple way to determine the fold of an extract. Figure 18.6 illustrates the linear relationship of between total phenol and the fold of an extract. A good one-fold vanilla extract usually contains total phenols at 350 to 450 µg chlorogenic acid equivalents mL^{-1} vanilla extract. Figure 18.7 shows the total phenol content in the tested samples. The graph shows that most samples contained 350 to 400 µg chlorogenic acid equivalent per 1 mL extract, but samples # 11 and 39 contain less than 50 µg mL^{-1}. This low level of total phenols suggests these vanilla extracts are less than one-fold or were produced from low quality beans. Samples 20, 40, 42, and 46 contain over 800 µg mL^{-1}. Samples 20 and 42 are likely synthetic based on the high vanillin content. Sample 46 is labeled as imitation vanilla extract. Sample 40 has very low levels of vanillin and the other flavor compounds. The high total phenol content suggests the presence of other compounds, such as ethyl vanillin or coumarin.

18.7 Conclusion and Recommendation

In addition to the analyses presented above, we tasted all the extracts with water and 5% sugar with a group of 10 trained non-professional individuals. The results revealed that vanilla samples labeled as being from the same sources were totally different in the way

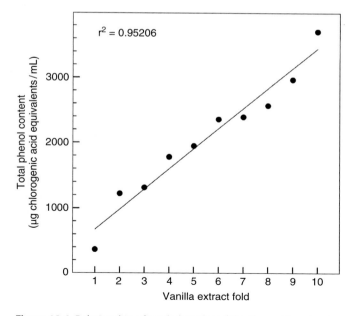

Figure 18.6 Relationship of total phenols to fold of a vanilla extract.

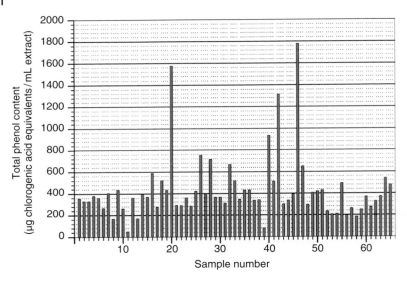

Figure 18.7 Total phenol analysis in samples 1 to 65.

they were evaluated. When a vanilla extract carries a label "100% Pure Madagascar Bourbon" or "100% Tahitian" we would expect that even when they are coming from different companies, the taste will be similar due to their common origins of the beans. The study indicated that there were no similarities among such retail products. This hints that there is no standardization in the retail business of vanilla extracts.

The standard of identity for vanilla extract does not require labeling of the strength, although it must be a minimum of one-fold to be called vanilla extract. There is no requirement for indicating the country of origin of the beans. Having a blend of extracts from beans of different origins or quality is allowed and can often result in a better product. Producers may even use different beans or blends in different bottling runs of the same product. However, it is expected that the producers are accurately labeling the products. For example, an extract labeled 100% Madagascar Bourbon should include only beans from Madagascar (only *Vanilla planifolia*). The quality of the beans is the main factor that affects the quality of the extract but there are no regulations regarding bean quality, so all the beans became top of the line and labeled as gourmet!

From our analysis of the flavor components in a large sample of vanilla extracts, it is obvious there are many companies that sell vanilla extract in the retail market that have very little knowledge of the subject or the FDA regulations regarding proper labeling. Retail vanilla extract is like expensive perfumes in many ways. For some it is more important that the bottle looks gourmet without paying enough attention to the product on the inside. If the vanilla on the retail market has no standard and the market is flooded with undefined vanilla extract, how would a consumer know what to buy and what to use? Can the average consumer tell the difference between good quality or poor quality, as an expert will? Does the industry need to teach people what is a good extract? Or it is a matter of taste?

References

Booth, D.Y.J., Sinha, A., Galasso, K.E. and Havkin-Frenkel, D. (2008) Discoloration of oregano post-distillation left-over leaf extract. *Acta Horticulturae (ISHS)*, **778**, 93–97.

Gassenmeier, K., Riesen, B. and Magyar, B. (2008) Commercial quality and analytical parameters of cured vanilla beans (*Vanilla planifolia*) from different origins from the 2006-2007 crop. *Flavour and Fragrance Journal*, **23**, 194–201.

Havkin-Frenkel, D. (2009) New vanilla routes. *The World of Food Ingredients*, **March**, 36–39.

Havkin-Frenkel, D., Podstolski, A. and Knorr, D. (1996) Effect of light on vanillin precursors formation by *in vitro* cultures of *Vanilla planifolia*. *Plant Cell, Tissue and Organ Culture*, **45**, 133–136.

19

Vanilla in Perfumery and Beverage Flavors
Felix Buccellato

19.1 Earliest Recorded Use of Vanilla

Vanilla beans and their extracts were used by the ancient Totonaco and Aztec Indians of Mexico, as first recorded in 1520 by Bernal Diaz, one of Hernan Cortes officers. The on-the-vine cured vanilla bean was used to make a drink called chocolatl, containing powdered cocoa beans, ground corn, and flavored with tlilxochitl (ground black vanilla pod and honey). Coffee emporiums of today offer similar combinations of chocolate blended with honey and vanilla.

What is the reason for the appeal of vanilla flavor manifested early on by the Aztecs and persisting to this day, as evidenced by the extensive use of vanilla and related products in foods, cosmetics, and personal products? Chemical analysis of vanilla flavor reveals an extremely complex natural product containing nearly 500 compounds and counting, and chief among them is vanillin. The richness and depth of vanilla flavor and aroma originates from the chemical complexity of the compounds present in vanilla extracts.

Vanillin is the characterizing compound in vanilla flavor. The molecular structure of vanillin indicates that the compound is a multifunctional aromatic with extraordinary olfactory properties, notably, extreme diffusivity and extreme odor intensity. These molecular properties of vanillin are the hallmark of potency and enchantment of flavors and fragrances.

Humans, as well as non-human primates, are capable of detecting extremely low levels of vanillin, well below the typical levels in vanilla extracts. In a comparison of the odor thresholds for vanillin by humans and the pygmy marmoset (*Cebuella pygmaea*), the detection threshold for humans was 11.8×10^{-14} M and that of the pygmy marmoset was 2.04×10^{-14} M (Glaser *et al.* 1995). These values approximate a threshold level of around 0.12 parts per trillion. This is a remarkably low odor threshold, indicating extraordinary detection levels for vanillin by humans and other primates.

An important reason for the appeal of vanilla might also arise from tolerance and, moreover, preference to vanilla flavor. There are many products in the flavor and fragrance field with low odor thresholds. Vanillin, a major component of vanilla, differs in a very important way. Most compounds with low odor thresholds including pyrazines, pyridines, thiols, various thiazoles, mercaptans, and thioesters, possess a very pungent

Handbook of Vanilla Science and Technology, Second Edition.
Edited by Daphna Havkin-Frenkel and Faith C. Belanger.
© 2019 John Wiley & Sons Ltd. Published 2019 by John Wiley & Sons Ltd.

and negative odor at high concentrations. Vanillin, by comparison, can be smelled and appreciated at extremely high levels and does not adversely impact the olfactory system. There are very few compounds that fall in this category. Finally, the flavor and aroma of vanilla or vanillin is perceived as very pleasant, and it seems people cannot get enough of it. Hence, complexity, low perception threshold, tolerance to vanilla flavor and aroma, even at high concentrations, and strong preference may have combined to create a universal fondness and craving for vanilla flavor.

Vanilla flavor is used extensively in food products, shown for example by the use of vanilla in non-carbonated and carbonated beverages (Tables 19.1 and 19.2). The universal popularity of vanilla is also indicated by its use in fragrances. About 50% of the over 10,000 fragrances make use of vanilla in some amount (Figure 19.1). Examples of the use of vanilla in oriental fragrances are given in Tables 19.3 and 19.4. The latter are often a combination of vanilla, balsamic resins such as olibanum (frankincense), balsam tolu, benzoin, and musks, which can be blended in an infinite number of ways. They have been described, not always correctly, as sweet, spicy, fruity, and animalic. The descriptor of "sweet" is a gustatory or taste description and does not directly apply to odor. In odor, the word "sweet" is used when there is an inability to attach a specific characterization, such as apple sweet, strawberry sweet, or caramel sweet. The term "sweet" for describing a fragrance property does not refer to a property of sugars or other artificial sweeteners, which do not give rise to odor. The "sweet" that people refer to in odor generally comes from association of the aroma and sugar. The odor of vanilla or vanillin that is mixed with other characters described as "sweet" stems from the association with the familiar sweetened and flavored products such as ice cream and cakes.

Oriental fragrances often incorporate an accord referred to as amber. It is a perfumery accord using vanilla, olibanum, balsamic resins, and citrus to varying degrees. They are often added as a complex or a foundation to an oriental fragrance. One of the most famous fragrances, based somewhat on vanilla, is a beautiful oriental blend called *Shalimar*. It has a foundation of vanilla and benzoin, topped off with bergamot and citrus notes with traces of many modifiers. The signature, however, is the beautiful and extremely long lasting and diffusive vanilla character.

While 50% of fragrances use vanilla, vanillin, or ethyl vanillin, other fragrances do not use vanilla or vanillin (Tables 19.5 and 19.6). The most common reason for non-use is stability and color requirements. There are several important fragrances that do not use vanilla because the accord or intended image is one that does not require any of the "sweet" character of vanillin.

What are the problems with the use of vanilla in fragrances? In a word: solubility. There are many carrier media in which vanilla extract, an alcohol and water-based dark brown material, is difficult or impossible to use. Other vanilla preparations also pose a problem. Vanilla absolute is a very expensive product. The dark color of vanilla absolute is problematic and, furthermore, it is a vanilla preparation that has lost the top notes that are normally present in an alcoholic solution of vanilla.

Extraction of vanilla beans into propylene glycol addresses the flammability issue of alcohol but does little to reduce the solubility problems where a highly lipophilic system is needed. Color for many applications is often a requirement for products such as candles and any wax based or silicon based system. There are also air freshener devices and systems where any hydrophilic glycols or carriers are extremely undesirable. This is also true for certain lipophilic encapsulation procedures.

Table 19.1 Examples of non-carbonated drinks containing vanilla.

Product	Brand	Mfg. Company
Vanilla Frappucino	Starbucks	Pepsi
Strawberries & Cream Frappucino	Starbucks	Pepsi
Cream Liqueur	Starbucks	Jim Beam Brands
Super Protein Vanilla Al'mondo	Odwalla	Coca Cola
Soy Smart Chai Soy milk Shake	Odwalla	Coca Cola
Vanilla Shake	Slim Fast	Unilever
Homemade Creamy Vanilla Shake	Ensure	Abbott
Strawberries & Cream Shake	Ensure	Abbott
Nutripals Vanilla Drink	PediaSure	Abbott
Vanilla Drink Powder	Nesquik	Nestle
Vanilla Twist Vodka	Smirnoff	Diageo
Vanilla Vodka	Svedka	Spirits Marque One, LLC
Vanilla Vodka	Absolut	V&S Vin & Spirits AB
The Activator Smoothie - Vanilla	Smoothie King	Smoothie King
The Hulk Smoothie - Vanilla	Smoothie King	Smoothie King
Low Carb Vanilla Smoothie	Smoothie King	Smoothie King
The Shredder Smoothie - Orange Vanilla	Smoothie King	Smoothie King
Slim & Trim Smoothie - Vanilla	Smoothie King	Smoothie King
Slim & Trim Smoothie - Orange Vanilla	Smoothie King	Smoothie King
Coca Cola Black Cherry Vanilla Icee	Icee	J&J Snack Foods Corp.
Orange Dream Icee	Icee	J&J Snack Foods Corp.
Strawberry Creme Icee	Icee	J&J Snack Foods Corp.
Belgian Blends French Vanilla Latte	Godiva	Coca Cola
Vanilla Bean Coolata	Dunkin Donuts	Dunkin Brands
Tazo Chai Tea Latte	Tazo	Starbucks
Tazo Decaffeinated Chai Tea Latte	Tazo	Starbucks
Tazo Vanilla Tea Latte	Tazo	Starbucks
Enriched Vanilla Rice Milk	Rice Dream	Hain Celestial Group
Heartwise Vanilla Rice Milk	Rice Dream	Hain Celestial Group
Vanilla Rice Milk	Rice Dream	Hain Celestial Group
Vanilla Hazelnut Rice Milk	Rice Dream Supreme	Hain Celestial Group
Enriched Refrigerated Vanilla Rice Milk	Rice Dream	Hain Celestial Group
Classic Vanilla Soy Milk	Soy Dream	Hain Celestial Group
Enriched Vanilla Soy Milk	Soy Dream	Hain Celestial Group
Enriched Refrigerated Vanilla Soy Milk	Soy Dream	Hain Celestial Group
Honey Vanilla Chamomile Herbal Tea	Celestial Seasonings	Hain Celestial Group

(Continued)

Table 19.1 (Continued)

Product	Brand	Mfg. Company
Canadian Vanilla Maple Decaf Black Tea	Celestial Seasonings	Hain Celestial Group
Vanilla Apple White Organic Tea	Celestial Seasonings	Hain Celestial Group
Honey Vanilla White Chai Tea	Celestial Seasonings	Hain Celestial Group
Vanilla Hazelnut Dessert Tea	Celestial Seasonings	Hain Celestial Group
Vanilla Strawberry Rose Ceylon Black Tea	Celestial Seasonings	Hain Celestial Group
Honey Vanilla Apple Cider	Celestial Seasonings	Hain Celestial Group
Vanilla Hazelnut Organic Coffee	Celestial Seasonings	Hain Celestial Group
Vanilla plus Soy milk	WestSoy	Hain Celestial Group
Vanilla Low fat Soy milk	WestSoy	Hain Celestial Group
Vanilla Lite Soy milk	WestSoy	Hain Celestial Group
Vanilla Non-fat Soy milk	WestSoy	Hain Celestial Group
Vanilla Soy Slender Soy milk	WestSoy	Hain Celestial Group
Vanilla Soy Shake	WestSoy	Hain Celestial Group
Vanilla Rice Drink	WestSoy	Hain Celestial Group

Table 19.2 Examples of carbonated drinks containing vanilla.

Product Name	Brand	Parent Company
Orange & Cream Soda	Jones Soda Co.	Jones Soda Co.
Cream Soda	Jones Soda Co.	Jones Soda Co.
Cream Soda	Mug Root beer	Pepsi
Diet Cream Soda	Mug Root beer	Pepsi
Crème Soda	Boylan's	Boylan's Bottle Co.
Orange Cream Soda	Boylan's	Boylan's Bottle Co.
Diet Cream Soda	Boylan's	Boylan's Bottle Co.
Natural Crème Soda	Boylan's	Boylan's Bottle Co.
Orange 'N Cream Soda	Stewart's	Cadbury Schweppes
Diet Orange 'N Cream Soda	Stewart's	Cadbury Schweppes
Cream Soda	Stewart's	Cadbury Schweppes
Cherries 'N Cream Soda	Stewart's	Cadbury Schweppes
Vanilla Cream Root beer	Barq's	Coca Cola
Vanilla Cream Ginger Ale	Canada Dry	Coca Cola
Vanilla Coke Zero	Coca Cola	Coca Cola
Diet Coke Black Cherry Vanilla	Coca Cola	Coca Cola
Cherry Vanilla	Dr Pepper	Cadbury Schweppes
Diet Cherry Vanilla	Dr Pepper	Cadbury Schweppes

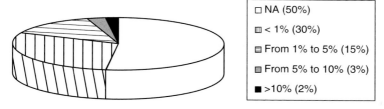

Figure 19.1 Pie chart showing percentage of vanilla or vanilla notes in the 50% of fragrances that include a vanilla component.

Table 19.3 Women's oriental fragrances where vanilla is a significant percentage and an aroma character.

Name of Fragrance	Company	Year	Materials used
Emeraude	Coty	1921	Vanilla, Florals, Musks
Shalimar	Guerlain	1925	Vanilla, Benzoin, Bergamot
Youth Dew	Lauder	1952	Vanilla, Balsams, Spices
Opium	St Laurent	1977	Vanilla, Citrus, Spices
Obsession	Calvin Klein	1985	Vanilla, Patchouli, Resins, Citrus, Musks
Coco Mademoiselle	Chanel	2001	Vanilla, Oakmoss, Florals, Amber
Burberry Brit	Burberry	2003	Icy Pear with base of Amber and Vanilla Bean
Joy	Jean Patou	1930	Bergamot, Roses, Jasmine, and Vanilla
Lovely	Sarah Jessica Parker	2005	Paper whites, Musk, and Amber (contains Vanilla)
Princess	Vera Wang	2006	Water Lily, Apricot, and Chiffon Vanilla
Midnight Fantasy	Britney Spears	2007	Vanilla, Amber, Musk

Table 19.4 Men's oriental fragrances where vanilla is a significant percentage and an aroma character.

Name of Fragrance	Company	Year	Materials used
Fougere Royal	Houbigant	1882	Vanilla, Oakmoss, Lavender
Canoe	Dana	1935	Vanilla, Oakmoss, Bergamot, Lavender
Brut	Faberge	1964	Vanilla, Oakmoss, Lavender, Patchouli
Paco Rabanne	Paco Rabanne	1973	Vanilla, Lavender, Oakmoss, Citrus
Obsession	Calvin Klein	1986	Vanilla, Patchouli, Citrus, Musk
Canoe II	New Dana	2001	Vanilla, Patchouli, Oakmoss, Musk

What can be done? Alternatives would be to extract vanilla beans with solvents less polar than water and alcohol. Alternatives might include types of esters or salicylates that are used in the flavor themselves or materials that can be used in the potential systems. However, there are limitations regarding what can be done in extraction of vanilla for flavors due to the standard of identity. The negative side of alternative

Table 19.5 Women's fragrances without vanilla or vanilla components.

Fragrance	Company	Year	Materials used
Muguet de Bois	Coty	1936	Bergamot, Citrus, Lily, Cyclamen
L'air du Temps	Ricci	1947	Bergamot, Rose, Lily, Muguet
Chanel 19	Chanel	1971	Bergamot, Rose, Woods, Muguet
Lauren	Cosmair	1978	Citrus, Ylang, Lily, Mandarin, Tropical fruits and flowers, Mimosa, Orange Flower
Calyx	Prescriptives	1986	Lily, Mandarin, Tropical Fruits and Flowers

Table 19.6 Men's fragrances without vanilla or vanilla components.

Fragrance	Company	Year	Materials used
Imperiale	Guerlain	1850	Citrus, Bergamot, Mandarin, Lemon, Verbena, Cedarwood
Aqua Velva Ice Blue	Williams Hispania	1935	Bergamot, Lemon, Petitgrain, Lavender, Mint, Spices, Sage, Rose
Eau Sauvage	Dior	1966	Jasmin, Bergamot, Citrus, Herbal
Eau d'Hadrien	Annick Goutal	1980	Citrus, Bergamot, Cypress
	Reintroduced	1995	

extraction methods is that vanillin, a highly polar material, and other polar constituents will not be extracted in the same way as they would be in alcohol-water systems. The resultant material will be application specific, not a real article of commerce or meant as a replacement for vanilla extract of any kind. However, this would create an opportunity for vanilla to be used in new ways that do not currently exist. If vanillin is needed, it is always available to bolster the body without color problems.

Some of the problems with using vanilla extracts have been solved by using vanillin. However, I espouse the view that vanillin does not offer the same richness and depth in flavor and aroma found in vanilla extract. I believe there is an opportunity to develop new types of vanilla bean extracts that would expand the use of vanilla, particularly for use in non-food items.

The real beauty of vanilla lies in its extreme complexity. It is not due merely to vanillin or percentage content of vanillin. If this were true, we could replace vanilla with synthetic vanillin in a suitable solvent or diluent. There are a wide variety of functional groups and molecules that make vanilla such a wonderful material (Table 19.7). Chapter 11 by Toth *et al.* presents a comprehensive summary of the volatile compounds that have been identified in vanilla. While little attention has been paid to the sesquiterpene and hydrocarbon compounds of vanilla, it is this author's view that the sesquiterpene and unsaturated hydrocarbon compounds play an extremely important role in the longevity and natural warm woody aroma that is intrinsic to a good quality vanilla.

Table 19.7 Compound classes contributing to the descriptors of vanilla aroma.

Descriptor	Compound Class
Vanilla	Vanillin
Fruity	Ethyl esters and other esters
Caramel/Cooked	Diacetyl, hydroxyl, ketones, cyclic compounds, lactones
Woody	Sequiterpenes, lactones
Spicy	Phenols, substituted phenols

When we look at flavor claims, there is little need to describe vanilla to the public. Only when there is an unfamiliar product such as Chai Tea, does a marketing department develop a complex explanation of the flavor using the terms like "exotic" and "spicy" to embellish the already "plain vanilla", which is exotic in its own right. As can be seen, there are a great many coffee and tea drinks using vanilla and sugar and a combination of other ingredients. These drinks are almost exactly like the drink types used 500 years ago by the Aztec chiefs. The New Chiefs now all go to Starbucks!

The problem of adulteration of vanilla with vanillin and a few other components has always plagued the vanilla market and the pricing. After 35 years of work and study of natural products, I am convinced that we are not yet able to do a better job than Nature. There is really nothing like the pure unadulterated vanilla extract or various folded and concentrated vanilla extracts that are available. It is always a shame when users try to stretch the use and performance of vanilla with the adulteration by vanillin or other mixtures to make an economic price point or performance point. When this happens, it affects the overall market by introducing unfair competitive pricing that can drive honest growers and manufacturers of vanilla extracts to their breaking points, or at worst put them out of business permanently. The supply is then cut, the prices go up and the demand for good quality extracts goes down, thus creating a spiral downwards in supply and upwards in pricing of good quality material. We have seen this happening in a variety of other natural products. The most striking examples are cinnamon bark oil, supplanted or replaced by cinnamic aldehyde, or bitter almond oil replaced by benzaldehyde. Our world would be much worse off without the quality of real vanilla extract in our daily lives.

Lastly, I hope the continued EU regulations do not prevent or eliminate use of natural products due to very weak contact allergens. This continues to affect the overall quality of items we use daily, as well as to impact the farmers and growers of botanical products.

Reference

Glaser, D., Etzweiler, F., Graf, R., Neuner-Jehle, N., Calame, J. and Mueller, P. (1995) The first odor threshold measurement in a non-human primate (*Cebuella pygmaea*; Callitrichidae) with a computerized olfactometer. In: *Chemical signals in vertebrates VII*, Apfelbach, R. and Mueller-Schwarze, D. (Eds), Pergamon, Oxford, pp. 445–455.

Part III

Biology of Vanilla

20

Vanilla Phylogeny and Classification
Kenneth M. Cameron

Reviews of plant systematics and economic botany regularly list *Vanilla planifolia* as a species unique in Orchidaceae (the orchid family), and it is true that the plant exhibits some features uncommon among orchids. For example, mature plants of this species are succulent climbing vines that begin life as terrestrials rooted in the soil, but may eventually become epiphytes attached to tree trunks and branches. Roots are used for both water and mineral absorption as well as for clinging to supports. New flowers open on a daily basis but last for only a few hours. Unlike most other orchids, the flowers shed individual pollen grains (monads) rather than packaging their pollen into complex pollinia. If pollinated, then each flower may develop into a long slender fruit, which often remains fleshy as a "berry" rather than drying out and dehiscing at maturity, as in a more typical orchid capsule. Whereas the vast majority of orchids produce microscopic dust-like seeds containing an undifferentiated embryo lacking endosperm and surrounded by only a thin transparent seed coat, the seeds of *Vanilla planifolia* are visible to the naked eye, black, hard, crustose, and may contain at least a few cells of endosperm to nourish the embryo within.

To the orchidologists all of the features listed above are exceptional, even within a family as diverse as Orchidaceae, but the question remains: Is *Vanilla planifolia* one-of-a-kind? The answer is "no". *Vanilla planifolia* may be the only species of more than 25,000 naturally occurring orchid species that is of significant agricultural value as a crop plant, and it may possess a number of unusual features compared to the majority of orchids, but it is not unique in the family. In fact, the genus *Vanilla* contains as many as 110 different species, which are distributed throughout tropical and subtropical areas of North America, South America, Africa, and Asia (Box 20.1). All are climbing vines, but differ considerably in their habitat preference, foliage, floral structure, and relationships with animal pollinators. Furthermore, *Vanilla* is only 1 genus out of 15 genera that are classified within the orchid subfamily Vanilloideae (the "vanilloid orchids"). Some of *Vanilla*'s closest relatives (e.g. *Pseudovanilla* and *Clematepistephium*) are also climbing vines. Others (e.g. *Cyrtosia*) also produce fleshy fruits containing spherical crustose seeds not unlike those of *Vanilla*. Still others, indigenous to temperate areas of northern North America and Asia (e.g. *Pogonia*), shed their pollen also as monads from flowers that share similar morphology with those of tropical *Vanilla* species (Cameron 2003).

Handbook of Vanilla Science and Technology, Second Edition.
Edited by Daphna Havkin-Frenkel and Faith C. Belanger.

Box 20.1 Checklist of All Currently Recognized Species of Vanilla Arranged By Native Continental Distribution. Based on Data By Govaerts *et al.* 2008

AFRICA

1) *Vanilla acuminata* Gabon
2) *Vanilla africana* West Africa
3) *Vanilla bampsiana* Zaïre
4) *Vanilla chalottii* Gabon
5) *Vanilla coursii* Madagascar
6) *Vanilla cucullata* Cameroon to Gabon
7) *Vanilla decaryana* Madagascar
8) *Vanilla francoisii* Madagascar
9) *Vanilla grandifolia* Príncipe to Zaïre
10) *Vanilla hallei* Gabon
11) *Vanilla heterolopha* Gabon
12) *Vanilla humblotii* Comoros
13) *Vanilla imperialis* West Africa to SW. Ethiopia and Angola
14) *Vanilla madagascariensis* Madagascar
15) *Vanilla nigerica* Nigeria to Cameroon
16) *Vanilla ochyrae* Cameroon
17) *Vanilla perrieri* Madagascar
18) *Vanilla phalaenopsis* Seychelles
19) *Vanilla polylepis* Kenya
20) *Vanilla ramosa* Ghana to Tanzania
21) *Vanilla roscheri* Ethiopia to Natal
22) *Vanilla seretii* West Africa

ASIA

23) *Vanilla abundiflora* Borneo
24) *Vanilla albida* Taiwan to Malesia
25) *Vanilla andamanica* Andaman Island
26) *Vanilla annamica* China to Vietnam
27) *Vanilla aphylla* Assam to Java
28) *Vanilla borneensis* Borneo
29) *Vanilla calopogon* Philippines
30) *Vanilla diabolica* Sulawesi
31) *Vanilla giulianettii* New Guinea
32) *Vanilla griffithii* West Malesia
33) *Vanilla havilandii* Borneo
34) *Vanilla kaniensis* New Guinea
35) *Vanilla kempteriana* New Guinea
36) *Vanilla kinabaluensis* Penninsular Malaysia to Borneo
37) *Vanilla moonii* Sri Lanka
38) *Vanilla ovalis* Philippines

39) *Vanilla palembanica* Sumatra
40) *Vanilla pierrei* Indo-China
41) *Vanilla pilifera* Assam to Borneo
42) *Vanilla platyphylla* Sulawesi
43) *Vanilla seranica* Maluku
44) *Vanilla shenzhenica* China
45) *Vanilla siamensis* China to Thailand
46) *Vanilla sumatrana* Sumatra
47) *Vanilla utteridgei* New Guinea
48) *Vanilla walkeriae* India & Sri Lanka
49) *Vanilla wariensis* New Guinea
50) *Vanilla wightii* India

NORTH AMERICA (including Mexico and Caribbean)

51) *Vanilla barbellata* South Florida to Caribbean
52) *Vanilla bakeri* Cuba
53) *Vanilla bicolor* Caribbean to South America
54) *Vanilla claviculata* Caribbean
55) *Vanilla correllii* Bahamas
56) *Vanilla dilloniana* South Florida to Caribbean
57) *Vanilla inodora* Mexico to Central America
58) *Vanilla mexicana* South Florida, Mexico to South America
59) *Vanilla odorata* Mexico to South America
60) *Vanilla palmarum* Cuba, South America
61) *Vanilla phaeantha* South Florida to Caribbean, Central America
62) *Vanilla planifolia* Caribbean, Mexico to Paraguay; widely cultivated
63) *Vanilla poitaei* Caribbean
64) *Vanilla pompona* Mexico to Central America

CENTRAL & SOUTH AMERICA (excluding Mexico and Caribbean)

65) *Vanilla acuta* Northern South America
66) *Vanilla angustipetala* Brazil
67) *Vanilla appendiculata* Guyana
68) *Vanilla bahiana* Brazil
69) *Vanilla barrereana* French Guiana
70) *Vanilla bertoniensis* Paraguay to Southern Brazil
71) *Vanilla bicolor* Caribbean to Southern Tropical America
72) *Vanilla bradei* Brazil
73) *Vanilla calyculata* Colombia
74) *Vanilla carinata* Brazil
75) *Vanilla chamissonis* French Guiana to NE Argentina
76) *Vanilla columbiana* Colombia
77) *Vanilla cristagalli* Brazil
78) *Vanilla cristatocallosa* Guyana to Brazil

79) **Vanilla denticulata** Brazil
80) **Vanilla dubia** Brazil
81) **Vanilla dungsii** Brazil
82) **Vanilla edwallii** Brazil to Argentina
83) **Vanilla fimbriata** Guyana
84) **Vanilla gardneri** E. Brazil
85) **Vanilla grandiflora** Southern Tropical America
86) **Vanilla hamata** Peru
87) **Vanilla hartii** C. America, Trinidad
88) **Vanilla helleri** C. America
89) **Vanilla hostmannii** Suriname
90) **Vanilla inodora** Mexico to Central America
91) **Vanilla insignis** Honduras
92) **Vanilla latisegmenta** Guyana
93) **Vanilla leprieurii** French Guiana
94) **Vanilla lindmaniana** Brazil
95) **Vanilla marowynensis** Suriname
96) **Vanilla methonica** Colombia
97) **Vanilla mexicana** S. Florida, Mexico to Trop. America
98) **Vanilla odorata** S. Mexico to Trop. America
99) **Vanilla organensis** Brazil
100) **Vanilla oroana** Ecuador
101) **Vanilla ovata** French Guiana to Brazil
102) **Vanilla parvifolia** S. Brazil to Paraguay
103) **Vanilla penicillata** Venezuela
104) **Vanilla perexilis** Paraguay to S. Brazil
105) **Vanilla phaeantha** S. Florida to Caribbean, C. America
106) **Vanilla planifolia** Caribbean, Mexico to Paraguay; widely cultivated
107) **Vanilla pompona** Mexico to C. America
108) **Vanilla porteresiana** French Guiana
109) **Vanilla purusara** Brazil
110) **Vanilla ribeiroi** Brazil
111) **Vanilla rojasiana** Paraguay to NE. Argentina
112) **Vanilla ruiziana** Peru
113) **Vanilla schwackeana** Brazil
114) **Vanilla sprucei** Colombia to N. Brazil
115) **Vanilla surinamensis** Suriname
116) **Vanilla trigonocarpa** Costa Rica to N. Brazil
117) **Vanilla uncinata** N. Brazil
118) **Vanilla weberbaueriana** Peru

Vanilla and its relatives are surviving members of what is likely an ancient lineage of flowering plants (Cameron 1999, 2000). Many are restricted to remote localities, and some are threatened with extinction. As such, a review of vanilloid orchid systematics (the scientific study of the diversity and classification of organisms) is critical to a comprehensive understanding of the biology of *Vanilla planifolia* and these exceptional orchids.

Vanilla planifolia may not be unique in Orchidaceae, but this wild species, domesticated centuries ago, provides humankind with a commodity cherished as a flavor and fragrance that is unique among spices.

20.1 Vanilloideae Among Orchids

The vanilloid orchids, Vanilloideae, have been recognized as a formal subfamily of Orchidaceae only during the past 10 years or so as DNA data have been used to re-evaluate relationships within the orchid family (Cameron *et al.* 1999; Cameron 2004, 2006). For a detailed review of this DNA-driven revolution in orchid taxonomy, see Cameron (2007). Current systems of orchid classification divide the family into 5 subfamilies (Chase *et al.* 2003). The largest, with approximately 650 genera and 18,000 species, is Epidendroideae. This subfamily includes the vast majority of tropical epiphytic orchids and those most highly prized as ornamental plants. In contrast, the second subfamily, Orchidoideae, contains almost exclusively terrestrial species classified within approximately 200 genera. These 2 subfamilies, both characterized by flowers containing only a single fertile anther (monandrous), are known to have evolved from a common ancestor. All species within the third subfamily, Vanilloideae, also have flowers with just a single fertile anther, but this condition is considered to have evolved independently from the two other aforementioned monandrous subfamilies (Freudenstein *et al.* 2002). In other words, the reduction in stamen/ anther number from several to one occurred at least twice within Orchidaceae. The hypothetical ancestor of all orchids, as for most monocotyledons in general, probably possessed six stamens within a radially symmetrical flower similar to what we would see in a daylily, onion, or daffodil today. Through the process of evolution, orchid flowers are thought to have undergone significant structural modifications resulting in flowers with pronounced bilateral symmetry, loss of stamens, and fusion of the remaining stamen(s) with the pistil to form a central column. A snapshot of this evolutionary process, frozen in time, can be seen today within the fourth subfamily, Apostasioideae, which contains only two extant genera: *Neuwiedia* and *Apostasia*. Species of *Neuwiedia* possess flowers with three fertile anthers. These are only partially fused with the base of the pistil, and the perianth of the flower is only slightly bilateral in symmetry. These are the most "primitive" of all orchids, in that they show the fewest modifications from the basic blueprint of a hypothetical pre-orchid ancestor. Species of *Apostasia* are similar in floral morphology, but have lost an additional anther so that the mature flowers contain just two. A pair of fertile anthers also defines the fifth orchid subfamily, Cypripedioideae. This group of species, most commonly referred to as "lady's slipper orchids", are classified within just four genera. The labellum or lip of these flowers is always inflated into a sack or pouch. In terms of relative size, Cypripedioideae (ca. 120 species) is more diverse than Apostasioideae (15 species), but less diverse than Vanilloideae (ca. 200 species).

20.2 Diversity Within Vanilloideae

Prior to being classified as their own subfamily of Orchidaceae, the vanilloid orchids were moved around from one orchid group to another. Robert Dressler's (1993) comprehensive, well respected, pre-molecular system of orchid classification listed many of the vanilloid orchids under the category *incertae cedis* (meaning of uncertain status). It is a mix of presumably primitive and advanced floral features among vanilloid orchids that has caused confusion for orchid taxonomists as to how these species should be best classified. At one time, it was suggested that they might be best treated as a family of plants, Vanillaceae, closely related to but separate from Orchidaceae (Lindley 1835). Throughout most of the nineteenth and twentieth centuries, however, they have been placed among the lower branches of the orchid subfamily Epidendroideae (Dressler 1979). The controversy of their position among orchids was eventually laid to rest using comparisons of DNA sequence information. In fact, one of the most unexpected results of the first molecular systematic studies of orchids was the segregation of *Vanilla* and its relatives from the other orchids with a single fertile stamen, and their position near the base of the orchid family tree (Cameron *et al.* 1999). Recognition of the group as a subfamily, therefore, helped to solve one of the better known enigmas of orchid systematics.

20.2.1 Tribe Pogonieae

The most recent systems of classification for subfamily Vanilloideae (Chase *et al.* 2003; Cameron 2003) considers 15 genera divided into 2 tribes: Vanilleae and Pogonieae. A phylogenetic reconstruction of the subfamily is shown in Figure 20.1. Within tribe Pogonieae are both tropical and temperate species. *Duckeella* contains one or possibly

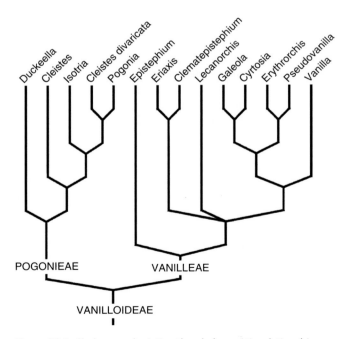

Figure 20.1 Cladogram depicting the phylogenetic relationships among genera of Vanilloideae based on a combination of nuclear, mitochondrial, and plastid DNA sequence data. The subfamily is divided into two tribes: Pogonieae and Vanilleae. Note that *Vanilla* shares a common ancestor with a clade of four genera including *Galeola* and *Pseudovanilla*.

two species from Venezuela and northern Brazil. The genus produces long linear leaves and bright yellow flowers that rise above wet grassland and savanna habitats. It may occasionally be found rooted in mats of floating vegetation. Closely related is the genus *Cleistes* (>30 species), which contains species mostly from tropical South America. These plants are also found in open savannas, but may survive predictable periods of drought by entering an annual state of dormancy and storing food within underground tubers. One species of this genus, *Cleistes divaricata*, is native to the southeastern United States. Detailed systematic studies of the tribe indicate that this species might be better treated as a separate genus (Cameron and Chase 1999). Also native to the United States is the genus *Isotria*, which contains two species. The flowers of *Isotria* are typical of other vanilloid orchids, but the arrangement of leaves into a single whorl is distinctive. These plants are spring ephemerals that emerge and reproduce quickly within their deciduous forest habitat before the tree canopy closes fully during the summer months. They also reproduce asexually via long underground roots, and are strongly mycorrhizal. The fourth genus within Pogonieae is *Pogonia*, which exhibits an interesting geographic disjunction between eastern North America (with one species, *P. ophioglossoides*) and eastern Asia (with 3–5 species). These plants are found most commonly in acidic bogs, around the edges of lakes, and within wet savannas. The fruits of all Pogonieae are true capsules and there is no evidence that they produce vanillin. Because of their dormancy requirements and strong relationships with mycorrhizal fungi in the soil, they are rarely cultivated and should be left undisturbed in their natural habitats.

20.2.2 Tribe Vanilleae

In addition to *Vanilla* itself, the tribe Vanilleae contains eight other genera that are exclusively tropical or subtropical, but wide in distribution, and diverse in morphology. Two of these, *Eriaxis* and *Clematepistephium*, are endemic to the isolated Pacific island of New Caledonia, 1,400 km northeast of Australia. Both genera are monotypic, containing only a single species each. *Eriaxis rigida* is an upright herb with stiff leathery leaves that grows within the full sun, metal-rich, nutrient poor soils of New Caledonia's maquis. The pink flowers of *Eriaxis* are highly attractive, but the genus is nearly impossible to cultivate outside of its native habitat. Its close relative, *Clematepistphium smilacifolium*, could not be more different. This species is restricted to the dense shade habitats of the New Caledonian rainforests, and grows as a climbing vine. In contrast to *Vanilla*, however, the species produces no aerial roots, but instead climbs by means of twining around the trunks of small trees. Its large, leathery leaves exhibit prominent venation patterns that are reticulate (netlike) rather than exclusively parallel as is the norm for monocotyledons (Cameron and Dickinson 1998). Fresh flowers of *Clematepistephium* are primarily green in color, but have been seen by very few people.

At one time, the two New Caledonian endemics described above were classified within the genus *Epistephium*, but that genus of ca. 20 species is now considered to be exclusively South American in distribution. Some species (e.g. *Epistephium ellipticum*) are diminutive, growing only a few centimeters tall, whereas others can grow to a height of nearly 2 m. They are erect herbs of the open savannas, often found in nutrient poor, white sand areas of Brazil and Venezuela, and some even scramble loosely through surrounding vegetation, although none are actual climbers. The leaves of *Epistephium* exhibit reticulate venation like their New Caledonian relatives, and the stunning flowers are mostly dark pink or violet. Like so many other vanilloid orchids, however, they strongly resist cultivation attempts, usually dieing after just a year or two. The fruits of these orchids are capsules that dehisce to release distinctive seeds with circular wings – highly unusual for Orchidaceae (Cameron and Chase 1998).

Winged seeds also are found in three other genera of vanilloid orchids (*Pseudovanilla, Erythrorchis,* and *Galeola*) that are closely related and native to tropical Southeast Asia, northeastern Australia, and a few Pacific islands. All three of these genera are leafless climbing vines, two of which (*Erythrorchis* and *Galeola*) completely lack chlorophyll. These non-photosynthetic genera are exclusively parasitic on fungi, a lifestyle referred to as mycoheterotrophic. The leafless genus *Pseudovanilla* is similar to the other two in most regards, but does eventually develop green pigment within its stems, even if it may persist in a presumably non-photosynthetic state during the juvenile stages of its lifecycle. Recent studies have shown that these orchids are the closest living relatives of *Vanilla* (Cameron and Molina 2006). They climb by means of aerial roots produced at each node of the stem, just like *Vanilla,* and their flowers are structurally similar to those of *Vanilla* species. However, the fruits of all three genera are dry and dehiscent at maturity, in order to release winged seeds for dispersal by air.

There are two other genera of Vanilloideae that persist as non-photosynthetic mycoheterotrophs: *Cyrtosia* and *Lecanorchis.* Both grow as erect herbs within forested areas of Southeast Asia, and both share a number of floral features with *Vanilla.* For example, the fruits of *Cyrtosia* are fleshy and contain small, black, spherical, crustose seeds. In contrast to *Vanilla,* however, the fleshy fruits of *Cyrtosia* are typically bright red, presumably to attract birds or mammals for dispersal (Nakamura and Hamada 1978). The plants themselves are yellow, dull orange, or light brown in color, and may grow more than 1 m tall. *Lecanorchis* plants are more characteristically black or dark brown in color, and less than 20 cm tall. The small flowers of *Lecanorchis* are similar in structure to many species of *Vanilla* in that the labellum is fused with the column along its margins to produce a floral tube. Also like many species of *Vanilla,* the labellum of *Lecanorchis* is ornamented with characteristic bristles and hairs, but *Lecanorchis* fruits are dry capsules lacking odor, and containing numerous dust-like seeds with long slender appendages of unknown adaptive advantage.

20.3 Origins and Age of Vanilloideae

A dogmatic notion has persisted among scholars that the orchid family is a relative newcomer on the evolutionary stage of flowering plants. To support this hypothesis, botanists cite the relatively low levels of genetic diversity among orchid genera and species, many of which can be hybridized easily with one another. They cite the fact that the geologically young Andes of South America and mountains of New Guinea are major centers of orchid diversity. They cite the close relationships between orchids and social bees, which are thought to have evolved much later than other insects, and they cite the fact that most orchid subtribes and genera are found in either the Old World or the New World, but rarely across oceans, suggesting that they evolved long after the separation of the continents.

Studies of Vanilloideae challenge the opinion that all orchid lineages are recently evolved, however, and new information concerning the systematics of Orchidaceae contradict some of the facts mentioned above. For example, *Vanilla* is one of a few orchid genera with a transoceanic distribution. Species are native to North America, South America, Africa, and Asia. Of course, this pattern could be explained by recent long distance dispersal of seeds (from Africa to the Caribbean, for example), but may also be indicative of a much older origin prior to the complete separation of Gondwana. The fact that vanilloid orchids survive as relics in the Guyana Shield of South America, tropical Australia and Africa, Madagascar, and on the island of New Caledonia (a non-volcanic island that separated from Gondwana around 65 million years ago) may also provide evidence of their considerable age. Furthermore, the vanilloid orchids show high rates of molecular divergence between genera as well as between species in some cases.

Vanilloideae are positioned near the base of the orchid family tree, and Orchidaceae is the basal family within the large monocot order Asparagales (which includes the onion, iris, agave, and hyacinth families among others). Recent estimates of the evolutionary age of these plants have calculated that Orchidcaeae may trace their origins back at least 76 million years (Ramirez *et al.* 2007) or even as much as 119 million years (Janssen and Bremer 2004). Vanilloid orchids, in turn, are at least 62 million years old, and probably older. These dates are based on a "molecular clock" approach to determining divergence times among branches within an evolutionary tree constructed with DNA data. Critical to this approach is the use of a calibration point for the "clock", which, in the case of Orchidaceae, has been provided by a 15- to 20-million-year-old fossil specimen of orchid pollen attached to an extinct species of bee preserved in amber (Ramirez *et al.* 2007).

20.4   Diversity Within *Vanilla*

Of the 15 genera now classified within Vanilloideae and already discussed, *Vanilla* is the most diverse. A formal monograph of the genus has never been published, and the last treatment considering all the species was written over 50 years ago by Portères (1954), who based his work on a revision by Rolfe (1896). A comparison of alternative systems of classification is presented in Table 20.1. The current worldwide checklist of all orchid species today recognizes 110 species of *Vanilla* (Govaerts *et al.* 2008). Half of these (ca. 61 species) are native to tropical South America, Central America, subtropical North America, and the Caribbean (i.e. they are Neotropical plants). Africa claims ca. 23 native species, with at least 5 of these restricted to Madagascar. The remaining species of *Vanilla* are found on the Indian subcontinent and throughout tropical Southeast Asia. No species of *Vanilla* are native to Australia or to the Pacific Islands. *Vanilla tahitensis* is grown throughout French Polynesia, but this "species" recently was shown to be a hybrid between Neotropical *V. planifolia* (the maternal parent) and *V. odorata* (the paternal parent). It is not a natural native of Polynesia (Lubinsky *et al.* 2008).

Species of *Vanilla* were formally classified by Rolfe (1896) into subgeneric sections based primarily on vegetative features and secondarily on flower morphology. *Vanilla* section *Aphyllae* was proposed in order to accommodate all of the leafless species in the genus (e.g. *V. aphylla*, *V. barbellata*, *V. roscheri*, and others). All of these are fully

Table 20.1 Comparison of infrageneric classification systems for *Vanilla*, including an informal system based on recent molecular phylogenetic studies. Number of species classified by Rolfe (1896) and Portères (1954) within each group are given.

Rolfe (1896)	Portères (1954)	DNA-based phylogeny (2009)	Representative species
Section *Aphyllae* (13 spp.)	Section *Aphyllae* (18 spp.)		
Section *Foliosae* (37 spp.)	Section *Foliosae*		
	Subsection *Membranaceae* (15 spp.)	Membranaceous clade	*V. mexicana*
	Subsection *Papillosae* (28 spp.)		
	Subsection *Lamellosae* (49 spp.)		
		Neotropical, leafy, fragrant clade	*V. planifolia*
		Paleotropical clade	
		V. africana subclade	*V. africana*
		African subclade	*V. roscheri*
		Caribbean leafless subclade	*V. barbellata*
		Asian subclade	*V. aphylla*

photosynthetic plants with green stems, but have either lost their leaves entirely or have modified them into rigid hook-like structures (e.g. *V. poitaei*) to assist with climbing. They are found most commonly in hot, xeric, full sun habitats where leaf reduction is probably an adaptation for water retention. Species within this section can be found in the Caribbean, continental Africa, Madagascar, and Asia. Although some of them produce fleshy fruits, there is no evidence that these are aromatic or worthy of cultivation as crop plants. Rolfe's classification of these species together implies that they share a recent common ancestor, but molecular studies have demonstrated that this is not the case (Cameron 2005). Instead, there appear to be at least three independent cases of leaf reduction/leaf loss in *Vanilla* – at least once in Africa, in the Caribbean, and in Asia. The section, therefore, is not monophyletic, but an artificial grouping of species with shared vegetative morphology derived by convergent evolution.

For the remaining species not classified in *Vanilla* subgenus *Aphyllae*, Rolfe created section *Foliosae*. As the name indicates, all of these are leafy. Since this is such a large group, Portères (1954) later divided the section further into subsections. *Vanilla* section *Foliosae* subsection *Membranaceae* is a small cluster of species characterized by wiry stems, thin leaves, short aerial roots, and flowers in which the labellum is not fused with the column. The labellum also lacks the complex bristles, hairs, and scales characteristic of other *Vanilla* species, and the fruits tend to dry on the vines and split lengthwise. In some species of this section, flowers are born solitary within leaf axils rather than in congested racemes. *Vanilla mexicana* exemplifies this subsection, and molecular systematic studies have demonstrated that the group is basal relative to all other species within the genus. Plants tend to grow in wet, shady habitats with rich organic soils, and probably survive in close association with mycorrhizal fungi. They are very difficult to transplant or cultivate, and there is no evidence that the fruits produce aromatic vanillin.

 Vanilla planifolia, *V. pompona*, and many other species of the genus were classified by Portères into *Vanilla* section *Foliosae* subsection *Lamellosae*. The group is so named because species within this group are characterized by flowers with flattened scale-like appendages (lamellae), and complex ornamentation on their labella. A third subsection, *Papillosae*, was erected for the 28 or so species characterized by fleshy leaves and flowers usually with thick trichomes running up the center of the labellum, but without lamellate scales. The labellum of all these flowers is always fused to the column along its margins to form a floral tube. Species within *Vanilla* section *Foliosae* are pantropical in distribution, but recent molecular systematic studies have demonstrated that this group is also artificial. Instead, species of *Vanilla* cluster by geography with leafy and leafless species intermixed, as can be seen in Figure 20.2. Specifically, all Old World species

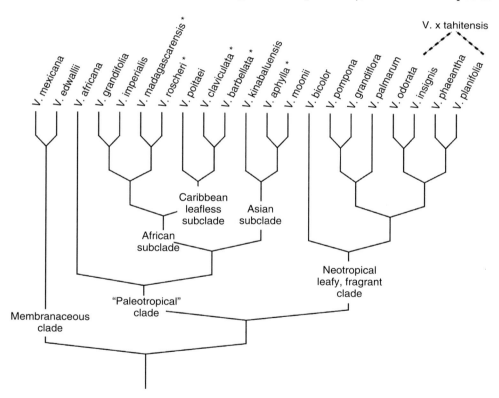

Figure 20.2 Phylogenetic relationships among select species of *Vanilla*. The cladogram is based on molecular sequence data from different genes including nuclear ribosomal ITS, plastid *rbcL*, *matK*, *rpoC*1, and others. Fully leafless species are marked with an asterisk (*). The hybrid origin of *V. tahitensis* from a cross between *V. odorata* and *V. planifolia* is highlighted by the dashed lines. Informal clade and subclades are labeled on the branch representing the common ancestor of each major species group.

(from Africa and Asia) share a common ancestor together with the leafless New World species. The latter were probably dispersed from Africa to the Caribbean via Atlantic hurricanes. All remaining Neotropical species, including *V. planifolia*, share a different common ancestor. It is within this group that aromatic fruits producing significant levels of vanillin are found. As such, the group has informally been named the "American fragrant species". Note that molecular studies position *V. tahitensis* (the only known tetraploid within the genus) inside this group of Neotropical relatives, thereby confirming the hybrid origin of Old World Tahitian vanilla from New World species. Whether the hybrids were man-made by Europeans in the Old World or were naturally occurring in Mesoamerica prior to European contact is uncertain.

20.5 Systematic Conclusions and Implications

Considering that *Vanilla* and its relatives have been surrounded by systematic controversy and uncertainly for such a long time, it is encouraging to witness the increased level of understanding regarding their classification, evolution, and origins that has come about in recent years. The relationship of these orchid species to one another, and to other orchids, has only been clarified during the past decade, especially as DNA data has been incorporated into systematic studies. Up to this point the vanilloid orchids resisted being shoehorned into any particular sub-tribe, tribe, or subfamily of Orchidaceae. We now realize that the single fertile anther of the *Vanilla* flower's column arose via a different evolutionary process compared to other orchid subfamilies with similar floral morphology. Future genetic studies into the structure and development of the *Vanilla* flower/fruit would be well advised to consider other genera of Vanilloideae with shared ancestry, rather than making direct comparisons to more distantly related orchids. Such comparisons could be misleading in their homology assumptions.

In a similar vein, it has become clear that the orchid family is older than previously hypothesized, and that living vanilloid orchids can trace their shared common ancestry back to around 65 million years ago. Through the process of evolution they have become adapted to a variety of habitats, pollinators, and seed dispersal strategies. Yet they share a common suite of genes. Fundamental differences in gene expression and regulation ultimately determine whether a vanilloid orchid is tropical or can survive subzero temperatures, whether it grows as an erect herb in open savannas or as a vine in the forest under-story, and whether it will produce a dry capsule or an aromatic fleshy fruit. As we move forward into the age of genomics, the manipulation of genes will become easier and more common. Future genomic studies, especially those targeting the improvement of *Vanilla* as a crop plant, may also want to study other closely related genera of tribe Vanilleae or even the entire subfamily Vanilloideae. Imagine the possibilities for increased cultivation potential, for example, if the *Vanilla planifolia* genome was modified to incorporate genes from *Cyrtosia*, a closely related erect herb capable of surviving in the cool climate of temperate China. Alternatively, the closely related *Epistephium* genome might provide clues as to the manner in which many vanilloid orchids exhibit drought tolerance and thrive in the full sun. Future *Vanilla* plantations might be established in areas with a variety of different climatic conditions as a result of genetic modification, and without a need for structures to support climbing vines.

At the same time that our understanding of Vanilloideae systematics has improved, systematic studies within the genus *Vanilla* have also broadened our view of its species, both wild and cultivated. Recent publications into the hybrid origin and population genetics of *Vanilla tahitensis* are the most notable. The genus is more diverse than people in the vanilla industry realize, and there are certainly a number of species other than *Vanilla planifolia* that produce vanillin and characteristic chemical profiles within their aromatic fruits. Focused breeding among species of the American fragrant *Vanilla* group would be a logical place to begin, if the development of novel flavors and/or fragrances was a desired goal.

Systematics is a unifying science that draws on tools from a variety of disciplines to address questions of species status, geographic distribution, taxonomy, nomenclature, and evolution – that is, all aspects of plant diversity. Many systematic questions centered around *Vanilla* have been addressed in recent years. For example, we now understand that classification systems emphasizing the leafy versus leafless habit of *Vanilla* species do not reflect actual patterns of evolution. Instead, geography is a better indicator of species relationships. At the same time that long standing questions are answered, however, new questions present themselves. For example, is *Vanilla pompona* a single species or a complex of several closely related species? Did *Vanilla tahitensis* ever persist as a naturally occurring hybrid in Mesoamerica or is it entirely man-made? How many native species of leafless *Vanilla* actually exist in Madagascar? Are reports of wild *Vanilla planifolia* vines in Mexico nothing more than examples of cultivated plants that have escaped to become feral? These and other systematic questions focused on *Vanilla* and its relatives eventually will be taken up by future generations of botanists. *Vanilla planifolia* may not be absolutely unique within Orchidaceae, but the vanilloid orchids offer a number of unique opportunities to better understand the basic biology of the largest family of plants on Earth, and humankind's singularly most popular flavor and fragrance.

References

Cameron, K.M. (1999) Biogeography of Vanilloideae (Orchidaceae). *XVI International Botanical Congress, Abstracts, 749*. St Louis, Missouri.

Cameron, K.M. (2000) Gondwanan Biogeography of Vanilloideae (Orchidaceae). *Southern Connections Congress, Programme and Abstracts, Lincoln*, New Zealand, pp. 25–26.

Cameron K.M. (2003) Vanilloideae. In: *Genera Orchidacearum Vol 3*, Pridgeon, A., Cribb, P, Chase, M. and Rasmussen, F. (Eds), Oxford University Press, Oxford, 281–334.

Cameron, K.M. (2004) Utility of plastid *psaB* gene sequences for investigating intrafamilial relationships within Orchidaceae. *Molecular Phylogenetics and Evolution*, 31, 1157–1180.

Cameron, K.M. (2005) Recent advances in the systematics biology of *Vanilla* and related orchids (Vanilloideae, Orchidaceae). In *Vanilla: First International Congress*. Allured Publishing, Carol Stream, IL.

Cameron, K.M. (2006) A comparison of plastid *atpB* and *rbcL* gene sequences for inferring phylogenetic relationships within Orchidaceae. In *Monocots: Comparative Biology and Evolution*, Columbus, J.T., Friar, E.A., Porter, J.M., Prince, L.M. and Simpson, M.G. (Eds), Rancho Santa Ana Botanic Garden, Claremont, CA, pp. 447–464.

Cameron, K.M. (2007) Molecular phylogenetics of Orchidaceae: the first decade of DNA sequencing. In *Orchid Biology Reviews and Perspectives Vol. IX*, Cameron, K., Arditti, J. and Kull, T. (Eds), The New York Botanical Garden Press, Bronx, NY, pp. 163–200.

Cameron, K.M. and Chase, M. (1998) Seed morphology of the vanilloid orchids. *Lindleyana*, 13, 148–169.

Cameron, K.M. and Chase, M.W. (1999) Phylogenetic relationships of Pogoniinae (Vanilloideae, Orchidaceae): an herbaceous example of the eastern North America-eastern Asia phytogeographic disjunction. *Journal of Plant Research*, 112, 317–329.

Cameron, K.M. and Dickinson, W.C. (1998) Foliar architecture of vanilloid orchids: insights into the evolution of reticulate leaf venation in monocotyledons. *Botanical Journal of the Linnean Society*, 128, 45–70.

Cameron, K. and Molina, M.C. (2006) Photosystem II gene sequences of *psbB* and *psbC* clarify the phylogenetic position of *Vanilla* (Vanilloideae, Orchidaceae). *Cladistics*, 22, 239–248.

Cameron, K.M., Chase, M., Whitten, M. et al. (1999) A phylogenetic analysis of the Orchidaceae, evidence from *rbcL* nucleotide sequences. *American Journal of Botany*, 86, 208–224.

Chase, M.W., Cameron, K.M., Barrett, R. and Freudenstein, J.F. (2003) DNA Data and Orchidaceae systematics: a new phylogenetic classification. In *Orchid Conservation*, ed. K.W. Dixon, Kell, S.P., Barrett, R.L. and Cribb, P.J. (Eds), Natural History Publications, Kota Kinabalu, Sabah, pp. 69–89.

Dressler, R.L. (1979) The subfamilies of the Orchidaceae. *Selbyana*, 5, 197–206.

Dressler, R.L. (1993) *Phylogeny and Classification of the Orchid Family*. Dioscorides Press, Portland, OR.

Freudenstein, J., Harris, E. and Rasmussen, F. (2002) The evolution of anther morphology in orchids: incumbent anthers, superposed pollinia, and the vandoid complex. *American Journal of Botany*, 89, 1747–1755.

Govaerts, R., Campacci, M.A., Holland Baptista, D. et al. (2008) World Checklist of Orchidaceae. The Board of Trustees of the Royal Botanic Gardens, Kew. http://www.kew.org/wcsp/ (accessed February 26, 2008)

Janssen, T. and Bremer, K. (2004) The age of major monocot groups inferred from 800 + *rbcL* sequences. *Botanical Journal of the Linnaean Society*, 146, 385–398.

Lindley, J. (1835) *Key to Structural, Physiological and Systematic Botany*. Longman, London.

Lubinsky, P., Cameron, K.M., Molina, M.C. *et al.* (2008) Neotropical roots of a Polynesian spice: the hybrid origin of Tahitian vanilla, *Vanilla tahitensis* (Orchidaceae). *American Journal of Botany*, 95, 1040–1047.

Nakamura, S.J. and Hamada, M. (1978) On the seed dispersal of an achlorophyllous orchid, *Galeola septetrionalis*. *Journal of Japanese Botany*, 53, 260–263.

Portères, R. (1954) Le genere *Vanilla* et ses especes. In *Le Vanillier et la Vanille dans le Monde*, ed. Bouriquet, G.,(Ed.), Paul Lechevalier, Paris, pp. 94–290.

Ramirez, S.R., Gravendeel, B., Singer, R.B., Marshall, C.R. and Pierce, N.E. (2007) Dating the origin of the Orchidaceae from a fossil orchid with its pollinator. *Nature*, 448, 1042–1045.

Rolfe, R.A. (1896) A revision of the genus *Vanilla*. *Journal of the Linnaean Society*, 32, 439–478.

21

Molecular Analysis of a Vanilla Hybrid Cultivated in Costa Rica

Faith C. Belanger and Daphna Havkin-Frenkel

There are 110 species in the genus *Vanilla*, but only two, *V. planifolia* and *V. tahitensis*, are permitted by the United States FDA (21CFR169.175) to be used in food products. *V. planifolia* is known to have originated in Central America and is currently widely grown commercially in Mexico, Madagascar, India, Papua New Guinea, Uganda, and Indonesia. The origin of *V. tahitensis*, which is cultivated in Tahiti and Papua New Guinea, has been the subject of much speculation and folklore, since it is not known to exist outside of cultivation. One common speculation has been that it is a hybrid originating from a cross between *V. planifolia* and *V. pompona*. Recently DNA sequence analysis revealed that in fact *V. tahitensis* is a likely hybrid between the Central American species *V. planifolia* and *V. odorata*, with *V. planifolia* as the maternal parent (Lubinsky *et al.* 2008). How *V. tahitensis* came to Tahiti has not been documented and why no wild plants have been found in Central America is still not known.

In addition to *V. planifolia* and *V. tahitensis*, some of the other Central American *Vanilla* spp. also produce aromatic flavor compounds in their pods, and have been used in some areas of Central and South America (Weiss 2002). The history of *Vanilla* use in South America is reviewed in Chapter 8 by Lubinsky *et al.* Also, the evolutionary relationships of some *Vanilla* spp. are reviewed in Chapter 14 by Cameron.

There is a vanilla cultivar widely grown in Costa Rica that is commonly considered to be a hybrid between *V. planifolia* and *V. pompona* (Schluter *et al.* 2007). In Chapter 3, Quirós describes the cultivar "Vaitsy" that was brought from Madagascar to Costa Rica. The cultivar is said to have been propagated through tissue culture and then widely disseminated to farmers throughout Costa Rica. Vaitsy is widely grown in Costa Rica because it has good resistance to the fungal disease caused by *Fusarium oxysporum* f. sp. *vanillae*. To our knowledge, the origin in Madagascar of Vaitsy has not been documented.

In Chapter 2, Hernandez-Hernandez discusses the *Fusarium* resistance of the cultivar "Tsy Taitry" developed in Madagascar. Tsy Taitry was developed from an interspecific hybrid between *V. planifolia* and *V. pompona* that was then backcrossed to *V. planifolia* (Bory *et al.* 2008a). Is the cultivar Vaitsy in Costa Rica the same as the interspecific hybrid cultivar Tsy Taitry developed in Madagascar? There is similarity in the names. However, without documentation from the individuals involved, we cannot answer this question.

Handbook of Vanilla Science and Technology, Second Edition.
Edited by Daphna Havkin-Frenkel and Faith C. Belanger.
© 2019 John Wiley & Sons Ltd. Published 2019 by John Wiley & Sons Ltd.

In this chapter we present DNA sequence data confirming the hybrid origin of a *Vanilla* cultivated in Costa Rica. The plant used for this analysis was from a plantation in Costa Rica and was considered to be a hybrid plant.

21.1 Methods

21.1.1 PCR Amplification, Cloning, and DNA Sequencing

Genomic DNA from leaf tissue of the putative hybrid *Vanilla* from Costa Rica, *V. planifolia*, *V. pompona*, and *V. tahitensis* was isolated by using a modification of the CTAB (hexadecyltrimethylammonium bromide) procedure (Doyle and Doyle 1987). The leaf tissue was first freeze dried and then 0.3 g of dried tissue was ground to a powder with 100 mg of PVPP in a mortar. CTAB extraction buffer (100 mM Tris-HCl, ph 8.0, 1.4 M NaCl, 20 mM EDTA, 1% 2-mercaptoethanol, 2% CTAB) (9 mls) was added to the powder and the mixture transferred to a centrifuge tube. The sample was incubated at 65 °C for 1 h, with mixing by inversion every 10 min. The sample was cooled for 10 min and then extracted by inversion for 5 min with an equal volume of chloroform:isoamyl alcohol (24:1, v/v). The sample was centrifuged for 10 min at 3,000 rpm and the aqueous layer was removed to a new tube and incubated with 5 ul of 100 mg/ml RNAse for 30 min at room temperature. The sample was extracted with chloroform:isoamyl alcohol (24:1 v/v) as before, the DNA in the aqueous layer precipitated by adding 2/3 volume of isopropanol and incubated at −20 °C overnight. The DNA was collected by centrifugation at 10,000 rpm for 15 min. and dissolved in 300 ul of TE. The DNA was again precipitated by adding 150 ul of 7.5 M ammonium acetate and 900 ul of ethanol, incubated at −20C overnight, and collected by centrifugation. The final DNA pellet was dissolved in 200 ul of TE.

The primers used for amplification of the nuclear 5.8S rDNA gene and the flanking internal transcribed spacer (ITS) regions were 5′TATGCTTAAAYTCAGCGGGT3′ and 5′AACA AGGTTTCCGTAGGTGA3′ (Cameron 2009). Approximately 0.5 μg of genomic DNA was used per 50 μl PCR reaction, which contained 10 mM Tris-HCl, pH 8.3, 50 mM KCl, 1.5 mM MgCl$_2$, 0.2 mM each dNTP, 20 pmoles each primer, and 1 μl of Taq polymerase. The initial denaturation was conducted at 94 °C for 4 min, followed by 30 cycles of 30 s denaturation at 94 °C, 30 s annealing at 52 °C, and 1 min extension at 72 °C, followed by a final extension at 72 °C for 10 min. The PCR products were directly cloned into the pGEM-T Easy vector and plasmids from individual transformed colonies were sequenced. GenBank accession numbers for the ITS clones generated in this study are GQ867234 to GQ867274.

The primers used for amplification of the chloroplast *psaB* gene were 5′ACGC GTCGTATTTGGTTTGGTATTGC3′ and 5′CAATGCCAATAAAAAGTAACCCATCC3′ (Cameron 2004). The PCR conditions were as described above with an annealing temperature of 60 °C. The PCR products were gel purified and sequenced using the amplification primers and one additional primer (5′ATGACCAATTCCAAAATTAGTTCTATACAT3′ (Cameron 2004). GenBank accession numbers for the *psaB* clones generated in this study are GQ888507 to GQ888510.

21.1.2 Phylogenetic Analysis

The CLUSTAL-X (Thompson *et al.* 1997) program was used to align DNA sequences. For the ITS analysis, the sequences included the ITS-1, 5.8S rDNA, and ITS-2 regions. For the phylogenetic analyses, the sequences were trimmed to include only the regions of sequence overlap for all the sequences in the analysis. Phylogenetic analysis was performed with the PAUP* program (version 4.0b10 for Macintosh; Swofford 2002). Phylogenetic trees were generated by using the maximum parsimony full heuristic search option set to simple sequence addition, tree-bisection-reconnection (TBR) branch swapping, and MulTrees on, with 1,000 bootstrap replications. Gaps were treated as missing data.

21.1.3 Preparation of Vanilla Extracts

Ten grams of cured beans were ground and mixed with 100 mL of 40% ethanol and placed in the oven at 60 °C overnight. The mix was filtered, brought up to 100 mL and the filtrate was used for HPLC analysis of four flavor components as described previously (Podstolski *et al.* 2002). This is the standard procedure commonly used in the industry for extraction.

21.2 Results and Discussion

Natural interspecific hybridization is recognized as an important factor in plant speciation (Soltis and Soltis 2009), and was the origin of many of the world's crop species (Hancock 1992; Hughes *et al.* 2007). Humans have intentionally used interspecific hybridization to improve crops, and in some cases have developed new species (Goodman *et al.* 1987; Lukaszewski and Gustafson 1987). Interspecific, and even intergeneric hybridization, is commonly used in developing new ornamental orchid varieties (Vainstein 2002).

As discussed above, interspecific hybridization between *V. planifolia* and *V. odorata* is considered to be the origin of the commercially important species *V. tahitensis* (Lubinsky *et al.* 2008). Since the hybridization event is believed to predate the documented understanding of how to hand pollinate *Vanilla* (Lubinsky *et al.* 2008), it is likely to have occurred naturally. Efforts have been made to improve cultivated *V. planifolia* through intentional interspecific hybridization. A breeding program in Madagascar produced *V. planifolia* × *V. tahitensis* and *V. planifolia* × *V. pompona* hybrids (Bory *et al.* 2008a). The cultivar Tsy Taitry mentioned above was developed by backcrossing a *V. planifolia* × *V. pompona* hybrid to *V. planifolia* (described by Bory *et al.* 2008a). An interspecific hybridization program was also carried out in Puerto Rico, where hybrids between *V. planifolia*, *V. phaentha*, and *V. pompona* were developed (Childers *et al.* 1988; Knudson 1950; Theis and Jimenez 1957). More recently, production of interspecific hybrids between *V. planifolia* and *V. aphylla*, a leafless species, was reported (Divakaran *et al.* 2006). The trait of particular interest that was driving these projects is resistance to *Fusarium*, since losses to the disease can be devastating. *V. planifolia* is susceptible to the disease, whereas *V. aphylla*, *V. pompona*, and *V. phaentha* are reported to be tolerant to the disease (Divakaran *et al.* 2006; Weiss 2002; Childers *et al.* 1988). Introgression of disease resistance into a crop species from related species is a commonly used approach in plant breeding. As described above, the cultivars Tsy Taitry and Vaitsy are reported to have better tolerance to *Fusarium* than does *V. planifolia*.

The sequences of the nuclear 5.8S ribosomal DNA (rDNA) gene and the flanking internal transcribed spacers (ITS) are frequently used in plant phylogenetic studies to deduce the evolutionary relationships of different species. Although the rDNA genes are present in high copy within the genome, in diploid species concerted evolution has been found to result in homogenization of the sequences, such that within a species a single sequence often predominates, although some sequence variants can be maintained (Baldwin *et al.* 1995; Alvarez and Wendel 2003; Bailey *et al.* 2003; Small *et al.* 2004). As already discussed, past natural interspecific hybridization is the origin of many plant species. Such allopolyploid species may have distinct ITS sequence variants originating from their ancestral parental species. Phylogenetic placement of such divergent sequences into different clades can be used to infer the identities of the subgenome donors (Gaut *et al.* 2000; Sang *et al.* 1995). This was the approach used to identify the parental species of *V. tahitensis* (Lubinsky *et al.* 2008). However, in some polyploid species, over evolutionary timescales, concerted evolution has also occurred between homoeologous loci and has resulted in homogenization of the ITS regions to one of the parental types (Wendel *et al.* 1995; Wang *et al.* 2000; Fortune *et al.* 2008, Rotter *et al.* 2010).

F1 hybrids and recent backcross progenies are expected to have ITS copies from both parental species, and sequence analysis of cloned ITS fragments can be used to confirm the hybrid status of the plants. Such an analysis was used to determine the parental species of a recently formed natural *Citrus* hybrid (Xu *et al.* 2006). We used analysis of ITS sequences to confirm the hybrid nature of a *Vanilla* widely cultivated in Costa Rica. The ITS region of *V. planifolia*, *V. pompona*, and the presumed hybrid, was PCR amplified, cloned into a bacterial plasmid, and sequences of 6 to 28 clones per plant sample were obtained as described in the Methods Section. Maximum parsimony phylogenetic analysis of the *Vanilla* spp. ITS sequences is shown in Figure 21.1. The tree was based upon 686 total characters, of which 342 were constant, 163 variable characters were parsimony uninformative, and 181 characters were parsimony informative. Figure 21.1 includes previously reported ITS sequences, which were generated by direct sequencing of PCR products, and the sequences of the cloned ITS PCR products from *V. planifolia*, *V. pompona*, and the hybrid plant generated in this study. The accession numbers of the sequences used in the analysis presented in Figure 21.1 are listed in Table 21.1. The sequence from *Clematepistephium smilacifolium*, a species also in the tribe Vanilleae, was also included in the analysis. The *Pogonia japonica* sequence was chosen to root the tree, since the tribe Pogonieae is sister to the tribe Vanilleae (Cameron 2009).

In this analysis the *V. planifolia* and *V. pompona* sequences grouped into separate clades. Sequences from the Costa Rican hybrid were found in both the *V. planifolia* and *V. pompona* clades, suggesting both of these species were the parental species of the hybrid. For simplicity of presentation, in the tree presented in Figure 21.1, only five hybrid clone sequences are shown in each of the two clades. The phylogenetic tree produced when all the hybrid clones generated from this study were included, had the identical topology, with 6 hybrid sequences in the *V. planifolia* clade and 22 hybrid sequences in the *V. pompona* clade (not shown). Also, neighbor-joining analysis of the data generated a tree of similar topology (not shown).

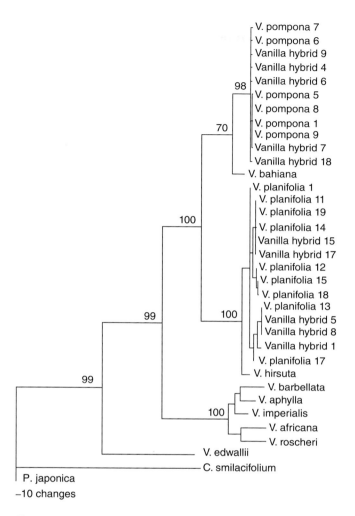

Figure 21.1 Rooted maximum parsimony phylogenetic tree of nuclear ITS sequences. The tree length is 872 and the consistency index is 0.727. The *Pogonia japonica* sequence was designated as the outgroup for rooting the tree. The numbers at the nodes are the bootstrap percentages based on 1,000 replications. For simplicity, the bootstrap percentages at the minor nodes are not labeled, but all were 60% or greater. Accession numbers of the sequences are given in Table 21.1.

In contrast to the biparental inheritance of nuclear genes such as the 5.8S rDNA genes, chloroplast genes are maternally inherited. Phylogenetic analysis of chloroplast genes can therefore be used to determine the maternal parent in an interspecific hybrid. We chose to analyze the chloroplast DNA encoded *psaB* gene, since sequences from other *Vanilla* spp. have been reported. The *psaB* gene encodes one of the subunits of the photosystem I protein complex (Shimada and Sugiura 1991). Maximum parsimony phylogenetic analysis of the *Vanilla* spp. *psaB* sequences is shown in Figure 21.2. The tree was based upon 1,627 total characters, of which 1,484 were constant, 88 variable characters were parsimony uninformative, and 55 variable characters were parsimony

Table 21.1 Accession numbers of DNA sequences used in the phylogenetic analyses presented in Figures 21.1 and 21.2.

Sequence name	Accession number	Source
ITS Sequences		
C. smilacifolium	FJ425838	Cameron 2009
P. japonica	AF151011	Cameron and Chase 1999
V. africana	FJ425834	Cameron 2009
V. aphylla	AF151006	Cameron and Chase 1999
V. bahiana	EU498163	Pansarin *et al.* 2008
V. barbellata	FJ425835	Cameron 2009
V. edwalii	EU498165	Pansarin *et al.* 2008
V. hirsuta	AF391786	Clements *et al.* 2002
V. imperialis	FJ425830	Cameron 2009
V. planifolia 1	AF391786	Clements *et al.* 2002
V. planifolia 11	GQ867239	This study
V. planifolia 12	GQ867240	This study
V. planifolia 13	GQ8672241	This study
V. planifolia 14	GQ867242	This study
V. planifolia 15	GQ867243	This study
V. planifolia 17	GQ867244	This study
V. planifolia 18	GQ867245	This study
V. planifolia 19	GQ867246	This study
V. pompona 1	EU498164	Pansarin *et al.* 2008
V. pompona 5	GQ867234	This study
V. pompona 6	GQ867235	This study
V. pompona 7	GQ867236	This study
V. pompona 8	GQ867237	This study
V. pompona 9	GQ867238	This study
V. roscheri	FJ425840	Cameron 2009
Vanilla hybrid 1	GQ867248	This study
Vanilla hybrid 4	GQ867271	This study
Vanilla hybrid 5	GQ867273	This study
Vanilla hybrid 6	GQ867265	This study
Vanilla hybrid 7	GQ867266	This study
Vanilla hybrid 8	GQ867261	This study
Vanilla hybrid 9	GQ867257	This study
Vanilla hybrid 15	GQ867253	This study
Vanilla hybrid 17	GQ867254	This study
Vanilla hybrid 18	GQ867247	This study

Table 21.1 (Continued)

Sequence name	Accession number	Source
psaB Sequences		
C. smilacifolium	AY380964	Cameron 2004
P. japonica	AY381061	Cameron 2004
V. africana	AY381089	Cameron 2004
V. aphylla	AY381088	Cameron 2004
V. barbellata	AY381090	Cameron 2004
V. edwallii	EU498093	Pansarin et al. 2008
V. imperialis	AY381091	Cameron 2004
V. inodora	AY381092	Cameron 2004
V. palmarum 1	AY381093	Cameron 2004
V. palmarum 2	EU498092	Pansarin et al. 2008
V. planifolia	GQ888507	This study
V. pompona	GQ888509	This study
V. roscheri	AY381095	Cameron, 2004
V. tahitensis	GQ888508	This study
Vanilla hybrid	GQ888510	This study

informative. Phylogenetic analysis of the *psaB* DNA sequences revealed the close relationship of the sequence from the Costa Rican hybrid plant with that of *V. planifolia* and *V. tahitensis*. Neighbor-joining analysis of the data generated a tree of identical topology (not shown). *V. planifolia* was previously determined to be the maternal parent of *V. tahitensis* (Lubinsky *et al.* 2008), and now we can conclude it was also the maternal parent of the Costa Rican hybrid plant.

Overall, the DNA sequence analysis of the nuclear ITS region and the chloroplast *psaB* gene presented here confirms that the *Vanilla* hybrid widely grown in Costa Rica originates from interspecific hybridization, with *V. planifolia* as the maternal parent and *V. pompona* as the paternal parent. However, it is not possible to determine from this analysis if the plant is the F1 hybrid or the progeny of a backcross.

A sample of cured beans from the hybrid and from medium-grade beans from Madagascar was extracted and the main flavor compounds analyzed by HPLC. The amounts and the ratios of the measured compounds were within the expected ranges for beans from Madagascar (Tables 21.2 and 21.3).

In summary, by using DNA sequence analysis, we have confirmed the hybrid origin of a *Vanilla* cultivar widely grown in Costa Rica. The analysis of nuclear and plastid genes from the hybrid revealed that it originated from an interspecific cross, with *V. planifolia* as the maternal parent and *V. pompona* as the paternal parent. Analysis of four marker compounds from cured beans indicated the extract from the hybrid is similar to extracts from *V. planifolia*. That a *Vanilla* originating from interspecific hybridization is being widely cultivated and is reported to have tolerance to *Fusarium* underscores the potential for hybridization and breeding in developing improved cultivars of *Vanilla*.

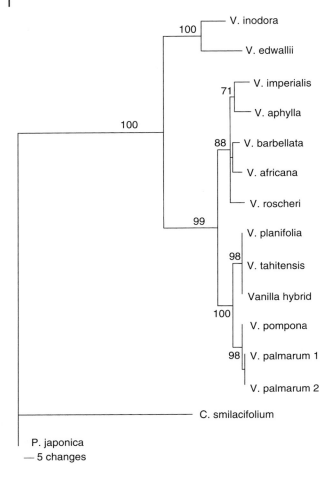

Figure 21.2 Rooted maximum parsimony phylogenetic tree of chloroplast *psaB* sequences. The tree length is 164 and the consistency index is 0.921. The *Pogonia japonica* sequence was designated as the out-group for rooting the tree. The numbers at the nodes are the bootstrap percentages based on 1,000 replications. For simplicity, the bootstrap percentages at the minor nodes are not labeled, but all were 50% or greater. Accession numbers of the sequences are given in Table 21.1.

Table 21.2 Analysis of marker compounds from cured beans from the Costa Rican hybrid and medium-grade beans from Madagascar.

Compound	Total mg	mg/mL	% Dry Wt
Costa Rican hybrid beans			
p-OH-Benzoic acid	1.91	0.01	0.0076
p-OH-Benzaldehyde	28.28	0.12	0.18
Vanillic acid	13.70	0.06	0.08
Vanillin	280.07	1.23	1.76

Table 21.2 (Continued)

Compound	Total mg	mg/mL	% Dry Wt
Madagascar beans			
p-OH-Benzoic acid	5.84	0.02	0.03
p-OH-Benzaldehyde	16.53	0.07	0.09
Vanillic acid	25.76	0.10	0.15
Vanillin	254.49	1.08	1.45

Table 21.3 Comparison of ratios of marker compounds from the Costa Rican hybrid beans and medium-grade beans from Madagascar with expected ratios.

	Costa Rican Hybrid	Madagascar	Expected Ratios[a]
Vanillin/pHB ald	10.25	15.42	10–20
Vanillin/Vanillic acid	20.5	10.8	12–29
pHB acid/pHB ald	0.08	0.28	0.15–0.35
Vanillic acid/pHB ald	0.5	1.43	0.53–1.5

a) *From Gassenmeier et al.* 2008.

References

Alvrez, I. and Wendel, J.F. (2003) Ribosomal ITS sequences and plant phylogenetic inference. *Molecular Phylogenetics and Evolution*, 29, 417–434.

Bailey, C.D., Carr, T.G., Harris, S.A. and Hughes, C.E. (2003) Characterization of angiosperm nrDNA polymorphism, paralogy, and pseudogenes. *Molecular Phylogenetics and Evolution*, 29, 435–455.

Baldwin, B.G., Sanderson, M.J., Porter, J.M., Wojciechowski, M.F., Campbell, C.S. and Donoghue, M.J. (1995) The ITS region of nuclear ribosomal DNA: a valuable source of evidence on angiosperm phylogeny. *Annals of the Missouri Botanical Garden*, 82, 247–277.

Bory, S., Lubinsky, P., Risterucci, A-M. *et al.* (2008a) Patterns of introduction and diversification of *Vanilla planifolia* (Orchidaceae) in Reunion Island (Indian Ocean). *American Journal of Botany*, 95, 805–815.

Bory, S., Grisoni, M., Duval M-F. and Besse, P. (2008b) Biodiversity and preservation of vanilla: present state of knowledge. *Genetic Resources and Crop Evolution*, 55, 551–571.

Cameron, K.M. (2004) Utility of plastid *psaB* gene sequences for investigating intrafamilial relationships within Orchidaceae. *Molecular Phylogenetics and Evolution*, 31, 1157–1180.

Cameron, K.M. (2009) On the value of nuclear and mitochondrial gene sequences for reconstructing the phylogeny of vanilloid orchids (Vanilloideae, Orchidaceae). *Annals of Botany*, 104, 377–385.

Cameron, K.M. and Chase, M.W. (1999) Phylogenetic relationships of Pogoniinae (Vanilloideae, Orchidaceae): an herbaceous example of the eastern North America – eastern Asia phytogeographic disjunction. *Journal of Plant Research*, 112, 317–329.

Childers, N.F., Cibes, H.R. and Hernandez-Medina, E. (1988) *Vanilla* – the orchid of commerce. In: The Orchids: a Scientific Survey, Withner, C.L. (Ed), Robert E. Krieger Publishing Company, Malabar, FL.

Clements, M.A., Jones, D.L., Sharma, I.K. *et al.* (2002) Phylogenetic systematics of the Diruideae (Orchidaceae) based on the ITS and 5.8S coding region of nuclear ribosomal DNA. *Lindleyana*, 17, 135–171.

Divakaran, M., Babu, K.N., Ravindran, P.N. and Peter, K.V. (2006) Interspecific hybridization in vanilla and molecular characterization of hybrids and selfed progenies using RAPD and AFLP markers. *Scientia Horticulturae*, 108, 414–422.

Doyle, J.J. and Doyle, J.L. (1987) A rapid DNA isolation procedure for small quantities of fresh leaf tissue. *Phytochemical Bulletin* 19, *11–15. irc.igd.cornell.edu/Protocols/ DoyleProtocol.pdf*

Fortune, P.M., Pourtau, N., Viron, N. and Ainouche, M.L. (2008) Molecular phylogeny and reticulate origins of the polyploid *Bromus* species from section *Genea* (Poaceae). *American Journal of Botany*, 95, 454–464.

Gassenmeier, K., Riesen, B. and Magyar, B. (2008) Commercial quality and analytical parameters of cured vanilla beans (*Vanilla planifolia*) from different origins from the 2006–2007 crop. *Flavour and Fragrance Journal*, 23, 194–201.

Gaut, B.S. Tredway, L.P., Kubik, C., Gaut, R.L. and Meyer, W. (2000) Phylogenetic relationships and genetic diversity among members of the *Festuca-Lolium* complex (Poaceae) based on ITS sequence data. *Plant Systematics and Evolution*, 224, 33–53.

Goodman, R.M., Hauptli, H., Crossway, A. and Knauf, V.C. (1987) Gene transfer in crop improvement. *Science*, 236, 48–54.

Hancock, J.F. (1992) Plant Evolution and the Origin of Crop Species. Prentice Hall, Englewood Cliffs, NJ.

Hughes, C.E., Govindarajulu, R., Robertson, A., Filer, D.L., Harris, S.A. and Bailey C.D. (2007) Serendipitous backyard hybridization and the origin of crops. *Proceedings of the National Academy of Sciences*, 104, 14, 389–14, 394.

Knudson, L. (1950) Germination of seeds of *Vanilla. American Journal of Botany*, 37, 241–247.

Lubinsky, P., Cameron, K.M., Molina, M.C. *et al.* (2008) Neotropical roots of a polynesian spice: the hybrid origin of Tahitian vanilla, *Vanilla* tahitensis (Orchidaceae). *American Journal of Botany*, 95, 1040–1047.

Lukaszewski, A.J. and Gustafson, J.P. (1987) Cytogenetics of triticale. *Plant Breeding Reviews*, 5, 41–93.

Pansarin, E.R., Salatino, A. and Salatino, M.L.F. (2008) Phylogeny of South American Pogonieae (Orchidaceae, Vanilloideae) based on sequences of nuclear ribosomal (ITS) and chloroplast (*psaB, rbcL, rps16,* and *trnL-F*) DNA, with emphasis on *Cleistes* and discussion of biogeographic implications. *Organisms Diversity & Evolution*, 8, 171–181.

Podstolski, A., Havkin-Frenkel, D., Malinowski, J., Blount, J.W., Kourteva, G. and Dixon, R.A. (2002) Unusual 4-hydroxybenzaldehyde synthase activity from tissue cultures of vanilla orchid *Vanilla planifolia. Phytochemistry*, 61, 611–620.

Rotter, D., Ambrose, K.V. and Belanger, F.C. (2010) Velvet bentgrass (*Agrostis canina* L.) is the likely ancestral diploid maternal parent of allotetraploid creeping bentgrass (*Agrostis stononifera* L.). *Genetic Resources and Crop Evaluation*, 57, 1065–1077.

Sang, T., Crawford, D.J. and Stuessy, T.F. (1995) Documentation of reticulate evolution in peonies (Paeonia) using internal transcribed spacer sequences of nuclear ribosomal DNA: implications for biogeography and concerted evolution. *Proceedings of the National Academy of Sciences USA*, 92, 6813–6817.

Schluter, P.M., Soto Arenas, M.A. and Harris, S.A. (2007) Genetic variation in *Vanilla planifolia* (Orchidaceae). *Economic Botany*, 61, 328–336.

Shimada, H. and Sugiura, M. (1991) Fine structural features of the chloroplast genome: comparison of the sequenced chloroplast genomes. *Nucleic Acids Research*, 19, 983–996.

Small, R.L., Cronn, R.C. and Wendel, J.F. (2004) Use of nuclear genes for phylogeny reconstruction in plants. *Australian Systematic Botany*, 17, 145–170.

Soltis, P.S. and Soltis, D.E. (2009) The role of hybridization in plant speciation. *Annual Review of Plant Biology*, 60, 561–588.

Swofford, L. (2002) PAUP*: phylogenetic analysis using parsimony (*and other methods), Version 4. Sinauer Associates, Sunderland.

Theis, T. and Jimenez, F.A. (1957) A *Vanilla* hybrid resistant to *Fusarium* root rot. *Phytopathology*, 47, 579–581.

Thompson, J.D., Gibson, T.J., Plewniak, F., Jeanmougin, F. and Higgins, D.G. (1997) The CLUSTAL-X windows interface: flexible strategies for multiple sequence alignment aided by quality analysis tools. *Nucleic Acids Research*, 25, 4876–4882.

Vainstein, A. (2002) *Breeding for Ornamentals: Classical and Molecular Approaches*, Kluwer Academic Publishers, The Netherlands.

Wang, J-B., Wang, C., Shi, S-H. and Zhong, Y. (2000) Evolution of parental ITS regions of nuclear rDNA in allopolyploid *Aegilops* (Poaceae) species. *Hereditas*, 133, 1–7.

Weiss, E.A. (2002) *Spice Crops*. CABI Publishing, Wallingford, UK.

Wendel, J.F., Schnabel, A. and Seelanan, T. (1995) Bidirectional interlocus concerted evolution following allopolyploid speciation in cotton (Gossypium). *Proceedings of the National Academy of Sciences USA*, 92, 280–284.

Xu, C.J., Bao, L., Zhang, B. *et al.* (2006) Parentage analysis of huyou (*Citrus changshanensis*) based on internal transcribed spacer sequences. *Plant Breeding*, 125, 519–522.

22

Root Cause: Mycorrhizal Fungi of Vanilla and Prospects for Biological Control of Root Rots

Paul Bayman, María del Carmen A. González-Chávez, Ana T. Mosquera-Espinosa, and Andrea Porras-Alfaro

22.1   Introduction

Relationships between *Vanilla* and mycorrhizal fungi have been known for over a century, but are still mysterious in many ways. In this chapter we summarize what is known about mycorrhizal fungi of *Vanilla*, from two perspectives: phylogenetic diversity and function. We also review potential applications of mycorrhizal fungi for biocontrol of root rots in *Vanilla*, based on studies on other crops and diseases.

22.1.1   Orchids and Their Mycorrhizæ

Mycorrhizæ (from the Greek *mykes* = mushroom + *rhiza* = root) are mutualistic associations between plants and fungi. About 85% of plant species form mycorrhizæ and many plants cannot grow without them (Smith and Read 2002). The plants benefit because they receive nutrients, especially phosphorus, from the fungi. The enormous surface area of their hyphæ makes fungi far better at extracting nutrients from the soil than plant roots. In some cases they also protect plants from diseases, toxic soils and drought. The fungi benefit because they receive sugars from the plants.

The Orchidaceae is the largest family of flowering plants, with over 28,000 known species (Dressler 1993; Ackerman 2014). Orchids are found on every continent except Antarctica. Most orchids have minute, dust-like seeds with no stored nutrients to support germination. In nature, orchid seeds can germinate only if they establish a relationship with a fungus from which they can get energy (Rasmussen 1995).

Orchids in temperate areas are terrestrial, whereas tropical and subtropical orchids, which include about 75% of species, can be either terrestrial or epiphytic. In adult stages, terrestrial and epiphytic orchids differ in the extent of their relationships with mycorrhizal fungi: most terrestrial orchids are obligately mycorrhizal, whereas many epiphytic orchids appear to be facultatively mycorrhizal (Rasmussen 1995). *Vanilla* is an unusual orchid in being both terrestrial and epiphytic: most plants have roots in the soil or leaf litter and roots attached to tree bark (Figure 22.1).

Handbook of Vanilla Science and Technology, Second Edition.
Edited by Daphna Havkin-Frenkel and Faith C. Belanger.
© 2019 John Wiley & Sons Ltd. Published 2019 by John Wiley & Sons Ltd.

(A)   (B)

Figure 22.1 Roots of *Vanilla planifolia*. A: epiphytic roots attached to tree bark; B: extensively branched roots on surface of soil.

Mycorrhizæ of orchids are distinct from those of other plants in two ways: first, the fungi appear to receive little or nothing from the orchids. The fungus can receive sugars from the plant (Cameron *et al.* 2006), but it is unclear how often this happens in nature. It appears that orchids usually parasitize their fungi, in contrast to the mutualism typical of other mycorrhizæ (Smith and Read 2002). The second distinctive feature of orchid mycorrhizæ is their choice of partners: most orchids associate with fungi in the form-genus *Rhizoctonia* (Basidiomycota), though many terrestrial orchids have switched to other groups of fungi, mostly other Basidiomycota (Rasmussen 1995).

Knowledge of mycorrhizal fungi associated with orchids is growing (Shefferson *et al.* 2007, Roy *et al.* 2009). Techniques applied to study orchid-fungal interactions include DNA sequencing to identify fungi, based on both cultures and direct amplifications (Taylor and McCormick 2008; Roy *et al.* 2009), next-generation sequencing (Johnson *et al.* 2016) and symbiotic germination experiments both in the lab (Otero *et al.* 2004; 2005; Porras and Bayman 2007) and in the field (Rasmussen and Whigham 1993).

22.1.2 The Fungi: *Rhizoctonia* and Related Taxa

The *Rhizoctonia* fungi associated with orchids rarely produce sexual spores in culture, so they have traditionally been identified based on the morphology of their hyphæ and mycelia (Figure 22.2). The hyphæ form knots inside root cortex cells, called pelotons, which are digested by the plant (Figure 22.2A, B). These characters are convergent, causing unrelated fungi to be grouped together (Gonzalez *et al.* 2016). Molecular systematics studies have shown that these fungi actually comprise several sexual genera, of which three have been found in *Vanilla*: *Ceratobasidium*, *Thanatephorus* and *Tulasnella* (Porras-Alfaro and Bayman 2007; Mosquera-Espinosa *et al.* 2013; Johnson *et al.* 2016; Valdovinos Ponce *et al.* 2016). (*Rhizoctonia sensu strictu* is an asexual genus whose sexual stage belongs in *Thanatephorus*.) These fungi were formerly placed in two orders (*Thanatephorus* and *Ceratobasidium* in Ceratobasidiales and *Tulasnella* in Tulasnellales)

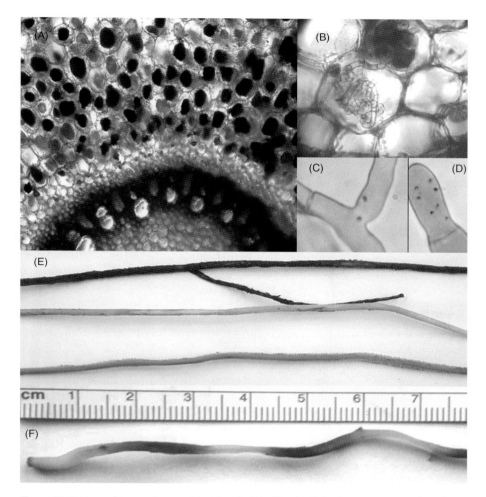

Figure 22.2 Mycorrhizæ and mycorrhizal fungi of *Vanilla planifolia*. A: transverse section of root showing cortex cells with extensive colonization by fungal pelotons (irregular inclusions in cells above the vascular cylinder); B: a single peloton in a root cortex cell; C and D: hyphæof *Ceratobasidium* and *Thanatephorus* showing characteristic binucleate and multinucleate cells, respectively, after staining with Safranin O (Sneh *et al.* 1991); E: roots from soil, air and tree bark (from top); F: soil root with the velamen and part of the cortex removed; areas with mycorrhizæ are indicated by brown color.

but are currently placed in Cantharellales (which includes the chanterelles) (Hibbett 2006; Gonzalez *et al.* 2016).

Fungi in the *Rhizoctonia* complex are best known as plant pathogens. In particular, *Thanatephorus cucumeris* (more widely known by its asexual name, *R. solani*) includes fungi that cause significant losses of rice, wheat, potato, beans and other crops (Vilgalys and Cubeta 1994). Some *R. solani* strains parasitize a wide range of crops; it is not known why orchids appear to be able to parasitize them in turn.

Relationships between plant pathogens and orchid mycorrhizal fungi are not clear; in one of the few studies involving *Vanilla*, a single strain of *R. solani* was both a pathogen and a mycorrhizal symbiont (Alconero 1969). Similarly, a pathogenic strain of *R. solani* was capable of stimulating seed germination in the orchid *Comparettia falcata* (Chávez *et al.* 2014). This subject is discussed in more detail below.

Although tropical orchids comprise the majority of species, knowledge of orchid mycorrhizal relationships comes mostly from temperate, terrestrial orchids. Despite the importance of *Vanilla* from economic, cultural and historical perspectives, surprisingly little is known about its mycorrhizal fungi. Mycorrhizae of *Vanilla* were first described over a century ago (Decordenoy 1904) but few studies have been published until recently.

22.2 Phylogenetic Diversity of Mycorrhizal Fungi of Vanilla

We have applied the resolving power of molecular systematics to identify mycorrhizal fungi of *Vanilla* (Porras-Alfaro and Bayman 2007; Mosquera-Espinosa *et al.* 2013; Valdovinos Ponce *et al.* 2016; Johnson *et al.* 2016; González-Chávez *et al.* 2017). These studies have varied in objectives and sampling strategies, but taken together they can give a preliminary picture of *Vanilla*'s symbionts in Puerto Rico, Cuba, Costa Rica, Colombia and Mexico and including both wild and cultivated plants. Table 22.1 summarizes the *Vanilla* species surveyed and the fungal genera identified that are presumably mycorrhizal. However, the total sample size is still small, exciting new fungi are still being found and further studies are clearly needed.

These studies suffer from a type of circularity: other fungi were isolated from and directly amplified from, *Vanilla* roots (Valdovinos Ponce *et al.* 2016; Johnson *et al.* 2016). The majority of these fungi are ascomycetes or 'imperfect' fungi. However, it is assumed that they are not mycorrhizal because they fall outside the range of basidiomycete fungi that are reported in the literature to have mycorrhizal associations with orchids. For simplicity and in the absence of evidence of a functional relationship, we make the same assumption here.

22.2.1 Methods

Since methods and sampling strategies varied among studies, they are summarized here. Porras-Alfaro and Bayman (2007) collected roots of wild *V. planifolia* and *V. poitæi* in Puerto Rico, roots of cultivated *V. planifolia* × *phæantha* in Costa Rica and roots of wild *V. aphylla* in Cuba (Figures 22.1, 22.3). Samples included both roots in soil and aerial roots attached to tree bark and both cultures from root pieces with pelotons and direct sequencing from root pieces. The nuclear ribosomal ITS region was sequenced. This region was chosen because it is easy to amplify, phylogenetically informative and well-represented in GenBank; it is one of the most widely used sequences for DNA barcoding of fungi. DNA sequences were used for BLAST searches in GenBank to identify the most similar sequences in the database. The same methods were used to identify a mycorrhizal fungus from *V. trigonocarpa* in Colombia (Mosquera-Espinosa *et al.* 2013).

Valdovinos Ponce *et al.* (2016) sampled *V. planifolia* and *V. pompona* in three states in Mexico. Both wild and cultivated plants were sampled, including both roots in soil and

Table 22.1 Mycorrhizal fungi reported from *Vanilla*: taxonomic and geographic distribution of genera. References: 1) Porras-Alfaro and Bayman, 2007; 2) Mosquera-Espinosa *et al.* 2013; 3) Johnson *et al.* 2016; 4) Valdovinos Ponce *et al.* 2016.

	Puerto Rico	Costa Rica	Cuba	Colombia	Mexico
V. planifolia	*Ceratobasidium,* *Tulasnella*				*Sebacina,* *Tulasnella*
V. pompona					*Ceratobasidium*
V. planifolia × *phæantha*		*Ceratobasidium*			
V. poitæi	*Ceratobasidium,* *Tulasnella*				
V. aphylla			*Ceratobasidium*		
V. trigonocarpa				*Ceratobasidium*	
Reference	1	1	1	2	3, 4

roots on vertical supports (trees and posts). A variety of ascomycete and basidiomycete fungi were isolated. Individual pelotons from these species were extracted from roots in soil and the ITS region was sequenced directly (González-Chávez *et al.* 2017). Roots with pelotons were also visualized with scanning electron microscopy (Figure 22.3).

High-throughput (amplicon) sequencing was used to compare fungal communities of soil vs. support root types in *V. planifolia* in Mexican *Vanilla* farms (Johnson *et al.* 2016). Farms with different cultivation systems were compared. Again, the ITS region was used.

22.2.2 Diversity of Mycorrhizal Fungi

Vanilla roots were colonized by four genera known to include mycorrhizal symbionts of orchids: *Ceratobasidium*, *Thanatephorus*, *Tulasnella* and *Sebacina* (Table 22.1). Many sequences were similar to sequences from other orchid-associated fungi in GenBank, suggesting that orchids tend to exploit certain groups. (Since it is not clear that the fungi receive benefits from the orchids, we prefer not to refer to these fungi as "specialized" for the interaction.)

Other basidiomycetes were found by direct amplification of pelotons of soil roots from four *Vanilla* Mexican agrosystems and are in the process of being identified (Johnson *et al.* 2016). The most interesting of these is *Scleroderma* (Boletales), known as an ectomycorrhizal symbiont of trees but not previously reported from orchids (González-Chávez *et al.* 2017).

Six million fungal ITS sequences were amplified from *V. planifolia* in Mexico and were grouped in 144 operational taxonomic units (OTUs) (Johnson *et al.* 2016). Many of these OTUs had not been previously reported from *Vanilla*. They included basidiomycetes such as ectomycorrhizal fungi (e.g. *Inocybe*) and saprotrophs (*Mycena*), as well as the previously known mycorrhizal genera mentioned above (Figure 22.4). On average, nine fungal OTUs colonized a single root segment. Of course, it is not clear how many of these fungi are mycorrhizal.

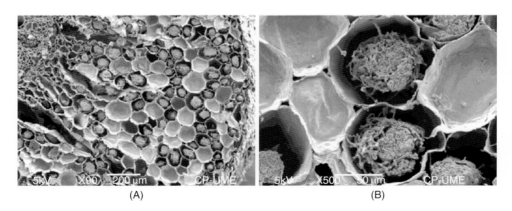

(A) (B)

Figure 22.3 Scanning electron micrographs of mycorrhizal pelotons in a root of *Vanilla pompona*. A: transverse section of root showing part of the vascular cylinder (upper left), the cortext and part of the velamen (lower right). B: close-up of cortical cells, several of which contain large pelotons in which fungal hyphae are visible.

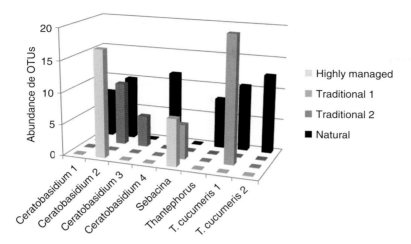

Figure 22.4 Relative abundance of fungi from *Vanilla* roots differs among agrosystems in Mexico. *Vanilla* roots from soil were sampled from farms that were highly managed, intermediate ("Traditional") or natural (i.e., *Vanilla* planted on forest trees). OTUs were defined by ITS2 sequencing and BLASTn searches with a 97% similarity threshold. All OTUs shown belong to genera known to be mycorrhizal symbionts of orchids. Redrawn from Johnson *et al.* (2016).

22.2.3   Fusarium

Many other fungi have been isolated from *Vanilla* roots, as would be expected (Bayman and Otero 2006). *Fusarium* is of particular interest for several reasons: it is both common and diverse in *Vanilla* roots (Johnson *et al.* 2016; Valdovinos Ponce *et al.* 2016; Torres-Cruz *et al.* 2017); it was once claimed to be mycorrhizal in orchids (Bernard 1904); and a more recent study showed that a *Fusarium* isolate stimulated orchid seed germination (Vujanovic *et al.* 2000). Also, some species may have potential as biological control agents against root rots (see section 22.4, below).

22.2.4   Distribution of Mycorrhizæ and Colonization of Roots

Colonization of *Vanilla* roots by mycorrhizal fungi is very variable. A technique was developed to expedite the search for mycorrhizæ in extensive *Vanilla* root systems: the velamen and part of the cortex were stripped off roots and areas with mycorrhizæ could be seen as brown spots on the cortex and vascular cylinder (Figure 22.2; Porras and Bayman 2003).

The extent of mycorrhizal colonization was compared in *V. planifolia*, *V. planifolia* cv. Rayada, *V. insignis* and *V. pompona* in five Mexican commercial agrosystems (González-Chávez *et al.* 2017). All *Vanilla* species presented pelotons (Figure 6.3), similar to those of orchid mycorrhizal fungi and those from *Vanilla* roots from other countries (see above).

However, the type of agrosystem influenced colonization rate and density of hyphal coils of pelotons. Mexican agrosystems use various types supports of tutors to grow *Vanilla* (González-Chávez *et al.* 2017). The traditional systems use live trees including

Citrus sinensis, Erythrina americana, Gliricidia sepium, *Pachira macrocarpa* and *Tabenaemontana* sp. These agrosystems had the highest levels of mycorrhizal colonization (70%) of *Vanilla* roots in soil. *Vanilla* grown in a shade house system with dead branches of *G. sepium* and concrete posts as tutors had lower colonization in roots in the soil (23% and 48%, respectively). The most intact pelotons were observed in the traditional agrosystem.

22.2.5 Roots in Soil vs. Roots on Bark

Are *Vanilla* roots in soil and leaf litter functionally equivalent to roots on tree bark or posts? It seems logical that the former are for absorption and the latter are primarily for support; if this is true, these two types of roots should differ in quantity and type of mycorrhizal fungi and soil roots should contribute more to plant nutrition. Roots in soil had significantly more mycorrhizæ than roots on tree bark (Porras-Alfaro and Bayman 2007). In *V. planifolia* a few roots attached to rough-barked trees were heavily colonized, but most roots on bark had no mycorrhizal pelotons. In *V. poitæi*, no mycorrhizæ were found in roots on bark. Similarly, fungal communities (and frequency of mycorrhizal fungi) differed between roots in soil and roots on bark or posts in cultivated *V. planifolia* (Johnson *et al.* 2016).

Vanilla roots on tree bark are important for support and contribute to plant nutrition through photosynthesis, but the lack of mycorrhizæ suggests that their contribution to mineral nutrition is limited in most cases. In forests we have seen healthy *Vanilla* plants dangling from branches with no roots in contact with any substrate, either bark or soil. These unattached roots have no mycorrhizæ, suggesting that adult plants do not require mycorrhizæ to survive.

22.2.6 Differences in Mycorrhizæ Among Agrosystems

Fungi in general and mycorrhizal fungi as well, differed among *V. planifolia* roots from the agrosystems in Mexico described above (Johnson *et al.* 2016; González-Chávez *et al.* 2017). *Vanilla* planted on forest trees had a richer community of orchid mycorrhizal fungi than in traditionally managed farms and an intensively managed farm was poorer still (Figure 22.4). However, some taxa of *Ceratobasidium* were more common in the intensively managed farm. Until functional differences between these taxa are determined (as discussed below) it will be difficult to infer which agroecosystem is best in terms of mycorrhizal relationships.

22.2.7 Limitations of Methods and Sources of Bias

Choice of sampling method determines which fungi are detected (Porras-Alfaro and Bayman 2007). Culture-independent studies (e.g., Johnson *et al.* 2016) may reveal additional taxa.

It is possible that other mycorrhizal fungi were present in *Vanilla* but were not amplified with the primers used; the most widely-used ITS primers are biased against Tulasnellaceae (Porras-Alfaro *et al.* 2003). Primers that amplify these fungi are available (Porras-Alfaro and Bayman 2007; Taylor and McCormick 2008).

22.3   Mycorrhizal Fungi of Vanilla Stimulate Seed Germination and Seedling Growth

The data discussed above show that *Vanilla* has a range of fungi that we believe to be mycorrhizal. But since the definition of mycorrhizæ is based on function, the occurrence of a fungus in *Vanilla* roots and pelotons is not sufficient to establish it as mycorrhizal. Which fungi are most efficient at promoting *Vanilla* germination, growth and yield? Here we summarize existing data, discuss what additional studies are needed and propose potential uses.

22.3.1   Seedling Germination Experiments

The fungi identified in roots of adult *Vanilla* plants are not necessarily the same as those that stimulate seed germination, as has been shown in other orchids (Bayman *et al.* 2016). To relate phylogeny to function, symbiotic germination of *V. planifolia* seeds was compared after inoculation with different mycorrhizal fungi (Porras-Alfaro and Bayman 2007). The growth medium contained cellulose as the sole carbon source; cellulose can be digested by fungi but not by plants, which obliges the seeds to get their carbon from the fungi (Otero *et al.* 2004). Five *Ceratobasidium* isolates and three *Tulasnella* isolates from *Vanilla* were tested, along with three *Ceratobasidium* isolates obtained from the epiphytic orchid *Ionopsis utricularioides* in Puerto Rico (Otero *et al.* 2005). Control plates had *V. planifolia* seeds without fungi (Porras-Alfaro and Bayman 2007).

Five of eight *Ceratobasidium* isolates stimulated germination of *Vanilla* seeds (Figure 22.5A). Germination frequency was highest with *Ceratobasidium* isolates from the epiphytic orchid *I. utricularioides* rather than with isolates from *Vanilla* itself. Seeds inoculated with *Tulasnella* and control seeds without fungi did not germinate.

Inoculation of seed germination plates with *Ceratobasidium* also reduced contamination by fungi and bacteria (Figure 22.5B, C). In contrast, plates with *Tulasnella* and control plates without fungi were often contaminated after two months' growth. Orchid seed germination and seedling growth are slow and in tropical environments it is hard to avoid contamination when petri plates are incubated for long periods; fungus mites often invade cultures, carrying undesirable bacteria and fungi. Symbiotic germination has the additional advantage of reducing this contamination, presumably through competition or antibiosis.

Vanilla seeds are difficult to germinate and this limits development of new *Vanilla* cultivars. *Vanilla* seeds have been found germinating on rotten wood (Childers *et al.* 1959), which suggests that fungi may be required to break the seed coat and allow water to reach the embryo. Symbiotic germination techniques may be useful in overcoming this barrier, both by stimulating germination and by reducing contamination.

The range of potential mycorrhizal symbionts of *Vanilla* has been expanded by the recent studies in Mexico cited above. These fungi should be included in experiments like those described here, to find the most effective fungi for promoting *Vanilla* seedling production.

22.3.2   Seedling Growth and Survival Experiments

Seeds and mericlones of *Vanilla* and other orchids are usually propagated in sterile culture under laboratory conditions. When plantlets are removed from culture flasks

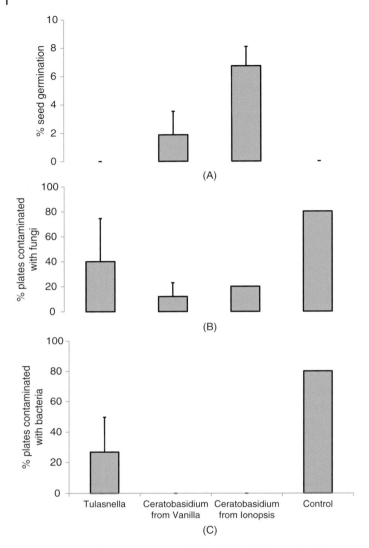

Figure 22.5 Symbiotic germination of *Vanilla planifolia* seeds and contamination of cultures. A: % germination of seeds after two months' incubation with different fungi (and control seeds and plates without fungi); B: contamination of seed germination plates by fungi: C: contamination of seed germination plates by bacteria. *Ceratobasidium* strains isolated from the orchid *Ionopsis* were most effective at stimulating seed germination and also at inhibiting contamination. Each bar shows the average of 3–5 strains with 5 plates per strain. Error bars = s.d. Redrawn from Porras-Alfaro (2004).

and moved to the greenhouse they are suddenly exposed to environmental challenges and potential pathogens. During this transition plants become very susceptible to infections and desiccation and many plantlets die or lose vigor in the process.

Inoculation with mycorrhizal fungi aided growth and survival of sterile *Vanilla* plantlets (Porras-Alfaro and Bayman 2007). Axenic mericlones of *V. planifolia* × *V. pompona* in flasks were inoculated with the same fungi used for seed germination experiments (above) and increase in plant growth was measured.Significant differences were found among treatments after two months' growth (Figure 22.6). Plants inoculated with a *Ceratobasidium* isolate from *Vanilla* and three isolates from the orchid *I. utriculari-oides* grew as much as control plants without fungi and produced as many or more new leaves than control plants (data not shown). However, plants inoculated with four iso-lates of *Ceratobasidium* and three isolates of *Tulasnella* grew significantly less than uninoculated control plants and produced significantly fewer new leaves.

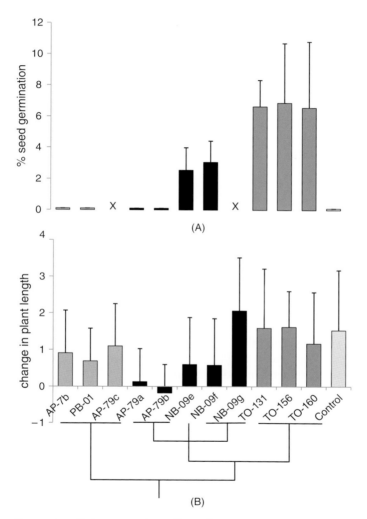

Figure 22.6 Phylogeny predicts effects of mycorrhizal fungi on seed germination and seedling growth in *Vanilla*. Relationships between fungal isolates are shown with a maximum parsimony tree below the graphs. A: Fungal stimulation of seed germination (%). B: change in length of mericloned plants growing in peat moss after inoculation with fungi. Bar colors indicate fungi: from left, *Tulasnella*, *Ceratobasidium* isolated from *Vanilla*, *Ceratobasidium* isolated from the orchid *Ionopsis utricularioides* and controls without fungi. Redrawn from Porras-Alfaro (2004); Porras-Alfaro and Bayman (2007).

Closely related fungi tended to have similar effects on growth of *Vanilla* seeds and plantlets (Figure 22.6; Porras-Alfaro 2004). The two *Ceratobasidium* isolates that inhibited both seed germination and reduced plant growth were very closely related to each other. The three isolates from *Ionopsis*, which stimulated both germination and growth, were also very closely related to each other. A similar finding was reported for *Ceratobasidium* isolates from epiphytic orchids in Puerto Rico (Otero *et al.* 2005). This association between phylogeny and function is useful, because it suggests a strategy for bioprospecting new isolates: choose those that are closely related to known, successful isolates.

The isolates from *Ionopsis* that stimulated germination and growth of *Vanilla* (Figures 22.5, 22.6) belong to *Ceratobasidium* Clade B as defined by Otero *et al.* (2004, 2005). Clade B was also the most successful *Ceratobasidium* clade at stimulating germination and seedling growth of *Ionopsis* (Otero *et al.* 2004).

Clade B is very widely distributed: isolates related to those found in mycorrhizæ of *I. utricularioides* were isolated from the same species in Cuba, Costa Rica, Panamá and Trinidad (Otero *et al.* 2007). This is an enormous range considering that this clade was previously unknown and doesn't even have a species name. Its ubiquity suggests that *Ceratobasidium* Clade B will not be difficult to find where *Vanilla* is grown, facilitating the use of local isolates. All these data suggest that this clade is potentially useful for orchid biotechnology.

22.4 Can Mycorrhizal Fungi Protect Vanilla Plants from Pathogens?

Root rots of *Vanilla* caused by *Fusarium* and other pathogens are serious problems for cultivation, as addressed in the chapter by Juan Hernández-Hernández in this volume (Chapter 2). One of the reasons for studying mycorrhizal fungi in *Vanilla* is the possibility that mycorrhizæ can protect plants from these pathogens. In this section we argue that although this ability has not been demonstrated, it is plausible and worth studying. Our argument is based on evidence that other types of mycorrhizal fungi can protect plants from diseases and that *Rhizoctonia* fungi (i.e., *Ceratobasidium*) are useful for biocontrol of a variety of plant pathogens, including *Fusarium*.

22.4.1 Biocontrol of Plant Diseases Using Arbuscular Mycorrhizal Fungi

Most crop plants have arbuscular mycorrhizæ (AM, also called vesicular-arbuscular mycorrhizæ or endomycorrhizæ), so this is the best-known type of mycorrhiza in terms of biocontrol. In many cases AM have been shown to increase plants' resistance or tolerance to diseases, but only if the symbiosis is well-established before exposure to the pathogen (Azcón-Aguilar *et al.* 2002). There is no single mechanism of protection: the AM fungi may affect the pathogen, the plant, or the rhizosphere (Azcón-Aguilar and Barea 1997).

AM have been used to control *Fusarium* root rots in several plants. AM fungi reduced damage in *Passiflora edulis* var. Flavicarpa caused by a combination of *F. oxysporum* f. *passiflorae* and several nematodes (Tofiño and Sánchez de Prager 1997). Co-inoculation with AM fungi and rhizosphere bacteria controlled *F. oxysporum* f. sp. *lycopersici* in tomato (Akköprü and Demir 2005). The AM fungus *Glomus macrocarpum* protected bean roots from *F. solani*, apparently by excluding the pathogen from the rhizosphere (Muchovej *et al.* 1991). Fungicides that controlled the pathogen *F. oxysporum* in the grass *Vulpia ciliata* also reduced growth of AM fungi, resulting in no net change in plant fecundity (Newsham *et al.* 1995). In this case the AM fungi had no visible beneficial effect on the plant in the absence of the pathogen, illustrating the complexity of microbial interactions in the rhizosphere.

22.4.2 Biocontrol of Plant Diseases Using *Ceratobasidium*

Ceratobasidium isolates from *Vanilla* and other orchids were able to protect rice plants from sheath blight, a serious disease (Mosquera-Espinosa *et al.* 2013). The sheath blight pathogen is *Rhizoctonia solani*, closely related to *Ceratobasidium*. Rice plants in the greenhouse were inoculated with different *Ceratobasidium* isolates and then with *R. solani* two days later. Disease severity was significantly reduced compared to control plants inoculated with *R. solani* alone.

Ceratobasidium has been used for biocontrol of diseases in a variety of crops. *Ceratobasidium* is often called "binucleate *Rhizoctonia*" in the older literature; cells are typically binucleate, distinguishing them from the multinucleate hyphæ of *Thanatephorus* (Figure 22.2; Sneh *et al.* 1991). There are several possible modes of action: directly inhibiting growth of the pathogen, competing with the pathogen for space or nutrients, or strengthening the plant's defense through production of growth promoters or systemic induced resistance (Cartwright and Spurr 1998; Burns and Benson 2000; González *et al.* 2000; Sneh *et al.* 2004). In orchids there is another potential mode of action which has not been explored: improving the nutritional status of the plant, making it less susceptible to pathogens.

Apathogenic and hypovirulent strains of *Ceratobasidium* are common in the soil, representing 10–30% of total *Rhizoctonia* isolates in a study in Israel (Ichielevich-Auster *et al.* 1985) and almost half the isolates in a study in New Zealand (Sneh *et al.* 2004). Both these studies found that some isolates were highly effective at protecting seedlings of a variety of crops from damping-off diseases. Both studies suggested systemic induced resistance as the mechanism of protection.

Particularly relevant to prospects for *Vanilla* are studies that used *Ceratobasidium* to control *Fusarium* diseases. *F. oxysporum* causes vascular wilts in spinach and tomato (*F. oxysporum* f. sp. *spinaciae* and f. sp. *lycopersici*, respectively). In both crops, prior inoculation with a hypovirulent *Ceratobasidium* significantly reduced disease severity (Muslim *et al.* 2003a, b). Inoculum levels of the pathogens were also reduced and germination of *Fusarium* propagules was delayed. A newly described species from Spain, *C. albasitensis*, protected a wide range of crop plants from pathogens, including *F. solani* (González *et al.* 2000). *Ceratobasidium* has also been used successfully for control of *Thanatephorus* pathogens in a variety of crops, including melons, tomato, lettuce, radish, chillies, potato, beans and sugar beet (González *et al.* 2006).

In some cases effective control of pathogens has been achieved by joint inoculation with *Ceratobasidium* and bacteria. A combination of *Ceratobasidium* and *Burkholderia cepacia* controlled *R. solani* root rot in poinsettia, especially when *B. cepacia* was inoculated first (Hwang and Benson 2002).

Unfortunately, it is difficult to compare fungi in different studies, because few studies have identified *Ceratobasidium* isolates to species level and few have provided DNA sequence data which would allow comparison. However, all these studies lead us to conclude that experiments using mycorrhizal fungi of *Vanilla* to control *Fusarium* root rots are worthwhile.

Effective, low-cost formulations for applying *Ceratobasidium* for biocontrol in the field have been demonstrated. For example, grain colonized with the fungus can be used to inoculate the soil around the crop (Cartwright and Spurr 1998). Most of the studies cited above contend that *Ceratobasidium* must be inoculated before exposure of the plant to the pathogen. A low-cost, low-tech method for inoculating orchids with *Ceratobasidium* is available (Bayman 2012).

22.4.3 Are *Rhizoctonia* Strains Used for Biocontrol also Potential Pathogens?

Before using *Rhizoctonia* isolates for biocontrol of root rots in *Vanilla*, we must ask whether these isolates have the potential to become pathogenic on *Vanilla* or other nearby plants. This question is difficult because lifestyles of orchid mycorrhizal fungi are not well understood. Do the orchid fungi primarily make their living as plant pathogens, saprotrophs or mycorrhizal symbionts of orchids or other plants (Gonzalez *et al.* 2016)? No one knows.

Some *Rhizoctonia* groups are clearly more pathogenic than others. *Thanatephorus* includes serious pathogens, most notably *R. solani*, which devastates many crops worldwide (Agrios 1997). Orchid mycorrhizal fungi may also be pathogens. *Thanatephorus* isolates from the orchid *Pterostylis acuminata* were moderately pathogenic on lettuce and severely pathogenic on cauliflower and radish (Carling *et al.* 1999). An *R. solani* isolate capable of forming mycorrhizæ on *Vanilla planifolia* and *V. phæantha* also formed pathogenic lesions on the same roots (Alconero 1969). These isolates are too pathogenic to be used for biocontrol of root rots in *Vanilla*; the cure may turn out to be more harmful than the disease. Some *Ceratobasidium* isolates are also pathogenic.

In practice, isolates of *Rhizoctonia* are often tested for pathogenicity on seedlings of radish or cucumber before they are used for biocontrol. In one such study, 248 *Rhizoctonia* isolates from crop residues in Japan were tested and five *Ceratabasidum* and one *Thanatephorus* isolate were found to be hypovirulent in both radish and cucumber seedlings (Villajuan *et al.* 1996). Similarly, a quarter of 153 *Rhizoctonia* isolates from soil in USA were hypovirulent or apathogenic on cabbage and several protected cabbage and cucumber seedlings from pathogens (Sneh and Ichielevich-Auster 1998).

The same *Ceratobasidium* that protects a plant from fungal pathogens may itself become mildly pathogenic when no other pathogen is present (Burns and Benson 2000). This is true for many types of microbial biocontrol agents and is compatible with the view that mutualisms are balanced antagonisms, in which either partner may become parasitic under favorable conditions (Saikkonen *et al.* 1998; Schultz *et al.* 1999). There may be a tradeoff between the protective function of the *Ceratobasidium* against pathogens and the cost of maintenance at expense of the plant.

The pathogenicity of *Ceratobasidium* biocontrol isolates under certain conditions and their relationship to the more pathogenic genus *Thanatephorus*, will complicate their importation for biocontrol purposes. The plant health inspection service of any country will be reluctant to permit their importation, even if the same species or clade already exists in the country. Instead, it is more practical to find local *Ceratobasidium* isolates with no or minimal pathogenicity for domestic use. The isolates can be identified by DNA sequencing and the ones most closely related to the desired clade can be selected. Collaboration among universities and commercial agriculture are necessary to advance research in *Vanilla* fungal symbionts with potential to be used for biological control in the field.

In many areas a single clone of *Vanilla* is cultivated, making the crop more susceptible to any pathogen that can attack that particular genotype. Changing mycorrhizal partners may allow adaptation to changing conditions when no variation is present in the population of the *Vanilla* plant itself – a sort of extended phenotype, or a hologenome in the sense of Rosenberg *et al.* (2007).

22.4.4   *Fusarium* Species as Potential Biocontrol Agents to Protect Vanilla from *Fusarium oxysporum* Root Rots

As mentioned above, *Fusarium* is common in *Vanilla* roots. We believe it is possible some that *Fusarium* species may be able to protect *Vanilla* against the root rot pathogen *F. oxysporum* f. sp. *vanillæ*. The evidence comes from maize. *Fusarium* is ubiquitous in maize; under field conditions, there is no maize plant or seed from which it cannot be isolated. Pre-inoculation with *F. moniliforme* can protect maize seedlings from the pathogen *F. graminearum* (Van Wyck *et al.* 1988). Both species produce mycotoxins and each can inhibit mycotoxin production by the other.

Similarly, *F. graminearum* (as an endophyte) protected *Physalis peruviana* from *F. verticillioides* (a pathogen) *in vitro* (Manosalva and Mosquera-Espinosa 2014). However, experiments to test such interactions in *Vanilla* have not been done.

22.5   Conclusions

We have summarized current knowledge about mycorrhizal fungi of *Vanilla*. This knowledge may have applications for commercial production of *Vanilla* in several ways:

- improving plant nutrition;
- stimulating seed germination for plant propagation and production of new hybrids;
- acclimatizing seedlings and mericlones to increase survival and growth after transplanting;
- protecting plants from root rots.

Further work is needed before any of these ideas can be applied commercially, but the experiments needed are relatively simple.

Interactions between organisms are complex (Saikkonen *et al.* 1998; Schultz *et al.* 1999), so three-way interactions, such as those between *Vanilla*, *Fusarium* and mycorrhizal fungi, will be still more complex. The argument that mycorrhizal fungi (and endophytes) are potentially useful for biocontrol of *Fusarium* root rots in *Vanilla* is complicated by the fact that the lifestyles and basic biology of these mycorrhizal fungi are still largely unknown.

References

Ackerman, J.D. (and collaborators) (2014) *Orchid Flora of the Greater Antilles*. Memoirs of the New York Botanical Garden, v. 109.

Agrios, G.N. (1997) *Plant Pathology*, 4th edn. Harcourt Academic Press, San Diego.

Akköprü, A. and Demir, S. (2005) Biological control of *Fusarium* wilt in tomato caused by *Fusarium oxysporum* f. sp. *lycopersici* by AMF *Glomus intraradices* and some *Rhizobacteria*. *Journal of Phytopathology* 153, 544–550.

Alconero, R. (1969) Mycorrhizal synthesis and pathology of *Rhizoctonia solani* in *Vanilla* orchid roots. *Phytopathology* 59, 426–430.

Azcón-Aguilar, C., Jaizme-Vega, M.C. and Calvet, C. (2002) The contribution of arbuscular mycorrhizal fungi to the control of soil-borne plant pathogens. In: *Mycorrhizal Technology in Agriculture: From Genes to Bioproducts*. Gianinazzi, S.,

Azcón-Aguilar, C. and Barea, J.M. (1997) Arbuscular mycorrhizas and biological control of soil-borne plant pathogens – an overview of the mechanisms involved. *Mycorrhiza* 6, 457–464.

Bayman, P. (2012) Growing epiphytic orchids from seed: a simple, nonsterile, symbiotic method. *Orchids/Lindleyana* 81, 564–566.

Bayman, P., Mosquera-Espinosa, A.T., Saladini-Aponte, C.M., Hurtado-Guevara, N.C. and Viera-Ruiz, N.L. (2016) Age-dependent mycorrhizal specificity in an invasive orchid, *Oeceoclades maculata*. *American Journal of Botany* 103, 1880–1889.

Bayman, P. and Otero, J.T. (2006) Endophytes of orchid roots. In: *Microbial Root Endophytes*. B. Schulz, C. Boyle and T. Sieber (eds). Springer Verlag, Berlin, Germany, pp 153–177.

Bernard, N. (1904) Recherches Experimentales sur les Orchidees. *Revue Générale de Botanique* 16, 405–451, 458–476.

Burns, J. and Benson, M. (2000) Biocontrol of damping-off of *Catharanthus roseus* caused by *Pythium ultimum* with *Trichoderma virens* and binucleate *Rhizoctonia* fungi. *Plant Disease* 84, 644–648.

Cameron, J.R.L. and Read, D.J. (2006) Mutualistic mycorrhiza in orchids: Evidence from plant–fungus carbon and nitrogen transfers in the green-leaved terrestrial orchid *Goodyera repens*. *New Phytologist* 171, 405–416.

Carling, D., Pope, E., Brainard, K. and Carter, D. (1999) Characterization of mycorrhizal isolates of *Rhizoctonia solani* from an orchid, including AG-12, a new anastomosis group. *Phytopathology* 64, 492–496.

Cartwright, K. and Spurr, H. (1998) Biological control of *Phytophthora parasitica* var. *nicotianae* on tobacco seedlings with non-pathogenic binucleate *Rhizoctonia* fungi. *Soil Biology and Biochemistry* 30, 1879–1884.

Chávez, H.K., Mosquera-Espinosa, A.T. and Otero-Ospina, J.T. (2014) Propagación in vitro de semillas de la orquídea *Comparettia falcata* Poepp. & Endl. (Orchidaceae) mediante técnicas simbióticas y asimbióticas. *Acta Agronómica* 64, 125.

Childers, N., Cibes, H. and Hernandez-Medina, E. (1959) Vanilla – The Orchid of Commerce. In: *The Orchids*. Withner, C., ed. The Ronald Press Company. NY, pp. 477–508.

Decordenoy, J.H. (1904) Contribution a la biologie du vanillier. *Journal d'Agriculture Tropicale* 4, 104–106.

Dressler, R.L. (1990) *The Orchids: Natural History and Classification.* Harvard University Press, Cambridge MA, 332 pp.

González-Chávez, M.C.A., Torres-Cruz, T.J., Albarrán-Sánchez, S., Carrillo-González, R., Carrillo-López, L.M. and Porras-Alfaro, A. (2018) Microscopic characterization of orchid mycorrhizal fungi: *Scleroderma* as a putative novel orchid mycorrhizal fungus in *Vanilla. Mycorrhiza* 28, 147–157.

Gonzalez, D., Rodriguez-Carres, M., Boekhoutc, T., Stalpersc, J., Kuramaed, E.E., Nakatanie, A.K., Vilgalys, R. and Cubeta, M.A. (2016) Phylogenetic relationships of *Rhizoctonia* fungi within the Cantharellales. *Fungal Biology* 120, 603–619.

González, V., Portal, M., Acero, J., Sánchez-Ballesteros, J. and Rubio, V. (2000) Biological control properties of a new *Rhizoctonia*-like species (BNR), *Ceratobasidium albasitensis* isolated in Spain. *Third International Symposium on Rhizoctonia.* National Chung Hsing University, pp 21–23.

González, V., Portal, M. and Rubio, V. (2006) Review. Biology and systematics of the form genus *Rhizoctonia. Spanish Journal of Agricultural Research* 4, 55–79.

Hibbett, D.S. (2006) A phylogenetic overview of the Agaricomycotina. *Mycologia* 98, 917–925.

Hwang, J. and Benson, D.M. (2002) Biocontrol of *Rhizoctonia* stem and root rot of poinsettia with *Burkholderia cepacia* and binucleate *Rhizoctonia. Plant Disease* 86, 47–53.

Ichielevich-Auster, M., Sneh, B., Koltin, Y. and Barash, I. (1985) Pathogenicity, host specificity and anastomosis groups of *Rhizoctonia* spp. isolated from soils in Israel. *Phytoparasitica* 13, 103–112.

Johnson, L., Gónzalez-Chávez, M.C.A., Carrillo-González, R., Porras-Alfaro, A. and Mueller, G. (2016) Amplicon sequencing reveals differences between root microbiomes of a hemiepiphytic orchid, *Vanilla planifolia* at four Mexican farms. *Inoculum* 67(4), 26.

Manosalva, J.L. and Mosquera-Espinosa, A.T. (2014) Diagnóstico de hongos patógenos en el cultivo de uchuva (*Physalis peruviana* L.) y la evaluación in vitro de algunos hongos con actividad biocontroladora. *Fitopatología Colombiana* 38, 1–7.

Mosquera-Espinosa, A.T., Bayman, P., Prado, G., Gómez-Carabalí, A. and Otero, J.T. (2013) The double life of *Ceratobasidium*: orchid mycorrhizal fungi and their potential for biocontrol of *Rhizoctonia solani* sheath blight of rice. *Mycologia* 105, 141–150.

Muchovej, J.J., Muchovej, R.M.C. and Goncalves, E.J. (1991) Effect of kind and method of fungicidal treatment of bean seed on infections by the VA mycorrhizal fungus *Glomus macrocarpum* and by the pathogenic fungus *Fusarium solani*. 2. Temporal-spatial relationships. *Plant and Soil* 132, 47–51.

Muslim, A., Horinouchi, H. and Hyakumachi, M. (2003a) Suppression of *Fusarium* wilt of spinach with hypovirulent binucleate *Rhizoctonia. Journal of General Plant Pathology* 69, 143–150.

Muslim, A, Horinouchi, H. and Hyakumachi, M. (2003b) Biological control of *Fusarium* wilt of tomato with hypovirulent binucleate *Rhizoctonia* in greenhouse conditions. *Mycoscience* 44, 77–84.

Newsham K.K., Fitter, A.H. and Watkinson, A.R. (1995) Multi-functionality and biodiversity in arbuscular mycorrhizas. *Trends in Ecology and Evolution* 10, 407–411.

Otero, J.T., Bayman, P. and Ackerman, J.D. (2005) Variation in mycorrhizal performance in the epiphytic orchid *Tolumnia variegata in vitro*: the potential for natural selection. *Evolutionary Ecology* 19, 29–43.

Otero, J.T., Ackerman. J.D. and Bayman, P. (2004) Differences in mycorrhizal specificity between two tropical orchids. *Molecular Ecology* 13, 2393–2404.

Otero, J.T., Flanagan, N.S., Herre, E.A., Ackerman, J.D. and Bayman, P. (2007) Widespread mycorrhizal specificity correlates to mycorrhizal function in the epiphytic orchid *Ionopsis utricularioides. American Journal of Botany* 94, 1944–1950.

Porras, A. and Bayman, P. (2003) Mycorrhizal fungi of *Vanilla*: root colonization patterns and fungal identification. *Lankesteriana* 7, 147–150.

Porras-Alfaro, A. (2004) *Mycorrhizal fungi of Vanilla: An Integral View of the Symbiosis.* M.S. thesis, University of Puerto Rico – Río Piedras.

Porras-Alfaro, A. and Bayman, P. (2007) Mycorrhizal fungi of *Vanilla*: diversity, specificity and effects on seed germination and plant growth. *Mycologia* 99, 510–525.

Rasmussen, H.N. (1995) *Terrestrial Orchids: From Seed to Mycotrophic Plant.* Cambridge University Press, NY, 456 pp.

Rasmussen, H.N. and Whigham, D.F. (1993) Seed ecology of dust seeds in situ: a new study technique and its application in terrestrial orchids. *American Journal of Botany* 80, 1374–1378.

Rosenberg, E., Koren, O., Reshef, L., Efron, R. and Zilber-Rosenberg, I. (2007) The role of microorganisms in coral health, disease and evolution. *Nature Reviews Microbiology* 5, 355–362.

Roy, M., Watthana, S., Stier, A., Richard, F., Vessabutr, S. and Selosse, M.A. (2009) Two mycoheterotrophic orchids from Thailand tropical dipterocarpacean forests associate with a broad diversity of ectomycorrhizal fungi. *BMC Biology* 7, 51. doi:10.1186/1741-7007-7-51.

Saikkonen, K., Faeth, S.H., Helander, M. and Sullivan, T.J. (1998) Fungal endophytes: a continuum of interactions with host plants. *Annual Review of Ecology and Systematics* 29, 319–343.

Schuepp, H., Barea, J.M. and Haselwandter, K. and Birkhauser, A.L.S., eds. Vertag, Switzerland, pp. 187–196.

Schulz, B., Rommert, A.K., Dammann, U., Aust, H.J. and Strack, D. (1999) The endophyte–host interaction: a balanced antagonism? *Mycological Research* 103, 275–1283.

Shefferson, R.P., Taylor, D.L., Weiss, M., Garnica, S., McCormick, M.K., Adams, S., Gray, H.M., McFarland, J.W., Kull, T., Tali, K., Yukawa, T., Kawahara. T., Miyoshi, K. and Lee, Y.I. (2007) The evolutionary history of mycorrhizal specificity among lady's slipper orchids. *Evolution* 61, 1380–1390.

Smith, S.E. and Read, D.J. (2002) *Mycorrhizal Symbiosis*, 2nd edn. Academic Press, San Diego, 605 pp.

Sneh, B., Burpee, L. and Ogoshi, A. (1991) *Identification of Rhizoctonia species.* APS Press. USA.

Sneh, B. and Ichielevich-Auster, M. (1998) Induced resistance of cucumber seedlings caused by some non-pathogenic *Rhizoctonia* (np-R) isolates. *Phytoparasitica* 26, 27–38.

Sneh, B., Yamoah, E. and Stewart, A. (2004) Hypovirulent *Rhizoctonia* spp. isolates from New Zealand soils protect radish seedlings against damping-off caused by *R. solani. New Zealand Plant Protection* 57, 54–58.

Taylor , D.L. and McCormick, M.K. (2008) Internal transcribed spacer primers and sequences for improved characterization of basidiomycetous orchid mycorrhizas. *New Phytologist* 177, 1020–1033.

Tofiño, R. and Sánchez de Prader, M. (1997) *Relación de hongos micorrizógenos con secadera del maracuyá. Problema especial.* Universidad Nacional de Colombia – Sede Palmira. Pronatta, 20 pp.

Torres-Cruz, T.J., Porras-Alfaro, A., Carrillo-González, R. and González Chávez M.C.A. (2017) Fungal colonization and *Fusarium* diversity across *Vanilla* cultivation systems. *Meeting of Mycological Society of America*, Athens, Georgia (Poster).

Valdovinos Ponce, G., Nava Díaz C., Carrillo González and González Chávez, M.C.A. (2016) *Buenas prácticas de manejo para favorecer la sanidad y nutrición de las plantas de vainilla. Colegio de Postgraduados, Montecillo, Texcoco, Mexico.* ISBN 978-607-715-318-24.

Van Wyck, P.S., Scholtz, D.J. and Marasas, W.F.O. (1988) Protection of maize seedlings by *Fusarium moniliforme* against infection by *Fusarium graminearum* in the soil. *Plant and Soil* 107, 251–257.

Vilgalys, R. and Cubeta, M.A. (1994) Molecular systematics and population biology of *Rhizoctonia*. *Annual Review of Phytopathology* 32, 135–155.

Villajuan-Abgona, R.V., Katsuno, N., Kageyama, K. and Hyakumachi, M. (1996) Isolation and identification of hypovirulent *Rhizoctonia* spp. from soil. *Plant Pathology* 45, 896–904.

Vujanovic, V., St-Arnaud, M., Barabe, D. and Thibeault, G. (2000) Viability testing of orchid seed and promotion of coloration and germination. *Annals of Botany* 86, 79–86.

23

Enzymes Characterized From Vanilla
Andrzej Podstolski

Plants produce numerous secondary metabolites in organized pathways of metabolite processing facilitated by proteinaceous biocatalysts – enzymes. For many practical and scientific reasons it is necessary to know enzymatic and metabolite pathways leading to final products of interest. Vanillin is the most desired compound synthesized in *Vanilla planifolia*. The components of vanilla aroma are well known; however, the steps of their biosynthesis, including that of vanillin, remain controversial. Some of the different proposed routes leading to vanillin have been thoroughly reviewed by Walton *et al.* (2003). In this chapter we will focus on particular enzymes catalyzing the chain of reactions beginning from *L*-phenylalanine and ending with vanillin. Attention will be given to the chain-shortening enzyme operating at the divergence point between general phenylpropanoid and benzoic acid derivative pathways.

23.1   *L*-Phenylalanine Ammonia-Lyse (Pal) and Cinnamate-4-Hydroxylase (C4h)

It is generally accepted that *L*-phenylalanine is a common precursor of numerous secondary metabolites synthesized in the phenylpropanoid (C_6–C_3) pathway. *L*-phenylalanine ammonia-lyase (PAL; EC 4.3.1.5) is the entry, regulatory enzyme coupling primary metabolism with secondary phenylpropanoid metabolism, producing a vast array of C_6–C_3 compounds including their derivatives (Jones 1984). Having such a regulatory position, the enzyme responds to several physical and biological elicitors, among them wounding, light, UV, temperature, and pathogen infections (Dixon and Paiva 1995). PAL converts *L*-phenylalanine to *trans*-cinnamic acid, the precursor of numerous phenolic compounds, such as 4-coumaric, caffeic, ferulic, and sinapic acids and their corresponding monolignols, structural elements of lignin, flavonoids and isoflavonoids, coumarins and benzoic acid (C_6–C_1) derivatives (Dixon and Paiva 1995; Dixon *et al.* 2002). In grasses, which are monocots, PAL is accompanied by tyrosine ammonia-lyase activity (TAL), resulting in formation of 4-coumaric acid, the first phenolic compound in the pathway, directly from tyrosine. In vanilla, which is a monocot, both PAL and TAL activities were identified (Havkin-Frenkel and Belanger

Handbook of Vanilla Science and Technology, Second Edition.
Edited by Daphna Havkin-Frenkel and Faith C. Belanger.
© 2019 John Wiley & Sons Ltd. Published 2019 by John Wiley & Sons Ltd.

2007). Feeding of green vanilla beans with ^{14}C L-tyrosine resulted in formation of vanillin, vanillic acid, vanillyl alcohol, protocatechuic alcohol and protocatechuic aldehyde, 4-hydroxybenzyl alcohol, 4-hydroxybenzaldehyde, 4-coumaric acid, and ferulic acid. Qualitatively similar results were obtained with ^{14}C L-phenylalanine but comparison of quantities of accumulated 4-coumaric and cinnamic acids in both cases showed that PAL produced 10 times more *tran*-cinnamic acid than TAL produced 4-coumaric acid. Also, the amount of vanillin and its precursors formed was 10 times higher in the case of ^{14}C L-phenylalanine feeding. Similar relations in vanillin precursor formation were found when ^{14}C L-phenylalanine and ^{14}C L-tyrosine were fed to V. planifolia tissue culture and on-the-vine green beans (Havkin-Frenkel and Belanger 2007).

PAL has a tetrameric structure (Havir and Hanson 1973, Bolwell *et al.* 2005) with the subunits encoded in the majority of plants by multigene families (Cramer *et al.* 1989; Wanner *et al.* 1995). The size of PAL subunits may vary a little, depending on the plant species and the enzyme isoform. PAL purified from French bean suspension culture was composed of 83 kDa subunits in the constitutively expressed enzyme and 77 kDa subunits in the elicitor or wound induced isoform. The enzyme isoform containing 77 kDa subunits was shown to be involved in cytokinin-induced xylogenesis and stimulation of biosynthesis of phenolics under stress conditions (Bolwell and Rodgers 1991).

PAL, despite its solubility, may associate with endomembrane bound cinnmate-4-hydroxylase (C4H; EC.1.14.13.11), a cytochrome P450 and the second enzyme in the phenylpropanoid pathway, creating a channel facilitating flux of intermediates at the entrance to phenylpropanoid pathway. C4H is the most abundant cytochrome P450 enzyme in plants (Potts *et al.* 1974). There are several indications that PAL and C4H, both phenylpropanoid pathway entry enzymes are regulated coordinately (Hahlbrock *et al.* 1976; Czichi and Kindl 1975, 1977; Hahlbrock and Scheel 1989). There have been many reports indicating that both PAL and C4H are concomitantly induced in response to stress (Mizutani *et al.* 1997; Koopmann *et al.* 1999; Blount *et al.* 2000). In tobacco (*Nicotiana tabacum*), there are two PAL isoforms, PAL 1 and PAL2, encoded by closely related gene families. PAL1 is localized in both cytosolic and microsomal fractions, while PAL2 is found only in the cytosol. In plants over-expressing C4H however, PAL2 has been also found associated with a microsomal fraction, indicating C4H capability to organize an endomembrane bound complex with PAL (Achnine *et al.* 2004). Existence of these two enzymes as a complex facilitating channeling of *trans*-cinnamic acid, the PAL activity product, may explain why the first accumulated product of the phenylpropanoid pathway is often 4-coumarate, but rarely *trans*-cinnamate. Athough there is no published data concerning C4H in *Vanilla*, it is obvious that its activity has to account for the observed 4-coumarate accumulation in this plant (Havkin-Frenkel *et al.* 1996).

23.2 Chain-shortening Enzymes

Downstream of the biosynthesis of 4-coumarate, there is a divergence point between the general phenylpropanoid pathway leading to C_6–C_3 metabolites, such as lignin components and flavonoids (C_6–C_3–C_6), and the pathway leading to C_6–C_1 metabolites, such as benzoic acid derivatives, including vanillin. Generally, biosynthesis of benzoic acids in plants may proceed by the following routes:

i)   The oxidative pathway proceeds through formation of hydroxycinnamic acid CoA-esters and subsequent NAD-dependent oxidation of the side chain in reactions, similar to β-oxidation of fatty acids, leading from ferulic acid to vanillic acid and vanillin (Zenk 1965) and from 4-coumaric acid to 4-hydroxybenzoic acid (Loschler and Heide 1994).

ii)   The non-oxidative pathway proceeds through a cofactor independent action of a hydrolyase type enzyme, 4-hydroxybenzaldehyde synthase (4HBS), on 4-coumaric acid producing 4-hydroxybenzaldehyde (French *et al.* 1976; Yazaki *et al.* 1991; Schnitzler *et al.* 1992; Podstolski *et al.* 2002), and on 2-coumaric acid producing salicylic aldehyde (Malinowski *et al.* 2007).

iii) The shikimate pathway proceeds via isochorismic acid producing salicylic acid (2-hydroxybenzoic acid) and 2,3-dihydroxybenzoic acid (Wildermuth *et al.* 2001; Muljono *et al.* 2002). This pathway is not dependent on phenylalanine as a precursor.

In the bacterium *Pseudomonas fluorescens*, there is yet another non-oxidative pathway in which feruloyl-CoA ester undergoes chain shortening by an enoyl-SCoA hydratase/isomerase type enzyme, ending with the respective aldehyde (vanillin) formation (Gasson *et al.* 1998; Mitra *et al.* 1999).

A very different pathway to vanillic acid biosynthesis in cell cultures of *V. planifolia* has been proposed by Funk and Brodelius (1990). It involves conversion of cinnamic acid glucose ester to the corresponding caffeic acid glucose ester, which then undergoes two steps of methylation, first in position 4 forming isoferulic acid and then in position 3 forming 3,4-dimethoxycinnamic acid glucose ester. The last compound then serves as a substrate for a chain-shortening enzyme requiring methylation of position 4 of hydroxycinnamic acid. As a result, 3,4-dimethoxybenzoic acid is formed, which upon demethylation in position 4 yields vanillic acid. This pathway still awaits direct enzymatic evidence.

A tempting idea presented by Zenk (1965) that vanillin in *V. planifolia* may be produced through the β-oxidative pathway from ferulic acid, a compound having the identical substitution pattern in the ring as does vanillin, has so far gained no support on the enzymatic side. None of the proposed enzymes of a β-oxidation-like-pathway leading from ferulic acid to vanillin have been characterized.

On the other hand, a non-oxidative chain shortening enzyme, 4HBS (Mr 28 kDa), converting 4-coumaric acid to 4-hydroxybenzaldehyde was isolated, purified, and characterized from *V. planifolia* embryo culture (Podstolski *et al.* 2002). 4HBS activity was highly specific for 4-coumaric acid. Cinnamic acid or further metabolites of the phenylpropanoid pathway, 2-coumaric, caffeic and sinapic acids, were not substrates. The activity with ferulic acid was very low, about 2% of that obtained with 4-coumaric acid. 4HBS required thiol reagents for its activity. In this respect, equally effective were dithiothreitol, CoA, and to a lesser extent cysteine and the reduced form of glutathione (Podstolski *et al.* 2002). 4HBS catalyzed a hydrolyase type reaction. This reaction presumably proceeds via an unstable intermediate, 4-hydroxyphenyl-β-hydroxypropionic acid, formed after hydration of the side chain double bond of 4-coumaric acid followed by side chain cleavage and release of acetate and 4-hydroxybenzaldehyde (Figure 23.1)

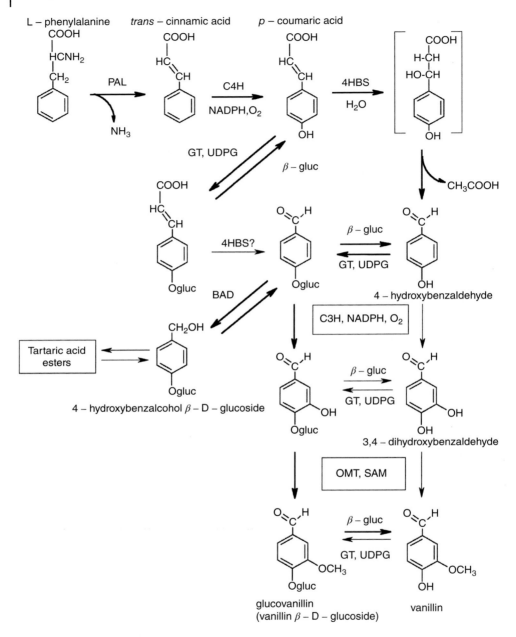

Figure 23.1 Proposed vanillin biosynthetic pathway (modified from Kanisawa *et al.* 1994 and Podstolski *et al.* 2002) in *Vanilla planifolia*. Bold arrows indicate suggested main flow of metabolites. Abbreviations: PAL, L-phenylalanine ammonia-lyase; C4H, cinnamic acid 4-hydroxylase; C3H, 4-hydroxybenzaldehyde 3-hydroxylase; 4HBS, 4-hydroxybenzaldehyde synthase; β-gluc, β-D-glucosidase; GT, glucosyltransferase; UDPG, uridine diphosphate glucose; BAD, 4-hydroxybenzoyl alcohol dehydrogenase; OMT, O-methyltransferase; SAM, S-adenosylmethionine; NADPH, nicotinamide adenine dinucleotide phosphate, reduced; Tartrate esters: bis[4-(β-D-glucopyranosyloxy)-benzyl]-2-isopropyltartrate and bis[4-(β-D-glucopyranosyloxy)-benzyl]-2-(2-butyl)tartrate.

(Yazaki *et al.* 1991; Schnitzler *et al.* 1992). The corresponding activity converting 2-coumaric acid to salicylic aldehyde (2-hydroxybenzaldehyde) was characterized from tobacco (Malinowski *et al.* 2007).

In addition to activity in the embryo culture, 4HBS activity was also observed in leaves, stems, roots, and vanilla pods. The highest specific activity was found in the pods where it was present during the whole developmental period with a peak of activity 7 to 8 months after pollination. The peak of 4HBS activity coincided with the maximum content of 4-coumaric acid and preceded the highest accumulation of 4-hydroxybenzaldehyde that is observed 10 to 11 months post-pollination (Podstolski *et al.* 2002). The presence of 4HBS was confirmed (using antibodies) in the cytoplasm of secretory hair-like cells located in the inner part of V*anilla* fruit. This inner part of the fruit contains 95% of the total vanillin, as well as all proposed vanillin precursors, 4-coumaric acid, 4-hydroxybenzaldehyde and 3,4-dihydroxybenzaldehyde (Joel *et al.* 2003).

23.3 4-Coumaric Acid 3-Hydroxylase (C3H)

In the proposed sequence of reactions leading to vanillin (Figure 23.1), the step following 4-hydroxybenzaldehyde formation is its hydroxylation in position 3 to protocatechuic aldehyde (3,4-dihydroxybenzaldehyde). This step involves introduction of the second hydroxyl group in the position *ortho* to the already existing one. Enzymes catalyzing such reactions, for example 4-coumarate 3-hydroxylase (C3H), which catalyzes formation of 3,4-dihydroxycinnamic acid (caffeic acid) from 4-coumaric acid, have not been fully characterized. It has been proposed that this hydroxylation might be carried out by monophenolase activity in the presence of an electron donor. The activity of a monophenolase oxidizing 4-coumaric acid, 4-hydroxybenzyl alcohol, 4-hydroxybenzaldehyde, and 4-hydroxybenzoic acid was found in *V. planifolia* shoot primordial culture (Debowska and Podstolski 2001), but its role in formation of 3,4-dihydroxy-derivatives was not confirmed. C3H has been characterized as a mixed function oxidase containing copper in the active site (Vaughan and Butt 1970) and requiring for its activity as electron donors either ascorbate (Kojima and Tekeuchi 1989), or NADPH (Vaughan and Butt 1970), or when in the chloroplast, plastoquinone or ferredoxin (Bartlett *et al.* 1972). On the other hand, there is data showing that phenolase and 3-hydroxylase activities could be completely separated, giving strong evidence for the existence of two independent enzymes (Duke and Vaughn 1982). More recent data obtained with *Arabidopsis* (Nair *et al.* 2002) showed that in this model plant cytochrome P450 monooxygenase, encoded by *CYP98A3*, was capable of 3-hydroxylation of 4-coumaric acid. This evidence strongly supports the view that C3H is a cytochrome P450, NADPH-dependent type monooxygenase, rather than a phenolase.

There is no available data concerning plant 4-hydroxybenzaldehyde 3-hydroxylase or 4-hydroxybenzoic acid 3-hydroxylase activities. Nevertheless, such activity (preferring NADPH to NADH as a co-substrate) was described in the bacterium *Corynebacterium glutamicum* growing on 4-hydroxybenzoic acid as the sole carbon source (Huang *et al.* 2008). This finding may create an opportunity for transgenic plant engineering.

23.4 *O*-Methyltransferase (OMT)

Protocatechuic aldehyde (3,4-dihydroxybenzaldehyde) is believed to be the immediate precursor of vanillin. Occurrence of this compound has been confirmed in vanilla pods (Joel *et al.* 2003) and in *Vanilla* tissue cultures as well (Havkin-Frenkel *et al.* 1996). Methylation of the hydroxyl group at position 3 catalyzed by an *O*-methltransferase (OMT) is considered the final step in the biosynthesis of vanillin (3-methoxy-4-hydroxybenzaldehyde). Reactions catalyzed by OMTs are very common in plants. These enzymes catalyze the transfer of the methyl group from S-adenosyl-*L*-methionine (SAM) to a hydroxyl group of various substrates.

In *V. planifolia*, at least five different OMTs have been characterized. From the tissue culture a cDNA encoding a multifunctional OMT was isolated and functionally characterized. Studies on its substrate specificity showed that it preferentially methylated the typical caffeic acid *O*-methyltransferase (COMT) substrates, caffeoylaldehyde and 5-hydroxyconiferylaldehyde, but also had activity with 3,4-dihydroxybenzaldehyde, the proposed immediate vanillin precursor (Pak *et al.* 2004). Its presence was found in leaves, stems, and roots but not in pods of *V. planifolia*. Although the enzyme was active on 3,4-dihydroxybenzaldehyde, its tissue location and substrate specificity indicated it was unlikely to be involved in vanillin biosynthesis and is most likely a COMT involved in lignin biosynthesis. Based on the relative activities of this enzyme, it is most probably connected with methylation of the 3-OH of caffeoylaldehyde and 5-OH of 5-hydroxyconiferylaldehyde during S-lignin formation (Dixon *et al.* 2001; Parvathi *et al.* 2001). Thus, in pods there must be another distinct 3,4-dihydroxybenzaldehyde *O*-methyltransferase (DOMT) responsible for vanillin biosynthesis. Strong DOMT activity was found in extracts of the inner part of pod at 5 to 11 months post-pollination (Pak *et al.* 2004), which is the pod developmental period when glucovanillin accumulation begins (Havkin-Frenkel *et al.* 1999).

In cell suspension culture of *V. planifolia*, two kinetin-inducible COMTs have been described. One was caffeic acid 4-*O*-methyltransferase, apparently connected with vanillic acid biosynthesis in cell suspension culture in the pathway proposed by Funk and Brodelius (1990). The other kinetin-inducible enzyme was a typical caffeic acid 3-*O*-methyltransferase connected to lignin biosynthesis (Xue and Brodelius 1998).

Two other *O*-methyltransferases from *V. planifolia*, Van OMT-2, and Van OMT-3, which have novel substrate preferences, have also been described (Li *et al.* 2006). These enzyme are capable of efficient methylation of the outer hydroxyl group in substrates having a 1,2,3-trihydroxybenzene structure, such as in methylgallate or in the B ring of the flavonol myricetin. Although the DNA sequences of Van OMT-2 and Van Omt-3 are 52% identical with the *V. planifolia* COMT described above (Pak *et al.* 2004), their activity with the usual COMT substrates was negligible. Based on phylogenetic analysis it was proposed that these enzymes evolved from the *V. planifolia* COMT (Li *et al.* 2006). The role of these enzymes in *V. planifolia* requires further elucidation.

23.5 Benzyl Alcohol Dehydrogenase (Bad)

4-Hydroxybenzylalcohol is one of the most abundant potential vanillin precursors found in green vanilla beans (Kanisawa *et al.* 1994). In *V. planifolia* embryo culture, it accumulates up to 3% of dry weight (Havkin-Frenkel *et al.* 1996). It was suggested that this alcohol is formed from 4-hydroxybenzaldehyde as a result of aromatic alcohol NADPH-dependent dehydrogenase activity (BAD). Although this enzyme in *V. planifolia* was not sufficiently investigated, there is data indicating its high activity in vanilla embryo culture. The enzyme was also found capable of fast and efficient reduction of vanillin to vanillyl alcohol (Havkin-Frenkel and Podstolski, unpublished).

23.6 Glycosyltransferases (GTS)

Many secondary metabolites including derivatives of benzoate, salicylate, phenylpropene, coumarins, flavonoids, anthocyanidins, terpenoids, and alkaloids occur in plant tissues as glucosides (Vogt and Jones 1997). Some of the presumed intermediates of the vanillin biosynthetic pathway downstream of 4-coumaric acid, such as 4-hydroxybenzaldehyde, 4-hydroxybenzyl alcohol, vanillin, vanillic acid, vanillyl alcohol, bis[4-(β-*D*-glucopyranosyloxy)-benzyl]-2-isopropyltartrate and bis[4-(β-*D*-glucopyranosyloxy)-benzyl]-2-(2-butyl) tartrate (in Figure 23.1 referred to as "tartrate esters") have been identified in extracts of green vanilla beans (Kanisawa *et al.* 1994; Dignum *et al.* 2004). Due to their more hydrophilic nature, glucosides are as a rule more soluble in water in comparison to their aglycones, the non-sugar partners. Higher solubility of glucosides allows for their higher concentration in cell storage compartments, usually vacuoles. Their high solubility may also facilitate transport of these metabolites between cell compartments and plant tissues and organs. As discussed below, vanillin accumulates in the green beans as a glucoside.

Glycosylation is carried out by glycosyltransferases (GTs), which transfer nucleotide-diphosphate activated sugars (UDP-sugar) to low molecular weight substrates. These enzymes occur in minute quantities in plant tissues as a set of related enzymes and thus are hard to separate and purify. GTs are very labile, making purification quite a difficult task. Those involved in secondary metabolism are usually soluble proteins of molecular weight between 45 and 60 kDa (Vogt and Jones 1997). Because a variety of secondary metabolites occur in the form of glycosides of different sugars, there are also several UDP-activated-sugars donating a sugar moiety to these low molecular substrates. The most frequent is UDP-glucose (in this case the respective **gly**cosides are called **gluco**sides), but UDP-galactose, UDP-rhamnose, and others are also found (Vogt and Jones 1997). A very important feature of glycosides is that they are not volatile. Glycosylation of a volatile aglycone makes it nonvolatile and allows for its accumulation at high concentrations in specialized tissues and cell compartments. Release of toxic aglycones by glycosylhydrolases is often induced as part of the defense response to microbial, fungal, or animal attack.

Although glucosides of some of the proposed vanillin biosynthesis intermediates have been found in green vanilla beans, it is still unknown if the glucosylated compounds actually serve as intermediates in the pathway. Addition of the sugar moiety to the intermediates may be a side reaction of the glucosyltransferase, "trapping" them from further modification. The added sugar moiety can be expected to change the shape of the substrate molecule and its interaction with the biosynthetic enzyme. For example, Van OMT-3, discussed above, was highly active with myricetin but not with a glycosylated derivative, myricitrin (Li *et al.* 2006). The answer to this question will require the isolation and characterization of the enzymes involved.

The concentration of glucovanillin (vanillin-β-*D*-glucoside) found in green vanilla pods is around 14% on a dry weight basis. Vanillin, the aglucone moiety in glucovanillin, comprises 45.8% by weight of the compound, which is also the percentage yield of vanillin when glucovanillin is hydrolyzed to completion. For example, 14 g of glucovanillin when hydrolyzed to completion, yields 6.4 g of vanillin. An actual recovery of 6 g vanillin therefore represents around 94% of the theoretically possible recovery. This level of vanillin recovery has been achieved after 7 days of curing under controlled laboratory conditions (Havkin-Frenkel *et al.* 2003). In traditional curing methods, however, the vanillin losses are considerably larger. A typical vanillin content of 2% on a dry weight basis of cured beans, represents a recovery of about 30%.

23.7 β-Glycosyl Hydrolases and Curing

Breakdown of a glucoside requires hydrolysis of the glucosidic bond between sugar and aglycone. In most natural glucosides, the bond is of the β-type. *In vitro*, this can be achieved non-enzymatically, especially at low pH and at increased temperature. In biological conditions the reaction is catalysed by β-glucosidase, a hydrolytic enzyme splitting β-glucosidic bonds between sugar and aglycone. β-Glucosidase usually occurs in plant tissues that accumulate or store bound secondary metabolites in the form of β-glucosides. Activity of β-glucosidase is the most important factor in formation of vanilla aroma during the curing of vanilla beans (Havkin-Frenkel *et al.* 2003). Although glucovanillin has long been postulated as the source of vanillin formed during the curing of the beans (Goris 1924), there is not much data concerning vanilla bean β-glucosidase itself (Arana 1943; Ranadive *et al.* 1983; Wild-Altamirano 1969; Hanum 1997). The enzyme was purified and characterized by Odoux *et al.* (2003). It consists of a 201 kDa tetramer composed of four identical 50 kDa subunits. It exhibits very narrow optimum activity at pH 6.5 and a temperature optimum of 40 °C. The enzyme was stable at neutral and alkaline pH, but is very unstable at pH below 6. Surprisingly, this β-glucosidase, despite purification to electrophoretic homogeneity, exhibited broad substrate specificity. It was active with several natural glycosides such as glucovanillin, prunasin, esculin, and salicin, as well as the artificial substrates 4-nitrophenyl-β-*D*-glucopyranoside, 4-nitrophenyl-β-*D*-fucopyranoside, 4-nitrophenyl-β-*D*-galactopyranoside, and 4-nitrophenyl-β-*D*-xylopyranoside. Looking at its broad specificity, it can be described rather as β-*D*-glycosidase than a β-*D*-glucosidase (Odoux *et al.* 2003). However, since the gene encoding the enzyme was not cloned and the activity of the recombinant enzyme characterized, it is possible that the activity characterized by Odoux *et al.* 2003 was originating from multiple enzymes. Activity of other glycosyl hydrolases, such as α-galactosidase, β-galactosidase, and β-mannosidase, has been reported in *V. planifolia* (Havkin-Frenkel and Belanger 2007). It will require future research to determine if the activities described above originate from separate enzymes, or rather are a result of the broad substrate specificity of β-glucosidase.

During the curing process, activity of glycosyl hydrolases liberate vanilla flavor components from their bound glycosidic forms. These hydrolytic enzymes and their substrates show different spatial distribution in the cell and in the tissues. The hydrolytic enzymes are located predominantly in the cytosolic fraction of the cell, whereas the glucoside substrates are stored in vacuolar compartments. In vanilla beans, β-glucosidase is predominantly located in the outer part of the fruit tissue and glucovanillin is in the inner placental part of the fruit (Joel *et al.* 2003; Havkin-Frenkel *et al.* 2003). Thus, after partial killing of the beans during the curing process, disorganization of the cellular compartments enables contact of β-glucosidase with the accumulated glucosides and the release of aroma constituents. Curing is a complex process also involving activity of other enzymes such as proteases, peroxidases, and polyphenoloxidases. The oxidative enzymes, peroxidases, and polyphenoloxidases, are responsible for browning of the pod tissue and for the formation of volatile ketones, aldehydes, and hydrocarbon derivatives (Adedji *et al.* 1993), which perhaps may also add to the final vanilla flavor.

References

Achnine, L., Blancaflor, E.B., Rasmussen, S. and Dixon, R.A. (2004) Colocalization of L-phenylalanine ammonia-lyase and cinnamate 4-hydroxylase for metabolic channeling in phenylpropanoid biosynthesis. *The Plant Cell*, 16, 3098–3109.

Adedji, J., Hartman, T.G. and Ho, C.-T. (1993) Flavor characterization of different varieties of Vanilla beans. *Perfumer & Flavorist*, 18, 25–33.

Arana, F.E. (1943) Action of a β-glucosidase in the curing of vanilla. *Food Research*, 8, 343–351.

Bartlett, D.J., Poulton, J.E. and Butt, V.S. (1972) Hydroxylation of *p*-coumaric acid by illuminated chloroplasts from spinach beet leaves. *FEBS Letters*, 23, 265–267.

Blount, J.W., Korth, K.L., Masoud, S.A., Rasmussen, S., Lamb, C. and Dixon R.A. (2000) Altering expression of cinnamic acid 4-hydroxylase in transgenic plants provides evidence for a feedback loop at the entry into the phenylpropanoid pathway. *Plant Physiology*, 122, 107–116.

Bolwell, G.P., Bell, J.N., Cramer, C.L., Schuch, W., Lamb, C.J. and Dixon, R.A. (2005) L-Phenylalanine ammonia-lyase from *Phaseolus vulgaris*. Characterisation and differential induction of multiple forms from elicitor-treated cell suspension cultures. *European Journal of Biochemistry*, 149, 411–419.

Bolwell, G.P. and Rodgers, M.W. (1991) L-phenylalanine ammonia-lyase from French bean. *Biochemical Journal*, 279, 231–236.

Cramer, C.L., Edwards, K., Dron, M. *et al.* (1989) Phenylalanine ammonia-lyase gene organization and structure. *Plant Molecular Biology*, 12, 367–383.

Czichi, U. and Kindl, H. (1975). Formation of *p*-coumaric acid and *o*-coumaric acid from L-phenylalanine by microsomal membrane fractions from potato: evidence of membrane –bound enzyme complexes. *Planta*, 125, 115–125.

Czichi, U. and Kindl, H. (1977) Phenylalanine ammonia-lyase and cinnamic acid hydroxylases as assembled consecutive enzymes on microsomal membranes of cucumber cotyledons: cooperation and subcellular distribution. *Planta*, 134, 133–143.

Debowska, R. and Podstolski, A. (2001) Properties of diphenolase from *Vanilla planifolia* (Andr.) shoot primordial cultured *in vitro*. *Journal of Agricultural Food Chemistry*, 49, 3432–3437.

Dignum, M.J.W., Van der Heijden, R., Kerler, J., Winkel, C. and Verpoorte, R. (2004). Identification of glucosides in green beans of *Vanilla planifolia* Andrews and kinetics of vanilla β-glucosidase. *Food Chemistry*, 85, 199–205.

Dixon, R.A., Achnine, L., Kota, P., Liu, C.-J., Reddy, M.S. and Wang, L. (2002) The phenylpropanoid pathway and plant defense: A genomics perspective. *Molecular Plant Pathology*, 3, 371–390.

Dixon, R.A., Chen, F., Guo, D. and Parvathi, K. (2001) The biosynthesis of monolignols: a "metabolic grid", or independent pathways to guaiacyl and syringyl units? *Phytochemistry*, 57, 1069–1084.

Dixon, R.A. and Paiva, N.L. (1995) Stress-induced phenylpropanoid metabolism. *Plant Cell*, 7, 1085–1097.

Duke, S.O. and Vaughn, K.C. (1982) Lack of involvement of polyphenol oxidase in ortho-hydroxylation of phenolic compounds in mung bean seedlings. *Physiologia Plantum*, 54, 381–385.

French, C.J., Vance, C.P. and Towers, G.H.N. (1976) Conversion of *p*-coumaric acid to *p*-hydroxybenzoic acid by cell free extracts of potato tuber and *Polyporus hispidus*. *Phytochemistry*, 15, 564–566.

Funk, C. and Brodelius, P.E. (1990) Phenylpropanoid metabolism in suspension cultures of *Vanilla planifolia*, Andr. III. Conversion of 4-methoxycinnamic acids into 4-hydroxybenzlic acids. *Plant Physiololgy*, 94, 102–108.

Gasson, M.J., Kitamura, Y., McLauchlan, W.R. *et al.* (1998) Metabolism of ferulic acid to vanillin. *Journal of Biological Chemistry*, 273, 4163–4170.

Goris, M.A. (1924) Sur la composition chimique des fruits verts de vanilla et le mode de formation du parfum de la vanilla. *Acad. Des Sci. Colon Paris, Compt. Rend.*, 179, 70–72.

Hahlbrock, K., Knobloch, K.H., Kreuzaler, F., Potts, J.R. and Wellmann, E. (1976) Coordinated induction and subsequent activity changes of two groups of metabolically interrelated enzymes. *European Journal of Biochemistry/FEBS*, 61, 199–206.

Hahlbrock, K. and Scheel, D. (1989) Physiology and molecular biology of phenylporopanoid metabolism. *Annual Review of Plant Physiology and Plant Molecular Biology*, 40, 347–369.

Hanum, T. (1997) Changes in vanillin and activity of β-glucosidase and oxidases during post harvest processing of vanilla beans (*Vanilla planifolia*). *Bulletin Teknologia dan Industri Pangan*, 8, 46–52.

Havir, E.A. and Hanson, K.R. (1973) L-phenylalanine ammonia-lyase (maize and potato). Evidence that the enzyme is composed of four subunits. *Biochemistry*, 12, 1583–1591.

Havkin-Frenkel, D. and Belanger, F.C. (2007) Application of metabolic engineering to vanillin biosynthetic pathways in *Vanilla planifolia*. In: Verpoorte, R., Alferman, A.W. and Johnson, T.S. (eds) *Application of plant metabolic engineering*. Springer Netherlands, pp. 175–196.

Havkin-Frenkel, D., French, J.C., Graft, N.M., Joel, D.M., Pak, F.E. and Frenkel, C. (2003) Interrelation of curing and botany in vanilla (*Vanilla planifolia*) bean. *Acta Horticulturae*, 629, 93–102.

Havkin-Frenkel, D., Podstolski, A. and Knorr, D. (1996) Effect of light on vanillin precursors formation by *in vitro* cultures of *Vanilla planifolia*. *Plant Cell, Tissue and Organ Culture*, 45, 133–136.

Havkin-Frenkel, D., Podstolski, A., Witkowska, E., Molecki P. and Mikolajczyk, M. (1999) Vanillin biosynthetic pathways: an overview. In: *Plant Cell and Tissue Culture for the Production of Food Ingredients*, Fu, T.J., Singh, G. and Curtis W.R. (Eds), Kluwer Academic, Plenum Publisher, New York, pp. 35–43.

Huang, Y., Zhao, K.X., Shen, X. H., Jiang, C. Y. and Liu, S. J. (2008) Genetic and biochemical characterization of a 4-hydroxybenzoate hydroxylase from *Corynebacterium glutamicum*. *Applied Microbiology and Biotechnology*, 78, 75–83.

Joel, D.M., French, J.C., Graft, N., Kourteva, G., Dixon, R.A. and Havkin-Frenkel, D. (2003) A hairy tissue produces vanillin. *Israel Journal of Plant Sciences*, 51, 157–159.

Jones, D.H. (1984) Phenylalanine ammonia-lyase: regulation of its induction, and its role in plant development. *Phytochemistry*, 23, 1349–1360.

Kanisawa, T., Tokoro, K. and Kawahara, S. (1994) Flavor development in the beans of *Vanilla planifolia* In: *Olfaction Taste XI*, Kurihara, K., Suzuki, N. and Ogawa, H. (Eds), Proceeding of the International Symposium. Springer, Tokyo, pp. 268–270.

Kojima, M. and Takeuchi, W. (1989) Detection and characterization of *p*-coumaric acid hydroxylase in mung bean, *Vigna mungo*, seedlings. *Journal of Biochemistry*, 105, 265–270.

Koopmann, E., Logemann, E. and Hahlbrock, K. (1999) Regulation and functional expression of cinnamate 4-hydroxylase from parsley. *Plant Physiology*, 119, 49–55.

Li, H.M., Rotter, D., Hartman, T.G., Pak, F.E., Havkin-Frenkel, D. and Belanger, F.C. (2006) Evolution of novel *O*-methyltransferases from the *Vanilla planifolia* caffeic acid *O*-methyltransferase. *Plant Molecular Biology*, 61, 537–552.

Loscher, R. and Heide, L. (1994) Biosynthesis of *p*-hydroxybenzoate from *p*-coumarate and *p*-coumaryl-coenzyme A in cell-free extracts of *Lithospermum erythrorhizon* cell cultures. *Plant Physiology*, 106, 271–279.

Malinowski, J., Krzymowska, M., Godon, K., Hennig, J. and Podstolski, A. (2007) A new catalytic activity from Tobacco converting 2-coumaric acid to salicylic aldehyde. *Physiologia Plantarum*, 129, 461–471.

Mitra, A., Kitamura, Y., Gasson, M.J. *et al.* (1999) 4-Hydroxycinnamoyl-CoA hydratase/lyase (HCHL) – an enzyme of phenylpropanoid chain cleavage from *Pseudomonas*. *Archives of Biochemistry and Biophysics*, 365, 10–16.

Mizutani, M., Ohta, D. and Sato, R. (1997) Isolation of a cDNA and genomic clone encoding cinnamate 4-hydroxylase from Arabidopsis and its expression manner in planta. *Plan Physiology*, 11, 755–763.

Muljono, R.A.B., Scheffer, J.J.C. and Verpoorte, R. (2002) Isochorismate is an intermediate in 2,3-dihydroxybenzoic acid biosynthesis in *Catharanthus roseus* cell cultures. *Plant Physiology and Biochemistry*, 40, 231–234.

Nair, R.B., Xia, Q., Kartha, C.J. *et al.* (2002) *Arabidopsis CYP98A3* mediating aromatic 3-hydroxylation developmental regulation of the gene and expression in yeast. *Plant Physiology*, 130, 210–220.

Odoux, E., Chauwin, A. and Brillouet, JM. (2003) Purification and characterization of vanilla bean (*Vanilla planifolia* Andrews) β-D-glucosidase. *Journal of Agricultural and Food Chemistry*, 51, 3168–3173.

Pak, F.E., Gropper, S., Dai, W.D., Havkin-Frenkel, D. and Belanger, F.C. (2004) Characterization of a multifunctional methyltransferase from the orchid *Vanilla planifolia*. *Plant Cell Reports*, 22, 959–966.

Parvathi, K., Chen, F., Guo, D., Blount, J.W. and Dixon, R.A. (2001) Substrate preferences of *O*-methyltransferases in alfalfa suggest new pathways for 3-*O*-methylation of monolignols. *Plant Journal*, 25, 193–202.

Podstolski, A., Havkin-Frenkel, D., Malinowski, J., Blount, J.W., Kourteva, G. and Dixon, R.A. (2002) Unusual 4-hydroxybenzaldehyde synthase activity from tissue cultures of vanilla orchid *Vanilla planifolia*. *Phytochemistry*, 61, 611–620.

Potts, J.R.M., Weklych, R. and Conn, E.E. (1974) The hydroxylation of cinnamic acid by Sorghum microsomes and requirement for cytochrome P450. *The Journal of Biological Chemistry*, 249, 5019–502.

Ranadive, A.S., Szkutnica, K., Guerrera, J.G. and Frenkel C. (1983) Vanillin biosynthesis in vanilla beans. In: Proceedings of the 9th International Congress on Essential Oils, Singapore, March 13–17, 1983; Essential Oils Association of Singapore: Singapore 1983; Book 2 pp 147–154.

Schnitzler, J.P., Madlung, J., Rose, A. and Seitz H.U. (1992) Biosynthesis of *p*-hydroxybenzoic acid in elicitor–treated carrot cell cultures. *Planta*, 188, 594–600.

Vaughan, P.F.T. and Butt, V.S. (1970) The action of *o*-dihydric phenols in the hydroxylation of *p*-coumaric acid by a phenolase from leaves of spinach beet (*Beta vulgaris* L.). *Biochemical Journal*, 119, 89–94.

Vogt, T. and Jones, P. (1997) Glucosyltransferases in plant natural product synthesis: characterization of a supergene family. *Plant Physiology*, 113, 755–763.

Walton, N.J., Mayer, M.J. and Narbad, A. (2003) Molecules of interest, vanillin. *Phytochemistry*, 63, 505–515.

Wanner, L.A., Li, G., Ware D., Somssich, I.E. and Davis, K.R. (1995) The phenylalanine ammonia-lyase gene family in *Arabidopsis thaliana*. *Plant Molecular Biology*, 27, 327–338.

Wild-Altamirano, C. (1969) Enzymic activity during growth of vanilla fruit. 1. Proteinase, glucosidase, peroxidase and polyphenoloxudase. *Journal of Food Science*, 34, 235–238.

Wildermuth, M. C., Dewdney, J., Wu, G. and Ausubel, F.M. (2001) Isochorismate synthase is required to synthesize salicylic acid for plant defense. *Nature*, 414, 562–565.

Xue, Z-T. and Brodelius, P.E. (1998) Kinetin-induced caffeic acid *O*-methyltransferases in cell suspension cultures of *Vanilla planifolia* Andr. and isolation of caffeic acid *O*-methltransferase cDNAs. *Plant Physiology and Biochemistry*, 36, 779–788.

Yazaki, K., Heide, L. and Tabata, M. (1991) Formation of *p*-hydroxybenzoic acid from *p*-coumaric acid by cell free extract of *Lithospermum erythrorhizon* cell cultures. *Phytochemistry*, 30, 2233–2236.

Zenk, M.H. (1965) Biosynthese von vanillin in *Vanilla planifolia* Andr. *Zeitschrift fur Pflanzenphysiologie*, 53, 404–414.

24

Vanillin Biosynthesis – Still Not as Simple as it Seems?

Richard A. Dixon

24.1 Introduction

Vanillin is the world's most popular flavor, and as such is probably the world's most popular plant natural product. It is also an extremely simple molecule. Why then, at a time when the biosynthesis of increasingly complex plant secondary metabolites is being elucidated at both the chemical and molecular genetic levels, should vanillin biosynthesis still be so controversial? Why do we know most of the steps involved in taxol biosynthesis (Heinig and Jennewein 2009), all of the steps involved in lignin (monolignol) biosynthesis (a pathway that shares similarities to the vanillin pathway(s)) (Humphreys and Chapple 2002; Vanholme *et al.* 2013), many of the steps involved in the formation of complex nitrogen-containing alkaloids (Kutchan 2002; Zeigler *et al.* 2006), but not how plants make 3-methoxy, 4-hydroxy-benzaldehyde? To be fair to the small body of researchers who have investigated vanillin biosynthesis, this question should probably be re-phrased to ask why we are still somewhat confused about the biosynthesis of most C_6-C_3 benzenoid derivatives in plants.

Vanillin is made in the "pods" of an orchid, *Vanilla planifolia*, a species that lacks genetic or genomic resources, and is stored as its 4-O-glucoside, glucovanillin. It is made in specialized cells within the pod (Joel *et al.* 2003), although there is still some disagreement as to exactly which cell types do or do not produce vanillin (Joel *et al.* 2003; Odoux and Brillouet 2009). The nature of the plant species and the restricted cellular location of its famous product should not present insurmountable problems for understanding vanillin biosynthesis, however, since many studies have addressed biosynthetic routes to more complex natural products through the application of molecular genetic approaches to specialized tissues in genetically recalcitrant plant species. Some of the best examples concern the biosynthesis of defensive compounds in glandular trichomes (Gang *et al.* 2002; Weathers *et al.* 2006; Nagel *et al.* 2008). My contention is that the simplicity of vanillin itself poses the major problem, because the structure lends itself to multiple theoretical biosynthetic pathways (Figure 24.1), and, because of a general promiscuity of many enzymes of plant phenolic metabolism, it is possible to find evidence to support any of these pathways from in vitro biochemical approaches. This certainly seems to be the case from a brief overview of

Handbook of Vanilla Science and Technology, Second Edition.
Edited by Daphna Havkin-Frenkel and Faith C. Belanger.
© 2019 John Wiley & Sons Ltd. Published 2019 by John Wiley & Sons Ltd.

the history of studies on the biosynthesis of vanillin and related compounds (Table 24.1), from which it is clear that our "understanding" of vanillin biosynthesis has not proceeded in a sequential manner. Rather, each new "advance" his provided an alternative model without effectively disproving existing models. Remarkably, seven years after publication of the first version of this chapter, this situation has changed little.

Table 24.1 A timeline for the development of concepts related to vanillin biosynthesis.

System and approach	Concept	Reference
Radiolabeling of *V. planifolia* pods	Vanillin is formed directly from ferulic acid	Zenk, 1965
Radiolabeling of *V. planifolia* tissue cultures	Intermediacy of isoferulic acid (which is subsequently demethylated)	Funk and Brodelius, 1990a,b
Enzyme assay in cell free extracts from *Lithospermum erythrorhizon*	Non-oxidative chain-shortening of coumaric acid to 4-hydroxybenzaldehyde	Yazaki *et al.*, 1991
Measuring metabolite levels in *V. planifolia* pods	Intermediacy of tartrate esters	Kanisawa *et al.*, 1994
Enzyme isolation and assay from cell cultures of *Hypericum androaemum*	Involvement of a cinnamoyl CoA hydratase/lyase in non-oxidative chain shortening	El-Mawla *et al.*, 2002
Enzyme isolation from *V. planifolia* cell cultures	Thiol-dependent non-oxidative conversion of 4-coumarate to benzaldehyde	Podstolski *et al.*, 2002
Gene cloned from *V. planifolia* embryo cultures	A cysteine-protease like protein catalyzing weak thiol-dependent non-oxidative conversion of 4-coumarate to benzaldehyde	Havkin-Frenkel *et al.*, 2013
Partial purification of an enzyme from pods of *V. planifolia*.	Iron-dependent dioxygenase for thiol-dependent chain-shortening of 4-coumaric and ferulic acids	Negishi and Negishi, 2014, 2015, 2017
Gene cloned from *V. planifolia* pods	Vanillin synthase, converting ferulic acid to vanillin; the same protein as in Havkin-Frenkel *et al.*, 2013, above	Gallage *et al.*, 2014

◄──

Figure 24.1 Scheme of potential pathways to vanillin, in comparison to monolignol and ferulic acid formation. The pathway on the left hand side of the figure shows the formation of ferulic acid from *trans*-cinnamic acid, according to recent studies on monolignol biosynthesis and the formation of ferulate in *Arabidopsis*. Vanillin is shown arising from two mechanistically different routes: directly from coumaric or ferulic acids by non-oxidative chain shortening, or via any one of three Coenzyme A esters by β-oxidation. The enzymes involved in the formation of ferulate from cinnamate have all be functionally identified; it is assumed that similar types of enzymes (or even possibly the same enzymes) could be involved in the hydroxylation, *O*-methylation and reduction of benzoyl CoA or benzaldehyde intermediates. The recently discovered route from coumaroyl shikimate to coumaroyl CoA via caffeoyl shikimate esterase is shown; previously, it was assumed that HCT could also operate in the "reverse direction" to convert caffeoyl shikimate to caffeoyl CoA. Both routes could potentially operate in parallel. Enzymes are: C4H, cinnamate 4-hydroxylase; 4CL, 4-coumarate (hydroxycinnamate) Coenzyme A ligase; HCT, hydroxycinnamoyl CoA shikimate hydroxycinnamoyl transferase; CSE, caffeoyl shikimate esterase; CCoAOMT, caffeoyl Coenzyme A 3-*O*-methyltransferase; CCR, cinnamoyl Coenzyme A reductase; AldH, aldehyde dehydrogenase; 4-HBS, 4-hydroxybenzaldehyde synthase; P450, cytochrome P450; OMT, *O*-methyltransferase.

24.2   Multiple Pathways to Vanillin Based on Biochemistry?

Past work on the vanillin pathway, and pathways leading to related benzenoids, has been reviewed in more detail elsewhere (Dignum *et al.* 2001; Walton *et al.* 2003; Wildermuth 2006; Kundu 2017). It is generally agreed that vanillin is a product of the phenylpropanoid pathway from L-phenylalanine, and that the hydroxyl group at the 4-position of the aromatic ring (*para* to the side chain) therefore originates through the action of the cytochrome P450 enzyme cinnamate 4-hydroxylase (C4H, Figure 24.1). This model is supported by labeling studies (Zenk 1965). The conversion of coumarate (4-hydroxycinnamate) to vanillin then "simply" requires four steps. These are: (1) shortening of the side chain by 2 carbons, catalyzed by a "chain shortening" enzyme or enzyme complex; (2) introduction of the aldehyde function to the side chain (in some models this may occur as an integral part of chain shortening; (3) introduction of the 3-hydroxyl group; (4) 3-*O*-methylation (Figure 24.1). Clearly the *O*-methylation reaction has to occur after the 3-hydroxylation, but other than that, these reactions could theoretically occur in any order. However, the number of possible theoretical pathways to vanillin is increased beyond three factorial by the fact that there is more than one mechanism for chain shortening of hydroxycycinnamic acids, and these lead to products with different oxidation states of the terminal group of the side-chain. Furthermore, if the model assumes a shared pathway to that involved in monolignol biosynthesis in which the first reactions are the ring modifications, additional reactions associated with formation of different types of ester intermediates could likely also be involved (see below) (Figure 24.1).

Similar complexities have been encountered in studies on related molecules. For example, salicylic acid (SA, 2-hydroxy-benzoic acid) was long thought to be synthesized through the phenylpropanoid pathway via L-phenylalanine, followed by chain shortening to a benzoic acid followed by subsequent 2-hydroxylation (Yalpani *et al.* 1993; Leon *et al.* 1995), and this was supported by genetic studies in which modification of expression of L-phenylalanine ammonia-lyase (the first enzyme of the phenylpropanoid pathway) gave disease response phenotypes predictably associated with modification of SA levels (Pallas *et al.* 1996). The subsequent demonstration that, at least in *Arabidopsis*, defense-associated SA formation occurs directly from the shikimate pathway via isochorismate (Wildermuth *et al.* 2001) came as a total surprise. Similarly, labeling and genetic studies have demonstrated that the formation of benzoic acids in Petunia flowers occurs by multiple pathways involving both oxidative and non-oxidative chain shortening (Boatright *et al.* 2004; Orlova *et al.* 2006). This complexity makes it quite difficult to interpret labeling studies, particularly if (as in the case of most studies on vanillin to date) multiple tissue types are being labeled and the labeling is only carried out over a short period relative to the period of biosynthesis and accumulation. It has been argued that the existence of multiple pathways to benzenoid natural products within one plant might reflect a biological need for flexible responses to different environmental conditions (Wildermuth 2006). This is quite plausible, but it seems to the present author that constitutive vanillin biosynthesis during the development of the vanilla pod is more likely to occur via a single major pathway. The question is how to elucidate that pathway when enzyme promiscuity can mislead *in vitro* studies.

Early labeling experiments suggested that vanillin biosynthesis in plants occurs via ferulic acid, a molecule known to be synthesized via the phenylpropanoid/monolignol pathway (Zenk 1965). More recent labeling studies came to the same conclusion, but did not actually measure incorporation of ferulate itself into vanillin (Gallage *et al.* 2014). Although a number of studies have suggested other alternatives (Table 24.1), it is instructive to consider this model for the formation of vanillin because it allows discussion of the types of enzymes that may be involved in the ring modification reactions, and their identification through functional genomics approaches.

At least in *Arabidopsis*, ferulate is formed from 4-coumarate by six or possibly seven enzymatic steps in a pathway, shared with monolignol biosynthesis; that is considerably more complex than envisaged at the time that the first labeling studies on vanillin biosynthesis were performed. The first step is the formation of a Coenzyme A ester through the action of 4-coumarate: CoA ligase (4CL), an enzyme generally encoded by multiple genes in plants (Ehlting *et al.* 1999) (Figure 24.1). The subsequent coumaroyl CoA ester is potentially a substrate for β-oxidative chain shortening (Figure 24.1) but, in the monolignol pathway, is directly converted to the corresponding shikimate ester by the action of hydroxycinnamoyl CoA: hydroxycinnamoyl transferase (HCT) (Hoffmann *et al.* 2003); it is this shikimate ester that undergoes hydroxylation of the aromatic ring at the 3-position by a second cytochrome P450 monooxygenase (Schoch *et al.* 2001). The subsequent 3-*O*-methylation does not happen at the shikimate ester stage, however; rather, the shikimate ester is converted back to the CoA ester through HCT acting in the reverse direction, and the resulting caffeoyl CoA is then methylated via caffeoyl CoA 3-*O*-methyltransferase (CCoAOMT) to yield feruloyl CoA. At least this was the accepted route until 2013, when a caffeoyl shikimate esterase (CSE) was discovered and ascribed a role in monolignol biosynthesis; this role requires caffeic acid to then be re-esterified to caffeoyl CoA (Vanholme *et al.* 2013). This compound is reduced to coniferaldehyde by the action of a cinnamoyl CoA reductase (CCR), another enzyme that is encoded by multiple genes in plants (Escamilla-Treviño *et al.* 2009). Finally, coniferaldehyde is converted to ferulic acid by the action of an aldehyde dehydrogenase (Nair *et al.* 2004) (Figure 24.1). It is important to note that detailed biochemical and genetic studies support the operation of this complex pathway over the simple mechanism whereby coumarate is converted to ferulate in two steps by 3-hydroxylation followed by 3-*O*-methylation, at least in dicotyledonous plants. However, early enzymatic work with crude and partially purified plant extracts did indeed suggest that this simpler pathway might operate, and the action of CSE (Vanholme *et al.* 2013) places caffeic acid (one methylation step away from ferulic acid) as a new component of the pathway.

The alternative and much simpler pathway to vanillin involves non-oxidative chain shortening. At least *in vitro*, 4-coumarate can be converted to 4-hydroxybenzaldehyde through a non-oxidative process requiring the presence of a thiol reagent but no other cofactor (Podstolski *et al.* 2002) (Figure 24.1). This type of reaction has been shown in extracts from both embryo cultures and pods of *V. planifolia* (Podstolski *et al.* 2002; Negishi and Negishi 2014), and in both cases, 4-coumarate is the preferred substrate over ferulate. A similar reaction, but requiring iron as a co-factor, has also been proposed (Negishi and Negishi 2015, 2017). Conversion to vanillin then simply requires 3-hydroxylation and *O*-methylation. Classical COMT enzymes are able to catalyze this methylation at the level of the benzaldehyde (Kota *et al.* 2004).

In 2013, we cloned a gene from embryo cultures of *V. planifolia* that encoded a protein that was part of a complex of proteins associated with the thiol-dependent chain shortening of 4-coumaric acid, designated 4-hydroxybenzaldehyde synthase (4-HBS). Remarkably, this protein exhibited extremely high sequence identity to cysteine proteases. When expressed in yeast, the protein appeared to confer a weak ability to chain shorten 4-coumaric acid, but not ferulic acid (Havkin-Frenkel *et al.* 2003). At the time, we were unable to confirm whether or not this protein was involved in vanillin biosynthesis.

24.3 Elucidation of Vanillin Biosynthesis via Molecular Biology?

What did it take to establish the above pathway for ferulate formation? Interestingly, labeling experiments played only a small part, as these can be difficult to interpret. Rather, the major paradigm shifts came about through the application of molecular genetic approaches in *Arabidopsis*, coupled with substrate specificity studies with recombinant enzymes. Unfortunately, *V. planifolia* does not appear to be a genetically tractable system at this stage. However, various tools of functional genomics that are now commonly applied to other systems provide alternative approaches to throw light on the biosynthetic pathway, and an increasing number of molecular biological studies addressing vanillin biosynthesis have be reported (e.g. Pak *et al.* 2004; Havkin-Frenkel and Belanger 2007; Wiedez *et al.* 2011; Fock-Bastide *et al.* 2014). These examples have, however, targeted known enzyme classes. In the absence of a clear biochemical mechanism for chain-shortening in vanillin biosynthesis that would infer involvement of one particular class of protein, the best molecular approach to obtain candidate genes involves gene expression profiling, both temporally and spatially.

The so-called "next generation" sequencing techniques have made it relatively simple to obtain massive RNA sequence datasets from relatively small amounts of tissue. Such datasets can easily be obtained from dissected tissues from vanilla pods throughout their period of development. As a control, similar datasets can be obtained from tissues shown not to accumulate significant amounts of vanillin, such as stems, roots, and leaves. After assembly and initial annotation of the sequences, the data can be mined for sequences matching the enzyme types predicted for involvement in vanillin biosynthesis based on all potential pathway models in Figure 24.1, or be interrogated in non-biased ways, for example through gene expression correlation analysis. Apart from the side chain shortening reaction (the most problematical part as it is not immediately clear what the chain-shortening enzyme might look like), the targeted enzyme types will include those known to catalyze aromatic hydroxylation and subsequent *O*-methylation, and possibly CoA ester reduction (analogous to CCR). It is more than likely that the hydroxylation reaction will be catalyzed by a cytochrome P450 enzyme, and that this will exhibit a significant degree of substrate specificity (Chapple 1998). Plant phenolic *O*-methyltransferases fall into two major classes, the type members being the so-called caffeic acid 3-*O*-methyltransferase (COMT, type I) which should properly be referred to as 5-hydroxyconiferaldehyde 3-*O*-methyltransferase based on its preferred substrate in the lignin pathway, and the type II CCoAOMT that is also involved in monolignol biosynthesis (Noel *et al.* 2003). Either type could potentially be involved in vanillin biosynthesis.

In contrast to most plant biosynthetic P450 enzymes, COMT is relatively promiscuous. In fact, the enzyme from alfalfa shows high activity against 3,4-dihydroxybenzaldehyde to form vanillin (Kota *et al.*, 2004), although this is unlikely to be a function for the enzyme in alfalfa. Because vanillin accumulation occurs over a long time period, high activity may not be critical for candidate enzymes. For example, the formation of a major strawberry aroma compound involves the activity of a COMT, even though this enzyme is much more active with monolignol precursors than it is with the precursor of the 2,5-dimethyl-4-methoxy-3(2H)-furanone flavor compound (Wein *et al.* 2002). Thus, *in vitro* biochemistry will ultimately need to be confirmed by either genetic approaches or detailed flux analysis measurements. Rapid techniques for reverse genetics based on virus-induced gene silencing are now being developed, and work well in some monocot systems (Lu *et al.* 2003; Ding *et al.* 2006). Likewise, techniques for precursor labeling and metabolic flux analysis are becoming increasingly sophisticated (Boatright *et al.* 2004).

As of the time of writing, two groups have reported RNA sequencing datasets for *V. planifolia* (Gallage *et al.* 2014; Rao *et al.* 2014). These have been mined for potential chain shortening enzymes, but the difficulties that have dogged the biochemical approach to vanillin biosynthesis still persist. Gallage *et al.* found a sequence in their dataset that was identical to the sequence of the cysteine protease-like protein that we reported in 2003. They claimed that this protein was a vanillin synthase, localized to the cells that synthesize vanillin, which could convert ferulate directly to vanillin (Gallage *et al.* 2014). However, our group has been unable to confirm this finding, either for the recombinant enzyme alone or when expressed in bacteria, yeast or plants (Yang *et al.* 2017). Analysis of our spatially and temporally detailed RNA sequencing data, along with immunolocalization studies, also showed that this protein was likely present in similar amounts in all tissues and organs of the plant, irrespective of whether or not they made or accumulated vanillin (Rao *et al.* 2014; Yang *et al.* 2017). However, a recent paper claims that expression of the cysteine protease-like protein in callus cultures of pepper results in increased accumulation of vanillin (Chee *et al.* 2017). As pepper naturally produces vanillin (as a part of the capsaicin molecule), it is possible that the cysteine protease-like protein co-acts with other proteins in the tissue to assist the chain-shortening reaction, consistent with our reports of the protein being part of a complex (Podstolski *et al.* 2002; Havkin-Frenkel *et al.* 2003; Yang *et al.* 2017).

A further factor adding to the mystery of vanillin biosynthesis relates to cell biology. The pathways shown in Figure 24.1 occur either in the cytosol or on cytochrome P450 enzymes anchored to the cytosolic side of the endoplasmic reticulum. But the cysteine-protease like protein has a chloroplast targeting sequence (Havkin-Frenkel *et al.* 2003; Gallage *et al.* 2014). Furthermore, it has been suggested, on the basis of cytological studies, that vanillin accumulates in specialized organelles termed phenyloplasts which are derived from plastids (Brillouet *et al.* 2014). Although controversial, similar organelles have been proposed to be involved in the biosynthesis of a different class of phenolic compounds, the condensed tannins (Brillouet *et al.* 2012). If vanillin biosynthesis, rather than just accumulation, occurs in re-differentiating plastids, perhaps associated with extensive protein breakdown, the environment in which the chain shortening reaction occurs might be difficult to re-create in *in vitro* biochemical studies.

Two factors are currently limiting the final assault on the vanillin pathway; the lack of a good experimental model system (e.g. a highly inducible cell or tissue culture) to simplify labeling experiments, and the lack of an efficient genetic transformation system for the plant. The ability to create loss of function mutants in *V. planifolia* would allow unequivocal testing of candidate genes for involvement in vanillin biosynthesis. There is also currently no clear economic benefit from understanding how the vanillin molecule is assembled in *V. planifolia*, since the idea of using such information to engineer the pathway, at least in *V. planifolia*, goes against the concept of natural vanilla, and synthetic vanillin is so cheap that introducing this molecule alone into other plants as an extractable flavor component also does not make much economic sense. These factors should not, however, be used to argue against supporting research on vanillin biosynthesis. The pathways and mechanisms uncovered could in the future prove critical for the development of more complex bioactives in plants with wide applications in agriculture, food science and biomedicine.

References

Boatright, J., Negre, F., Chen, X., Kish, C.M., Wood, B., Peel, G., Orlova, I., Gang, D., Rhodes, D. and Dudareva, N. (2004) Understanding in vivo benzenoid metabolism in petunia petal tissue. *Plant Physiol.*, 135, 1993–2011.

Brillouet J.-M., Romieu C., Schoefs B., Solymosi K., Cheynier, V., Fulcrand, H., Verdeil, J.-L. and Conéjéro. G. (2013) The tannosome is an organelle forming condensed tannins in the chlorophyllous organs of Tracheophyta. *Ann. Bot.*, 112, 1003–1014.

Brillouet, J.-M., Verdeil, J.-L., Odoux, E., Lartaud, M., Grisoni, M. and Conéjéro, G. (2014) Phenol homeostasis is ensured in vanilla fruit by storage under solid form in a new chloroplast-derived organelle, the phenyloplast. *J. Exp. Bot.*, 65, 2427–2435.

Chapple, C. (1998) Molecular-genetic analysis of plant cytochrome P450-dependent monooxygenases. *Annu. Rev. Plant Physiol. Plant Mol. Biol.*, 49, 311–343.

Chee, M.J.Y., Lycett, G.W., Khoo, T.J. and Chin, C.F. (2017) Bioengineering of the plant culture of *Capsicum frutescens* with vanillin synthase gene for the production of vanillin. *Mol Biotechnol.*, 59, 1. https://doi.org/10.1007/s12033-016-9986-2

Dignum, M.J.W., Kerler, J. and Verpoorte, R. (2001) Vanilla production: technological, chemical, and biosynthetic aspects. *Food Rev. Int.*, 17, 199–219.

Ding, X.S., Schneider, W.L., Chaluvadi, S.R., Mian, M.A.R. and Nelson, R.S. (2006) Characterization of a *Brome mosaic* virus strain and its use as a vector for gene silencing in monocotyledonous hosts. *Mol. Plant-Microbe Interact.*, 19, 1229–1239.

Ehlting, J., Buttner, D., Wang, Q., Douglas, C.J., Somssich, I.E. and Kombrink, E. (1999) Three 4-coumarate: coenzyme A ligases in *Arabidopsis thaliana* represent two evolutionarily divergent classes in angiosperms. *Plant J.*, 19, 9–20.

El-Mawla, A.M.A.A. and Beerhues, L. (2002) Benzoic acid biosynthesis in cell cultures of *Hypericum androsaemum*. *Planta*, 214, 727–733.

Escamilla-Treviño, L.L., Shen, H., Uppalapati, S.R., Ray, T., Tang, Y., Hernandez, T., Yin, Y., Xu, Y. and Dixon, R.A. (2009) Switchgrass (*Panicum virgatum* L.) possesses a divergent family of cinnamoyl CoA reductases with distinct biochemical properties. *New Phytol.*, 185, 143–155.

Fock-Bastide, I., Palama, T.L., Bory, S., Lécolier, A., Noirot, M., and Joët, T. (2014) Expression profiles of key phenylpropanoid genes during *Vanilla planifolia* pod development reveal a positive correlation between PAL gene expression and vanillin biosynthesis. *Plant Physiol. Biochem.*, 74, 304–314.

Funk, C. and Brodelius, P.E. (1990a) Phenylpropanoid metabolism in suspension cultures of *Vanilla planifolia*. Andr. II. Effects of precursor feeding and metabolic inhibitors. *Plant Physiol.*, 94, 95–101.

Funk, C. and Brodelius, P.E. (1990b) Phenylpropanoid metabolism in suspension cultures of *Vanilla planifolia* Andr. III. Conversion of 4-methoxycinnamic acids into 4-hydroxybenzoic acids. *Plant Physiol.*, 94, 102–108.

Gallage, N.J., Hansen, E.H., Kannangara, R., Olsen, C.E., Motawia, M.S., Jørgensen, K., Holme, I., Hebelstrup, K., Grisoni, M. and Møller, B.L. (2014) Vanillin formation from ferulic acid in *Vanilla planifolia* is catalysed by a single enzyme. *Nat. Commun.*, 5, DOI: 10.1038/ncomms5037.

Gang, D.R., Beuerle, T., Ullmann, P., Werck-Reichhart, D. and Pichersky, E. (2002) Differential production of *meta* hydroxylated phenylpropanoids in sweet basil peltate glandular trichomes and leaves is controlled by the activities of specific acyltransferases and hydroxylases. *Plant Physiol.*, 130, 1536–1544.

Havkin-Frenkel, D. and Belanger, F.C. (2007) Application of metabolic engineering to vanillin biosynthetic pathways in *Vanilla planifolia*. In: *Applications of Plant Metabolic Engineering.* Verpoorte, R. ed. Springer, pp. 175–196.

Havkin-Frenkel, D., Podstolski, A. and Dixon, R.A. (2003) *Vanillin biosynthetic pathway enzyme from Vanilla planifolia*. US Patent US20030070188 A1 (US Patent Office: United States, 2003).

Heinig, U. and Jennewein, S. (2009) Taxol: A complex diterpenoid natural product with an evolutionarily obscure origin. *African J. Biotechnol.*, 8, 1370–1385.

Hoffmann, L., Maury, S., Martz, F., Geoffroy, P. and Legrand, M. (2003) Purification, cloning, and properties of an acyltransferase controlling shikimate and quinate ester intermediates in phenylpropanoid metabolism. *J. Biol. Chem.*, 278, 95–103.

Humphreys, J.M. and Chapple, C. (2002) Rewriting the lignin roadmap. *Curr. Opin. Plant Biol.*, 5, 224–229.

Joel, D.M., French, J.C., Graft, N., Kourteva, G., Dixon, R.A. and Havkin-Frenkel, D. (2003) A hairy tissue produces vanillin. *Israel J. Plant Sci.*, 51, 157–159.

Kanisawa, T., Tokoro, K. and Kawahara, S. (1994) In: *Olfaction Taste XI* (Proceedings of the International Symposium). Kurihara, K., Suzuki, N. and Ogawa, H., eds. Springer, Tokyo, p. 268.

Kota, P., Guo, D., Zubieta, C., Noel, J.P. and Dixon, R.A. (2004) O-Methylation of benzaldehyde derivatives by "lignin-specific" caffeic acid 3-O-methyltransferase. *Phytochemistry*, 65, 837–846.

Kundu, A. (2017) Vanillin biosynthetic pathways in plants. *Planta* 245, 1069–1078.

Kutchan, T.M. (2002) Sequence-based approaches to alkaloid biosynthesis gene identification. In: *Phytochemistry in the Genomics and Post-Genomics Eras.* Romeo, J.T. and Dixon, R.A., eds. Elsevier, Oxford. pp. 163–178.

León, J., Shulaev, V., Yalpani, N., Lawton, M.A. and Raskin, I. (1995) Benzoic acid 2-hydroxylase, a soluble oxygenase from tobacco, catalyzes salicylic acid biosynthesis. *Proc. Natl. Acad. Sci. USA*, 92, 10413–10417.

Lu, R., Martin, H.A.M., Peart, J.R., Malcuit, I. and Baulcombe, D.C. (2003) Virus-induced gene silencing in plants. *Methods*, 30, 296–303.

Nagel, J., Culley, L.K., Lu, Y., Liu, E., Matthews, P.D., Stevens, J.F. and Page, J.E. (2008) EST analysis of hop glandular trichomes identifies an *O*-methyltransferase that catalyzes the biosynthesis of xanthohumol. *Plant Cell* 20, 186–200.

Nair, R.B., Bastress, K.L., Ruegger, M.O., Denault, J.W. and Chapple, C. (2004) The *Arabidopsis thaliana* Reduced Epidermal Fluorescence1 gene encodes an aldehyde dehydrogenase involved in ferulic acid and sinapic acid biosynthesis. *Plant Cell*, 16, 544–554.

Negishi, O. and Negishi, Y. (2014) Biosynthesis of vanillin: enzyme involved in the conversion of ferulic acid to vanillin in *Vanilla planifolia*. In: *Flavour Science*. Ferreira, V. and Lopez, R. eds. Elsevier Inc, Oxford, pp. 205–209.

Negishi, O. and Negishi, Y. (2015) Enzyme catalyzing cleavage of the ferulic acid side chain in *Vanilla planifolia*. In: *Flavour Science: Proceedings from the XIV Weurman Flavour Research Symposium*. Taylar, A.J. and Mottram, D.S., eds. Context Products Ltd., Leicestershire, UK, pp. 233–236.

Negishi, O. and Negishi, Y. (2017) Phenylpropanoid 2,3-dioxygenase involved in the cleavage of the ferulic acid side chain to form vanillin and glyoxylic acid in *Vanilla planifolia*. *Biosci. Biotechnol. Biochem.* 81, 1732–1740.

Noel, J.P., Dixon, R.A., Pichersky, E., Zubieta, C. and Ferrer, J.-L. (2003) Structural, functional, and evolutionary basis for methylation of plant small molecules. *Rec. Adv. Phytochem.*, 37, 37–58.

Odoux, E. and Brillouet, J.-M. (2009) Anatomy, histochemistry and biochemistry of glucovanillin, oleoresin and mucilage accumulation sites in green mature vanilla pod (*Vanilla planifolia*; Orchidaceae): a comprehensive and critical reexamination. *Fruits*, 64, 1–21.

Orlova, I., Marshall-Colon, A., Schnepp, J., Wood, B., Varbanova, M., Fridman, E., Blakeslee, J.J., Peer, W.A., Murphy, A.M., Rhodes, D., Pichersky, E. and Dudareva, N. (2006) Reduction of benzenoid synthesis in petunia flowers reveals multiple pathways to benzoic acid and enhancement in auxin transport. *Plant Cell*, 18, 3458–3495.

Pak, F.E., Gropper, S., Dai, W.D., Havkin-Frenkel, D. and Belanger, F.C. (2004) Characterization of a multifunctional methyltransferase from the orchid *Vanilla planifolia*. *Plant Cell Rep.*, 22, 959–966.

Pallas, J.A., Paiva, N.L., Lamb, C.J. and Dixon, R.A. (1996) Tobacco plants epigenetically suppressed in phenylalanine ammonia-lyase expression do not develop systemic acquired resistance in response to infection by tobacco mosaic virus. *Plant J.*, 10, 281–293.

Podstolski, A., Havkin-Frenkel, D., Malinowski, J., Blount, J.W., Kourteva, G. and Dixon, R.A. (2002) Unusual 4-hydroxybenzaldehyde synthase activity from tissue cultures of the vanilla orchid *Vanilla planifolia*. *Phytochemistry*, 61, 611–620.

Rao, X., Krom, N., Tang, Y., Widiez, T., Havkin-Frenkel, D., Belanger, F.C., Dixon, R.A. and Chen, F. (2014) A deep transcriptomic analysis of pod development in the vanilla orchid (*Vanilla planifolia*). *BMC Genomics* 15, 964.

Schoch, G., Goepfert, S., Morant, M., Hehn, A., Meyer, D., Ullmann, P. and Werck-Reichart, D. (2001) CYP98A3 from *Arabidopsis thaliana* is a 3'-hydroxylase of phenolic esters, a missing link in the phenylpropanoid pathway. *J. Biol. Chem.*, 276, 36566–36574.

Vanholme, R., Cesarino, I., Rataj, K., Xiao, Y., Sundin, L., Goeminne, G., Kim, H., Cross, J., Morreel, K., Araujo, P., Welsh, L., Haustraete, J., McClellan, C., Vanholme, B., Ralph, J., Simpson, G.G., Halpin, C. and Boerjan W. (2013) Caffeoyl shikimate esterase (CSE) is an enzyme in the lignin biosynthetic pathway. *Science* 341, 1103–1106.

Walton, N.J., Mayer, M.J. and Narbad, A. (2003) Vanillin. *Phytochemistry*, 63, 505–515.

Weathers, P.J., Slkholy, S. and Wobbe, K.K. (2006) Artemisinin: The biosynthetic pathway and its regulation in *Artemisia annua*, a terpenoid-rich species. *In Vitro Cell. Dev. Biol.-Plant*, 42, 309–317.

Wein, M., Lavid, N., Lunkenbein, S., Lewinsohn, E., Schwab, W. and Kaldenhoff, R. (2002) Isolation, cloning and expression of a multifunctional *O*-methyltransferase capable of forming 2,5-dimethyl-4-methoxy-3(2H)-furanone, one of the key aroma compounds in strawberry fruits. *Plant J.*, 31, 755–765.

Widiez, T., Hartman, T. G., Dudai, N., Yan, Q., Lawton, M., Havkin-Frenkel, D. and Belanger, F.C. (2011) Functional characterization of two new members of the caffeoyl CoA *O*-methyltransferase-like gene family from *Vanilla planifolia* reveals a new class of plastid-localized *O*-methyltransferases. *Plant Mol. Biol.*, 76, 475–488.

Wildermuth, M.C. (2006) Variations on a theme: synthesis and modification of plant benzoic acids. *Curr. Opin. Plant Biol.*, 9, 288–296.

Wildermuth, M.C., Dewdney, J., Wu, G. and Ausubel, F.M. (2001) Isochorismate synthase is required to synthesize salicylic acid for plant defence. *Nature*, 414, 562–565.

Yalpani, N., Leon, J., Lawton, M.A. and Raskin, I. (1993) Pathway of salicylic-acid biosynthesis in healthy and virus-inoculated tobacco. *Plant Physiol.*, 103, 315–321.

Yang, H., Barros-Rios, J., Kourteva, G., Rao, X., Chen, F., Shen, H., Liu, C., Podstolski, A., Belanger, F., Havkin-Frenkel, D. and Dixon, R.A. (2017) A re-evaluation of the final step of vanillin biosynthesis in the orchid *Vanilla planifolia*. *Phytochemistry* 139, 33–46.

Yazaki, K., Heide, L. and Tabata, M. (1991) Formation of *p*-hydroxybenzoic acid from *p*-coumaric acid by cell free extract of *Lithospermum erythrorhizon* cell cultures. *Phytochemistry*, 30, 2233–2236.

Zenk, M.H. (1965) Biosynthese von vanillin in *Vanilla planifolia* Andr. *Z. Pflanzenphysiol.*, 53, 404–414.

Ziegler, J., Voigtländer, S., Schmidt, J., Kramell, R., Miersch, O., Ammer, C., Gesell, A. and Kutchan, T.M. (2006) Comparative transcript and alkaloid profiling in *Papaver* species identifies a short chain dehydrogenase/reductase involved in morphine biosynthesis. *Plant J.*, 48, 177–192.

25

Vanilla planifolia – The Source of the Unexpected Discovery of a New Lignin
Fang Chen and Richard A. Dixon

25.1   Introduction

Lignification occurs mostly in vessels, tracheids, and fibrous tissues of vascular plants where the lignin provides mechanical support and enhances water transport. Lignin is a heterogeneous polymer produced by the oxidative polymerization of monolignols, i.e., *p*-coumaryl, coniferyl and sinapyl alcohols. The biosynthesis of these monolignols shares the general phenylpropanoid biosynthesis pathway with vanillin biosynthesis and diverges to monolignol-specific pathways after *p*-hydroxycinnamoyl CoA biosynthesis. When monolignols are polymerized, they give rise to the three lignin subunits, namely *p*-hydroxyphenyl (H), guaiacyl (G) and syringyl (S) units (Boerjan and Baucher, 2003; Vanholme *et al.* 2010). Lignin composition varies from species to species. Within the same plant species, lignin composition may vary depending on the tissues, cell types and developmental stages and is also influenced by environmental stress. In angiosperms, lignin is mainly made of guaiacyl and syringyl units, while in gymnosperms, lignin is predominantly of the guaiacyl type (Whetten and Sederoff, 1995; Higuchi, 1990). During lignin biosynthesis, the monolignol precursors are functionalized by aromatic hydroxylation and *O*-methylation to generate monolignols that differ in their aromatic substitution patterns (Figure 25.1). It was accepted for many years that natural lignins are composed only of *p*-hydroxyphenyl, guaiacyl, and syringyl units. Catechyl (C) and 5-hydroxyl guaiacyl (5H) units that may derive from polymerization of the corresponding caffeyl and 5-hydroxy coniferyl alcohols (Figure 25.1), although theoretically possible, were not believed to exist naturally in plants. However, our studies on lignin biosynthesis in the *Vanilla planifolia* seed coat demonstrated that caffeyl alcohol is indeed a natural monolignol. It can be polymerized and forms a unique lignin structure, in which the catechyl units are predominantly linked by benzodioxane moieties. Here, we provide a brief overview of the recent studies of lignin biosynthesis in *V. planifolia*, which drastically changed our understanding of lignin biosynthesis.

Handbook of Vanilla Science and Technology, Second Edition.
Edited by Daphna Havkin-Frenkel and Faith C. Belanger.
© 2019 John Wiley & Sons Ltd. Published 2019 by John Wiley & Sons Ltd.

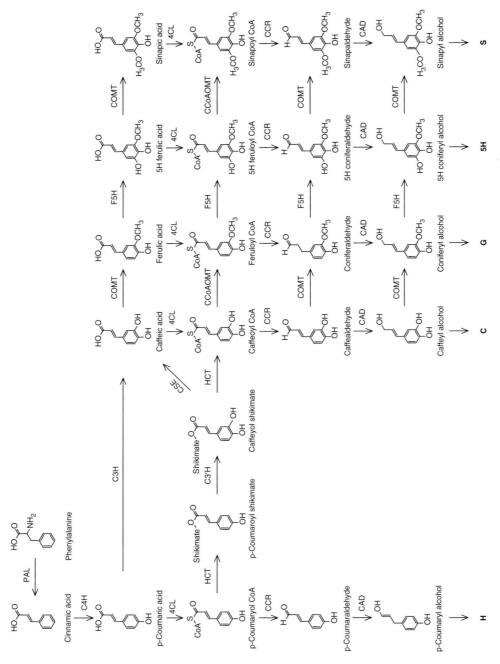

Figure 25.1 The theoretical lignin biosynthesis pathways.

25.2   Identification of C-lignin in *V. planifolia*

The research on *V. planifolia* in the Dixon lab had been focused, as in other labs, on vanillin biosynthesis. This simple mysterious molecule has attracted much attention in vanilla related research all over the world. Only after over 10 years of studies did we decide to take a look at the lignin deposited in the vanilla beans, essentially to satisfy the curiosity of Dr. Daphna Havkin-Frenkel who wanted to know what is the composition of lignin in vanilla beans, mostly because of the similarity between the vanillin biosynthetic pathway and lignin. There had been a notion that the vanilla pod contains less lignin then other plants because it accumulates vanillin. This did not sound like a promising avenue for further research at the beginning, since lignin in the vascular bundle of vanilla beans and seed coat unlikely will affect the biosynthesis of vanillin, which is made only in specific tissues/cell types during maturation of the vanilla bean (Joel *et al.* 2003). We decided to analyze the lignin composition of the ground vanilla bean powder by thioacidolysis, a chemical degradation method that breaks the lignin polymer into monomer structures (H, G and S monomers) via cleavage of uncondensed alkyl aryl ether (β-O-4) structures (Rolando *et al.* 1992; Lapierre *et al.* 1986). The monomers detected by GCMS after thioacidolysis give information on the composition of non-condensed alkyl aryl structures in the lignin polymer, and the S/G ratio can be calculated to represent the lignin composition.

Initial studies on the lignin of mature beans revealed the expected presence of regular G and S type lignin, with an S/G ratio of 0.4. At the same time, the gas chromatogram of vanilla bean samples showed many other smaller peaks that were not found in the analysis of *V. planifolia* stem cell wall materials (Figure 25.2). A small doublet peak in the gas chromatogram at a retention time after the G monomer caught our attention. At the beginning, we thought that this peak must represent an intermediate in the vanillin biosynthesis pathway, and identifying this compound may shed light on the vanillin biosynthesis pathway. However, close examination of mass spectra of this compound showed it to be consistent with the α, β, γ-trithioethylpropylcatechol derived from catechyl (C) units (Lapierre *et al.* 1992; Chen *et al.* 2012). This was a big surprise and very exciting, since (non-methylated) catechyl units had never be found in lignin polymers, even in transgenic plants downregulated in the caffeoyl-CoA 3-*O*-methyltransferase (CCoAOMT) or caffeic acid 3-*O*-methyltransferase (COMT) responsible for methyl substitution of catechyl units. More strikingly, after separating the vanilla seed from other parts of the bean, thioacidolysis revealed that the lignin in the isolated seeds was entirely composed of C units, with practically no release of G or S monomers (Figure 25.2). In contrast, thioacidolysis of the pod residue (after seed isolation), stem, leaf, and aerial roots released normal G and S monomers with essentially no C monomer, indicating that the lignins present in these tissues are typical G-rich G/S-lignins (Chen *et al.* 2012). Similar results were obtained from seed coats of two other *Vanilla* species, *V. pompona* and *V. tahitensis*. Consequently, we hand pollinated the flowers of *V. planifolia* in our greenhouse and investigated C-lignin deposition in vanilla seeds at different developmental stages. We found that C-lignin was first detected in the seed coat at around 8 weeks after pollination, when the color of the seed coat turns from transparent white to brown. This is about 2–3 months before the appearance in the pods of the flavor compound vanillin. In mature seed, thioacidolysis released about

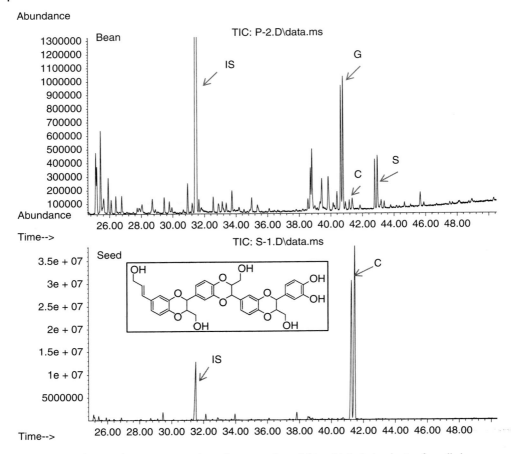

Figure 25.2 The gas chromatograms show the separation of thioacidolysis products of vanilla bean and seeds. Insert shows a structure of C-lignin. IS: internal standard; C: catechyl monomer; G: guaiacyl monomer; S: syringyl monomer.

70 μmol/g C monomers. We estimated that the absolute levels of the C-lignin in the vanilla seed are substantially higher than this because thioacidolysis is not very efficient in breaking down the unusual structure of the C-lignin polymer. Klason analysis of the seed indicated a very high level (~80%) of acid-insoluble lignin polymer (Table 25.1). The majority of the remaining material in the seed coat was crystalline cellulose. The chemical compositions of the pod residue and the stem were similar overall; these tissues were most rich in cellulose, with modest levels of hemicellulosic and pectic sugars, and Klason lignins.

C-lignin was then isolated from vanilla seed by cellulase treatment of ball milled *V. planifolia* seed followed by extraction with dimethylsulfoxide (DMSO). The isolated seed coat lignin was then acetylated and subjected to long-range ^{13}C-^{1}H correlation NMR analysis (Chen *et al.*, 2012). Volume integration of the contour signals confirmed that lignin from vanilla seed is almost exclusively composed of C units; no typical G and S lignin signals were detected in this polymer. High-field HSQC spectra of the

Table 25.1 Chemical composition of *V. planifolia* seed, pod and stem tissues.

Composition	Seed	Pod	Stem
Klason lignin (mg/g)	817	166	129
Crystalline cellulose (mg/g)	164	361	426
Rhamnose (%)	2.4	4.4	2
Arabinose (%)	7.7	17.9	3.4
Xylose (%)	8.7	35.5	76.9
Mannose (%)	2.7	5.3	1.1
Galactose (%)	27.8	18.1	4.2
Glucose (%)	49.5	16.5	12.1

side-chain (aliphatic) regions showed benzodioxanes were the dominant units in vanilla seed coat lignin, accounting for over 98% of the total identifiable dimeric units. The extensive benzodioxane linkages found in the C-lignin polymer indicate that, during lignin polymerization, the rearomatization of quinone methide (QM) intermediates after free radical coupling seems to be exclusively via the efficient internal trapping by the *o*-hydroxyl group in C units (Chen *et al.* 2012; Tobimatsu *et al.* 2013). The main benzodioxane backbones in the seed polymer are *trans/cis*-isomeric mixtures. The *trans/cis* composition of the benzodioxane rings in the seed polymer is 97:3, very similar to that of the dehydrogenation polymer (DHP) made from caffeyl alcohol *in vitro* (96:4). These results suggest that the stereochemistry of rearomatization of the QM is under simple kinetic chemical control. Therefore, it is most likely that caffeyl alcohol is enzymatically oxidized by plant oxidoreductases, such as peroxidases and laccases, but the cross coupling of free radical onto the growing polymer is independent of enzymes or other proteins, in the same way as conventional monolignols are coupled during lignin polymerization. The normal β-O-4 linkages, which are the predominant linkages in typical natural lignins, were absent in these polymers. In contrast, a G/S type lignin rich in β-aryl ether units, phenylcoumaran and resinol units can be found in lignin isolated from vanilla stems, as is typical for angiosperm stem lignins. In summary, thioacidolysis and 2D-NMR data clearly indicate that the C-polymer is essentially a homopolymer synthesized purely from caffeyl alcohol, and with benzodioxanes as essentially the only intermonomer unit in the polymer (Figure 25.2).

25.3 Identification of Genes Potentially Involved in Lignin and Vanillin Biosynthesis

To address potential pathways of vanillin and C-lignin biosynthesis, we have generated very large RNA sequence datasets from vanilla beans at different times of development, and representing different tissue types, including the seeds, hairs, placental and mesocarp tissues, using next-generation sequencing technologies (Rao *et al.* 2014). We used annotated lignin pathway genes in *Arabidopsis* as query sequences and identified the *V. planifola* genes homologous to genes potentially involved in the monolignol pathway. These included genes encoding L-phenylalanine ammonia-lyase (PAL), cou-maroyl shikimate 3'-hydroxylase (C3'H), cinnamate 4-hydroxylase (C4H), 4-coumarate: CoA ligase (4CL), cinnamoyl CoA reductase (CCR), hydroxycinnamoyl CoA: shiki-mate hydroxycinnamoyl transferase (HCT), caffeoyl-CoA 3-*O*-methyltransferase (CCoAOMT), caffeic acid 3-*O*-methyltransferase (COMT), cinnamyl alcohol dehy-drogenase (CAD), and caffeoyl shikimate esterase (CSE). Our transcriptome data indicate that C-lignin formation in the seed coat involves coordinate expression of monolignol biosynthetic genes without the expression of 3-*O*-methyltransferases, which would convert caffeoyl moieties to feruloyl moieties. Most of the predicted lignin-related genes shared similar expression patterns in vanilla bean tissues, with highest expression in seeds at 10 weeks post-pollination, and with lowest expression in the dark and light mesocarp tissues. Strikingly, CCoAOMT transcripts were virtually absent from developing seeds, in spite of the high expression level of the other lignin biosynthetic genes. The quantitative real-time RT-PCR analysis confirmed that PAL, 4CL, CCR, CAD and C4H displayed very high transcript levels in 8 week seeds and relatively high levels in 10 week seeds, and low transcript levels in dark and light meso-carp tissues, whereas CCoAOMT transcripts were detected in dark and light mesocarp tissues, but barely in seed tissues. The low level of CCoAOMT transcripts is likely the biochemical basis for the production of the C-lignin in the vanilla seed coats. This result is consistent with our finding of low OMT enzyme activities in the C-lignin containing seed coat of *Cleome hassleriana* (see below). It is currently not clear how the low expression specifically of the *O*-methyltransferases is regulated in vanilla seeds, because the enzymes of the whole monolignol pathway are normally coordi-nately regulated. Deep mining of transcription factors in the transcriptome database should help us to answer this question.

25.4 C-Lignin Biosynthesis in Other Plants

V. planifolia is a member of the *Orchidaceae* family in the order Asparagales. Although we also found C-lignin in the seed coats of *V. pompona* and *V. tahitensis*, C units were not detected in the seed coats of *Phalaenopsis* orchid species, nor in seeds of *Asparagus* and *Agave*, two other members of the monocot order Asparagales. However, similar C-lignins are found in the seed coats of a number of cactus species (family *Cactaceae*, order Caryophyllales) (Chen *et al.* 2013). Our data indicated that lignins in the seed coats of most cactus species were either of the C- or G/S-types, but never both. The taxonomy of the *Cactaceae* is constantly under revision, but our analysis showed that C-lignin or G/S-lignin types were sometimes found in very closely related species. This finding suggests that the formation of C-lignin is not an ancient trait, but has occurred recently and probably frequently within the plant kingdom. The genetic/biochemical mechanisms that allow for the monolignol pathway to be derailed into production of high concentrations of caffeyl alcohol are probably relatively simple. Interestingly, some cactus species possessed another novel natural lignin composed primarily of 5H units.

We discovered that *Cleome hassleriana* is another plant species which accumulates high levels of C-lignin in its seed coats (Tobimatsu *et al.* 2013). The seed coats of *C. hassleriana* not only contain C-lignin but also have a high content of G lignin. Shortly after pollination, thioacidolysis-derived G lignin can be detected. The lignin level increases rapidly and reaches a plateau at around 12 days after pollination (DAP). At this time, no C-lignin is detected. The C-lignin signal first appears at 14 DAP, when the seed coat starts to darken, and then rapidly develops over the next 6 days, by which time the seed coats became dark and hard. Deposition of C and G lignins are therefore at least temporally separated during *C. hassleriana* seed development.

Analysis of lignin monomer composition by thioacidolysis indicated that lignins in the seed coats of several members of the *Euphorbiaceae*, including *Jatropha curcas* (physic nut), *Aleurites moluccana* (candlenut) and *Ricinus communis* (castor bean), also contain high levels of caffeyl alcohol-derived C units (Tobimatsu *et al.* 2013). Unlike in the members of the *Orchidaceae* and *Cactaceae*, the seed coats of these plants also contain significant amounts of G and/or S units derived from coniferyl and/or sinapyl alcohols. However, there is no evidence for direct linkages between C lignin and G/S lignins present in plant seed coats. Analysis of cell wall preparations from stem, leaves, and roots from one *Euphorbiaceae* plant indicated, not surprisingly, that the lignin in vegetative tissues was composed of G and S units without any trace of C units.

25.5 Commercial Value of C-Lignin as a Novel Natural Polymer

The presence of naturally high levels of C-lignin in the seed coats of plant species like *Jatropha*, candlenut and castor bean suggests a near-term resource for exploitation of this polymer. Over the past 50 years, there has been extensive interest to convert lignin to valuable chemicals and materials. The most common products derived from lignin include vanillin and some deoxygenated aromatic hydrocarbons like benzene, toluene and xylenes (Rinaldi *et al.* 2016; Zakzeski *et al.* 2010; Kang *et al.* 2013). However, the inherent heterogeneity of lignin makes it difficult to selectively convert lignin to target products. C-lignin has a significant potential as a novel material for transformation into value-added chemicals and materials. The simplicity of C-lignin suggests that there would be few products from its depolymerization, making the separation of the desired products much easier. Research has shown that lignin extracted from candlenut nutshell contains high phenolic hydroxyl group content. The depolymerization of organosolv lignin extracted from candlenut over Cu-doped porous metal oxide catalyst showed that the lignin could be converted into simple mixtures of aromatic products in high yield, with the 4-(3-hydroxypropyl)-catechol being the major product that could be isolated in high purity, although the researchers did not know at the time that candlenut contains high C-lignin content (Klein *et al.* 2010; Barta *et al.* 2014).

C-lignin has a structure and properties that make it the ideal renewable source for biobased material manufacturing. For example, attempts have been made to use C-lignin isolated from vanilla seeds to make carbon fiber, a lightweight high performance material with broad industrial applications (Nar *et al.* 2016; Dixon *et al.* 2015). C-lignin can produce fine uniform fibers and processes unceasingly, compared to Kraft lignin which could only be electrospun for a short period of time. Furthermore, surface variation analysis by scanning electron microscopy indicated that the surface of the C-lignin fibers was much smoother than that of the Kraft lignin fibers. Carbonization at 900°C imparted more graphitic properties with higher carbon yield to the C-lignin derived carbon than to that from the Kraft lignin. The challenge now is to develop sources and efficient extraction and conversion processes for this natural biopolymer that can provide the necessary volumes for industrial processes.

The discovery of linear C-lignin homopolymer in nature is a historical milestone in lignin chemistry which drastically changes our view of lignification and represents important new opportunities for redesign of lignin synthesis in plants for sustainable biomass and biomaterial production in the future. The discovery of C-lignin in the vanilla plant demonstrated that lignification in nature is even more flexible than we had thought before (Eudes *et al.* 2014; Ragauskas *et al.* 2014). Plants can be engineered to produce lignin with differing compositions without major impacts on plant performance. In the future, dedicated bioenergy crops like poplar and switchgrass can be engineered to produce homogeneous linear C-lignin. Doing so will not only reduce the biomass recalcitrance to produce biofuels, but also facilitate the valorization of lignin for chemical and polymer production.

References

Barta K, Warner GR, Beach ES and Anastas PT (2014) Depolymerization of organosolv lignin to aromatic compounds over Cu-doped porous metal oxides. *Green Chemistry* 16(1), 191–196.

Boerjan W, Ralph J and Baucher M (2003) Lignin biosynthesis. *Annual Review of Plant Biology* 54(1), 519–546.

Chen F, Tobimatsu Y, Havkin-Frenkel D, Dixon RA and Ralph J (2012) A polymer of caffeyl alcohol in plant seeds. *Proceedings of the National Academy of Sciences USA* 109(5), 1772–1777.

Chen F, et al. (2013) Novel seed coat lignins in the Cactaceae: structure, distribution and implications for the evolution of lignin diversity. *The Plant Journal* 73(2), 201–211.

Dixon R, D'souza N, Chen F and Nar M (2015) *Carbon fibers derived from poly-(caffeyl alcohol)(pcfa)*. US Patent 20150354100 A1.

Eudes A, Liang Y, Mitra P and Loque D (2014) Lignin bioengineering. *Current Opinion in Biotechnology* 26, 189–198.

Higuchi T (1990) Lignin biochemistry: biosynthesis and biodegradation. *Wood Science and Technology* 24(1), 23–63.

Joel DM, et al. (2003) Research Briefs: A hairy tissue produces vanillin. *Israel Journal of Plant Sciences* 51(3), 157–159.

Kang S, Li X, Fan J and Chang J (2013) Hydrothermal conversion of lignin: a review. *Renewable and Sustainable Energy Reviews* 27, 546–558.

Klein AP, Beach ES, Emerson JW and Zimmerman JB (2010) Accelerated solvent extraction of lignin from Aleurites moluccana (Candlenut) nutshells. *Journal of Agricultural and Food Chemistry* 58(18), 10045–10048.

Lapierre C, Monties B and Rolando C (1986) Thioacidolysis of poplar lignins: identification of monomeric syringyl products and characterization of guaiacyl-syringyl lignin fractions. *Holzforschung-International Journal of the Biology, Chemistry, Physics and Technology of Wood* 40(2), 113–118.

Nar M, et al. (2016) Superior plant based carbon fibers from electrospun poly-(caffeyl alcohol) lignin. *Carbon* 103, 372–383.

Ragauskas AJ, et al. (2014) Lignin valorization: improving lignin processing in the biorefinery. *Science* 344(6185), 1246843.

Rao X, et al. (2014) A deep transcriptomic analysis of pod development in the vanilla orchid (*Vanilla planifolia*). *BMC Genomics* 15(1), 964.

Rinaldi R, et al. (2016) Paving the way for lignin valorisation: recent advances in bioengineering, biorefining and catalysis. *Angewandte Chemie International Edition* 55(29), 8164–8215.

Rolando C, Monties B and Lapierre C (1992) Thioacidolysis. *Methods in Lignin Chemistry.* Springer, New York, pp. 334–349.

Tobimatsu Y, et al. (2013) Coexistence but independent biosynthesis of catechyl and guaiacyl/syringyl lignin polymers in seed coats. *The Plant Cell* 25(7), 2587–2600.

Vanholme R, Demedts B, Morreel K, Ralph J and Boerjan W (2010) Lignin biosynthesis and structure. *Plant Physiology* 153(3), 895–905.

Whetten R and Sederoff R (1995) Lignin biosynthesis. *The Plant Cell* 7(7), 1001.

Zakzeski J, Bruijnincx PC, Jongerius AL and Weckhuysen BM (2010) The catalytic valorization of lignin for the production of renewable chemicals. *Chemical Reviews* 110(6), 3552–3599.

Part IV

Biotechnological Production of Vanillin

26

Biotechnology of Vanillin: Vanillin from Microbial Sources

Ivica Labuda

26.1   Introduction

Vanillin is one of the key additives to food products, perfumery, beverage, and an intermediate in the pharmaceutical industry. Vanillin can be produced synthetically or through biotechnology, and can also be obtained from vanilla beans. However, vanillin, which is a single chemical entity, should not be confused with vanilla extract, which contains more than 100 different components. Besides being an important flavor and food ingredient, vanillin is also gaining interest due to its biological activities such as antioxidant, antitumorigenic, tranquilizer and antidepressant.

The focus of this chapter is to compare different biotechnological advances leading to natural bio-vanillin and vanillin related compounds.

26.1.1   Why?

The world of microorganisms is vast and has been very productive for humankind. In addition to antibiotics, polymers, and bioremediation, microorganisms have been ever-present in foods. Fermentation and biotransformation give us the flavors of cheese, wine, beer and other foods. Therefore, it is only natural to consider this subject to find a way to produce a natural counterpart to a synthetic one.

Biotechnologically produced vanillin is not a replacement for vanilla extract. It is to replace synthetic vanillin with a natural flavor, at an affordable price. Rhovanil^R is the first such vanillin manufactured through biotechnology. Genetic manipulation offers tremendous possibilities and opportunities for strain and productivity improvements. At the same time customers look for more transparent regulation and labeling of genetically manipulated foods. rDNA modified strains (GMO) have been used in the food and flavor industry very carefully. That may change as CRISPR/Cas9 technology makes genetic changes more naturally so that they can even go undetected. This will undoubtedly trigger discussions and regulatory response.

Furthermore, fermentation products can be certified organic if they comply with the regulations for organic foods. In manufacturing of organic foods only non-GMO

Handbook of Vanilla Science and Technology, Second Edition.
Edited by Daphna Havkin-Frenkel and Faith C. Belanger.
© 2019 John Wiley & Sons Ltd. Published 2019 by John Wiley & Sons Ltd.

ingredients are allowed. Additionally, the requirements for kosher status need to be considered.

26.1.2   How?

Biotechnologically produced vanillin can be formed via microbial, enzymatic and cell culture processes. The processes can be also adapted for the formation of other vanillin related flavorings when they present either economic advantage or distinctive end products.

In general, biological processes are performed under gentle processing conditions and tend to have lower yields than chemical reactions. Typical hurdles are the initial concentration of precursor, its solubility, toxicity, and the accumulation of the finished product. To make good yields economically feasible, the engineering of the process must be coupled with a detailed understanding of the metabolic pathways. More advanced processing technologies such as whole-cell biocatalysis in biphasic systems, cofactor regeneration for *in vitro* oxidation, immobilization, membrane separation of the product, etc., are being introduced to improve the product yields.

Microbial or enzymatic processes have used any of the following precursors: lignin, curcumin, siam benzoin resin, phenolic stilbenes, isoeugenol, eugenol, ferulic acid, aromatic amino acids, and glucose via *de novo* biosynthesis (Figure 26.1). The most attention was given to the biotransformation of ferulic acid and eugenol, as they are the most readily available precursors. Many fungi and bacteria have the capacity to metabolize the above-mentioned precursors. Some of microbial metabolic pathways have been elucidated and cloned with the help of genetic engineering.

Over the past two decades many excellent papers and reviews have been written, e.g. Priefert *et al.* 2001; Xu *et al.* 2007; Kaur and Chakraborty 2013. Considerable progress has been made in characterizing the biochemistry and molecular genetics of the microbial catabolic and anabolic pathways leading to vanillin. Accumulated knowledge presents an opportunity for direct evolution of specific genes, improvement of fermentation conditions and successful commercialization.

26.2   Substrates

26.2.1   Ferulic Acid (4-Hydroxy 3-Methoxy Cinnamic Acid)

This phenolic acid has been the most attractive precursor for microbial formation of vanillin. Ferulic acid has a similar chemical structure and is abundant in nature. It is one of the hydroxycinnamates that are major constituents of plants such as grains, woods, grasses, fruits, vegetables, etc., bound as esters of polysaccharides, flavonoids, amides, lipids, and long chains alcohols. Ferulic acid forms a covalent bond with carbohydrates. Almost all ferulic acid in cereals is esterified with arabinose, glucose, xylose or galactose residues in the pectic or hemicellulosic component of cell walls. The ester bond is between the carboxylic group of the acid and the arabinoxylan. The molecular weight of these esters reaches over 50,000 Daltons. These esters behave as antioxidants.

Crystalline ferulic acid is relatively insoluble (0.8 g/L) in water at pH lower than pH 9. Therefore, some studies suggested using a solution of ferulic acid dissolved in 0.2 N

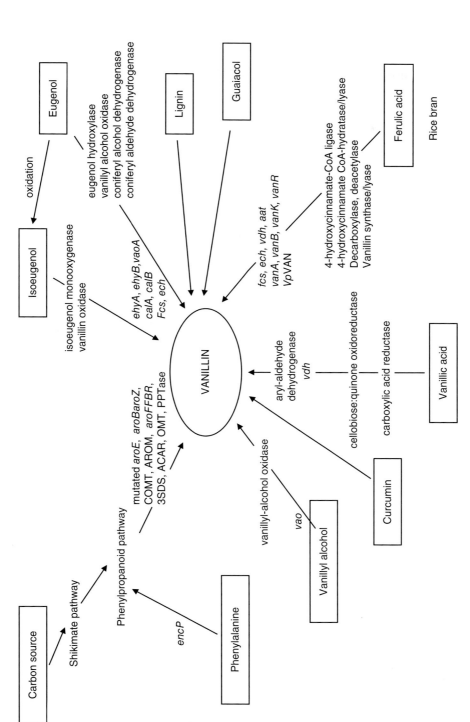

Figure 26.1 Different microbial routes to vanillin.

sodium hydroxide. The higher pH helps with biotransformation since phenolic compounds tend to oxidize at the high pH and higher temperature to their corresponding quinoid. Quinoid being an electrophilic intermediate, reacts immediately with nucleophiles. The pH impacts the solubility of ferulic acid and thus its availability to the biotransformation.

Highly purified ferulic acid is essential to elucidate metabolic pathways and to be used as an antioxidant in cosmetics, nutraceuticals and pharmaceuticals; however, it is not a viable substrate for commercial vanillin production. Agricultural waste products such as sugar beet pulp, brewer's spent grain, rice bran, wheat bran or corn bran were identified as potential sources of a cost-effective ferulic acid (Cheetham *et al.* 1996, 2005). Such hydrolytic processes need optimization to maximize the yield.

Chemicals and enzymes hydrolyze the ester bond. Alkaline and acid hydrolyses performed at higher temperatures (85–100°C) effectively break the covalent bond. The corn cob hydrolysis at high pH and temperature yields between 1–9 g/L of ferulic acid and 2.1 g/L of *p*-coumaric acid (e.g. Torre *et al.* 2008; Zhao *et al.* 2014). An inexpensive method to enrich and partially purify ferulic acid is an alkaline hydrolysis coupled with adsorption on active powdered charcoal (Ou *et al.* 2007). Even though efficient, such chemical hydrolyses may not be considered natural under the current US and EC regulations.

An enzymatic treatment of autoclaved plant material presents an economically feasible route to natural ferulic acid. Several microbial enzymes were identified as efficient in releasing ferulic acid from plant cell walls. Enzymes such as feruloyl esterase, cinnamoyl ester hydrolases, xylanases and carbohydrases have been investigated for their ability to release phenolic precursors from a variety of matrixes. Feruloyl esterase has been associated with many microorganisms such as *Aspergillus niger, Talaromyces stipitatus, Fusarium oxysporum, Humicola insolens, Streptomyces avermitilis,* etc. Two different feruloyl esterases (FAEA and FAEB) from *A. niger* released ferulic and caffeic acids from apple marc, coffee pulp, wheat straw, maize husk and sugar beet pulp (Benoit *et al.* 2006). Besides conventional batch fermentation, feruloyl esterase can be also produced by microencapsulated *Lactobacillus fermentum* 11976. Alginate encapsulated *L. fermentum* produced feruloyl esterase to de-esterify ethyl ferulate with a yield of 45.59% of ferulate (Bhathena *et al.* 2007).

Once a free ferulic acid is available, it is converted to vanillin in a series of enzymatic reactions. Enzymes involved in the biotransformation are typically induced by ferulic acid.

Five different catabolic pathways of ferulic acid have been proposed:

1) non-β-oxidative deacetylation (CoA-dependent)
2) β-oxidative deacetylation (CoA-dependent)
3) non-oxidative decarboxylation
4) CoA-independent deacetylation
5) side chain reduction.

26.2.1.1 Non-β-oxidative Deacetylation (CoA-dependent)
The genes coding the enzymes needed for non-β-oxidative deacetylation were identified and their activities confirmed (Gasson *et al.* 1998; Narbad and Gasson 1998, Figure 26.2). This pathway utilizes three enzymatic steps. It requires two enzymes,

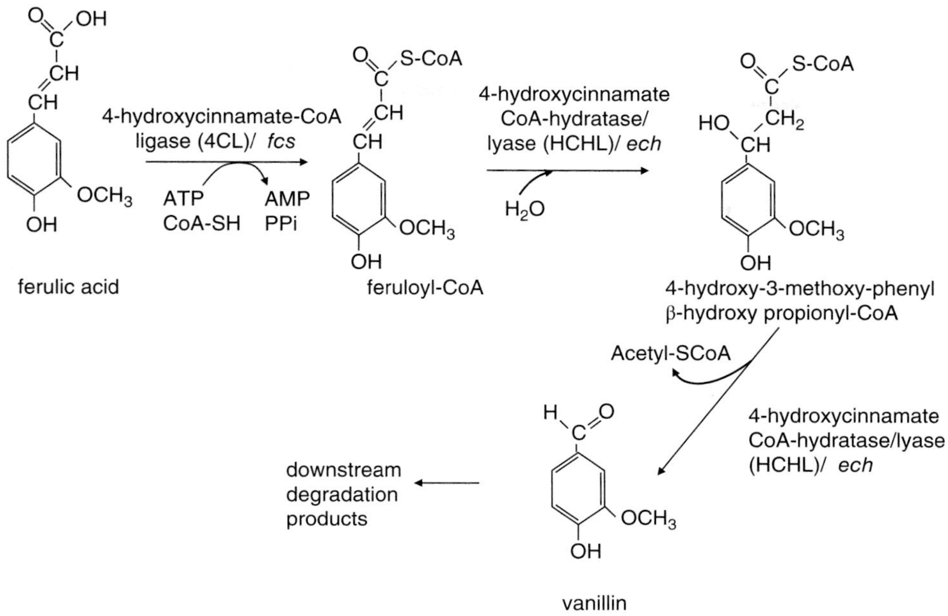

Figure 26.2 Non-β-oxidative deacetylation (CoA-dependent).

4-hydroxycinnamate-CoA ligase (*fcs*) (4-CL) and 4-hydroxycinnamate CoA-hydratase/ lyase (HCHL) (*ech*). First, the 4-CL transforms ferulic acid to feruloyl-SCoA. 4-CL also catalyzes the reaction with 4-coumaric and caffeic acids yielding 4-hydroxybenzaldehyde and protocatechuic aldehyde, respectively (Mitra *et al.* 1999). HCHL catalyzes hydration of feruloyl-SCoA into a transient intermediate 4-hydroxy-3-methoxyphenyl-β-hydroxypropionyl-SCoA and subsequently cleaves the side chain in a retro-aldol fashion into vanillin and acetyl-SCoA (Gasson *et al.* 1998; Narbad and Gasson 1998). The third gene associated with *fcs* and *ech* is vanillin-oxidoreductase, which is induced by ferulic acid (Gasson *et al.* 1998). The genes and enzymes involved in the corresponding reactions have been characterized in *P. fluorescens* AN103 (Gasson *et al.* 1998), *Pseudomonas* sp. HR199 (Overhage *et al.* 1999), *Streptomyces setonii* (Muheim and Lerch 1999), *Amycolatopsis* sp. HR167 (Achterholt *et al.* 2000), *Delftia acidovorans* (Plaggenborg *et al.* 2001), and *Pseudomonas putida* KT2440 (Plaggenborg *et al.* 2003). Identification and characterization of the genes coding for these enzymes offer new opportunities for metabolic engineering and for the construction of recombinant strains.

26.2.1.2 β-Oxidative Deacetylation (CoA-Dependent)

During the β-oxidation deacetylation analogous pathway the first step is the formation of feruloyl-SCoA by 4-hydroxycinnamate-SCoA ligase (*fcs*) (Figure 26.3). The second step involves a reduction catalyzed by 4-hydroxycinnamate-CoA-hydratase/lyase (*ech*) to give 4-hydroxy-3-methoxyphenyl-β-ketopropionyl-SCoA. The next step is a typical β-oxidation thiolytic cleavage yielding vanillyl-SCoA and acetyl-SCoA. Vanillyl-SCoA hydrates and splits off CoASH yielding vanillin. This step is catalyzed

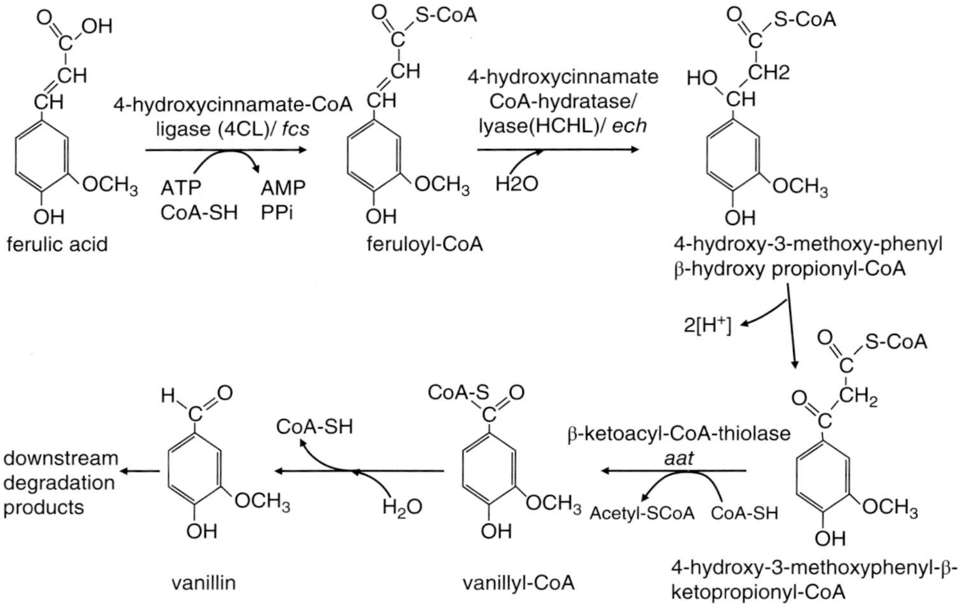

Figure 26.3 β-Oxidative deacetylation (CoA-dependent).

by β-ketoacyl-CoA-thiolase (*aat*). Limited experimental evidence is available to support this route to vanillin. Also, a differing experiment proving that the deletion of β-ketothiolase gene *aat* adjunct to other genes (*fcs, ech, vdh*) does not influence ferulic acid catabolism into vanillin in *Pseudomonas* sp. HR199 has been reported (Overhage *et al.* 1999).

26.2.1.3   Non-Oxidative Decarboxylation

The first step involves decarboxylase, which isomerizes ferulic acid to a quinoid intermediate and subsequently decarboxylates to 4-vinylguaiacol (Figure 26.4). The specificity of decarboxylase varies based on the production microorganism. For example, *Bacillus pumilus* has two decarboxylases active towards ferulic and p-coumaric, but with no activity against caffeic acid. On the other hand, *Lactobacillus plantarum* has two inducible decarboxylases, one for ferulic acid and one for caffeic and *p*-coumaric acids (Barthelmebs *et al.* 2000). In some microorganisms, such as the yeast *Brettanomyces anomalus*, the enzyme is constitutively produced at low levels and induced upon addition of the substrate (Edlin *et al.* 1998). *Rhodotorula glutinis* and *R. rubra* were reported to metabolize ferulic acid via decarboxylation (Huang *et al.* 1993a; Labuda *et al.* 1992a). *R. glutinis* converts 90% of ferulic acid to 4-vinylguaiacol within 2 hours (Labuda *et al.* 1992b) as reported also for a recombinant *E. coli* carrying pETD*fdc* from *B. pumilus* (Furuya *et al.* 2015). This mechanism was also reported for *Bacillus coagulans* (Karmakar *et al.* 2000). *Pestalotia palmarum* bioconversion broth contained also methoxyhydroquinone because of oxidative decarboxylation of vanillic acid besides vanillin, vanillic acid, vanillyl alcohol and protocatechuic acid (Rahouti *et al.* 1989).

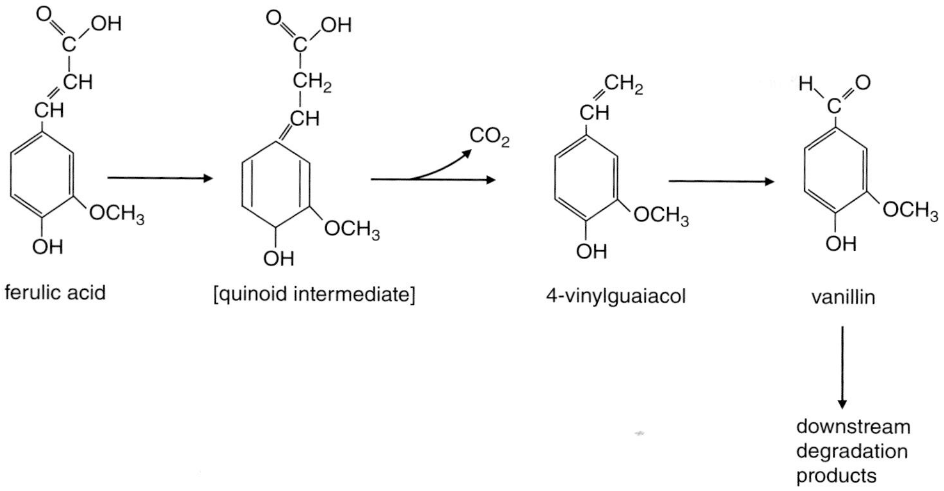

Figure 26.4 Non-oxidative decarboxylation.

26.2.1.4 CoA-Independent Deacetylation

During this degradation pathway, the trans double bond of ferulic acid is hydrated to yield a transient intermediate 4-hydroxy-3-methoxy-β-hydroxypropionic acid (Figure 26.5). This step is followed by aldolase cleavage of the acetate portion directly yielding vanillin.

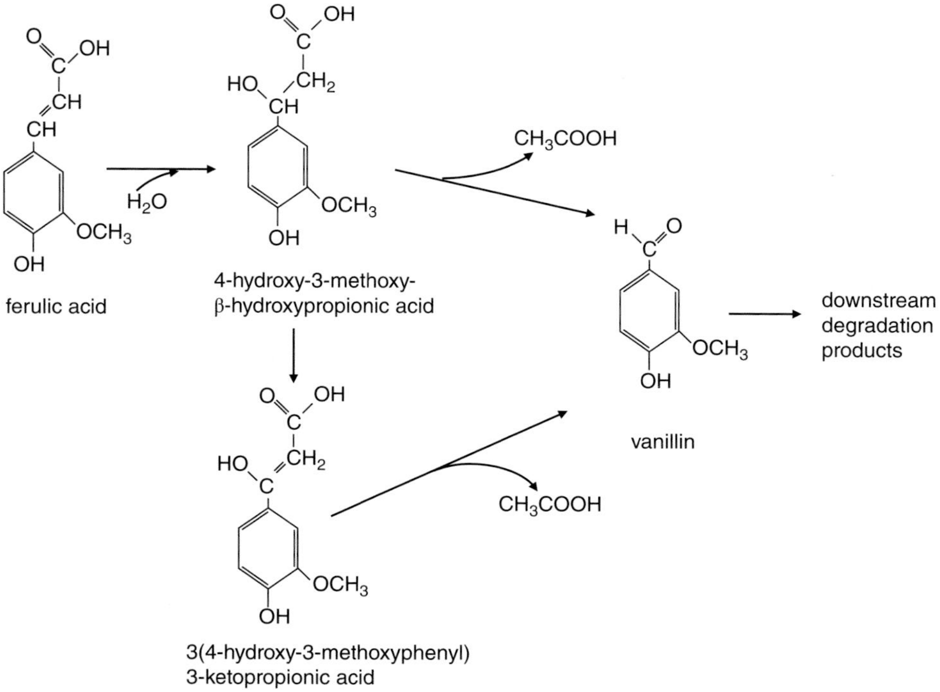

Figure 26.5 CoA-independent deacetylation.

The transient intermediate can be also reduced to 3(4-hydroxy-3-methoxyphenyl)3-ketopropionic acid and then deacetylated into vanillin. An enzyme ferulic acid deacetylase (*fca*) has been identified in *P. putida* strain WCS358 as an catalyst deacetylating ferulic acid yielding acetate and vanillin (Venturi *et al.* 1998). CoA-independent deacetylation was suggested for *Delftia acidovorans, Pseudomonas* sp., *P. mira, Bacillus subtilis, Corynebacterium glutamicum, E. coli, Streptomyces setonii, Aspergillus niger, Fomes fomentarius, Fusarium solani, Polysporus versicolor,* and *Rhodotorula rubra.*

26.2.1.5 Side-Chain Reductive Pathway

As with decarboxylation, this pathway involves isomerization of ferulic acid to a transient quinoid intermediate which is reduced to dihydroferulic acid (4-hydroxy 3-methoxy phenylpropionic acid) (Rosazza *et al.* 1995) (Figure 26.6). This pathway was reported for *S. cerevisiae* expressed under anaerobic conditions (Huang *et al.* 1993b) or *L. plantarum* (Barthelmebs *et al.* 2000). Depending on the microorganism, dihydroferulic acid is further metabolized to homovanillic, or vanillic acids. Vanillic acid can be further reduced to vanillin. Ferulic acid can be also directly reduced to coniferyl alcohol or demethylated to caffeic acid.

Based on the intermediates it seems logical that parallel pathways exist in several microorganisms. For example, intermediates vinylguaiacol and dihydroferulic acid found in *C. glutamicum* suggest that decarboxylation and reductive conversion take place at the same time. An experiment studying the *Rhodococcus* genome and its expression supports this phenomenon. Genome sequence data of *Rhodococcus* sp. I24

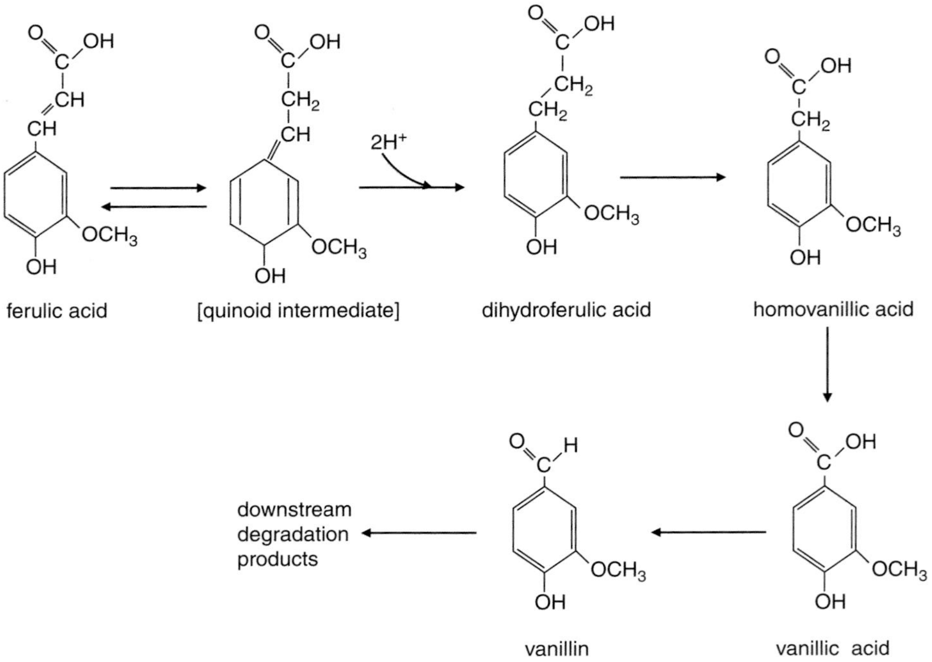

Figure 26.6 Side-chain reductive pathway.

suggested a CoA-dependent, non-β-oxidative pathway for ferulic acid bioconversion. However, the expression and functional characterization of corresponding structural genes from I24 suggested that degradation of ferulic acid in this strain proceeds via a β-oxidative pathway (Plaggenborg *et al.* 2006).

Besides being the most suitable precursor, ferulic acid exhibits many physiological functions, including antioxidant, antimicrobial, anti-inflammatory, anti-thrombosis, and anti-cancer activities. It also protects against coronary disease, lowers cholesterol and increases sperm viability. It is well regarded as a free radical scavenger that greatly reduces oxidative damage effect (e.g. Kanski *et al.* 2002).

It is ferulic acid that is utilized as a precursor in commercial processes to produce natural vanillin, by companies such as Firmenich, Symrise, Solvay and Comax.

26.2.2 Eugenol and Isoeugenol

Clove oil is an inexpensive source of eugenol (2-methoxy-4-(2-propenyl)-phenol), a main component of the essential oil of the clove tree *Syzygium aromaticum*. Both eugenol and isoeugenol are produced by plants as a part of their biodefense mechanisms. Eugenol's antiseptic properties are widely useful in dentistry. Although structurally similar to vanillin and economically viable, eugenol is the most challenging due to its toxicity and low water solubility. There are two strategies, which could decrease toxicity and improve solubility. The first strategy is to change the genetic make-up of the producing microorganisms to increase the tolerance to higher concentrations of eugenol. The second strategy is to use bi-phasic or immobilized processes where the substrate and the product are continuously removed. Some microorganisms such as *Pseudomonas* or *E. coli* are more tolerant to eugenol than other bacteria and fungi.

Isoeugenol is more readily metabolized. Although present in essential oils, it is usually prepared from eugenol via the chemical route. Isomerization of eugenol to isoeugenol is catalyzed by metal ions at elevated temperature in potassium hydroxide (KOH). Such a process is not considered natural under the USA and EU laws. A newer method using microwave technology requires less harsh conditions to perform this alkaline catalysis (Sinha *et al.* 2002). Pathways, intermediates, and enzymes involved in eugenol and isoeugenol metabolism have been identified (Hua *et al.* 2007; Xu *et al.* 2007; Zhang *et al.* 2007; Yamada *et al.* 2007).

The eugenol degradation pathway through a quinone intermediate and coniferyl alcohol was confirmed (Figure 26.7). The genes encoding the enzymes needed for the oxidative catabolism of eugenol were isolated: eugenol hydroxylase, vanillyl alcohol oxidase (*ehyA*, *ehyB* and *vaoA*), coniferyl alcohol dehydrogenase (*calA*), and coniferyl aldehyde dehydrogenase (*calB*) (Figure 26.7) (Priefert *et al.* 2001; Overhage *et al.* 2003, Xu *et al.* 2007). Eugenol hydroxylase and vanillyl alcohol oxidase catalyze the first and second steps during which a transient epoxide, eugenol oxide, is formed and then transformed into coniferyl alcohol. The major intermediates identified are coniferyl alcohol, coniferyl aldehyde, ferulic acid, vanillin, vanillin alcohol and vanillic acid. The intermediate ferulic acid is then transformed into vanillin following non-oxidative deacetylation, employing 4-hydroxycinnamate CoA ligase (4-CL) (*fcs*) and 4-hydroxycinnamate CoA-hydratase/lyase (HCHL) (*ech*) (Gasson *et al.* 1998; Overhage *et al.* 2003). Another epoxide-diol pathway suggested intermediates eugenol epoxide, eugenol diol and their further metabolism into coniferyl alcohol (Priefert *et al.* 2001).

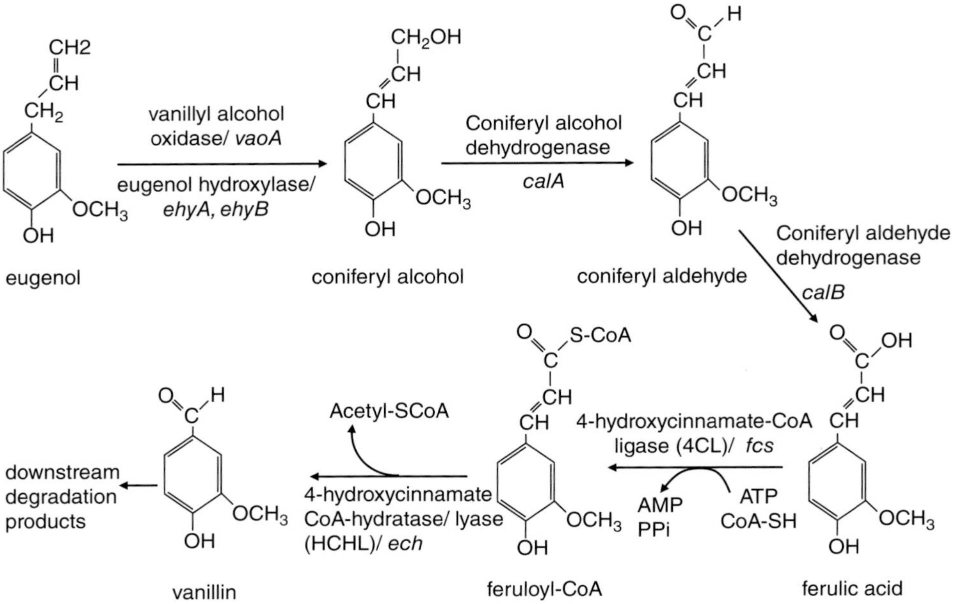

Figure 26.7 Suggested degradation pathway for eugenol.

Isoeugenol (2-methoxy-4-(1-propenyl)-phenol) is metabolized through an epoxide-diol pathway (Figure 26.8). It undergoes side-chain epoxidation, epoxide hydrolysis to a diol, and diol cleavage to vanillin. Intermediates were identified as isoeugenol-epoxide (IE) and isoeugenol-diol (ID) leading to vanillin or vanillyl alcohol. Vanillin undergoes downstream oxidation into vanillic acid. The biotransformation with resting cells of *B. pumillus* showed that vanillin was oxidized to vanillic acid and then to protocatechuic acid before the aromatic ring was broken. These findings suggest that isoeugenol is degraded through an epoxide-diol pathway (Hua *et al.* 2007). The same pathway was found in resting cells of *Nocardia iowensis* (Seshadri *et al.* 2008). The cell-free extract exhibiting activity of two enzymes, isoeugenol monooxygenase and vanillin oxidase, converted isoeugenol to vanillin and vanillic acid without cofactors.

26.2.3 Lignin

Lignin is a cross-linked phenylpropanoid polymer and is a major plant component. It is well known that vanillin is formed in small quantities from aromatic compounds known to be precursors in the biosynthesis and degradation of lignin. Enzymes needed for degradation of lignin are associated with white-rot fungi such as *Phanerochaete chrysosporium,* which delignify and decompose lignin efficiently. It is known that lignin peroxidase and laccase play a role in the depolymerization of lignin. Guaiacol intermediates released from lignin are further converted into many phenolic compounds such as coniferyl alcohol, coniferyl aldehyde, ferulic acid, vanillin, vanillic acid, *p*-hydroxycinnamic aldehyde, dehydrodivanillin, etc. This microbial process is slow and the vanillin levels are too low to be competitive.

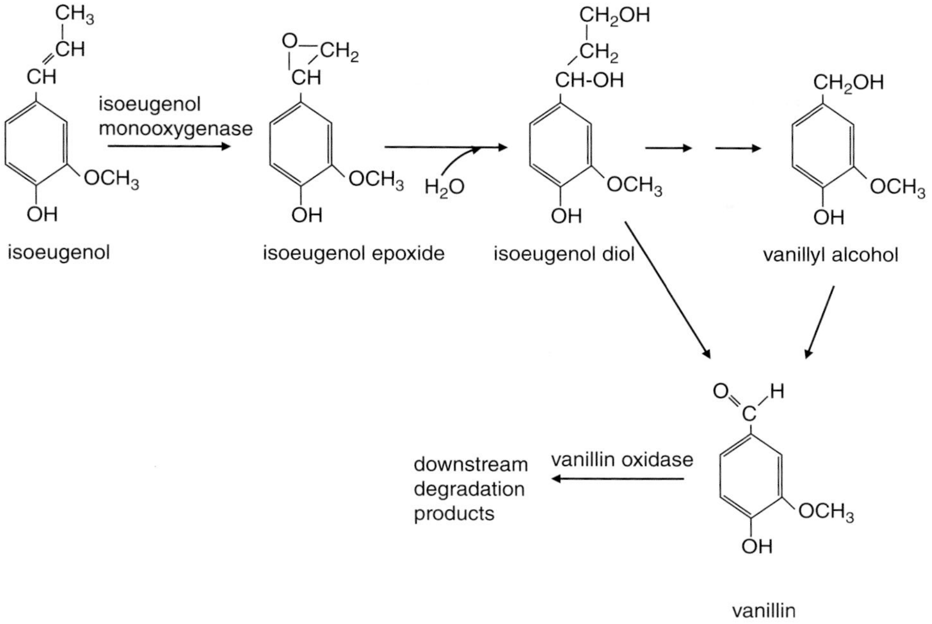

Figure 26.8 Suggested degradation pathway for isoeugenol.

26.2.4 Sugars

Sugars as carbon source play three important roles during the vanillin formation:

1) as an energy source for the producing organism;
2) as a precursor of vanillin; and
3) as a co-factor.

Glucose is typically employed as a carbon source fed together with ferulic acid, euge-nol or another precursor. In some instances the type of carbohydrate influences vanillin formation as demonstrated with cellobiose, glucose, fructose and maltose during *P. cinnabarinus* bioconversion (Oddou *et al.* 1999; Lessage-Messen *et al.* 1997, 2002). On the other hand, *A. niger* bioconversion of ferulic acid to vanillin was impacted very little by the type of used carbohydrate. Only slightly higher levels of vanillin accumulated with lactose and cellobiose than with glucose, galactose or sucrose (Labuda *et al.* 1992a).

Using glucose as a vanillin precursor is economically a very attractive proposition. A recombinant *E. coli* and *Saccharomyces cerevisiae* capable of forming vanillin from glucose were successfully constructed and are described below. In some recombinant *E. coli* strains with an arabinose inducible promoter, this pentose was used as an inducer. As arabinose is found in plant hydrolysates, it is important to establish its impact on vanillin generation. *S. setonii* formed roughly half of the vanillin in the presence of ara-binose (3.1 g/L) compared to that produced with glucose (6.1 g/L) (Gunnarsson and Palmqvist 2006). In some instances, a carbohydrate such as cellobiose acts as an elicitor of the vanillin pathway (Lessage-Messen *et al.* 2002).

26.3 Microorganisms

Microorganisms selected for vanillin formation have been isolated based on their ability to metabolize phenolic precursors to vanillin as well as to tolerate higher concentrations of phenolic compounds. The substrates and reaction conditions have been identified to maximize the yield of vanillin. It appears that genetically unrelated microorganisms share similar pathways to transform phenolic precursors.

26.3.1 Bacteria

26.3.1.1 *Pseudomonas*
Pseudomonas strains have a very broad and versatile metabolism of phenolic compounds, including eugenol, ferulic acid or vanillin.

It was *P. fluorescens* in which the genes coding the enzymes involved in the formation of vanillin from ferulic acid were identified as 4-CL (*fcs*) and HCHL (*ech*) enzymes for the first time. A close relationship of hydroxycinnamate catabolism to the β-oxidation of fatty acids was also demonstrated (Narbad and Gasson 1998; Gasson *et al.* 1998). In *P. fluorescens*, the HCHL belongs to the superfamily of fatty acid β-oxidation associated hydratases/isomerases. However, a NAD+ binding domain typical of enoyl-SCoA hydratase/isomerases was absent in the *P. fluorescense* gene.

The eugenol to ferulic acid pathway was characterized in *Pseudomonas* strain H199 and the genes encoding the enzymes were identified as eugenol hydroxylase (*ehyA* and *ehyB* genes), coniferyl alcohol dehydrogenase (*calA*) and coniferyl aldehyde dehydrogenase (Rabenhorst 1996; Priefert *et al.* 1999).

Under the optimized reaction conditions, *P. putida* IE27 cells produced 16.1 g/L vanillin from 150 mM isoeugenol, with a molar conversion yield of 71% at 20°C after a 24-h incubation in the presence of 10% (v/v) dimethyl sulfoxide (Yamada *et al.* 2007).

Vanillin was shown to be further oxidized through vanillic acid and protocatechuic acid in *Pseudomonas* sp. which expresses a complete set of genes (*vdh, vanA/vanB, pcaG/pcaH*) needed for this metabolism.

Partial insertional deactivation of vanillin dehydrogenase encoded by *vdh* in a non-pathogenic *P. fluorescens* BF13 prevented complete metabolism of vanillin. Vanillin yield was 20% of the ferulic acid (Di Gioia *et al.* 2011).

26.3.1.2 *Streptomyces*
Streptomyces setonii is a strain reported to give the highest bioconversion yields of vanillin. It was identified by Sutherland *et al.* (1983) and later showed an impressive tolerance towards high concentration of this aromatic aldehyde (Muheim and Lerch 1999). Typical by-products were vanillic alcohol, vanillic acid, guaiacol, 4-vinylguaiacol and 2-methoxy-4-ethyl-phenol. These intermediates suggest that *S. setonii* employs a non-oxidative decarboxylation pathway that involves initial enzymatic isomerization to the quinoid intermediate which subsequently decarboxylates to 4-vinylguaiacol (Figure 26.4).

Bioconversion outcomes heavily depend on the pH of the medium. The pH 8.2 was shown to be ideal for accumulation of vanillin by *S. setonii*, reaching 90% conversion, whereas lower pH favored reduction of vanillin to vanillyl alcohol and higher pH oxidation to vanillic acid (Gunnarsson and Palmqvist 2006). The same study also showed the influence of the carbohydrate on vanillin accumulation. The best yield of vanillin

(9.2 g/L) was achieved when ferulic acid (13.1 g/L) and glucose were used at pH 8.2. On the other hand, with arabinose as a carbon source and pH 8.2, vanillin reached only 4.7 g/L, although ferulic acid bioconversion was reported to be several fold higher than on glucose. There was no vanillyl alcohol formed when arabinose was used as a carbon source.

Another species, *S. sannanensis*, produced mostly vanillic acid (0.4 g/L) when grown only on ferulic acid (0.94 g/L) with no additional carbon source. Accumulation of vanillic acid slowed down after the inhibition of hydroxycinnamate-CoA-ligase, which suggested involvement of the CoA-dependant degradation pathway in *S. sannanensis* (Ghosh *et al.* 2007).

A soil isolate of *Streptomyces* sp. displayed activity to modify phenylpropanoids derived from lignin. Cloning of the genes *vanA* and *vanB* encoding vanillate demethylase into *S. lividans* demonstrated that veratric acid can be demethylated by vanillate demethylase into vanillic acid (Nishimura *et al.* 2006). Vanillic acid can be subsequently reduced to vanillin.

The marine actinomycete *Streptomyces maritimus* is one of few microorganisms that expresses the gene *encP* encoding bacterial phenylalanine ammonia lyase (PAL), an enzyme that catalyzes the conversion of the amino acid l-phenylalanine to *trans*-cinnamic acid. PAL is a part of phenylpropanoid pathway found in higher plants (Xiang and Moore 2005). This could provide an opportunity to combine the bacterial and plant phenylpropanoid pathways.

Streptomyces halstedii was selected among 535 actinomycetes from soil. It converted cinnamic acid in cinnamamide (95% molar conversion from 2 g/L of substrate) and ferulic acid in vanillic acid (8% molar conversion from 1 g/L of substrate) with low quantities of vanillin (0.10–0.15 g/L) (Brunati *et al.* 2004).

26.3.1.3 Bacillus

It was found that bacilli degrade a wide range of aromatic acids such as cinnamic, 4-coumaric, 3-phenylpropionic, 3-(*p*-hydroxyphenyl)propionic, ferulic, benzoic, and 4-hydroxybenzoic acids. Ferulic acid was metabolized through a non-β-oxidative mechanism with intermediates 4-hydroxy-3-methoxyphenyl-β-hydroxypropionic acid, vanillin, and vanillic acid (Peng *et al.* 2003). On the other hand, isoeugenol was metabolized through an epoxide-diol pathway (Hua *et al.* 2007).

Bacillus coagulans BK07 decarboxylates ferulic acid within 7 hours to 4-vinylguaiacol, which is immediately converted to vanillin and then oxidized to vanillic acid. Vanillic acid is further demethylated to protocatechuic acid (Karmakar *et al.* 2000). Very high conversion rate (95%) of ferulic acid to 4-vinylguaiacol is similar to that of *Rhodotorula glutinis*.

A soil isolate of *Bacillus subtilis* is able to grow on isoeugenol as a sole carbon source and converts it into vanillin. The bioconversion has been achieved with cells as well as with the cell free extract. It was reported that growing cells of *B. subtilis* B2 produced 0.61 g/L of vanillin whereas cell free extracts yielded 0.9 g/L vanillin from isoeugenol (Shimoni *et al.* 2000). Isoeugenol biotransformation by *B pumillus* led to vanillin as a main product. The growing cells of *B. pumillus* S-1 converted 10 g/L isoeugenol to 3.75 g/L vanillin in 150 hours, with a molar yield of 40.5% that is the highest up to now (Hua *et al.* 2007). Sequencing of this strain suggested that the monooxygenases and epoxide hydrolases were the most probable enzymes capable of converting isoeugenol to vanillin (Su *et al.* 2011).

26.3.1.4 *Corynebacterium*

Corynebacterium glutamicum utilizes many aromatic compounds such as ferulate, vanillate and vanillin as the sole carbon source. This bacterium has outstanding ability to switch between metabolic pathways based on the exposure of the cells to a specific aromatic compound. Induction of four genes, *vanA*, *vanB* and *vanK* genes and the divergently transcribed *vanR* gene, is needed for catabolism of ferulic acid (Merkens *et al.* 2005). Expression of these genes is not catabolically suppressed when ferulic acid is co-metabolized with glucose. The metabolic channeling of ferulate, vanillin, and vanillate into the protocatechuate branch of the β-ketoadipate pathway is controlled by a PadR-like transcriptional regulatory protein repressor (Brinkrolf *et al.* 2006).

Besides vanillin, 4-vinylguaiacol, vanillic acid, dihydroferulic acid were identified by LC/MS chromatography (Labuda *et al.* 1992b). These intermediates suggest that ferulic acid is metabolized in parallel via decarboxylation and side-chain decarboxylation. Other compounds such as 3,4-dihydroxybenzoic acid, 3,4-dihydroxybenzaldehyde, 3,4-dihydroxy cinnamic acid, and 3,4-dimethoxy benzaldehyde were formed by *C. glutamicum* from ferulic acid as well. It is assumed that these intermediates are the products of the downstream metabolism of vanillic acid.

26.3.1.5 *Escherichia coli*

Non-recombinant *E. coli* cannot form vanillin from glucose, ferulic acid, eugenol or isoeugenol. At the same time *E. coli* is easy to manipulate with highly developed genetic engineering tools and relatively easy to scale-up. The combination of these features makes it ideal to exploit *E. coli* for vanillin production.

Several attempts to create a recombinant *E. coli* for vanillin production were successful. A recombinant *E. coli* transforms glucose into vanillin in two steps: in the first step vanillic acid was formed through shikimate pathways; secondly vanillic acid was enzymatically reduced to vanillin. Recombinant *E.coli* required five enzymes: shikimate dehydrogenase (mutated *aroE*), 3-dehydroquinate synthase and 3-dehydroshikimate (*aroBaroZ*), catechol-*O*-methyltransferase (COMT) and 3-deoxy-D-arabino-heptulosonic acid 7-phosphate synthase (*aroFFBR*) (Frost 2002; Li and Frost 1998). The recombinant microbe performed dehydration of 3-dehydroshikimic acid and regio-selectively, methylated the resulting protocatechuic acid into vanillic acid. This methylation was a rate-limiting step, which was not overcome by doubling the amount of the plasmid with COMT. However, the addition of L-methionine improves the vanillate formation. In the last step, the reduction of vanillic acid to vanillin is done enzymatically by purified arylaldehyde dehydrogenase from *Neurospora crassa*. This enzymatic step requires cofactors ATP and NADP, which makes this process quite cost prohibitive. Nevertheless, the yield of 5 g/L of vanillin from glucose is impressive especially when *E. coli* does not produce any vanillin without the above described genetic changes.

Ferulic acid biotransformation needed a different strategy to develop recombinant *E. coli*. It was done by inserting the *fcs* (feruloyl-CoA synthetase) and *ech* (enoyl-CoA hydratase/aldolase) genes from *Amycolatopsis* sp. strain HR104 and *Delftia acidovorans* under the control of the arabinose-inducible promoter P_{BAD} into the pBAD24 expression vector. Optimized cell growing conditions led to the highest vanillin production with the *E. coli* strain carrying the *Amycolatopsis* genes, 0.58 g/L of vanillin from 1 g/L ferulic acid (Yoon *et al.* 2005). Another study by Furuya *et al.* (2015) suggested a yield of 7.8 g/L in a two-stage coenzyme independent process using two recombinant *E, coli* strains. In the first step ferulic acid was transformed into 4-vinylguaiacol, which served as a precursor for vanillin in the second step. The ferulic acid decarboxylase gene *fdc* was from *Bacillus pumilus* and the 4-vinylguaiacol oxygenase gene *Cso2* was from *Caulobacter segnis*.

Eugenol is another substrate for which recombinant *E. coli* strains were developed (Overhage *et al.* 2003). Two recombinant strains were constructed. The first step allowed eugenol to be transformed into ferulic acid by inserting the genes *calA* and *calB*, encoding coniferyl alcohol dehydrogenase and coniferyl aldehyde dehydrogenase of *Pseudomonas* sp. and the *Penicillium simplicissimum* vanillyl alcohol oxidase gene. The second step converted ferulic acid to vanillin by inserting the genes encoding feruloyl-coenzyme A (CoA) synthetase (*ech*) and enoyl-CoA hydratase/aldolase (*fcs*) of *Pseudomonas* sp. strain HR199 into another *E. coli* (Overhage *et al.* 1999). These two recombinant strains of *E. coli* when used together successfully converted eugenol to vanillin. First the resting cells of recombinant strain *E. coli* XL1-Blue (pSK*vaom*P-*calA*mcalB*) converted eugenol to ferulic acid with a molar yield of 91%. The maximum concentration of ferulic acid was 14.7 g/L after 30 hours fermentation. This corresponded to 93.3% molar biotransformation yield with respect to concentration of eugenol. The maximum production rate from ferulic acid was achieved by feeding ferulic acid at a rate 2.8 g/hour/L. This process led to 0.3 g/L of vanillin besides 0.1 g/L of vanillyl alcohol and 4.6 g/L of ferulic acid.

The attempt to use genes encoding eugenol hydroxylase and its electron acceptor from *Pseudomonas* sp. strain HR199, in *E. coli* XL1-Blue were shown to be unsuitable for performing the eugenol bioconversion.

The isoeugenol monooxygenase gene of *P. putida* IE27 was inserted into the recombinant plasmid in *E. coli* BL21. It enabled the transformed cells to produced vanillin with the yield of 28.3 g/L from 230 mM (37.7 g/L) isoeugenol. The molar conversion rate was 81% after 6 hours (Yamada *et al.* 2008). The recombinant cells converted isoeugenol to vanillin without generating undesirable by-products.

26.3.1.6 *Amycolatopsis* sp.

Using *Amycolatopsis* sp. strain HR167, 11.5 g of vanillin per liter was produced from 19.9 g/L of ferulic acid within 32 h on a 10-liter scale, corresponding to a molar yield of 77.8% (Rabenhorst and Hopp 2000).

Recombinant vanillin tolerant bacteria *Amycolatopsis sp.* HR167 was made by cloning a vanillyl alcohol oxidase (*vaoA*) from *Penicillium simplicissimum* CBS 170.90 on a hybrid plasmid. This new recombinant strain gained a capability to grow on eugenol (Overhage *et al.* 2006). Resting cells produced downstream intermediates from eugenol with the conversion rate of 0.36 g/L/hour. Intermediates were identified as coniferyl alcohol (4.7 g/L), coniferyl aldehyde, ferulic acid, guaiacol and vanillic acid. Vanillin was present only in minute amounts.

26.3.1.7 Lactic Acid Bacteria (LAB)

LABs are a main staple of human nutrition and therefore their application for vanillin production would be very logical.

De Revel *et al.* (2005) reported that LAB associated with wine produce small levels of vanillin when wine was made in wood barrels. This implies that phenolic compounds in the wood were utilized as vanillin precursor. LABs such as *Oenococcus oeni* or

Lactobacillus sp. used in wine making convert ferulic acid to vanillin at low levels. However, these two LABs are not able to form vanillin from eugenol, isoeugenol or vanillic acid. Other LABs such as *Lactobacillus* sp. and *Pediococcus* sp. strains could produce significant quantities of 4-vinylguaiacol from ferulic acid. It is suggested that LABs reduce vanillin non-enzymatically to the corresponding vanillyl alcohol. *Lactobacillus lactis* has been engineered to transform ferulic acid to vanillin (Bloem *et al.* 2007)

Heterologous expression of 4-hydroxy cinnamate:CoA ligase (4CL) from the plant *Arabidopsis thaliana* in LAB was successfully achieved (Martinez-Cuesta *et al.* 2005), however, the yield of the 4CL was very low in comparison with the formation of this enzyme in *Pseudomonas* when grown on ferulic acid. An easy expression of plant enzymes in bacterial cells is atypical.

26.3.1.8   *Clostridium*

Clostridium aerotolerans and *C. xylanolyticum*, closely related species, were reported to transform cinnamic acid to 3-phenylpropionic acid anaerobically. Both species also reduced a wide range of cinnamic acid derivatives, including *o*-, *m*- and *p*-coumaric, *o*-, *m*- and *p*-methoxycinnamic, *p*-methylcinnamic, caffeic, ferulic, isoferulic and 3,4,5-trimethoxy-cinnamic acids to their corresponding 3-phenylpropionic acid derivatives (Chamkha *et al.* 2001).

26.3.2   Fungi and Yeasts

Selected fungi and yeasts can convert ferulic acid, isoeugenol or eugenol to vanillin. Not as many pathways have been fully elucidated to confirm the existence of enzymes and their specific intermediates as has been done for bacteria. Decarboxylation, side-chain reduction, CoA-independent and β-oxidative decarboxylation were all reported as mechanisms of ferulic acid bioconversion by fungi and yeasts.

The major asset of fungi useful for vanillin production is their capacity to express enzymes such as feruloyl esterase and cinnamic acid hydrolases, peroxidases and decarboxylases. Esterases and hydrolases are instrumental in releasing ferulic acid from plant cell walls. Peroxidases, especially lignin peroxidases from the basidiomycete *Phanerochaete chrysosporium* are involved in the depolymerization of lignin type of materials during which some traces of vanillin are produced (ten Have *et al.* 1998).

Decarboxylases catalyze elimination of the carboxylic group of ferulic acid to form 4-vinylguaiacol (4-hydroxy-3-methoxystyrene). This has been reported in yeasts such as *Brettanomyces anomalus* (Edlin *et al.* 1998), *S. cerevisiae* (Clausen *et al.* 1994) and *Rhodotorula glutinis* (Labuda *et al.* 1992b). The corresponding genes from *S. cerevisiae* have been cloned and expressed in *E. coli* (Clausen *et al.* 1994; Ago and Kikuchi 1998).

The yeast *R. glutinis* transformed 1 g/L ferulic acid to 4-vinylguaiacol with the conversion rate of 90% within 24 h. Extended time led to accumulation of vanillic acid with about 14% after 95 hours of incubation. A vanillin conversion rate of about 8.5% was observed only in the presence of sulfhydro-reagents. Besides vanillin, vanillic acid and 4-vinylguaiacol, LC/MS chromatography identified dihydroferulic acid, 3,4-dihydroxy benzoic acid in the broth (Labuda, unpublished data). Based on the intermediates it appears that *R. glutinis* utilizes two parallel pathways, decarboxylation and side-chain reductive pathways.

β-Oxidation of ferulic acid employing hydroxycinnamate decarboxylase of *Rhodotorula rubra* led to accumulation of vanillic acid (Huang *et al.* 1993). This pathway, which is analogous to the β-oxidation of fatty acid, is thought to include a thiolytic cleavage of 4-hydroxy-3-methoxyphenyl-β-ketopropionyl.

As with *R. glutinis*, *A. niger* seem to utilize parallel metabolic pathways, decarboxylation and side-chain reductive pathways. LC/MS identified vanillin, vanillic acid, 4-vinylguaiacol, protocatechuic aldehyde, protocatechuic acid and 3,4-dimethoxy benzaldehyde (veratraldehyde) in *A. niger* medium after 150 hours continuous process with ferulic acid (Labuda, unpublished data).

A. niger deacetylates ferulic acid to vanillic acid very efficiently. In a two-step bioconversion process vanillic acid is subsequently reduced to vanillin by *Pycnoporus cinnabarinus* (Brunerie *et al.* 1996; Lessage-Messen *et al.* 2002). Cellobiose is suggested to enhance channeling of the bioconversion of vanillic acid into vanillin by controlling methoxyhydroquinone formation in *P. cinnabarinus*. This process could be further improved by introducing a chimeric bifunctional enzyme that increases the synergistic effect on the degradation of complex substrates. For example, a chimeric construct composed of the sequences encoding the feruloyl esterase A (FAEA) is fused to the endoxylanase B (XYNB) of *A. niger* which allows for efficient release of ferulic acid from plant materials (Levasseur *et al.* 2005). Another example is the improved secretion efficiency of a fused dockerin-feruloyl esterase protein which consists of feruloyl esterase of *A. niger* and dockerin of *Clostridium thermocellum* by introducing 514 amino acid sequences of glucoamylase and a dibasic proteolytic processing site upstream of the chimeric protein (Levasseur *et al.* 2004).

Mutation and genetic manipulation of *A. niger, A. flavus*, and *Penicillium chrysogenum* resulted in constructed strains that exhibited both ferulic acid esterase and alkene cleavage activities. These strains converted ferulic acid glycoside directly into vanillic acid (Cheetham *et al.* 2005).

Eugenol and isoeugenol are tolerated and metabolized by few fungi. The fungus *Byssochlamys fulva* V107 converted eugenol to 21.9 g/L of coniferyl alcohol within 36 hours representing 94.6% molar yield (Furukawa *et al.* 1999).

Induction of vanillyl alcohol oxidase was reported for *Penicillium simplicissimum* when grown on veratryl alcohol, anisyl alcohol, or 4-(methoxymethyl)-phenol (Fraaije *et al.* 1997). In addition to vanillyl-alcohol oxidase, an intracellular catalase peroxidase was induced. Induction of vanillyl-alcohol oxidase in *P. simplicissimum* was prevented by the addition of isoeugenol to veratryl alcohol-containing media, but growth is unaffected. Vanillyl alcohol oxidase gene (*vaoA*) from *P. simplicissimum* CBS 170.90 cloned into *Amycolapsis* enabled this bacterium to metabolize eugenol.

The fungus *Phanerochaete chrysosporium (Sporotrichum pruinosum)* a microorganism known to decompose lignin into vanillin was used to produce vanillin from green coconut husk (Barbosa *et al.* 2008).

Hansen *et al.* (2009) (Figure 26.9) describe a different approach by creating new transgenic *de novo* pathways to form vanillin from glucose in the yeasts *Saccharomyces cerevisiae* and *Schizosaccharomyces pombe*. The new pathway combines bacterial, mold, plant and human genes. The newly engineered pathway included 3-dehydroshikimate dehydratase from *Podospora pauciseta*, an aromatic carboxylic acid reductase (ACAR)

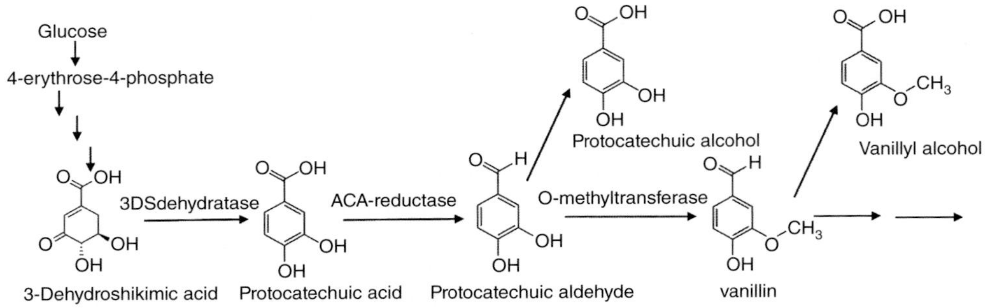

Figure 26.9 Suggested transgenic gene construct for production of vanillin from glucose (Hansen *et al.,* 2009).

from *Nocardia* sp., and a human *O*-methyl transferase. In *S. cerevisiae* co-expression of a *C. glutamicum* phosphopantetheinyl transferase (PPTase) was additionally required. Elimination of the yeast alcohol dehydrogenase gene (*adg6*) prevented the reduction of vanillin to vanillyl alcohol. Additionally, to eliminate vanillin toxicity and improve yield, a gene coding for UDP-glycosyltransferase from *Arabidopsis thaliana* was used to create vanillyl-β-D-glucoside. This work demonstrates the possibility of creating engineered pathways in organisms which normally do not posses the capacity to produce vanillin and other flavor compounds.

Similarly, patent application US 20170172184 (Goldsmith *et al.* 2017) (Figure 26.10) describes a transgenic *S. cerevisiae* with the shikimate pathways genes coding for enzymes caffeoyl-*O*-methyltransferase (COMT), arom multifunctional enzyme (AROM), 3-dehydroshikimate dehydratase (3DSD), aromatic carboxylic acid reductase (ACAR), 4-(hydroxymethyl)-2-methoxyphenol alcohol oxidase (VAO), *O*-methyltransferase (OMT), or phosphopantetheine transferase (PPTase) under one regulatory gene. The promise of this patent application is that this construct will produce only vanillin without any other derivatives of vanillin and therefore the process will not need complicated purification downstream processing.

Some adverse features such as slow growth, more viscous broth, pellet formation, uncontrolled fragmentation and lysis of the mycelium can complicate the

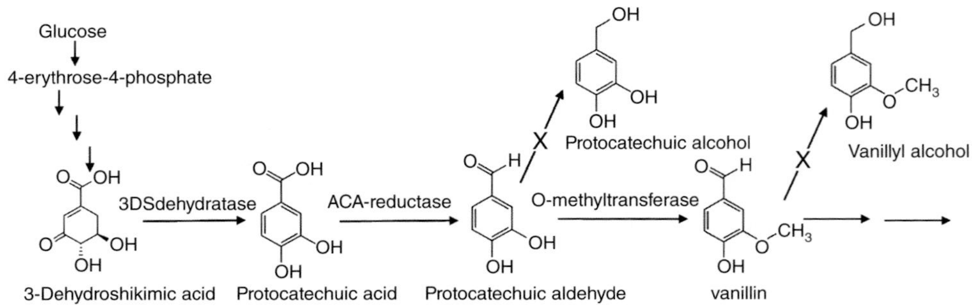

Figure 26.10 Suggested transgenic gene construct for production of vanillin from glucose without by-products (Goldsmith *et al.,* 2017).

bioconversion and result in a reduced productivity and more complex downstream processing.

26.4 Processes

Metabolic engineering offers several options on how to improve economics of the process, for example: (1) improvement of the expression of existing genes; (2) reduction of the downstream degradation of vanillin; (3) usage of alternative pathways such as shikimate; (4) creation of transgenic microorganisms or plants, etc.

The type of organism used for production, strain improvement, process parameters, cost of the media and substrate, its availability, toxicity, and solubility, downstream processing with product recovery are some of the equally crucial parameters. Other crucial factors are product inhibition, product down-stream metabolism, product toxicity to the cells, and product recovery and its purity. Several approaches have been attempted to increase the productivity. Table 26.1 summarizes the highest yields of vanillin produced by different microorganisms with the substrates such as ferulic acid, glucose, eugenol and isoeugenol.

26.4.1 Direct Bioconversion Process

In this process, the substrate is directly biotransformed into vanillin, which is recovered by subsequent organic solvent extraction or by hydrophobic adsorption. Economically feasible yields are achieved with pre-purified ferulic acid as a substrate. Its conversion yields about 6–13 g/L vanillin with a molar yield of about 75% within 40–50 hours fermentation. This was reported for a biotransformation with actinomycetes, such as *Amycolatopsis* sp. HR167 (Rabenhorts and Hopp 2000) and *S. setonii* ATCC 39116 (Müller *et al.* 1998). Ferulic acid (22.5 g/L) added in a NaOH solution was converted by *S. setonii* to vanillin (9–13 g/L) at 40–75% molar rate (Muheim *et al.* 2001). To achieve this high bioconversion rate, resting cells of *S. setonii* were used. Glucose was depleted. The pH of the bioconversion was maintained at pH 8.5.

Amycolatopsis sp. bioconversion was done in a stepwise manner with ferulic acid added at 19.9 g/L. After 47 hours, the culture produced 11.5 g/L vanillin with a molar conversion rate of 77% (Rabenhorst and Hopp 2000). On the other hand, *S. setonii* started to accumulate vanillin after glucose was exhausted and ferulic acid was added in a stepwise fed-batch manner. Productivity varied based on the fermentation volume between 51–75%. The best yield of 13.9 g/L was achieved in a 10-liter fermenter from 22.5 g/l of ferulic acid (Müller *et al.* 1998). Besides vanillin, several other phenolic metabolites such as vanillyl alcohol, vanillic acid, guaiacol, *p*-ethyl guaiacol and 2-methyl-4-ethylphenol were produced during this bioconversion. The side-products are structurally similar and may present a challenge with the purification of vanillin.

Isoeugenol (10 g/L) bioconversion with a growing culture of *B. pumillus* S-1, yielded 3.75 g/L vanillin after 150 hours, with a molar yield of 40.5% (Hua *et al.* 2007). This was the highest reported yield for isoeugenol up to now.

Solid state fermentation by *Phanerochaete chrysosporium* using autoclaved green coconut husk waste as a substrate yielded vanillin in the range between 44 to 52 μg/g of solid support. Sun dried husks were found suitable for vanillin production. Vanillin was extracted by organic solvent from the solid support (Barbosa *et al.* 2008).

Table 26.1 The highest microbial bioconversion yields of vanillin from various precursors.

Substrate	Microorganisms	Yield (g/L)	Molar yield (%)	Variable parameters	Reference
Ferulic acid	*S. setonii* ATCC39116	13.9	75	pH 8.5, glucose 17 h	Müller *et al.* 1998
Ferulic acid	*Amycolatopsis* sp. HR167	11.5	77.8	32 h	Rabenhorst and Hopp, 2000
Ferulic acid	*A. niger* I-1472 *P. cinnabarius* MUCL 38467	3.6	82	Continuous, 360 h phospholipids cellobiose	Lessage-Messen *et al.* 1999, 2002
Ferulic acid	*P. putida* Zyl581	2.2	73	pH 8.5 43 h	Cheetham *et al.* 2005
Ferulic acid	*S. setonii* ATCC 39116	9.3	90	pH 8.2, glucose	Gunnarsson and Palmkvist, 2006
Ferulic acid	*E. coli* BL21 pETDfdc *E. coli* BL21 pETDcso2 pGro7	7.8	40	pH 9.5 and 10.5 24 h water/butyl acetate	Furuya *et al.* 2015
Ferulic acid sugar beet	*P. putida* Zyl581	1.7	54	pH 8.5 40 h	Cheetham *et al.* 2005
Ferulic acid rice bran	*A. niger* CGMCC 0774 *P. cinnabarius* CGMCC 1115	2.8	61.6	pH 7.0 72 h, resin HD-8	Sun *et al.* 2008
Ferulic acid corn cob	*E. coli* JM109/pBBI	0.24	18	22 h	Rivas *et al.* 2009
Ferulic acid rice bran	*Amycolatopsis sp.* DSM 9991	93.6	46		Kindel *et al.* 2013

Substrate	Organism/Enzyme			Conditions	Reference
Vanillic acid	Micrococcus isabellinus Zyl849	1.5	80	pH 3.8; 20 h continuous process; oil extraction; 18.9 g product	Cheetham et al. 2005
Glucose	E. coli ATCC 98859/Neurospora crassa SY7A	0.3		20 g/l glucose 0.4 g/l methionine 48 h, pH 7.0	Frost, 2002
Glucose	S. cerevisiae transgenic	0.01–0.2		72–120 h	Goldsmith et al. 2017
Eugenol	Pseudomonas sp. HR199	0.3	89		Overhage et al. 2003
Eugenol	Lipoxygenase	0.14	1.1	36 h, charcoal	Wu et al. 2008
Isoeugenol	Pseudomona putida IE27	16.1	71	pH 20°C, 10% DMSO	Yamada et al. 2007
Isoeugenol	Pseudomona putida IE27	28.3	75	in 60% isoeugenol 24 h	Yamada et al. 2008
Isoeugenol	Pseudomonas chlororaphis	1.2	12.9		Kasana et al. 2007
Isoeugenol	Bacillus pumillus S-1	3.8	40.5		Hua et al. 2007
Isoeugenol	Bacillus fusiformis CGMCC1347	8.1	17.5	pH 7.0 72 h, resin H103	Zhao et al. 2006
Isoeugenol	Bacillus fusiformis SW-B9	32.5	5.8	60% isoeugenol 72 h	Zhao et al. 2005
Isoeugenol	lipoxygenase	2.46	13.3	H_2O_2, charcoal	Li et al. 2005

26.4.2   Bi-Phasic Fermentation

The use of an additional phase whether solid or liquid provides a three-fold benefit: it protects the cells against the substrate toxicity, prevents downstream product degradation and speeds up product recovery. A typical choice representing liquid phases would be either oil or a non-polar solvent. On the other hand, solid phase can be selected from active charcoal, or a variety of hydrophobic inert resins such as Amberlit.

For example, a biphasic bioconversion of isoeugenol using *Bacillus fusiformis* led to a yield of vanillin up to 32.5 g/L, when isoeugenol (60%) was used both as a substrate and as a non-polar solvent (Zhao *et al.* 2005). The disadvantage of this system is very high viscosity and difficulty of recovering vanillin from the isoeugenol phase.

A biphasic water/hexane system was employed in the production of 4-vinylguaiacol from ferulic acid by *B. pumillus*. The yield of 9.6 g/L of vinylguaiacol from 25 g/l of ferulic acid indicates 38% conversion yield, or 49.7% molar yield (Lee *et al.* 1998).

Use of a biphasic water/vegetable oil process with *Micromucor isabellinus* Zyl849 produced 18.9 g/L vanillin from vanillic acid after 20 hours (Cheetham *et al.* 2005). Vanillic acid concentration was maintained at 1.5 g/L throughout the experiment. Culture medium was continuously pumped through an extraction vessel and extracted by warm oil at pH 6.0. The aqueous portion containing vanillic acid, which was not extracted into oil phase was returned to the fermenter. A similar experiment was done with *A. fumigatus* Zyl747 with the yield of 16.26 g of vanillin after 665 hours (Cheetham *et al.* 2005).

Furuya *et al.* (2015) reported 7.8 g/L vanillin after 24 hours of two-stage fermentation using two recombinant *E. coli* strains. The second stage was in abi-phasic system of water and butyl acetate.

26.4.3   Mixed Culture Fermentation

This type of fermentation has received the most attention in recent years (e.g. Lesage-Meessen *et al.* 1996; 2002; Cheetham *et al.* 2005; Sun *et al.* 2008). The approach depends on the substrate and selected microorganisms.

If the precursor is plant material containing ferulic acid, the fermentation process consists of three steps: (1) ferulic acid is released from the plant materials either by chemical, microbial or enzymatic hydrolysis; (2) transformation of ferulic acid into vanillic acid; and (3) reduction of vanillic acid into vanillin.

Once ferulic acid is unbound, it is biotransformed to vanillic acid in the second step performed for example by Ascomycetes such as *A. niger*, Basidiomycetes such as *Bjerkandera adusta*, and Actinomycetes such as *S. setonii*. In the third step, vanillic acid is reduced to vanillin and vanillyl alcohol for example by *Pycnoporus cinnabarinus*.

As an example of such mixed culture bioconversion is a process using autoclaved maize bran and two fungi, *A. niger* CNCM I-1472 and *P. cinnabarinus* MUCL 38467. First *A. niger* released ferulic acid that was concomitantly deacetylated into vanillic acid. Afterwards *P. cinnabarinus* reduced vanillic acid into vanillin (Lesage-Meessen *et al.* 2002) with the yield of 767 mg/L vanillin. In another experiment, *A. niger* was fed ferulic acid continuously at 430 mg/L for 24 hours to a final concentration of 5.05 g/L. After culturing for 15 days, the final concentration of vanillic acid in the culture was 3.60 g/L with a molecular conversion rate of 82%. Cellobiose was used as elicitor and

Amberlit XAD2 resin was used to isolate vanillin from the broth. When only *A. niger's* extracellular enzymes were used in a combination with *P. cinnabarinus,* 584 mg/L of vanillin was obtained (Lessage-Messen *et al.* 1999, 2000). The entire process is quite long because it requires also preculture of both fungi and purification of the product.

A similar process using pre-purified ferulic acid as the starting substrate, with a combination of yeast *R. rubra* with bacterium *Nocardia* sp. was mentioned as a possible combination. First *R. rubra* IFO 889 produces vanillic acid by β-oxidation from ferulic acid. Vanillic acid is then converted by *Nocardia* sp. NRRL 5646 carboxylic acid reductase to vanillin (Li and Rosazza 2000).

A bacterial fermentation coupled with an enzymatic reduction serves as another example. A recombinant *E. coli* that contains a part of the plant shikimate pathway produces vanillic acid from glucose. Subsequently vanillic acid is reduced to vanillin by a purified aryl aldehyde dehydrogenase (Frost 2002).

26.4.4 Continuous Fermentation with Immobilized Cells

The highest bioconversion yields of vanillin formation from ferulic acid are achieved when the cells are in their resting growth phase. This warrants a continuous process. Cells, particularly unicellular microorganisms such as bacteria and yeasts, can utilize a membrane system where cells are supplied to the feed stock and the product is removed in a continuous mode. Another way is to immobilize the cells on porous polymeric materials. Ferulic acid or other substrates are then fed at a calculated rate to allow for the bioconversion to take place. The product is recovered either by a solvent extraction or by a hydrophobic adsorption. Physically immobilized *C. glutamicum* was fed 1 g/L/ hour of ferulic acid and it produced continuously 0.24 g/L/hour of vanillin, which was continuously removed by resin over 164-hour period (Labuda, unpublished data). Even though the productivity is quite low, the improvement of this set-up can result in a commercially viable process.

26.4.5 Enzymes

Vanillin production from eugenol and isoeugenol by enzymatic biotransformation is another way to produce vanillin. Styrene dioxygenase, vanillyl alcohol oxidase (dehydrogenase) and lipoxygenase were used to produce vanillin from eugenol and isoeugenol (van den Heuvel *et al.* 2004; Li *et al.* 2005). Creosol and vanillylamine were transformed to vanillyl alcohol and then to vanillin in a two-step process under the catalysis of vanillyl alcohol oxidase (van den Heuvel *et al.* 2004).

A crude enzyme preparation of lipoxygenase extracted from soybean, converted isoeugenol to vanillin. The process was assisted with powdered activated carbon and hydrogen peroxide, and yielded 2.46 g /L vanillin with a molar yield of 13.3% (Li *et al.* 2005).

On the other hand, when eugenol was used as precursor with soybean lipoxygenase in a silicon rubber membrane bioreactor, the vanillin yield after 36 hours was only 0.12 g/L. The conversion rate was 1.01%. Upon addition of charcoal, into the broth vanillin concentration in the receiving solution reached 0.14 g/L, and the conversion rate of clove oil increased to 1.14% (Wu *et al.* 2008). Similarly to microbial processes, lipoxygenase yielded more vanillin from isoeugenol than eugenol.

26.4.6 Cofactors

Sulfhydro compounds, e.g. dithiothreitol (DTT) and gluthathione were reported as important cofactors for accumulation of vanillin (Labuda *et al.* 1992b). Alliegro (2000) suggested that DTT can interact with cysteine-less protein domains and impact the activity of proteins. Recent patents on formation of vanillin used compounds such as thioacetic acid, as a part of their vitamin solutions without elaborating on their importance (Cheetham *et al.* 2005) or usage of sulfur dioxide. DTT binds to the active sites of shikimate dehydrogenase which impacts the catalysis of the reversible NADPH linked reduction of 3-dehydroshikimate to yield shikimate (Michel *et al.* 2003).

DMSO (10%) was used to dissolve isoeugenol in the water phase. It was effective in promoting vanillin formation by *P. putida* when isoeugenol was a precursor. The addition of DMSO inhibited the formation of vanillic acid (Yamada *et al.* 2007).

Addition of phospholipids accelerated bioconversion of ferulic acid to vanillic acid. Phospholipids such as soy phospholipids (phosphatidylcholine, lysophosphatidyl-choline, phosphatidyl-ethanolamine, acylphosphatidylethanolamine, phosphatidylino-sitol, phosphatidic acid, etc.) were suggested to be used at a concentration range of 0.1–20 g/L (Lessage-Messen *et al.* 1999).

Enzymatic reactions using peroxidase, lipoxygenase, laccase and some lyases require hydrogen peroxide as an electron donor to improve the conversion of isoeugenol into vanillin. As mentioned above, lipoxygenase needed as a cofactor H_2O_2 (0.1%) and 10 g/L of charcoal to form vanillin from isoeugenol (Li *et al.* 2005).

26.5 Downstream Processing and Recovery

Recovery of the final product from a fermentation or any other reaction mixture is critical and of high economic importance. Structural similarity of secondary metabolites formed during the vanillin processes presents a challenging engineering problem. Vanillin may be recovered by a solvent extraction followed by crystallization. This process typically results in a medium purity with a yield about 90%. This process has higher costs, which come from the usage of solvents and their environmental and toxicological impact. Another recovery process is a vacuum distillation as described by Gayet *et al.* (2017). Another method is organophilic pervaporation during which vanillin is directly recovered from the fermentation media, at the temperature of fermentation in a solvent-free process, assuring the quantitative recovery of pure vanillin free of other media constituents (Brazinha *et al.* 2011). Ultrafiltration, carbon dioxide extraction, membrane filtration or macroporous absorption can be considered for vanillin purification as well. All the above mentioned separation techniques will become unnecessary if the transgenic constructs of either *E. coli* or *S. cerevisiae* (Goldsmith *et al.* 2017) manage to express only the pathways for vanillin without the structurally similar by-products. Such processes would undoubtedly require simpler downstream processing to recover vanillin.

26.6 Conclusions

Recent changes in the US regulations allow vanillin produced by microbial bioconversion/fermentation to be considered natural. This change has been a crucial step, which allowed the extensive research from both private and academic laboratories to be transformed into a commercial product.

Even though metabolic engineering, process optimization and enhanced downstream processing improved the yields, the final cost depends on the raw materials and process productivity. Current cost is driven by the cost of ferulic acid. The cost of isoeugenol is lower, however one must question its naturalness since oxidation of eugenol to isoeugenol requires chemical catalysis.

Gene editing and the synthetic biology approach will undoubtedly bear fruit and deliver a viable construct with phenylpropanoid, chorismate and dehydro-shikimate pathways, whether in microorganisms or plants.

References

Achterholt, S., Priefert, H. and Steinbuchel, A. 2000. Identification of *Amycolatopsis* sp. strain HR167 genes, involved in the bioconversion of ferulic acid to vanillin. *Applied Microbiology and Biotechnology* 54, 799–807.

Ago, S. and Kikuchi, Y. 1998. *Ferulic acid decarboxylase.* US patent 5,955,137.

Alliegro, M.C. 2000. Effects of dithiothreitol on protein activity unrelated to thiol-disulfide exchange: for consideration in the analysis of protein function with Cleland's reagent. *Analytical Biochemistry* 282, 102–106.

Barbosa, E.S., Perrone, D., Vendramini, A.L.A. and Leite, S.G.F. 2008. Vanillin production by *Phanerochaete chrysosporium* grown on green coconut agro-industrial husk in solid state fermentation. *BioResources* 3(4), 1042–1050.

Barthelmebs, L., Divies, C. and Cavin, J.-F. 2000. Knockout of the p-coumarate decarboxylase gene from *Lactobacillus plantarum* reveals existence of two other inducible enzymatic activities involved in phenolic acid metabolism. *Applied and Environmental Microbiology* 66, 3368–3375.

Benoit, I., Navarro, D., Marnet, N., Rakotomanomana, N., Lesage-Meessen, L., Sigoillot, J.-C., Asther, M. and Asther, M. 2006. Feruloyl esterases as a tool for the release of phenolic compounds from agro-industrial by-products. *Carbohydrate Research* 341, 1820–1827.

Bhathena, J., Kulamarva, A., Urbanska A.M., Martoni, Ch. and Prakash, S. 2007. Microencapsulated bacterial cells can be used to produce the enzyme feruloyl esterase: preparation and in vitro analysis. *Applied Microbiology and Biotechnology* 75, 1023–1029.

Bloem, A., Bertrand, A., Lonvaud-Funel, A. and de Revel, G. 2007. Vanillin production from simple phenols by wine-associated lactic acid bacteria. *Letters in Applied Microbiology* 44, 62–67.

Brazinha, C., Barbosa, D.S. and Crespo, J.G. 2011. Sustainable recovery of pure natural vanillin from fermentation media in a single pervaporation step. *Green Chemistry* 13, 2197–2203.

Brinkrolf, K., Brune, I. and Tauch, A. 2006. Transcriptional regulation of catabolic pathways for aromatic compounds in *Corynebacterium glutamicum. Genetics and Molecular Research* 5, 773–789.

Brunerie, P. and Asther, M. 1996. A two-step bioconversion process for vanillin production from ferulic acid combining *Aspergillus niger* and *Pycnoporus cinnabarinus. Journal of Biotechnology* 50, 107–113.

Brunati, M., Marinelli, F., Bertolini, C., Gandolfi, R., Daffonchio, D. and Molinari, F. 2004. Biotransformations of cinnamic and ferulic acid with actinomycetes. *Enzyme and Microbial Technology* 34, 3–9.

Chamkha, M., Garcia, J. L. and Labat, M. 2001. Metabolism of cinnamic acids by some *Clostridiales* and emendation of the descriptions of *C. aerotolerans, C. celerecrescens* and *C. xylanolyticum. International Journal of Systematic and Evolutionary Microbiology* 51, 2105–2111.

Cheetham, P., Gradley, M. and Sime, J. 1996. *Enzymatic preparation of phenolic compounds from plant materials.* WO-A-96/39859.

Cheetham, P., Gradley, M. and Sime, J. 2005. *Flavor/aroma materials and their preparation.* US patent 6,844,019.

Clausen, M., Lamb, C.J., Megnet, R. and Doerner, P.W. 1994. *PAD1* encodes phenylacrylic acid decarboxylase which confers resistance to cinnamic acid in *Saccharomyces cerevisiae. Gene* 142, 107–112.

Di Gioia, D., Luziatelli, F., Negroni, A., Ficca, A.G., Fava, F. and Ruzzi, M. 2011. Metabolic engineering of *Pseudomonas fluorescens* for the production of vanillin from ferulic acid. *Journal of Biotechnology* 156, 309–316.

Edlin, D.A.N., Narbad, A., Gasson, M.J., Dickinson, J.R. and Lloyd, D. 1998. Purification and characterization of hydroxycinnamate decarboxylase from *Brettanomyces anomalus. Enzyme and Microbial Technology* 22, 232–239.

Fraaije, M.W., Pikkemaat, M.L. and van Berkel, W.J.H. 1997. Enigmatic gratuitous induction of the covalent flavoprotein vanillyl-alcohol oxidase in *Penicillium simplicissimum. Applied and Environmental Microbiology* 63(2), 435–439.

Frost, J.W. 2002. *Synthesis of vanillin from carbon source.* US patent 6,372,461.

Furukawa, H., Wieser, M., Morita, H. and Nagasawa, T. 1999. Microbial synthesis of coniferyl alcohol by the fungus *Byssochlamys fulva* V107. *Bioscience, Biotechnology and Biochemistry* 63, 1141–1142.

Furuya, T., Miura, M., Kuroiwa, M. and Kino, K. 2015. High-yield production of vanillin from ferulic acid by a coenzyme-independent decarboxylase/oxygenase two-stage process. *New Biotechnology* 32, 335–339.

Gasson, M., Kitamura, Y., McLauchlan, W.R., Narbad, A., Parr, A.J., Parsons, E.L.H., Payne, J., Rhodes, M.J.C. and Walton, N.J. 1998. Metabolism of ferulic acid to vanillin. A bacterial gene of the enoyl-SCoA hydratase/isomerase superfamily encodes an enzyme for the hydration and cleavage of a hydroxycinnamic acid SCoA thioester. *Journal of Biological Chemistry* 273, 4163–4170.

Gayet, H., Revelant, D. and Vibert, M. 2017. *Method for the purification of natural vanillin.* US Patent 9617198.

Ghosh, S., Sachan, A., Sen, S.K. and Mitra, A. 2007. Microbial transformation of ferulic acid to vanillic acid by *Streptomyces sannanensis* MTCC 6637. *Journal of Industrial Microbiology and Biotechnology* 34, 131–138.

Goldsmith, N., Hansen, E.H., Meyer, J.-P. and Brianza, F. 2017. *Methods of improving production of vanillin.* US patent application 20170172184.

Gross, B., Asther, M., Corrieu, G. and Brunerie, P. 1993. Production of vanillin by bioconversion of benzenoid precursors by *Pycnoporus.* US Patent 5,262,315.

Gunnarsson, N. and Palmqvist, E.A. 2006. Influence of pH and carbon source on the production of vanillin from ferulic acid by *Streptomyces setonii* ATCC 39116. In: *Flavor*

Science: Recent Advances and Trends. Bredie W.L.P. and Petersen M.A., eds. Elsevier, Oxford, pp. 73–78.

Hansen, E.H., Moller, B. L., Kock, G.R., Bunner, C.M., Kristensen, C., Jensen, O.R., Okkels, F.T., Olsen, C.E., Motawia, M.S. and Hansen, J. 2009. *De novo* biosynthesis of vanillin in fission and baker´s yeast. *Applied and Environmental Microbiology* 75, 2765–2774.

van den Heuvel, R.H.H., van den Berg, W.A.M., Rovida, S. and van Berkel, W.J.H. 2004. Laboratory-evolved vanillyl-alcohol oxidase produces natural vanillin. *Journal of Biological Chemistry* 279, 33492–33500.

Hua, D., Ma, C., Lin, S., Song, L., Deng, Z., Maomy, Z., Zhang, Z., Yu, B. and Xu, P. 2007. Biotransformation of isoeugenol to vanillin in a newly isolated *Bacillus pumilus* strain: identification of major metabolites. *Journal of Biotechnology* 130, 463–470.

Huang, Z., Dostal, L. and Rosazza, J.P.N. 1993a. Mechanism of ferulic acid conversion to vanillic acid and guaiacol by *Rhodotorula rubra*. *Journal of Biological Chemistry* 268, 23594–23598.

Huang, Z., Dostal, L. and Rosazza, J.P.N. 1993b. Microbial transformation of ferulic acid by *Sacharomyces cerevisiae* and *Pseudomonas fluorescens*. *Applied and Environmental Microbiology* 50, 2244–2250.

Kanski, J., Aksenova, M., Stoyanova, A. and Butterfield, D.A. 2002. Ferulic acid antioxidant protection against hydroxyl and peroxyl radical oxidation in synaptosomal and neuronal cell culture systems in vitro: structure-activity studies. *Journal of Nutritional Biochemistry* 13, 273–281.

Karmakar, B., Vohra, R.M., Nandanwar, H., Sharma, P., Gupta, K.G. and Sobti, C. 2000. Rapid degradation of ferulic acid via 4-vinylguaiacol and vanillin by a newly isolated strain of *Bacillus coagulans*. *Journal of Biotechnology* 80, 195–202.

Kasana, R.C., Sharma, U.K., Sharma, N. and Sinha, A.K. 2007. Isolation and identification of a novel strain of *Pseudomonas chlororaphis* capable of transforming isoeugenol to vanillin. *Current Microbiology* 54, 457–461.

Kindel, G., Hilmer, J.-M. and Gross, E.-E. 2013. *Vanillin*. EP appl. 20130157041.

Labuda, I.M., Keon, K.A. and Goers, S.K. 1992a. Microbial bioconversion process for the production of vanillin. In: *Progress in Flavour Precursor Studies*. Schreier, P. and Peter Winterhalter, P., eds. Wurzburg, Germany. pp. 477–482.

Labuda, I.M., Keon, K.A. and Goers, S.K. 1992b. *Bioconversion process for the production of vanillin*. US patent 5,128,253.

Lessage-Messen, L., Haon, M., Delattre, M., Thibault, J., Colonna-Ceccaldi, B. and Asther, M. 1997. An attempt to channel the transformation of vanillic acid into vanillin by controlling methoxyhydroquinone formation in *Pycnoporus cinnabarinus* with cellobiose. *Applied Microbiology and Biotechnology* 47, 393–397.

Lessage-Messen, L., Delattre, M., Haon, M. and Asther, M. 1999. *Methods for bioconversion of ferulic acid to vanillic acid or vanillin and for the bioconversion of vanillic acid to vanillin using filamentous fungi*. US patent 5,866,380.

Lessage-Messen, L., Delattre, M., Haon, M. and Asther, M. 2000. *Aspergillus niger which produces vanillic acid from ferulic acid*. US patent 6,162,637.

Lessage-Messen, L., Lomascolo, A., Bonnin, E., Thibault, J.-F., Buleon, A., Roller, M., Asther, M., Record, E., Colonna-Ceccaldi, B. and Asther, M. 2002. A biotechnological process involving filamentous fungi to produce natural crystalline vanillin from maize bran. *Applied Biochemistry and Biotechnology* 102–103, 141–153.

Levasseur, A., Pagès, S., Fierobe, H.-P., Navarro, D., Punt, P., Belaïch, P.-R., Asther, M. and Record, E. 2004. Design and production in *Aspergillus niger* of a chimeric protein associating a fungal feruloyl esterase and a clostridial dockerin domain. *Applied and Environmental Microbiology* 70, 6984–6991.

Levasseur, A., Navarro, D., Punt, P.J., Belaïch, J.-P., Asther, M. and Record, E. 2005. Construction of engineered bifunctional enzymes and their overproduction in *Aspergillus niger* for improved enzymatic tools to degrade agricultural by-products. *Applied and Environmental Microbiology* 71(12), 8132–8140.

Li, K. and Frost, J.W. 1998. Synthesis of vanillin from glucose. *Journal of the American Chemical Society* 120, 10545–10546.

Li, T. and Rosazza, J.P.N. 2000. Biocatalytic synthesis of vanillin. *Applied and Environmental Microbiology* 66(2), 684–687.

Li, Y.H., Sun, Z.H., Zao, L.Q. and Xu, Y. 2005. Bioconversion of isoeugenol into vanillin by crude enzyme extracted from soybean. *Applied Biochemistry and Biotechnology* 125:1–10.

Martinez-Cuesta, M.C., Gasson, M.J. and Narbad, A. 2005. Heterologous expression of the plant coumarate: CoA ligase in *Lactococcus lactis. Letters in Applied Microbiology* 40, 44–49.

Merkens, H., Beckers, G., Wirtz, A. and Burkovski, A. 2005. Vanillate metabolism in *Corynebacterium glutamicum. Current Microbiology* 51, 59–65.

Michel, G., Roszak, A.W., Sauve, V., Maclean, J., Matte, A., Coggins, J.R., Cygler, M. and Lapthorn, A.J. 2003. Structures of shikimate dehydrogenase AroE and its paralog YdiB. *Journal of Biological Chemistry* 278, 19463–19472.

Mitra, A., Kitamura, Y., Gasson, M.J., Narbad, A., Parr, A.J., Payne, J., Rhodes, M.J.C., Sewter, C. and Walton, N.J. 1999. 4-Hydroxycinnamoyl-CoA hydratase/lyase (HCHL) – an enzyme of phenylpropanoid chain-cleavage from *Pseudomonas. Archives of Biochemistry and Biophysics* 365, 10–16.

Muheim, A. and Lerch, K. 1999. Towards a high-yield bioconversion of ferulic acid to vanillin. *Applied Microbiology and Biotechnology* 51, 456–461.

Muheim, A., Müller, B., Munch, T. and Wetli, M. 2001. *Microbiological process for producing vanillin.* US patent 6,235,507.

Narbad, A. and Gasson, M.J. 1998. Metabolism of ferulic acid via vanillin using a novel CoA-dependent pathway in a newly-isolated strain of *Pseudomonas flluorescens. Microbiology* 44, 1397–1405.

Nishimura, M., Ishiyama, D. and Davies, J. 2006. Molecular cloning of *Streptomyces* genes encoding vanillate demethylase. *Bioscience, Biotechnology and Biochemistry* 70, 2316–2319.

Oddou, J., Stentelaire, C., Lesage-Meessen, L., Asther, M. and Colonna Ceccaldi, B. 1999. Improvement of ferulic acid bioconversion into vanillin by use of high-density cultures of *Pycnoporus cinnabarinus. Applied Microbiology and Biotechnology* 53, 1–6.

Ou, S., Luo, Y., Xue, F., Huang, C., Zhang, N. and Liu, Z. 2007. Separation and purification of ferulic acid in alkaline-hydrolysate from sugarcane bagasse by activated charcoal adsorption/anion macroporous resin exchange chromatography. *Journal of Food Engineering* 78, 1298–1304.

Overhage, J., Priefert, H. and Steinbüchel, A. 1999. Biochemical and genetic analysis of the ferulic acid catabolism in *Pseudomonas* sp. strain HR199. *Applied and Environmental Microbiology* 65, 4837–4847.

Overhage, J., Steinbüchel, A. and Priefert, H. 2003. Highly efficient biotransformation of eugenol to ferulic acid and further conversion to vanillin in recombinant strains of *Escherichia coli*. *Applied and Environmental Microbiology* 69(11), 569–576.

Overhage, J., Steinbuchel, A. and Priefert, H. 2006. Harnessing eugenol as a substrate for production of aromatic compounds with recombinant strains of *Amycolatopsis* sp. HR167. *Journal of Biotechnology* 125, 369–376.

Peng X., Misawa, N. and Harayama, S. 2003. Isolation and characterization of thermophilic bacilli degrading cinnamic, 4-coumaric, and ferulic Acids. *Applied and Environmental Microbiology* 69, 1417–1427.

Plaggenborg, R., Steinbüchel, A. and Priefert, H. 2001. The coenzyme A-dependent, non-β-oxidation pathway and not direct deacetylation is the major route for ferulic acid degradation in *Delftia acidovorans*. *FEMS Microbiology Letters* 205, 9–16.

Plaggenborg, R., Overhage, J., Steinbüchel, A. and Priefert, H. 2003. Functional analyses of genes involved in the metabolism of ferulic acid in *Pseudomonas putida* KT2440. *Applied Microbiology and Biotechnology* 61, 528–535.

Plaggenborg, R., Overhage, J., Loos, A., Archer, J.A.C., Lessard, P., Sinskey, A.J., Steinbüchel, A. and Priefert, H. 2006. Potential of *Rhodococcus* strains for biotechnological vanillin production from ferulic acid and eugenol. *Applied Microbiology and Biotechnology* 72, 745–755.

Priefert, H., Overhage, J. and Steinbüchel, A. 1999. Identification and molecular characterization of the eugenol hydroxylase genes (*ehyA/ehyB*) of *Pseudomonas* sp. strain HR199. *Archives of Microbiology* 172, 354–363.

Priefert, H., Rabenhorst, J. and Steinbuchel, A. 2001. Biotechnological production of vanillin. *Applied Microbiology and Biotechnology* 56, 296–314.

Rabenhorst, J. 1996. Production of methoxyphenol-type natural aroma chemicals by biotransformation of eugenol with a new *Pseudomonas* sp. *Archives of Microbiology and Biotechnology* 46, 470–474.

Rabenhorst, J. and Hopp, R. 2000. *Process for the preparation of vanillin and microorganisms suitable therefore*. US patent 6,133,003.

de Revel, G., Bloem, A., Augustin, M., Lonvaud-Funel, A. and Bertrand, A. 2005. Interaction of *Oenococcus oeni* and oak wood compounds. *Food Microbiology* 22, 569–575.

Rivas, B., Aliakbarian, B. Torre, P., Perego, P., Dominguez, J.M., Zilli, M. and Converti, A. 2009. Vanillin bioproduction frm alkaline hydrolysis of corn cob by *Escherichia coli* JM109/pBB1. *Enzyme and Microbial Technology* 44, 154–158.

Rosazza, J.P.N., Huang, Z., Dostal, L., Volm, T. and Rousseau, B. 1995. Review: Biocatalytic transformation of ferulic acid: an abundant natural product. *Journal of Industrial Microbiology* 15, 457–471.

Seshadri, R., Lamm, A.S., Khare, A. and Rosazza, J.P.N. 2008. Oxidation of isoeugenol by *Nocardia iowensis*. *Enzyme and Microbial Technology* 43, 486–494.

Shimoni, E., Ravid, U. and Shoham, Y. 2000. Isolation of a *Bacillus* sp. capable of transforming isoeugenol to vanillin. *Journal of Biotechnology* 29, 1–9.

Sinha, A.K., Joshi, B.P. and Dogra, R. 2002. *Microwave assistant process for the preparation of the substituted phenylaldehydes from trans and cis-phenylpropenes*. World patent WO/2002/079132.

Su, F., Hua, D., Zhang, Z., Wang, X., Tang, H., Tao, F., Tai, C., Wu, Q., Wu, G. and Xu, P. 2011. Genome sequence of *Bacillus pumilus* S-1, an efficient isoeugenol-utilizing producer for natural vanillin. *Journal of Bacteriology* 193, 6400–6401.

Sun, Z., Zheng, P., Guo, X., Lin, G., Yin, H., Wang, J. and Bai, Y. 2008. *Method for the producing vanillic acid and vanillin from waste residue of rice bran oil by fermentation and biotransformation.* US Patent 7,462,470.

Sutherland, J.B., Crawford, D.L. and Pometto III, A.L. 1983. Metabolism of cinnamic, p-coumaric and ferulic acids by *Streptomyces setonii. Canadian Journal of Microbiology* 29, 1253–1257.

ten Have, R., Rietjens, I.M.C.M., Hartmans, H., Swarts, H.J. and Field, J.A. 1998. Calculated ionization potentials determine the oxidation of vanillin precursors by lignin peroxidase. *FEBS Letters* 430, 390–392.

Torre, P., Allakbarian, B., Rivas, B., Dominguez, J.M. and Converti, A. 2008. Release of ferulic acid from corn cobs by alkaline hydrolysis. *Biochemical Engineering Journal* 40, 500–506.

Venturi, V., Zennaro, F., Degrassi, G., Okeke, B.C. and Brischi, C.V. 1998. Genetics of ferulic acid bioconversion to protocatechuic acid in plant-growth-promoting *Pseudomonas putida* WCS358. *Microbiology* 144, 965–973.

Wu, Y.-T., Feng, M., Ding, W.-W., Tang, X.-Y., Zhong, Y.-H. and Xiao, Z.-Y. 2008. Preparation of vanillin by bioconversion in a silicon rubber membrane bioreactor. *Biochemical Engineering Journal* 41, 193–197.

Xiang, L. and Moore, B.S. 2005. Biochemical characterization of a prokaryotic phenylalanine ammonia lyase. *Journal of Bacteriology* 187, 4286–4289.

Xu, P., Hua, D. and Ma, C. 2007. Microbial transformation of propenylbenzenes for natural flavour production. *Trends in Biotechnology* 25, 571–576.

Yamada, M., Okada, Y., Yoshida, T. and Nagasawa, T. 2007. Biotransformation of isoeugenol to vanillin by *Pseudomonas putida* IE27 cells. *Applied Microbiology and Biotechnology* 73, 1025–1030.

Yamada, M., Okada, Y., Yoshida, T. and Nagasawa, T. 2008. Vanillin production using *Escherichia coli* cells over-expressing isoeugenol monooxygenase of *Pseudomonas putida. Biotechnology Letters* 30, 665–670.

Yoon, S.-H., Li, C., Lee, Y.-M., Lee, S.-H., Kim, S.-H., Choi, M.-S., Seo, W.-T., Yang, J.-K., Kim, J.-Y. and Kim, S.-W. 2005. Production of vanillin from ferulic acid using recombinant strains of *Escherichia coli. Biotechnology and Bioprocess Engineering* 10, 1226–8372.

Zhang, Y. 2006. Metabolism of isoeugenol via isoeugenol-diol by a newly isolated strain of *Bacillus subtilis* HS8. *Applied Microbiology and Biotechnology* 73, 771–779.

Zhao, L.Q., Sun, Z.H., Zheng, P. and Zhu, L.L. 2005. Biotransformation of isoeugenol to vanillin by a novel strain of *Bacillus fusiformis. Biotechnology Letters* 27, 1505–1509.

Zhao, L.Q., Sun, Z.H. and Zheng, P. 2006. Biotransformation of isoeugenol to vanillin by *Bacillus fusiformis* CGMCC1347 with the addition of resin HD-8. *Process Biochemistry* 41, 1673–1676.

Zhao, S., Yao, S., Ou, S., Lin, J., Wang, Y. Peng, X., Li, A. and Yu, B. 2014. Preparation of ferulic acid from corn bran: Its improved extraction and purification by membrane separation. *Food and Bioproducts Process* 92, 309–313.

Index

Handbook of Vanilla Science and Technology, Second Edition.
Edited by Daphna Havkin-Frenkel and Faith C. Belanger.
© 2019 John Wiley & Sons Ltd. Published 2019 by John Wiley & Sons Ltd.